Sustainable Waste Management in the Tourism and Hospitality Sectors

Reshma Sucheran
Durban University of Technology, South Africa

IGI Global
Scientific Publishing
Publishing Tomorrow's Research Today

Vice President of Editorial	Melissa Wagner
Managing Editor of Acquisitions	Mikaela Felty
Managing Editor of Book Development	Jocelynn Hessler
Production Manager	Mike Brehm
Cover Design	Phillip Shickler

Published in the United States of America by
IGI Global Scientific Publishing
701 East Chocolate Avenue
Hershey, PA, 17033, USA
Tel: 717-533-8845
Fax: 717-533-8661
Website: https://www.igi-global.com E-mail: cust@igi-global.com

Library of Congress Cataloging-in-Publication Data

Names: Sucheran, Reshma, 1970- editor.
Title: Sustainable waste management in the tourism and hospitality sectors / edited by Reshma Sucheran.
Description: Hershey, PA : IGI Global Scientific Publishing, 2025. | Includes bibliographical references and index. | Summary: "This academic research book emerges as a pivotal resource for addressing the pressing global concerns surrounding climate change, particularly within waste management in the tourism and hospitality sectors"-- Provided by publisher.
Identifiers: LCCN 2024053014 (print) | LCCN 2024053015 (ebook) | ISBN 9798369361108 (hbk) | ISBN 9798369361115 (pbk) | ISBN 9798369361122 (ebk)
Subjects: LCSH: Refuse and refuse disposal. | Sustainable tourism. | Climatic changes.
Classification: LCC HD4482 .S87 2025 (print) | LCC HD4482 (ebook) | DDC 363.72/8--dc23/eng/20250204
LC record available at https://lccn.loc.gov/2024053014
LC ebook record available at https://lccn.loc.gov/2024053015

British Cataloguing in Publication Data
A Cataloguing in Publication record for this book is available from the British Library.

Table of Contents

Detailed Table of Contents

This chapter presents a bibliometric analysis of 280 research publications on waste management in the tourism and hospitality industry. Using Bibliometric R and the Web of Science database, this study examined several key factors, including publication and citation trends, author productivity and influence, leading organisations, countries, journals, and common themes in the field. The data show a steady rise in the number of publications and citations over time, reflecting the growing recognition of the importance of waste management in the tourism sector. This increase has been especially noted in recent years and is driven by a broader societal shift toward environmental awareness and the need for sustainable practices. The analysis of author productivity highlights varying levels of impact within the field, with some individuals making particularly significant contributions. Additionally, the study identifies key organizations, countries, and journals essential for advancing research on waste management in tourism.

Waste management significantly impacts the sustainable state of the environment, public health, and economic stability. This study thoroughly analyses the many waste streams produced, with particular attention on solid waste, wastewater, hazardous waste, e-waste, and food waste. Every category has different difficulties and consequences for waste management procedures and regulations. Food waste has serious negative effects on the environment and the economy and is frequently caused by overproduction and consumer behaviour. Solid waste, including various materials, needs efficient recycling and disposal techniques to reduce its impact. Because of the speed at which technology is developing, e-waste pollutes ecosystems with harmful materials. Understanding waste streams will help stakeholders create more long-lasting and efficient solutions to safeguard the environment and public health.

Waste management practices are crucial in mitigating the negative impacts and promoting sustainability in the tourism and hospitality sectors. This chapter uses relevant case studies to explore global perspectives on waste management education, regional variations, cultural considerations, innovative methodologies, obstacles, opportunities, and future recommendations. It highlights the challenges in different regions, from Asia to Europe's eco-conscious destinations and Africa's emerging markets. Challenges such as inadequate infrastructure, limited funding, and low stakeholder awareness persist, yet these also present opportunities for international cooperation and knowledge sharing. The chapter advocates leveraging technology, increasing stakeholder participation, and integrating waste management education into broader sustainability initiatives. Recommendations are provided for policymakers, business leaders, educators, and community leaders. Emphasizing cultural sensitivity, creativity, cooperation, and long-term vision, the chapter underscores the importance

The worldwide demand for wildlife tourism-based experiences has driven a global increase in demand for wildlife tourism experiences within protected areas. Tourism within protected areas has both positive and negative effects. The positive effects relate to the increased financial support for conservation initiatives that provide concrete substantiation for the long-term survival and conservation of protected areas. The negative effects, however, relate to the increased tourism infrastructure and the environmental impacts of accommodating guests and their hospitality staff within the protected areas. Of particular interest in this chapter is the issue of Waste management in protected areas, specifically recyclable waste. A mixed method single case study research design was employed to determine baselines and benchmarks of recyclable waste at a wildlife tourism lodge in the Greater Kruger Park area. Reflections on general waste management were also made regarding proposed principles for responsible waste management in protected areas.

This chapter examines sustainable waste management practices within the hospitality industry, a crucial segment of the global tourist sector that significantly impacts environmental health through waste production (Oluwole et al., 2020). Initially, it outlines the industry's structure and its role in waste generation, emphasizing the necessity for sustainable practices to mitigate environmental harm. The chapter highlights the environmental repercussions of poor waste management, including pollution and biodiversity loss, stressing the urgency for proactive measures. Additionally, it reviews the legal frameworks governing waste management and their importance in ensuring sustainability. Strategies for reducing, reusing, recycling, recovering, and disposing of waste are discussed, along with the role of technology in enhancing efficiency. Case studies illustrate successful initiatives, underscoring sustainable waste management's economic, social, and corporate benefits. These benefits include cost savings, improved public perception, and enhanced corporate responsibility.

Despite economic difficulties, global warming, and the ongoing recovery from the pandemic, the hospitality industry, a key player in the global economy, is under more urgency to adopt sustainable practices. This chapter examines 'Lean Hospitality,' an approach to the hospitality industry that borrows lean concepts from manufacturing operations to strengthen efficiency, reduce waste, and satisfy customers more. It examines how these concepts have changed and how they manage different kinds of waste, such as solids, liquids, gases, food choices, and potentially hazardous materials. The 5S Lean Philosophy, Value Stream Mapping, Standardized Work, and Poka-Yoke are some of the methodologies studied to see how they might promote operational excellence and continuous improvement, offering a promising future for the industry. Examples of real-world advantages and disadvantages are provided through case studies, including the Ritz-Carlton in Riyadh's effective lean implementation.

This chapter presents the specific environmental practices of Egyptian hotels and identifies potential barriers to implementing sustainable initiatives. The aim was to gain a holistic perspective on the factors affecting hotel environmental sustainability practices rather than merely discussing ecological problems. The analysis was based on qualitative data from 42 semi-structured interviews with top management from a sample of four- and five-star hotels in Egypt. The chapter reveals seven internal barriers: staff capabilities, physical limitations, the attitudes of top management, cost concerns, access to resources, owner attitudes, and day-to-day activities. External factors include government regulations, incentives, and guest attitudes. This chapter also examines the theoretical implications and discusses how future research might address environmental challenges in hospitality.

This chapter explores the motivations driving stakeholder engagement and the challenges faced in waste management initiatives within the tourism sector, focusing on the perspectives of self-help groups (SHGs). Through qualitative research methods, including in-depth interviews, this study examines the diverse motivations that prompt SHGs to participate in waste management projects, ranging from environmental concerns to socioeconomic empowerment. Moreover, it identifies and analyses the multifaceted challenges SHGs encounter, such as limited resources, inadequate infrastructure, and bureaucratic hurdles. This study contributes to a deeper understanding of stakeholder dynamics in tourism waste management by elucidating these motivations and challenges. The findings suggest that while SHGs play a crucial role in fostering sustainable practices, effective collaboration and support mechanisms are essential to address the challenges encountered and maximize the potential benefits of stakeholder engagement in waste management projects within the tourism sector.

initiatives, and innovative technologies. This will contribute to developing a more sustainable and environmentally conscious society.

Large cruise ships, capable of accommodating over 6,000 passengers and 1,500 crew members, generate substantial waste, necessitating effective management strategies to mitigate environmental impacts. This study employs a systematic literature review to identify key themes and research gaps in cruise waste management research. Using the PRISMA flow method, 47 relevant research articles were selected based on relevance. The data from these articles were analysed using bibliometric and content analysis techniques. The key themes identified include plastic waste, microplastic waste, food waste, wastewater, and waste management strategies. The study revealed significant research gaps that demand immediate attention and action concerning the effectiveness of waste management strategies, socio-economic considerations, interdisciplinary approaches, regulatory compliance, food waste disposal methods, regional collaborations, policy initiatives, and integrated waste management approaches.

The tourism industry contributes to and is impacted by plastic waste. The Walt Disney Company (Disney) is one of the largest media and entertainment companies contributing to global travel and waste. Disney theme parks represent a microcosm of society and the conflict between consumerism and environmentalism in the effort to reduce single-use plastic and waste to landfills. Disney has targets to reduce single-use plastic and waste at its wholly owned and operated parks and resorts in France and the United States (U.S.). As a multinational company that operates without a legally binding international agreement on plastic waste, Disney is subject to the plastic policies of each jurisdiction. The policy context in which Disney established its waste targets and the actions taken by Disney to reduce plastic waste in theme parks are discussed in this chapter. Moreover, opportunities to improve waste management practices and shared learning were identified.

The exponential increase in tourism escalates greywater production from laundry, gardening, bathing, and sewage treatment, leading to water scarcity. Simple physical and biological processes can purify greywater for reuse, leading to water conservation. Constructed wetlands offer new approaches to dealing with greywater, especially in polluted tourist areas, through plant, soil and microbial activity filtration processes. These systems are feasible for tropical regions with limited water resources. This chapter examines the suitability of constructed wetlands for greywater treatment, highlighting techniques, performance evaluations and environmental benefits. By integrating these systems, tourism facilities can reduce their ecological footprints with significant savings on water consumption. Constructed wetlands aid in improving greywater management and promoting sustainable water use because they have numerous advantages, such as low maintenance costs and adaptability, which are crucial in ecological preservation amidst rising tourism numbers.

Food waste presents a severe sustainability challenge. Of the food produced globally, 30% is wasted, and if no significant improvement is made, this figure is expected to double by 2050. With the continuous growth of the hospitality sector, an increasing amount of waste, particularly food waste, is generated. Globally, the hospitality industry is the third-largest contributor to food waste. Food waste mitigation is fundamental to the sustainable improvement of the food service industry, owing to its adverse consequences and ethical concerns. For this reason, there is an increasing demand for practical solutions in the food service industry. Establishments in the food service industry can make a significant contribution to global sustainability efforts by creating partnerships with stakeholders and providing educational programs aimed at waste reduction to employees and consumers. Several food service establishments actively implement measures to minimize waste and recognize and accept their responsibilities.

Increasing waste production presents substantial environmental and public health obstacles, necessitating inventive and sustainable waste management approaches. This chapter explores the fundamental concepts of circular economy and how it may be used to change waste management systems from linear disposal methods to sustainable and efficient models that make the most of resources. It explores the influence of efficient waste management on diminishing environmental deterioration, fostering resource preservation, and bolstering economic development through waste materials' reuse, recycling, and repurposing. The chapter highlights the importance of adopting comprehensive strategies that combine technology advancements, policy restructuring, and community involvement to enable a shift toward circular waste management practices that will support attaining global sustainability.

The convergence of waste-to-energy innovations and the tourism industry presents a dynamic frontier with significant potential for sustainable development. As tourism continues to grow globally, the environmental footprint of waste generated by this sector is becoming increasingly concerning. Waste-to-Energy (WtE) technologies, with the crucial role of technology providers, offer a promising solution by converting waste into usable energy, thereby reducing environmental impact and promoting a circular economy. This paper explores future trends and opportunities in integrating WtE systems within the tourism industry. By embracing WtE innovations, the tourism industry can significantly reduce its waste burden, lower energy costs, and contribute to a more sustainable future. This chapter underscores the necessity of multidisciplinary collaboration and supportive regulatory frameworks to fully realise the potential of WtE technologies in transforming the environmental landscape of tourism.

Preface

CHAPTER 1: GLOBAL TRENDS IN WASTE MANAGEMENT RESEARCH WITHIN THE TOURISM AND HOSPITALITY SECTOR: A BIBLIOMETRIC ANALYSIS

This paper conducts a bibliometric analysis of 280 studies on waste management in tourism and hospitality. Using Bibliometric R with the Web of Science database, it explores publication and citation trends, author productivity and influence, countries, journals, and themes. The data shows an increase in publications and citations, reflecting heightened attention to waste management in tourism, driven by growing environmental consciousness and sustainability imperatives. Author analysis reveals varying impact, with specific contributors making notable contributions. Examining leading organizations, countries, and journals provides insights into the global dedication and distribution of scholarly works in waste management in tourism. Thematic analysis identifies sustainability, waste management, and circular economy as central themes, aligning with Sustainable Development Goals (SDGs) 11 and 12. The study emphasizes interdisciplinary collaboration's importance in promoting sustainable practices, enhancing understanding of waste management studies in the tourism sector.

CHAPTER 2: SUSTAINABLE WASTE MANAGEMENT IN TOURISM AND HOSPITALITY SECTORS

The sustainable state of the environment, public health, and economic stability are all significantly impacted by waste management. This study offers a thorough analysis of the many waste streams produced, with particular attention on solid waste, wastewater, hazardous waste, e-waste, and food waste. Every category has different difficulties and consequences for waste management procedures and regulations. Food waste has serious negative effects on the environment and the economy and is frequently caused by overproduction and consumer behaviour. To reduce its impact, solid waste which includes a variety of materials needs efficient recycling and disposal techniques. Because of the speed at which technology is developing, e-waste pollutes ecosystems with harmful materials. Understanding waste streams will help stakeholders create more long-lasting and efficient solutions to safeguard the environment and public health.

CHAPTER 3: GLOBAL PERSPECTIVES ON WASTE MANAGEMENT EDUCATION IN TOURISM AND HOSPITALITY

Waste management is essential to mitigate the negative impacts and promote sustainability in Tourism and Hospitality. the chapter explores global perspectives on waste management education, regional variations, cultural considerations, innovative methodologies, obstacles, opportunities, and future recommendations using appropriate case studies. It highlights the challenges in different regions, from Asia to Europe's eco-conscious destinations and Africa's emerging markets. Challenges such as inadequate infrastructure, limited funding, and low stakeholder awareness persist, yet these also present opportunities for international cooperation and knowledge sharing. The chapter advocates leveraging technology, increasing stakeholder participation, and integrating waste management education into broader sustainability initiatives. Recommendations are provided for policymakers, business leaders, educators, and community leaders. Emphasizing cultural sensitivity, creativity, cooperation, and long-term vision, the chapter underscores the importance of collaboration among stakeholders.

CHAPTER 4: WASTE MANAGEMENT IN PROTECTED AREAS: BENCHMARKING RECYCLABLE WASTE EMANATING FROM TOURISM

The worldwide demand for wildlife tourism-based experiences has driven a global increase in demand for wildlife tourism experiences within protected areas. Tourism within protected areas has both positive and negative effects. The positive effects relate to the increased financial support for conservation initiatives that provide concrete substantiation for the long-term survival and conservation of protected areas. The negative effects however relate to the increased tourism infrastructure and environmental effects of accommodating guests and their hospitality staff within the protected areas. Of particular interest in this chapter is the issue of Waste management in protected areas, specifically recyclable waste. A mixed method single case study research design was employed to determine baselines and benchmarks of recyclable waste at a wildlife tourism lodge in the Greater Kruger Park area. Reflections on the general waste management at the were also made in terms of proposed principles for responsible waste management in protected areas.

CHAPTER 5: SUSTAINABLE WASTE MANAGEMENT PRACTICES IN THE HOSPITALITY INDUSTRY TOWARDS ENVIRONMENTAL RESPONSIBILITY AND ECONOMIC VIABILITY

The hospitality industry, a significant part of the global tourism sector, generates significant amounts of waste that contributes to environmental degradation. This chapter explores sustainable waste management techniques in the industry, focusing on waste generation, environmental consequences, legal frameworks, and the use of sustainable solutions. It examines waste generation in sectors like hotels, restaurants, and resorts, and the factors contributing to its development. The chapter also highlights the harmful environmental consequences of inadequate waste management, such as air, water, and soil contamination, negative impacts on biodiversity, and greenhouse gas emissions and climate change. Furthermore, it discusses the framework and standards regulating waste management in the hotel business, emphasizing the need for strict norms to promote environmental sustainability. Finally, the chapter advocates for a comprehensive approach to sustainable waste management, including strategies for reducing, reusing, recycling, recovering, and responsibly disposing of garbage.

CHAPTER 6: MINIMIZING WASTE IN THE HOSPITALITY INDUSTRY: THE OPPORTUNITIES AND CHALLENGES OF LEAN TOURISM

Lean Hospitality is a complete and efficient hospitality waste management solution. It prioritizes operational efficiency, customer satisfaction, waste reduction, and sustainability. This strategy uses lean manufacturing methods to improve service, resource use, and profitability in the hospitality industry. Cultural opposition, different service delivery architectures, and workforce lean competency are issues. These issues demand extensive training and strategic change management. Due to its complicated service delivery architecture and intangible client connections, the industry faces many challenges. Interfaces complicate trash identification and process standardization. Additionally, the focus on short-term financial rewards often obscures the long-term benefits of lean concepts, undermining their long-term adoption. To address these issues, a lean company culture must be actively fostered. Lean Hospitality can help the hospitality industry achieve environmental sustainability and operational efficiency with continuing education and strategic leadership.

CHAPTER 7: BEYOND BIN BASICS: CONFRONTING WASTE IN TOURISM AND HOSPITALITY

The tourism and hospitality sectors face significant challenges in managing waste, which pose environmental, economic, and social risks. This chapter explores waste management challenges within these sectors, examining the current state of waste management, environmental impacts of poor waste management, economic and social implications, innovative solutions and best practices, case studies of successful initiatives, and future outlooks. Key findings highlight the urgency for sustainable waste management practices, emphasizing the need for collaboration between businesses, governments, communities, and consumers. The chapter concludes with a call to action for integrating sustainability into operations, investing in infrastructure and technology, and advocating for supportive policies. Future outlooks focus on embracing innovation, increasing awareness, and fostering collaboration to build a more sustainable future for the tourism and hospitality sectors.

CHAPTER 8: IMPACT OF WASTE MANAGEMENT IN TRANSFORMING DHARAMSHALA TO SUSTAINABLY DEVELOPED TOURIST DESTINATION - AN EMPIRICAL INVESTIGATION: IMPACT OF WASTE MANAGEMENT IN TRANSFORMING DHARAMSHALA TO SUSTAINABLY DEVELOPED TOURIST DESTINATION

The burgeoning tourist influx in Dharamshala, exiled government capital of his holiness Dalai Lama and home to mighty Dhauladhaar, presents a pressing challenge for sustainable waste management, necessitating a comprehensive understanding of its impact on stakeholders. Through a nuanced analysis of stakeholder perspectives, including local communities, hospitality businesses, government bodies, non-government bodies and tourists, it elucidates the divergent interests, concerns, and aspirations shaping the waste management landscape. With the development in the infrastructure, growing tourist influx leading to mount of waste growing multi-fold and its need to manage the same in a sustainable manner. Furthermore, it underscores the imperative of fostering collaboration, innovation and stakeholder engagement to forge a sustainable path forward. By integrating international best practices, innovative technologies, and community-driven approaches.

CHAPTER 9: NAVIGATING THE PATH TO SUSTAINABILITY OVERCOMING ENVIRONMENTAL BARRIERS IN THE EGYPTIAN HOTEL INDUSTRY (EMPHASIZES THE JOURNEY AND SOLUTIONS)

This chapter presents the specific environmental practices facing Egyptian hotels and identifies potential barriers to implementing sustainable initiatives. The goal of this chapter was to gain a holistic perspective regarding the factors that impact environmental sustainability practices in hotels beyond simply discussing the environmental problems. Qualitative data analysis of 42 semi-structured interviews conducted among top management at a sample selected from four and five-star hotels in Egypt. This study reveals seven internal barriers: staff capabilities, physical limitations, the attitude of top management, cost concerns access to resources and no owner attitudes and day-to-day activities. The external factors are government regulation, incentives, and guest attitude. It subsequently examines the theoretical implications and discusses how future research might address environmental challenges in hospitality. Practical implications are established, including the necessity for staff training, management involvement, and partnerships with government and local stakeholders.

CHAPTER 10: EXPLORING STAKEHOLDER ENGAGEMENT IN TOURISM WASTE MANAGEMENT: SELF-HELP GROUPS' PERSPECTIVES

This chapter explores the motivations driving stakeholder engagement and the challenges faced in waste management initiatives within the tourism sector, focusing on the perspectives of self-help groups (SHGs). Through qualitative research methods, including in-depth interviews, the study examines the diverse motivations that prompt SHGs to participate in waste management projects, ranging from environmental concerns to socio-economic empowerment. Moreover, it identifies and analyzes the multifaceted challenges SHGs encounter, such as limited resources, inadequate infrastructure, and bureaucratic hurdles. This research contributes to a deeper understanding of stakeholder dynamics in tourism waste management by elucidating these motivations and challenges. Findings suggest that while SHGs play a crucial role in fostering sustainable practices, effective collaboration and support mechanisms are essential to address the encountered challenges and maximize the potential benefits of stakeholder engagement in waste management projects within the tourism sector.

CHAPTER 11: URBAN SOLID WASTE MANAGEMENT AND PUBLIC AWARENESS- FINDINGS FROM A STRATIFIED SURVEY IN BHUBANESWAR

The study aimed at accessing the current situation of solid waste management in the city and the awareness level and participation motive of the residents. A stratified random sampling method was used to collect 1447 responses from the residents of Bhubaneswar. The findings indicate that only 4.35% of the households are segregating waste. Since a whole lot of people are not habituated to such practices and many believe that it is of no use because the municipality workers are mixing it up again after collecting the waste are the two prominent reasons for non-segregation. One good thing is many people are selling most of the solid waste to the Kawadiwalas. The general awareness of the respondents was found to have a significant relationship with their willingness to participate in SWM. It has also been observed that female respondents have a lower level of basic awareness of SWM. senior citizens, and lower socio-economic groups, on the other hand, have a significantly higher basic awareness and are more willing to participate in SWM.

CHAPTER 12: WASTE MANAGEMENT PRACTICES IN KERALA'S PREMIER TOURIST DESTINATIONS A REVIEW

Kerala is known as "God's Own Country" because of its varied tapestry of lush landscapes, tranquil backwaters, and lively cultural heritage. The state's commitment to sustainable tourism demonstrates how serious it is about protecting its abundant natural and cultural resources for the enjoyment and benefit of future generations. Waste management in Kerala is a significant and urgent problem because of its high population density, fast urban growth, and flourishing tourism sector. It is crucial to tackle this problem to reduce environmental deterioration and protect public health. The waste management system in Kerala combines ancient practices with modern technology to effectively tackle the intricate difficulties brought about by urbanization and population expansion. Kerala aims to accomplish effective trash collection, segregation, recycling, and disposal through community engagement, decentralized initiatives, and innovative technologies. This will contribute to developing a more sustainable and environmentally conscious society.

CHAPTER 13: CRUISE SHIP WASTE MANAGEMENT A SYSTEMATIC REVIEW AND RESEARCH GAP ANALYSIS

LARGE CRUISE SHIPS generate substantial waste, necessitating effective management strategies to mitigate environmental impacts. This study employs a systematic literature review to identify key themes and research gaps in cruise waste management research. Using the PRISMA method, 47 relevant research articles were selected. The data from these articles were analysed using bibliometric and content analysis techniques. The key themes identified include plastic waste, microplastic waste, food waste, wastewater, and waste management strategies. The study revealed significant research gaps that demand immediate attention and action concerning the effectiveness of waste management strategies, socio-economic considerations, interdisciplinary approaches, regulatory compliance, food waste disposal methods, regional collaborations, policy initiatives, and integrated waste management approaches. These findings highlight the urgent need for further research to develop comprehensive and sustainable waste management solutions for the cruise industry.

CHAPTER 14: REDUCING SINGLE-USE PLASTIC AND WASTE: FANTASY OR REALITY AT DISNEY THEME PARKS

The tourism industry contributes to and is impacted by plastic waste. The Walt Disney Company (Disney) is one of the largest media and entertainment companies and a contributor to global travel and waste. Disney theme parks represent a microcosm of society and the conflict between consumerism and environmentalism in the effort to reduce single-use plastic and waste to landfill. Disney has targets to reduce single-use plastic and waste at its wholly owned and operated parks and resorts in France and the United States (U.S.). As a multinational company operating in the absence of a legally binding international agreement on plastic waste, Disney is subject to the plastics policies of each jurisdiction. The policy context in which Disney has established its waste targets and the actions taken by Disney to reduce plastic waste in the theme parks are discussed and opportunities to improve waste management practices and share learnings are identified.

CHAPTER 15: CONSTRUCTED WETLANDS AS A GREYWATER MANAGEMENT ALTERNATIVE IN THE TOURISM INDUSTRY

The exponential increase in tourism escalates greywater production from laundry, gardening, bathing, and sewage treatment, leading to water scarcity. Simple physical and biological processes can purify greywater for reuse, leading to water conservation. Constructed wetlands offer new approaches to dealing with greywater especially in polluted tourist areas through plant, soil and microbial activity filtration processes. These systems are feasible for tropical regions with limited water resources. This chapter examines the suitability of constructed wetlands for greywater treatment highlighting techniques, performance evaluations and environmental benefits. By integrating these systems, tourism facilities can reduce their environmental footprints thus realizing significant savings on water consumption. Constructed wetlands aids in improving greywater management and promoting sustainable water use because they have numerous advantages such as low maintenance costs and adaptability which are crucial in ecological preservation amidst rising tourism numbers.

CHAPTER 16: A COMPREHENSIVE REVIEW OF FOOD WASTE MANAGEMENT STRATEGIES IN THE FOOD SERVICE INDUSTRY

Of the food produced globally, 30% is wasted and if no significant improvement is made this figure is expected to double by 2050. Globally the hospitality industry is the third largest contributor to food waste presenting serious sustainability concerns. The mitigation of food waste is fundamental to the sustainable improvement of the industry, and it must establish effective solutions by creating partnerships with stakeholders and providing educational programs to employees and consumers aimed at waste reduction. The strategies encompass various operational elements including procurement, inventory management, menu design and portion control. This literature review provides an analysis of food waste management strategies and initiatives implemented by the food service industry to improve resource efficiency and promote sustainability. A literature search of academic journals and industry reports was utilised, serving as a valuable resource for the food service industry, policy makers and researchers to address the issue of food waste through effective management approaches.

CHAPTER 17: REVOLUTIONIZING RESPONSIBLE CONSUMPTION AND PRODUCTION OF FOOD IN HOSPITALITY INDUSTRY: MACHINE LEARNING AND DEEP LEARNING APPROACH TOWARDS ZERO FOOD WASTE

The hospitality industry plays a pivotal role in addressing challenges related to responsible consumption and food production. The hotel sector is leading the way in tackling important issues related to responsible food production and consumption to lead revolutionary solutions by using the capabilities of deep learning and machine learning technology. The integration of modern technical solutions becomes crucial in the current setting, where global poverty, malnutrition, and the complex dynamics of food adulteration considerably affect the industry's demand-supply mechanism. Keeping a close eye on the larger global scene, this study promotes practical actions that go beyond simple technology implementations. In the face of global issues such as hunger, malnutrition and food adulteration, the assimilation of advanced technological strategies becomes imperative. This chapter explores the role of ML and DL in achieving zero waste within the hospitality sector, addressing challenges and presenting innovative strategies for effective food loss-waste management.

CHAPTER 18: FOSTERING A CIRCULAR ECONOMY IN TOURISM: INSIGHTS FROM EGYPT

This chapter examines the state of the circular economy (CE) in Egypt's tourism sector, revealing both challenges and opportunities. The chapter uses a multi-faceted approach, combining secondary data analysis with a literature review. It systematically reviews and synthesizes existing data from government reports, industry statistics, and prior research studies. It analyses barriers, demographic influences, and organisational dynamics and explores how technology can aid in overcoming these challenges. The chapter also draws on best practices from successful transitions elsewhere. Findings highlight the need for robust policies, targeted educational campaigns, and technological adoption. Effective CE transition in Egypt will require structured approaches, international collaboration, and tailored practices. This chapter is novel for its detailed analysis of CE adoption in Egypt's tourism, offering strategic pathways for advancing circularity. It fills a critical gap by providing an integrated approach to an unexplored topic in a key developing country and Middle Eastern destination.

CHAPTER 19: TURNING THE WASTE INTO WEALTH: PROGRESSING TOWARD GLOBAL SUSTAINABILITY THROUGH THE CIRCULAR ECONOMY IN WASTE MANAGEMENT

Increasing waste production presents substantial environmental and public health obstacles, necessitating inventive and sustainable waste management approaches. This chapter explores the fundamental concepts of circular economy and how it may be used to change waste management systems from linear disposal methods to sustainable and efficient models that make the most of resources. It explores the influence of efficient waste management on diminishing environmental deterioration, fostering resource preservation, and bolstering economic development through waste materials' reuse, recycling, and repurposing. The chapter highlights the importance of adopting comprehensive strategies that combine technology advancements, policy restructuring, and community involvement to enable a shift toward circular waste management methods that support global sustainability objectives.

CHAPTER 20: WASTE-TO-ENERGY INNOVATIONS: EVIDENCE FROM TOURISM INDUSTRY

The convergence of waste-to-energy innovations and the tourism industry presents a dynamic frontier with significant potential for sustainable development. As tourism continues to grow globally, the environmental footprint of waste generated by this sector is becoming increasingly concerning. Waste-to-Energy (WtE) technologies, with the crucial role of technology providers, offer a promising solution by converting waste into usable energy, thereby reducing environmental impact and promoting a circular economy. This paper explores future trends and opportunities in integrating WtE systems within the tourism industry. By embracing WtE innovations, the tourism industry can significantly reduce its waste burden, lower energy costs, and contribute to a more sustainable future. This chapter underscores the necessity of multidisciplinary collaboration and supportive regulatory frameworks to fully realize the potential of WtE technologies in transforming the environmental landscape of tourism.

In concluding this exploration of *Sustainable Waste Management Within the Tourism and Hospitality Sectors*, we reflect on the profound insights and transformative ideas presented throughout this volume. The urgent need for environmental stewardship in these industries is undeniable, as they face significant challenges in managing the waste they generate. This book, *Sustainable Waste Management in Tourism and Hospitality Sectors*, stands as a testament to the collective effort required to address these issues.

From the historical context established in Chapter 1, where the rise in research activity underscores a growing commitment to sustainability, to the practical strategies outlined in subsequent chapters, the book delivers a comprehensive framework for tackling waste management. Each chapter, whether discussing global trends, regional practices, or innovative technologies, contributes to a nuanced understanding of how to mitigate the environmental impacts associated with tourism and hospitality waste.

We are reminded that sustainable waste management is not a singular task but a multifaceted challenge requiring diverse solutions. As explored in Chapter 2, the spectrum of waste—from food and e-waste to plastics and hazardous materials—necessitates targeted approaches and robust recycling methods. Chapters 3 and 4 highlight the importance of education and the unique demands of protected areas, while Chapters 5 through 7 delve into the economic and operational facets of implementing sustainable practices.

The real-world applications and case studies presented, such as those in Chapter 8's examination of Dharamshala and Chapter 12's review of Kerala's tourist destinations, illustrate the tangible impacts of effective waste management. Similarly, Chapters 13 and 14 provide insights into the cruise industry and large-scale efforts by companies like Disney, demonstrating how industry leaders can drive meaningful change.

Innovations such as waste-to-energy technologies discussed in Chapter 20 and the use of machine learning for zero food waste in Chapter 17 reflect the forward-thinking approaches essential for transforming waste management practices. The circular economy concepts explored in Chapters 18 and 19 offer a vision for a sustainable future where waste is minimized, resources are conserved, and economic benefits are maximized.

As the editor, we express my gratitude to the contributors who have provided their expertise, insights, and innovative solutions. Their dedication to advancing sustainable practices in tourism and hospitality is both inspiring and essential. To the readers, I hope this book serves as a catalyst for action, encouraging you to embrace and advocate for sustainable waste management strategies within your spheres of influence.

Together, through informed action and collaborative efforts, we can drive the necessary changes to create a more sustainable future for the tourism and hospitality sectors, ensuring their positive contribution to the global sustainability agenda.

Reshma Sucheran
Durban University of Technology, South Africa

Chapter 1
Global Trends in Waste Management Research Within the Tourism and Hospitality Sector:
A Bibliometric Analysis

Reshma Sucheran
Durban University of Technology, South Africa

ABSTRACT

This chapter presents a bibliometric analysis of 280 research publications on waste management in the tourism and hospitality industry. Using Bibliometric R and the Web of Science database, this study examined several key factors, including publication and citation trends, author productivity and influence, leading organisations, countries, journals, and common themes in the field. The data show a steady rise in the number of publications and citations over time, reflecting the growing recognition of the importance of waste management in the tourism sector. This increase has been especially noted in recent years and is driven by a broader societal shift toward environmental awareness and the need for sustainable practices. The analysis of author productivity highlights varying levels of impact within the field, with some individuals making particularly significant contributions. Additionally, the study identifies key organizations, countries, and journals essential for advancing research on waste management in tourism.

DOI: 10.4018/979-8-3693-6110-8.ch001

1. INTRODUCTION

Bibliometric techniques use statistical tools to assess and measure the growth of literature in a specific field. This approach is reliable for uncovering historical developments (Batistic & Kase, 2015) and identifying numerical patterns in publications on a particular topic. It also provides a way to connect the micro-level perspectives of journals and institutions with the macro-level view of national and global trends. This chapter aims to analyse and document the patterns of waste management research in tourism and hospitality from 1996 to 2023. Through a bibliometric analysis, this chapter offers a concise summary of the most noteworthy research achievements. Peer-reviewed articles on waste management were reviewed, focusing on factors such as the distribution of articles over time, the most relevant subject categories, leading scientific journals, and the most active countries and institutions. Additionally, a content analysis of the author's keywords was performed. This chapter also explores key metrics, including the total local citation score (TLCS), total global citation score (TGCS), and average local citation frequency of an item in the collection (ALCS).

2. LITERATURE REVIEW

2.1 Waste Management in Hospitality and Tourism

The global tourism industry is anticipated to expand significantly in the coming years, albeit accompanied by escalated waste generation. The amount of waste produced per tourist per day is approximately 1.6 kg, according to Obersteiner et al. (2021). Wealthier countries report higher waste generation per tourist and more advanced waste management practices, whereas poorer countries rely predominantly on landfill disposal methods (Ezeah et al., 2015). A study conducted among 120 hotels in Hoi An, Vietnam, revealed that hotels generate approximately 2.5 kg of solid waste per guest per day, with 58.5% being biodegradable, 25.8% recyclable, and 15.7% categorised as other waste (Pham et al., 2019). Various factors influencing the volume and composition of waste include hotel size, pricing range, type of restaurant, and the presence of a garden on the premises. According to Pirani and Arafat (2014), waste generated by tourism and hospitality comprises approximately 37–72% organic waste, 6–40% paper and cardboard, 5–15% plastic, and 3–14% glass. Around one-third of the hospitality sector's waste is food waste, of which almost one-third comes from plate waste, which refers to unused edible food on guests' plates after meals. Papargyropoulou et al. (2019) suggest that 92% of this plate waste could be avoided through more effective management practices.

The exponential growth of global tourism has created a need for inventive waste reduction measures owing to the association between tourist arrivals and waste generation (Han et al., 2020). Moreover, enhancing waste management aligns with Sustainable Development Goals (SDGs) and fosters environmental sustainability. Waste management strategies employed by the tourism sector differ worldwide, with sustainable waste management falling behind other environmental measures (Diaz-Farina et al., 2023). Insufficient waste management threatens the destination's sustainability, quality of tourism experiences, and overall well-being of the local community (Diaz-Farina et al., 2023; Dileep, 2007). Global attention to sustainable waste management solutions in the tourism industry emphasises the necessity for widespread strategies addressing social, economic, and environmental issues. Adopting the waste hierarchy principles, such as "reduce, reuse, recycle", is essential for attaining sustainable waste management objectives (Dileep, 2007). Recycling, alongside the strategies of reducing and reusing waste, is vital in Integrated Solid Waste Management (ISWM), which addresses waste-related challenges (Memon, 2010). A vital element for effectively preventing and treating waste in the tourism industry is a comprehensive understanding of waste generation through solid waste audits (Pirani & Arafat, 2014). Tourists are also ready to pay higher prices for ecologically friendly products and services, such as hotels prioritising sustainability (Kang et al., 2012). Consequently, numerous hotels are implementing eco-friendly policies and recycling processes to increase profitability, enhance customer satisfaction, and improve their corporate brand image (IHRA 1995).

Standard environmental practices implemented in the tourism and hospitality industry include water conservation and reuse, integral energy efficiency, reduction of material consumption, and solid waste mitigation. These practices often involve using environmentally friendly products and address pressing issues such as single-use plastics and increasing food waste, particularly during the COVID-19 pandemic (Filimonau & Tochukwu, 2020). Despite the importance of effective waste management in mitigating waste generation by the hospitality industry, sustainable waste management remains less advanced and widespread than other environmental initiatives (Pirani & Arafat, 2014). Given its significant waste generation, efforts to improve waste management in the hospitality sector are crucial. However, sustainable waste management practices require further development and adoption within the industry to achieve broader environmental objectives. To address these challenges, integrated waste reduction techniques must be used to benefit tourism facilities, host communities, and the environment (Ezeah et al., 2015).

2.2 An Overview of Bibliometric Analysis

The emergence of contemporary science can be attributed to the application of accurate measurement techniques to the inquiry of nature. Professionals and researchers need thorough theoretical and practical methods to analyse experimental data efficiently. Van-Raan (2004) stressed the crucial significance of measurement in science by referring to Onnes's assertion, "Measuring is knowing." This emphasises the importance of using quantitative analysis in scientific investigation. Bibliometrics is a systematic methodology that quantitatively analyses scientific publications' production, development, maturation, and usage. Pritchard introduced Bibliometrics in 1969 as an alternative to "statistical bibliographies." It has developed in tandem with the rapid expansion of scientific knowledge, which has created the need to store and arrange large volumes of published material in bibliographic databases (Gutiérrez-Salcedo et al., 2018). These databases store significant information such as citations, keywords, titles, journals, authors, H-Index, and institutions. This data is used to support scientific evaluation studies that adopt bibliometric approaches.

Bibliometric analysis is, therefore, an effective method for efficiently managing substantial amounts of data and recognising patterns within particular subjects across time (Xu et al., 2018).

This analytical methodology combines performance analysis and scientific mapping approaches (Donthu et al., 2021). Performance analysis evaluates the influence and ranking of different research components in a specific subject, such as journals, countries, authors, papers, and institutions, and uses multiple indicators to measure their impact and determine their success in the field (Cobo et al., 2011; Zupic & Cater, 2015; Donthu et al., 2021). Productivity may be quantified by the total or average number of publications, while impact is assessed through the yearly citation count or the H-Index. The H-Index, conceived by Hirsh, measures academics' productivity and influence, denoting the number of articles by a researcher that have garnered h citations or more (Hirsch et al., 2005).

Science mapping visually represents the structure and growth of scientific fields by analysing the connections between different research elements, both physical and conceptual (Zupic & Cater, 2015; Cobo et al., 2011). It uses techniques like bibliographic coupling, co-authorship, co-word analysis, co-citation, and citation patterns, often alongside network analysis, to describe a subject's intellectual and bibliographic landscape. Citation analysis measures a publication's influence based on how often it is cited (Donthu et al., 2021). Coauthorship analysis examines collaborative efforts and reveals the social networks within a field (Zupic & Cater, 2015).

Keyword co-occurrence analysis identifies terms that frequently appear together, suggesting conceptual links, and explores clusters of terms around common themes. Co-citation analysis tracks how often two publications are cited together, uncovering the conceptual framework of a research field (Rossetto et al., 2018). Bibliographic coupling identifies publications with similar citation patterns to highlight recurring themes in the literature (Donthu et al., 2020). Performance analysis provides descriptive insights into the contributions of researchers, organisations, and countries by measuring their productivity and impact (Donthu et al., 2021).

3. METHODOLOGY

Figure 1 illustrates the bibliometric analysis flow diagram utilised in this study. The search extracted various document types, authors, publication years, languages, countries, sources, popular articles, and institutions from the Web of Science (WOS) database. Web of Science (WoS) was selected for its comprehensive index coverage compared to other databases (Yakath Ali et al., 2021) and its widespread acceptance and frequent use (van Nunen et al., 2018). Developed by Thomson Reuters Scientific, Web of Science is a widely recognised and utilised database in research. It provides more reliable and uniform records and enhanced visual representations for citation analysis compared to other databases. Additionally, users can download complete citation records in a ".txt" file format compatible with various literature analysis tools, such as Bibexcel (Newell & Cousins, 2015). The database utilised in this chapter for searching scientific information included Science Citation Index Expanded (SCI-E), Social Sciences Citation Index (SSCI), Conference Proceedings Citation Index- Science (CPCI-S), and Conference Proceedings Citation Index- Social Science & Humanities (CPCI-SSH), which are all sub-databases within Web of Science.

Figure 1. Bibliometric analysis flow diagram

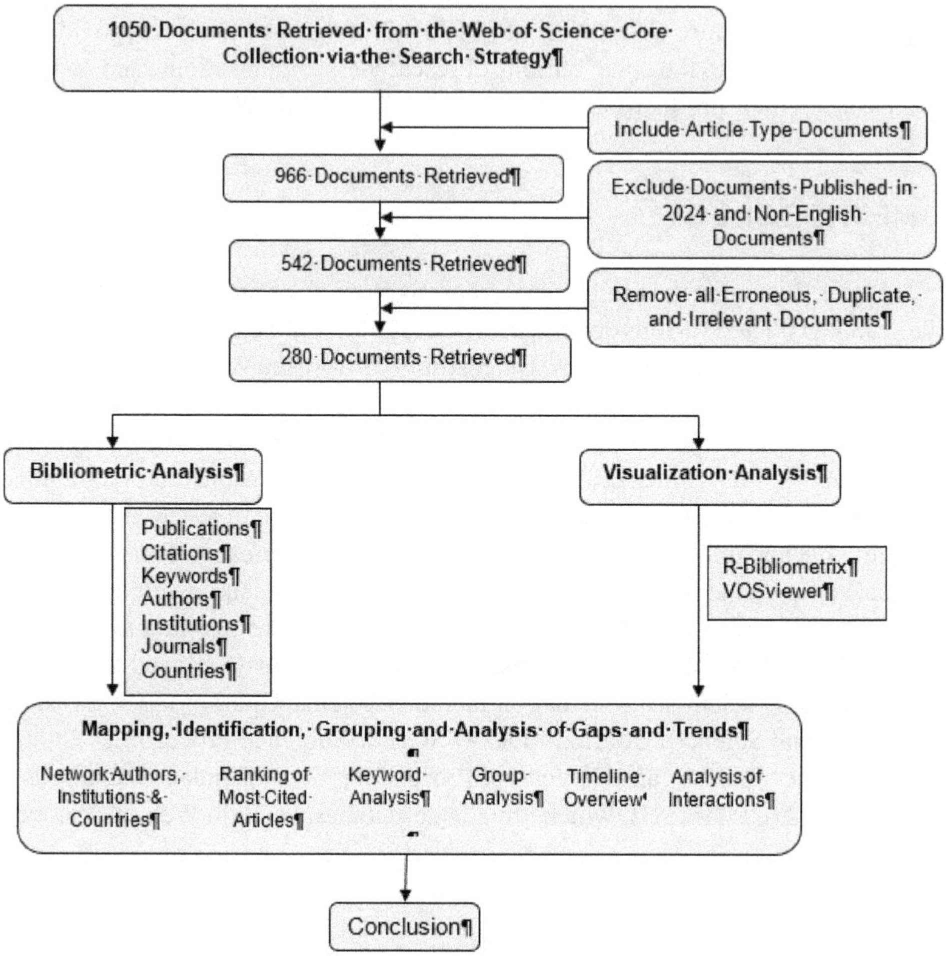

Source: Author's Compilation

For the bibliometric analysis in this chapter, R-based Bibliometrix was employed. Bibliometrix, a software based on the R programming language, was chosen for its capability to efficiently generate and present extensive bibliometric maps using a web browser interface. This software is commonly used to highlight research units based on scientific production and citation and to visualise graphs analysing co-citation, co-authorship, and co-occurrence (Aria & Cuccurullo, 2017). Biblioshiny, a web-based graphical interface powered by Bibliometrix, is a tool for non-coders to conduct bibliometric analysis (Aria & Cuccurullo, 2017). Initially developed using R, Bibliometrix was enhanced for compatibility with other R packages. The creators of Bibliometrix, Massimo Aria and Corrado Cuccurullo from the University of Naples and Luigi Vanvitelli from the University of Campania in Italy, designed this tool to interact with data from WoS, Scopus, and Dimensions. Its user-friendly interface is structured with a main menu that follows the SMA workflow, offering various analytics and graphs for metrics at three levels (source, author, document) and three knowledge structures (conceptual, intellectual, and social). The generated graphs and performance analyses can be exported in different file formats, including maps to Pajek and HTML, and tables to the clipboard or saved as Excel, PDF, or printed. Bibliometrix, a bibliometric analysis tool coded in R, operates within an ecosystem software environment consisting of open libraries, algorithms, and graphical software (Aria & Cuccurullo, 2017).

Table 1. Search strings and keywords

Group	Keywords
Group 1: Waste management-related words	(terms "waste management" OR "sustainable waste management" OR "waste reduction" OR "waste disposal" OR "waste generation")
Group 2: Tourism industry-related words	("tourism" OR "hospitality" OR "hotel*" OR "food service" OR "restaurant*" OR "tourism destinations" OR "travel industry")

Table 1 presents the search strategy employed to refine the documents for analysis in this study. The search strategy involved using specific keywords grouped into two main categories. The first group comprised waste management-related terms such as "waste management," "sustainable waste management," "waste reduction," "waste disposal," and "waste generation." These terms encompass various aspects of waste handling practices, including strategies to minimise waste generation, implement sustainable disposal methods, and manage waste materials effectively across different contexts. The second group of keywords targeted specific tourism industry segments to complement the waste management terms. These included terms like "tourism," "hospitality," "hotel*," "food service," "restaurant*," "tourism destinations," and "travel industry." The complete search string was structured as: "waste management" OR "sustainable waste management" OR "waste reduction" OR "waste disposal" OR "waste generation") AND ("tourism" OR "hospitality" OR "hotel*" OR "food service" OR "restaurant*" OR "tourism destinations" OR "travel industry"). The search string combined these two groups of keywords using logical operators (e.g., "OR") to refine the scope of the retrieved articles. By incorporating waste management-related terms and tourism industry-specific keywords, the search string effectively filtered out irrelevant articles and retrieved scholarly literature addressing waste management challenges and solutions within the tourism and hospitality sectors.

Table 2. Refinement through inclusion and exclusion criteria

Inclusion criteria	Exclusion criteria
Focus on waste in the tourism and hospitality industry	All studies not directly related to waste management in the tourism and hospitality industry
Articles	Review article, early access, proceeding paper, editorial material, retracted publication
English	Papers in languages other than English
1996-2023	Exclude articles published in 2024

To expand the scope of the investigation and incorporate the specified keywords detailed in Table 1, a search query was executed on April 16, 2024, encompassing all document types without imposing date restrictions. As shown in Figure 1, the query was applied to titles, abstracts, and keywords, initially retrieving 1050 documents. Filtering to include articles only, 966 documents were identified, with the further exclusion of articles from 2024, resulting in 914 documents. Only documents addressing waste management in the tourism and hospitality sectors were considered relevant, yielding 542 results. Subsequently, titles and abstracts were scrutinised to exclude entries that did not meet the search parameters, were irrelevant, contained errors, or were duplicates, resulting in a final dataset of 280 documents. Bibliometric graphs, tables, and maps were developed and evaluated using Biblioshiny and the Web of Science. Refinement through inclusion and exclusion criteria is critical in ensuring the relevance and quality of research articles extracted from the Web of Science for bibliometric analysis of waste management in tourism and hospitality.

4. RESULTS AND DISCUSSION

4.1 Performance Analysis

Table 3. Overview of data

Primary Information on the Data	Timespan	1996-2023
	Documents	280
	Document average age	5.46
	Average citations per document	23.18
	References	12981

continued on following page

Table 3. Continued

Document Contents	Keywords Plus (ID)	662
	Author's keywords (DE)	1057
Authors	Authors	839
	Authors of single-authored documents	32
Authors Collaboration	Single-authored documents	33
	Co-Authors per document	3.33
	International co-authorships (%)	38.21
Document Types	Article	280

The analysis reviewed performance data on documents, authors, institutions, countries, journals, and citations from 1996 to 2023. This timeframe aligns with discussions from the 1992 Rio Earth Summit on environmental impact, emphasising sustainable development and the integration of economic, social, and environmental considerations. The average age of the documents is relatively recent at 5.46 years, reflecting a focus on contemporary research developments in waste management within tourism and hospitality.

A key metric is the average number of citations per document, which is 23.18, indicating a significant scholarly impact and influence in the field. The dataset includes a substantial collection of 12,981 references, reflecting deep engagement with the existing literature on waste management in tourism and hospitality. Keywords Plus (662) and Author's keywords (1,057) provide terms used to categorise and describe the research articles, aid in thematic analysis, and identify key research topics and clusters.

The dataset featured 839 unique authors, with 32 documents authored by one individual, suggesting high collaboration among researchers. On average, each document has 3.33 co-authors, which reflects the collaborative nature of the research in this area. The percentage of international co-authorships was 38.2%, confirming significant global and cross-cultural collaboration. All 280 selected documents were articles because they focused on original research published in scholarly journals.

4.2 Publications Per Year

Figure 2. Number of publications and citations per year

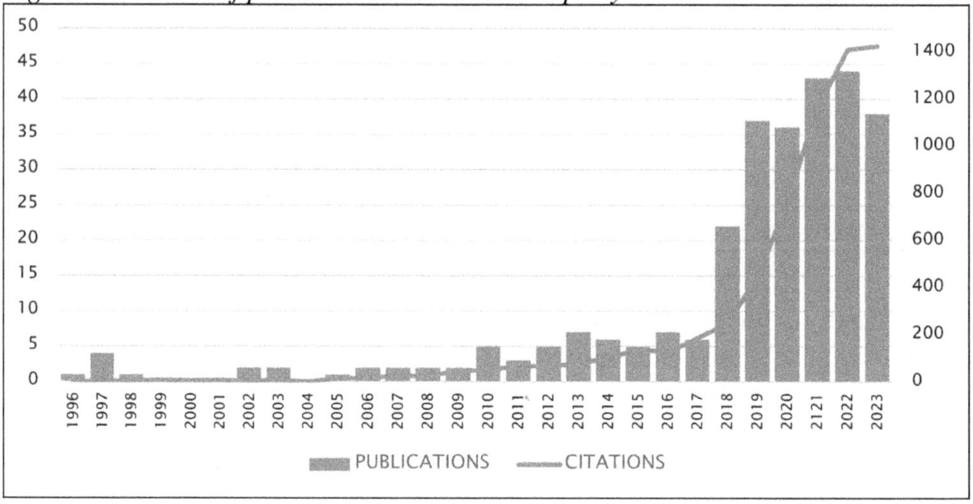

Figure 2 presents the publication and citation trends of waste management in the tourism sector from 1996 to 2023, showing how research output and its impact have changed over time. From 1996 to 2005, the number of publications was relatively low, with only an occasional increase. Despite this, citations have steadily grown, increasing interest in tourism-related waste management. In 2009, there was a noticeable increase in publications and citations. The number of publications has grown significantly from 2018 onwards, peaking in 2019, and citations have grown even faster than publications, especially in recent years. Data from 2020 show a sharp rise in both publications and citations.

These findings reflect the growing recognition of the importance of waste management in the tourism and hospitality sectors. The consistent increase in citations, even during times of lower publication output, suggests that early research in this area has continued to influence and inform current studies. The sharp increase in publications and citations after 2009 points to a global shift towards sustainability and environmental awareness, prompting more detailed investigations into waste management practices in the tourism sector and an increasing interest in developing innovative solutions to reduce waste and promote sustainable practices. Moreover, the rapid growth in citations suggests that research in this field is widely shared and influences academic discussions and practical applications in the industry.

4.3 Author Analysis

Table 4. Top contributing authors based on the number of studies, total citations and h-index

Author	Publications	h_index	Total Citations
Filimonau V	10	6	144
Han H	5	5	344
Kim W	5	5	319
Arbulu I	3	3	170
Fujiwara T	5	3	39
Hall CM	3	3	210
Kim MJ	3	3	210
Lozano J	3	3	170
Rey-Maquieira J	3	3	170
Wang L	3	3	264

Table 4 provides a detailed overview of the authors' contributions to waste management research in tourism using metrics such as publications, H-Index, and total citations. Filimonau stands out with ten publications, an H-Index of 6, and 144 citations. Although Filimonau has a solid publication record, the moderate H-Index and citation count suggest a steady, though not dominant, impact on the academic community. In contrast, authors such as Han and Kim W., each with five publications, exhibit a stronger influence, with H-Indexes of 5 and total citation counts of 344 and 319, respectively. These figures underline significant scholarly contributions to waste management research in tourism. Other authors, including Arbulu, Fujiwara, Hall, Kim MJ, Lozano, Rey-Maquieira, and Wang, have each published three papers. Their impact varies, with H-Index values ranging from three to five and total citations ranging from 39 to 264. While their output is consistent, these differences in metrics reflect the diverse ways in which they contribute to advancing knowledge in this specialised area of research.

Table 6. Total citations

Author	Year	Title of Article	Total Citations
Xiong et al.	2018	Sources and distribution of microplastics in China's largest inland lake - Qinghai Lake	359
Erdogan and Baris	2007	Environmental protection programs and conservation practices of hotels in Ankara, Turkey	206
Wang et al.	2017	The weight of unfinished plate: A survey-based characterisation of restaurant food waste in Chinese cities.	165
Mbaiwa	2003	The socio-economic and environmental impacts of tourism development on the Okavango Delta, north-western Botswana	158
Trung and Kumar	2005	Resource use and waste management in the Vietnam hotel industry	156
Johnson	2002	Environmentally sustainable cruise tourism: a reality check	152
Han et al.	2018	Water conservation and waste reduction management for increasing guest loyalty and green hotel practices	144
Arbulu et al.	2015	Tourism and solid waste generation in Europe: A panel data assessment of the Environmental Kuznets Curve	116
Singjai et al.	2018	Green initiatives and their competitive advantage for the hotel industry in developing countries	112
Kularatne et al.	2019	Do environmentally sustainable practices make hotels more efficient? A study of major hotels in Sri Lanka	107

Table 5 presents the most-cited documents in the bibliometric analysis of waste management in tourism. Xiong et al.'s (2018) study of microplastics in Qinghai Lake stands out with 359 citations, highlighting its significant influence and growing concern over microplastic contamination in inland water bodies. Erdogan and Baris' (2007) work on environmental protection programs in hotels in Ankara, Turkey, has received 206 citations, reflecting the importance of sustainability in the hospitality industry, especially in conservation practices. Other influential studies include Wang et al. (2021) on restaurant food waste in Chinese cities, Han et al. (2020) on water conservation and waste reduction in green hotel practices, and Kularatne et al. (2019) on sustainable practices in Sri Lankan hotels. These studies address crucial issues, such as food waste reduction, water conservation, and sustainable hotel management, highlighting the growing focus on environmental responsibility within tourism. The high citation counts of these papers confirm their academic- and industry-wide influence. They have shaped industry standards, informed policy discussions, and enriched academic discourse on sustainable waste management in tourism. However, relying solely on citation counts has limitations. Although high citations may reflect a study's academic significance, they do not always measure the research's practical impact in solving real-world waste management challenges. Factors such as publication platforms, author reputation, and citation practices within specific fields also played a role in these counts.

Figure 3. Authors' production over time

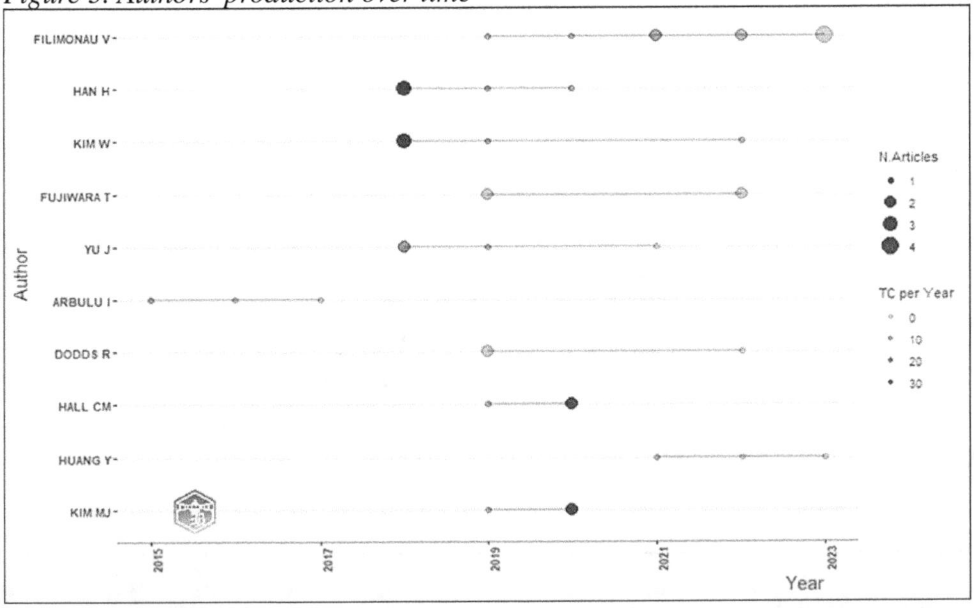

Figure 3 illustrates the productivity of authors in waste management research over time, highlighting key trends in the bibliometric analysis of this field. For example, authors such as Han H, Kim W, and Yu J showed increased productivity in 2018, marking a peak period of concentrated research efforts. Additionally, the data reveals that authors like Filimonau V, Fujiwara T, and Dodds R consistently produced numerous publications across multiple years, demonstrating their ongoing commitment and significant contributions to waste management in tourism.

Figure 4. Co-citation network by authors

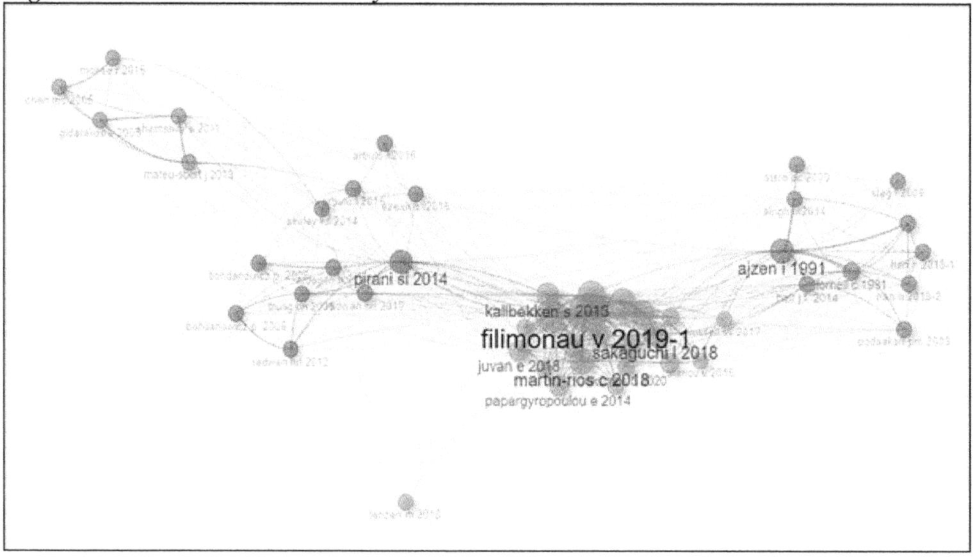

Figure 4 presents the co-citation network of papers from a bibliometric analysis of waste management in tourism, revealing key contributors and the relationships between their work. Authors like Pirani (2014), Ajzen (1991), and Filimonau (2019) stand out with high betweenness centrality values, signifying their important role in connecting various research themes within this field. Their work is foundational, extensively cited by other scholars, and continues to shape the discourse and theoretical frameworks surrounding waste management in tourism. The temporal trends shown in the network highlight the evolving nature of research in this area. While earlier seminal works, such as Ajzen's (1991) theory of planned behaviour, remain highly relevant, more recent contributions, such as Filimonau's (2019) research, point to ongoing advancements and the emergence of new topics. This combination of older, established works with newer studies reflects the field's dynamic growth and the continuous development of waste management strategies within tourism. The interdisciplinary connections revealed by the co-citation network, with authors appearing across different clusters, confirm the collaborative nature of this research area. Scholars from various disciplines, including behavioural sciences, environmental studies, and management, are working together to find innovative solutions for sustainable waste practices in tourism. This interdisciplinary approach is critical for addressing the complex challenges of waste management and developing integrated solutions that combine environmental sustainability with tourism industry development.

4.4 Analysis by Journals, Organizations and Countries

Table 7. Top contributing organizations, countries, journals

Affiliations	TS	Journal Title	TS	TC	Country	TS	TC
Cebu Technological University	10	Sustainability	27	322	China	23	823
University of Costa Rica	10	Journal of Cleaner Production	16	533	USA	22	307
Bournemouth University	9	Journal of Sustainable Tourism	12	341	Spain	18	526
North-West University	9	Marine Pollution Bulletin	12	524	Italy	17	326
Ryerson University	9	Waste Management Research	10	543	United Kingdom	17	669
University of Surrey	9	Resources Conservation and Recycling	8	249	India	12	126
Dong A University	8	Waste Management	8	543	Indonesia	10	36
IPB University	8	International Journal of Hospitality Management	7	370	Korea	10	573
San Jose State University	8	Environmental Science and Pollution Research	6	72	Thailand	10	231
University of La Laguna	8	Tourism Management	5	488	Germany	8	147

Table 7 provides a detailed overview of the leading organisations, countries, and journals contributing to waste management research in the tourism industry. Cebu Technological University and the University of Costa Rica are prominent, each contributing ten studies reflecting their strong focus on exploring sustainable waste practices in tourism destinations. The University of Costa Rica, known for its environmental expertise, likely provides valuable insights into managing waste in tourism settings. Bournemouth University, North-West University, Ryerson University, and the University of Surrey also play significant roles, each contributing nine studies. Other key institutions, such as Dong-A University, IPB University, San Jose State University, and the University of La Laguna, have each contributed eight studies. Leading journals in this field include *Sustainability*, *Journal of Cleaner Production*, *Journal of Sustainable Tourism*, *Marine Pollution Bulletin*, and *Waste Management Research*. These journals frequently publish influential research on waste management in tourism and have high citation counts.

Regarding geographic contribution, China leads in publication volume with 23 studies, but the United Kingdom, despite having fewer studies (17), has a much higher total citation count (669). This suggests that UK-based research is particularly influential and recognised within the academic community, reflected by its higher impact. Other countries, such as Spain, Italy, Germany, and the United States, also play key roles in advancing research on waste management in tourism, particularly in Europe and North America, where strong research networks and institutional support foster collaborations. The presence of Asian countries like India, Indonesia, Korea, and Thailand among the top contributors shows a growing research interest in waste management within fast-developing tourism markets. Understanding these regional contributions and collaborative dynamics is vital for policymakers, researchers, and practitioners. It helps shape effective policies, promote international partnerships, and advance sustainable waste management practices in the tourism sector globally.

Figure 5. Co-citation network by journals

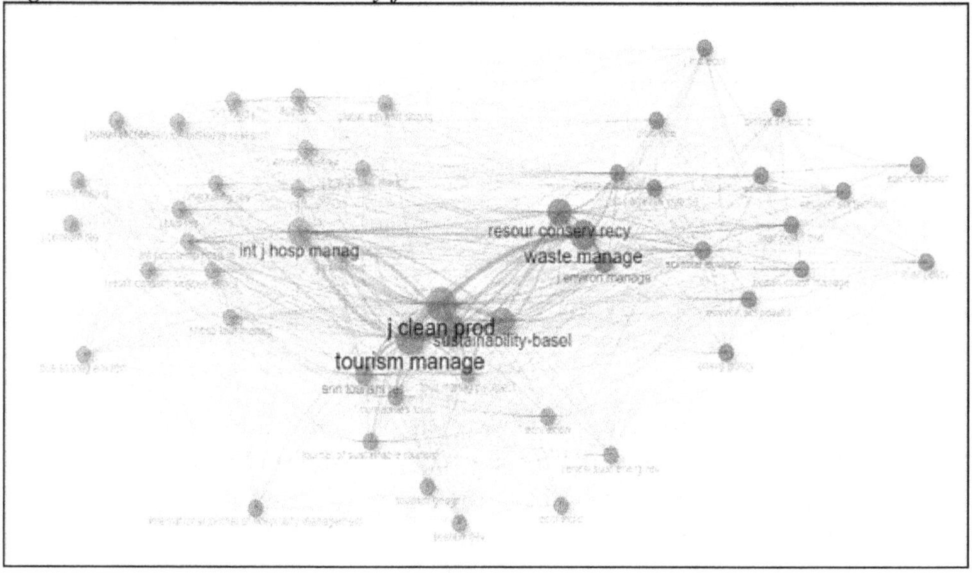

The data in Figure 5 illustrates the co-citation network of sources from a bibliometric analysis of waste management in tourism. Journals like *the Journal of Cleaner Production, Tourism Management, and the Journal of Sustainable Tourism* show high betweenness centrality values, indicating their crucial role in linking various research areas within waste management in tourism. The presence of journals from different fields, such as environmental sciences (*Journal of Environmental Management*), hospitality management (*International Journal of Hospitality Management*), and recycling (*Resources, Conservation & Recycling*), emphasises the interdisciplinary nature of this research. The high centrality values of these journals suggest that they act as key connectors between different research clusters, fostering collaboration and sharing ideas. By bridging disciplines, these journals help integrate knowledge and promote innovative solutions and best practices across academic fields related to waste management in tourism.

Figure 6. Collaboration network by country

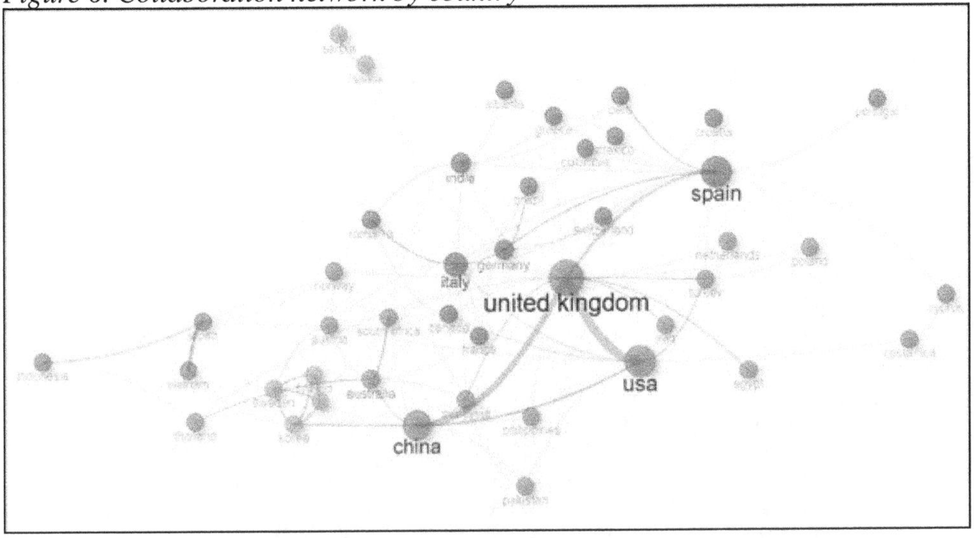

Figure 7 shows the country collaboration network in the bibliometric analysis of waste management in tourism. The United Kingdom, Spain, China, and the USA are central hubs with high betweenness centrality values, highlighting their key roles in fostering international collaborations and spreading knowledge across different research areas. These countries likely have strong research networks and are highly active in global partnerships, driving progress in sustainable waste management in tourism. The data also points to the significant global influence of countries like Germany, Italy, and India, which also display high centrality values. These nations are actively involved in international research collaborations, underscoring the importance of diverse regional contributions in tackling environmental challenges within the tourism industry. Additionally, the network reveals growing collaborations with countries like Russia, Thailand, Australia, Japan, and Malaysia, signalling their increasing participation in and impact on the global conversation about sustainable waste management in tourism.

4.5 Keyword Analysis

Figure 7. Word cloud

The findings from the word cloud analysis in Figure 7 provide several important insights into research on waste management in the tourism and hospitality sectors. The prominence of terms like "tourism," "sustainability," "waste management," and "food waste" shows a clear focus on addressing the environmental impacts of tourism activities. This suggests that research in this field aims to develop strategies to manage and reduce waste while integrating these efforts into broader sustainability initiatives. The frequent appearance of terms such as "sustainable tourism," "solid waste," and "recycling" highlights an inclusive approach to sustainability, where waste management is seen as a key part of the tourism industry's overall sustainability goals. The mention of "restaurant" and "hospitality industry" points to the significant role of food-related activities in waste management research. This suggests a growing need for focused studies on managing food waste within the hospitality sector, emphasising improving waste-handling practices in restaurants and hotels. Addressing food waste is crucial for creating targeted solutions to reduce waste in these settings.

The variety of terms in the word cloud reflects the multidisciplinary nature of waste management research in tourism. Effective waste management in this sector requires insights from environmental science, sustainability, hospitality management, and waste engineering. Researchers are increasingly adopting interdisciplinary approaches to tackle the complex challenges of waste management in tourism. Additionally, the prominence of specific terms helps highlight key research priorities and areas that may need further investigation. For example, if certain aspects of waste management are less frequently mentioned, this could indicate gaps in the research that require more attention. Understanding these priorities can help guide future studies and focus on areas needing further development.

Figure 8. Trend topics

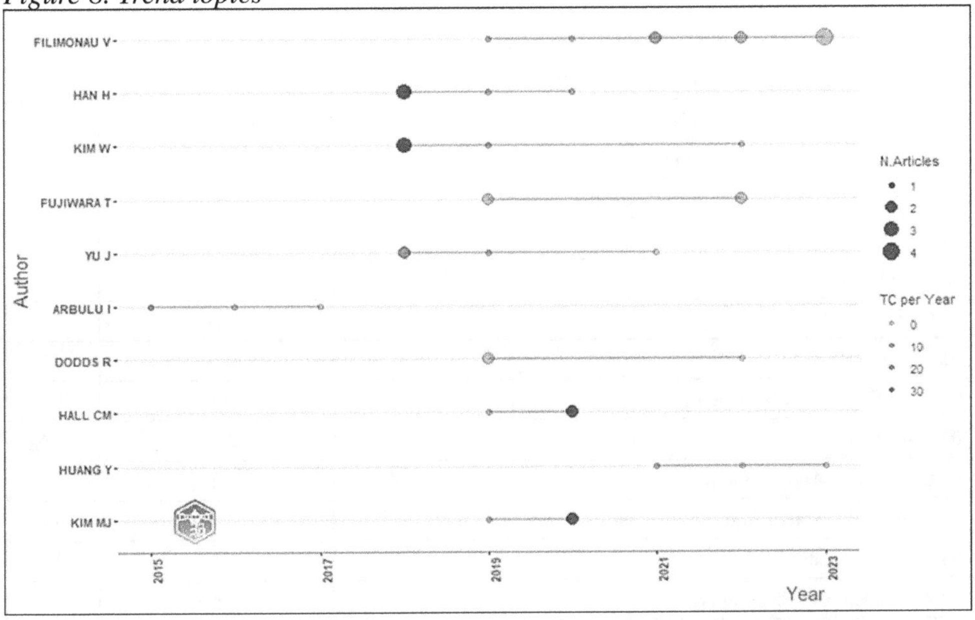

The data in Figure 6 shows essential trends in waste management research in the tourism and hospitality sector, revealing a clear shift towards sustainability and innovative practices. One key observation is the growing focus on sustainability, the circular economy, and effective waste management. The recent surge in interest, particularly in 2022, reflects increasing awareness of tourism's environmental challenges. This suggests that researchers and industry practitioners are increasingly committed to adopting sustainable practices to reduce waste and minimise tourism's environmental footprint. It marks a broader movement towards making sustainability a core part of tourism operations. The evolution of research topics over time unveils a deepening commitment to sustainable practices. For instance, the consistent focus on "sustainable tourism" since 2013, with a notable rise in interest in 2022, shows ongoing efforts to improve sustainability within the industry. This ongoing attention reflects the sector's dedication to meeting sustainability goals and addressing environmental concerns. Likewise, growing interest in topics like "food waste," "recycling," and "marine litter" confirms the complex nature of waste management. This trend highlights the need for an all-embracing approach to tackling waste generation, disposal, and environmental impacts, including reducing food waste, improving recycling, and addressing marine pollution.

4.6 Relevance to Sustainable Development Goals (SDGs)

Table 8. Sustainable development goals (SDGs)

Sustainable Development Goals	Record Count	%
11 Sustainable Cities And Communities	161	56.6
12 Responsible Consumption And Production	105	36.9
14 Life Below Water	42	14.7
15 Life On Land	14	4.9
13 Climate Action	11	3.8
01 No Poverty	5	1.7
03 Good Health And Well Being	4	1.4
02 Zero Hunger	3	1.0
06 Clean Water And Sanitation	2	0.7
09 Industry Innovation And Infrastructure	2	0.7
04 Quality Education	1	0.3
07 Affordable And Clean Energy	1	0.3
08 Decent Work And Economic Growth	1	0.3

The data provides valuable insights into the correlation between Sustainable Development Goals (SDGs) and waste management in the tourism industry. The most frequently referenced SDG is "11 Sustainable Cities and Communities," with 161 occurrences and a citation count of 56,690. The literature places significant emphasis on resolving urban sustainability challenges, including managing waste concerning tourism. The focus on sustainable cities and communities aligns to establish environmentally conscious and resilient towns and cities that facilitate sustainable tourism practices. The SDG cited the second most frequently is "12 Responsible Consumption and Production," with 105 occurrences and a citation count of 36,972. The statement highlights the importance of promoting sustainable consumption and production practices in the tourism sector, namely reducing waste generation, encouraging recycling, and minimising resource use. Additional Sustainable Development Goals (SDGs) noted in the data are "14 Life Below Water," "15 Life on Land," and "13 Climate Action," albeit they have lower frequency and citation counts. These goals highlight the interdependence between waste management in tourism and broader objectives of environmental sustainability, such as conserving marine and terrestrial ecosystems and reducing the impacts of climate change. In addition, the comparatively lower occurrences of Sustainable Development Goals (SDGs) such as "01 No Poverty," "03 Good Health and Well Being," and "02 Zero Hunger" indicate that waste management in tourism may have an indirect impact on these goals. However, they are not as explicitly emphasised in the literature as urban sustainability and responsible consumption. In general, the data suggests a clear correlation between waste management studies in tourism and specific Sustainable Development Goals (SDGs) related to urban sustainability, responsible consumerism, and environmental preservation. By prioritising these aims, researchers and practitioners can help achieve broader SDGs and tackle waste management issues in the tourism sector.

5. CONCLUSION

Examining publication and citation trends in waste management within tourism reveals evolving scholarly interest and impact. The increasing number of publications and citations indicates a growing recognition of waste management's importance in tourism. This trend, notable in recent years, reflects a societal shift toward environmental awareness and sustainability. The findings highlight a growing scholarly interest in waste management within tourism, the impact of key research contributions, and the need for interdisciplinary collaboration. These insights are essential for advancing knowledge, informing policy, and guiding practical efforts to achieve sustainable waste management in the tourism industry.

The future of waste management research in the tourism industry is expected to evolve significantly, with several key focus areas. A primary emphasis must be on integrating sustainability principles across various tourism sectors, including accommodation, transportation, and recreational activities, to foster more environmentally friendly practices. This involves developing strategies to minimise waste, improve recycling processes, and implement circular economy practices. Technological advancements must also play a crucial role in future research. Innovations such as smart waste management systems, enabled by Internet of Things (IoT) sensors, and improvements in waste treatment technologies are anticipated. Digital platforms for monitoring and managing waste will likely become central to enhancing efficiency in the sector's waste management practices. Another significant focus must be on understanding and influencing consumer behaviour. Upcoming studies must explore how tourists' attitudes and actions affect waste production and management. Moreover, research must aim to identify ways to encourage more sustainable behaviours among travellers and tourism operators, addressing the challenges associated with waste management.

The application of circular economy models will continue to gain importance, and therefore, future research must focus on integrating circular economy principles, emphasising waste reduction through recycling, reusing, and upcycling materials within the tourism industry. Future studies must also assess the development and impact of policies related to waste management in tourism in terms of evaluating the effectiveness of various regulations and identifying best practices for implementing policies that support sustainable tourism. And lastly, this field of research will need to adapt to emerging challenges such as climate change, new types of waste (e.g., electronic waste), and the effects of global events like pandemics. Addressing these evolving issues will be essential for developing resilient and effective waste management strategies in the tourism industry.

ACKNOWLEDGEMENT

This work is based on the research supported by the National Institute for the Humanities and Social Sciences REF NO: BRI22/1215.

REFERENCES

Ajzen, I. (1991). The Theory of planned behavior. *Organisational Behavior and Human Decision Processes*, 179-211. https://doi.org/ 90020.DOI: 10.1016/0749-5978(91)

Aria, M., & Cuccurullo, C. (2017). Bibliometrix: An R-Tool for Comprehensive Science Mapping Analysis. *Journal of Informetrics*, 11(4), 959–975. DOI: 10.1016/j.joi.2017.08.007

Arias-Hidalgo, D. (2019). *Sustainable tourism practices for water treatment and solid waste management in tourist lodges in Talamanca*. Editorial Tecnológica de Costa Rica.

Batistic, S., & Kase, R. (2015). The organisational socialisation field fragmentation: A bibliometric review. *Scientometrics*, 104(1), 121–146. DOI: 10.1007/s11192-015-1538-1

Cobo, M. J., López-Herrera, A. G., Herrera-Viedma, E., & Herrera, F. (2011). Science mapping software tools: Review, analysis, and cooperative study among tools. *Journal of the American Society for Information Science and Technology*, 62(7), 1382–1402. DOI: 10.1002/asi.21525

Diaz-Farina, E., Díaz-Hernández, J. J., & Padrón-Fumero, N. (2023). Analysis of hospitality waste generation: Impacts of services and mitigation strategies. *Annals of Tourism Research*, 4, 100083.

Dileep, M. R. (2007). Tourism and Waste Management: A Review of "Zero Waste" Implementation at Kovalam. *Asia Pacific Journal of Tourism Research*, 12, 377–392. DOI: 10.1080/10941660701823314

Donthu, N., Kumar, S., Mukherjee, D., Pandey, N., & Lim, W. M. (2021). How to Conduct a Bibliometric Analysis: An Overview and Guidelines. *Journal of Business Research*, 133, 285–296. DOI: 10.1016/j.jbusres.2021.04.070

Erdogan, N., & Baris, E. (2007). Environmental protection programs and conservation practices of hotels in Ankara, Turkey. *Tourism Management*, 28(2), 604–614. DOI: 10.1016/j.tourman.2006.07.003

Ezeah, C., Fazakerley, J., & Byrne, T. (2015). Tourism waste management in the European Union: Lessons learned from four popular EU tourist destinations. *American Journal of Climate Change*, 4(5), 431–445. DOI: 10.4236/ajcc.2015.45035

Filimonau, V., & Sulyok, J. (2021). 'Bin it and forget it!': The challenges of food waste management in restaurants of a mid-sized Hungarian city. *Tourism Management Perspectives*, 37, 100759. DOI: 10.1016/j.tmp.2020.100759

Filimonau, V., & Tochukwu, C. O. (2020). Exploring managerial approaches to mitigating solid waste in hotels of Lagos, Nigeria. *Journal of Cleaner Production*, 270, 122410. DOI: 10.1016/j.jclepro.2020.122410

Gutiérrez-Salcedo, M., Martínez, M. Á., Moral-Munoz, J. A., Herrera-Viedma, E., & Cobo, M. J. (2018). Some bibliometric procedures for analyzing and evaluating research fields. *Applied Intelligence*, 48, 1275–1287. DOI: 10.1007/s10489-017-1105-y

Han, H., Chua, B.-L., Hyun, S. S., & Hyun, S. S. (2020). Eliciting customers' waste reduction and water-saving behaviors at a hotel. *International Journal of Hospitality Management*, 87, 102386. DOI: 10.1016/j.ijhm.2019.102386

Hirsch, J. E. (2005). An Index to Quantify an Individual's Scientific Research Output. *Proceedings of the National Academy of Sciences of the United States of America*, 102(46), 16569–16572. DOI: 10.1073/pnas.0507655102 PMID: 16275915

International Hotel & Restaurant Association. (1995). Environmental Good Practice in Hotels: Case Studies From the International Hotel & Restaurant Association Environmental Award. Retrieved from http://www.pnuma.org/industria/documentos/Environmental%20Good%20Practice%20in%20Hotels.pdf

Kang, K. H., Stein, L., Heo, C. Y. H., & Lee, S. (2012). Views on environmentalism and consumers' willingness to pay for environmental sustainability in the hotel industry. *International Journal of Hospitality Management*, 31(2), 564–572. DOI: 10.1016/j.ijhm.2011.08.001

Kularatne, T., Wilson, C., Månsson, J., Hoang, V., & Lee, B. (2019). Do environmentally sustainable practices make hotels more efficient? A study of major hotels in Sri Lanka. *Tourism Management*, 71, 213–225. DOI: 10.1016/j.tourman.2018.09.009

Newell, J. P., Goldstein, B., & Foster, A. (2019). A 40-year review of food-energy-water nexus literature and its application to the urban scale. *Environmental Research Letters*, 14(7), 18. DOI: 10.1088/1748-9326/ab0767

Obersteiner, G., Gollnow, S., & Eriksson, M. (2021). Carbon footprint reduction potential of waste management strategies in tourism. *Environmental Development*, 39, 100617. DOI: 10.1016/j.envdev.2021.100617 PMID: 34513580

Obersteiner, G., & Gruber, I. (2017). Urban strategies for waste management in tourist cities D2.5 – Status quo (baseline) assessment report. Retrieved from https://ec.europa.eu/research/participants/documents/downloadPublic?documentIds=080166e5bc2d54d9&appId=PPGMS

Papargyropoulou, E., Steinberger, J. K., Wright, N., Lozano, R., Padfield, R., & Ujang, Z. (2019). Patterns and Causes of Food Waste in the Hospitality and Food Service Sector: Food Waste Prevention Insights from Malaysia. *Sustainability (Basel)*, 11(21), 6016. DOI: 10.3390/su11216016

Papargyropoulou, E., Wright, N., Lozano, R., Steinberger, J., Padfield, R., & Ujang, Z. (2016). A conceptual framework for studying food waste generation and prevention in the hospitality sector. *Waste Management (New York, N.Y.)*, 49, 326–336. DOI: 10.1016/j.wasman.2016.01.017 PMID: 26803473

Pham Phu, S. T., Hoang, M. G., & Fujiwara, T. (2019). Analysing solid waste management practices for the hotel industry. *Global Journal of Environmental Science Management*, 4(1), 19–30.

Pirani, S. I., & Arafat, H. A. (2016). Solid waste management in the hospitality industry: A review. *Journal of Environmental Management*, 146, 320–336. DOI: 10.1016/j.jenvman.2014.07.038 PMID: 25194519

Rossetto, D. E., Bernardes, R. C., Borini, F. M., & Gattaz, C. C. (2018). Structure and evolution of innovation research in the last 60 years: Review and future trends in the field of business through the citations and co-citations analysis. *Scientometrics*, 115(3), 1329–1363. DOI: 10.1007/s11192-018-2709-7

van Nunen, K., Li, J., Reniers, G., & Ponnet, K. (2018). Bibliometric Analysis of Safety Culture Research. *Safety Science*, 108, 248–258. DOI: 10.1016/j.ssci.2017.08.011

van Raan, A. F. J. (2004). Sleeping Beauties in Science. *Scientometrics*, 59(3), 467–472. DOI: 10.1023/B:SCIE.0000018543.82441.f1

Wang, L., Filimonau, V., & Li, Y. (2021). Exploring the patterns of food waste generation by tourists in a popular destination. *Journal of Cleaner Production*, 279, 123890. DOI: 10.1016/j.jclepro.2020.123890

Xiong, X., Zhang, K., Chen, X., Shi, H., Luo, Z., & Wu, C. (2018). Sources and distribution of microplastics in China's largest inland lake – Qinghai Lake. *Environmental Pollution*, 235, 899–906. DOI: 10.1016/j.envpol.2017.12.081 PMID: 29353805

Yakath Ali, N. S., Yu, C., & See, K. F. (2021). Four Decades of Airline Productivity and Efficiency Studies: A Review and Bibliometric Analysis. *Journal of Air Transport Management*, 96, 102099. DOI: 10.1016/j.jairtraman.2021.102099

Zupic, I., & Cater, T. (2015). Bibliometric Methods in Management and Organization. *Organizational Research Methods*, 18(3), 429–472. DOI: 10.1177/1094428114562629

Chapter 2
Sustainable Waste Management in Tourism and Hospitality Sectors

Amrik Singh
https://orcid.org/0000-0003-3598-8787
Lovely Professional University, India

Vipan Kumar
https://orcid.org/0009-0001-4765-1797
Hyatt Hotels, India

ABSTRACT

Waste management significantly impacts the sustainable state of the environment, public health, and economic stability. This study thoroughly analyses the many waste streams produced, with particular attention on solid waste, wastewater, hazardous waste, e-waste, and food waste. Every category has different difficulties and consequences for waste management procedures and regulations. Food waste has serious negative effects on the environment and the economy and is frequently caused by overproduction and consumer behaviour. Solid waste, including various materials, needs efficient recycling and disposal techniques to reduce its impact. Because of the speed at which technology is developing, e-waste pollutes ecosystems with harmful materials. Understanding waste streams will help stakeholders create more long-lasting and efficient solutions to safeguard the environment and public health.

DOI: 10.4018/979-8-3693-6110-8.ch002

1. INTRODUCTION AND BACKGROUND OF STUDY

For public health and water resources protection, wastewater management is crucial. Hazardous garbage requires special handling and disposal since it contains toxic, combustible, and reactive materials. This study thoroughly reviews the sources, environmental effects, and creative mitigation techniques for these wastes' adverse effects. The conversation highlights the significance of integrated waste management systems and the demand for a multidisciplinary strategy to handle the complexity of trash in a world that is changing quickly. The need for a multidisciplinary approach to waste management is emphasised, as it underlines the complexity of the issue and the necessity for collaboration. Responsible waste treatment in the tourism and hospitality business is one of the critical aspects of sustainable operations, as waste generation creates significant environmental impacts in these industries. The waste management scheme must be comprehensive since a wide range of waste types are brought into being - food waste, solid waste, e-waste, etc. This study gives an in-depth look at the most frequently seen waste types in restaurant and hotel sectors, also fulfilling a detailed analysis of their origins, difficulties, and strategies for their sustainable disposal. The hospitality industry leaves much solid waste behind, with the packaging materials, single-use items, in-room guest amenities, and office equipment accounting for it. According to Pirani and Arafat (2014), hotels need to manage solid waste well in order to curb environmental damage and improve business performance (Talukder et al., 2024; Singh et al., 2024; Singh et al., 2024a; Ansari & Singh, 2024; Singh & Singh,2024a; Singh & Singh, 2024b; Sharma & Singh,2024; Bhalla et al., 2023; Singh & Bathla, 2023; Singh & Hassan, 2024). Ultimately, this idea is also made in the study conducted by Camilleri-Fenech et al. (2020), who give a detailed case study of a five-star hotel waste generation very tactfully, putting in the effort to underline the immense necessity of adequate waste segregation and recycling programs. One of the most common types of waste, often the most extensive, is food waste in the hospitality industry. On a practical note, Dhir (2020) posits the importance of dealing with food waste through systematic literature reviews and framework development to achieve the global goals of improving the efficiency of food waste management and making the world greener and healthier. The COVID-19 pandemic has also impacted waste management, and Chick et al. (2020) addressed the perspectives and challenges of waste management in the hospitality sector after the COVID-19 pandemic. The pandemic has increased stress on exploring innovative waste management approaches and awareness of the general cleanliness of waste.

Another essential aspect in terms of sustainability is the effective wastewater management since the amounts of water used in hospitality operations including guest rooms, kitchens, laundry services, and swimming pools are very high(Talukder et al., 2024; Singh et al., 2024; Singh et al., 2024a; Ansari & Singh, 2024; Singh & Singh,2024a; Singh & Singh, 2024b; Sharma & Singh,2024; Bhalla et al., 2023; Singh & Bathla, 2023; Singh & Hassan, 2024). In their entitled 'Nature-based solutions implemented in constructed wetlands for wastewater treatment', an article published in Science of the Total Environment, Justino et al. (2023) endorse the use of constructed wetlands as an alternative to advanced methods in wastewater treatment. Gautam et al. (2015) argued that water management systems and reutilization are achievable in hospitality outlets and demonstrated a real-life application of the concepts using the example of Delhi. E-waste, which includes used goods and appliances with circuits and connectors, is a particular type of waste because of the toxicity of some of its parts and the recycling process, which is difficult and hazardous. There is a need to control e-waste pollution and to encourage sustainable practices within the sector to reduce negative impacts on the environment (Pirani & Arafat, 2014; Singh et al., 2024; Singh et al., 2024a; Ansari & Singh, 2024; Singh & Singh,2024a; Singh & Singh, 2024b; Sharma & Singh,2024; Bhalla et al., 2023; Singh & Bathla, 2023; Singh & Hassan, 2024). The hazardous wastes generated include cleaning compounds, pesticides, and pharmaceuticals, which have health and environmental implications. Triantafyllou and Cherrett (2010) discuss the aspects of retail sector supply chain management with specific reference to hazardous waste; nonetheless, their findings could help understand the issue in the hospitality industry context. The management, storage and disposal of hazardous waste are essential to avoid the harm these wastes may cause to the community and the extended environment. Having identified that mixed wastes are constantly generated in tourism and hospitality settings, proper management of the wastes in their various forms is a sustainable practice.

Through varied literature reviews, this chapter will provide detailed insight into the most typical troubles and possible opportunities in managing food waste, solid waste, e-waste, wastewater, and hazardous waste. Given the expansive scope of this research, which seeks to bridge theory and practice, a postpositivist paradigm is deemed suitable. The study predominantly relies on secondary data to fulfil its objectives. These secondary sources include previous studies, research articles, industry reports, journals, magazines, and other relevant publications.

If proper waste management strategies are applied, the environmental impact of the operations and waste produced by the industry can be minimised. Challenges are that food waste management can improve efficiency by reducing methane emissions related to composting (Singh et al., 2024; Singh et al., 2024a; Ansari & Singh, 2024; Singh & Singh,2024a; Singh & Singh, 2024b; Sharma & Singh,2024; Bhalla et al., 2023; Singh & Bathla, 2023; Singh & Hassan, 2024). A sound recycling system can assist in the reclaiming of various senior materials for reuse, hence promoting the circular economy. At the same time, proper e-waste management will help to contain toxin pollution. The constructed wetland as another efficient wastewater treatment is another innovative technology. By implementing sustainable procedures in waste management, benefits such as efficient sorting, better management of resources, and lower expenses are achieved. Furthermore, they enhance guests' satisfaction and business image, as customers are beginning to pay attention to the sustainability performances of businesses. Implementing recycling solutions in the hospitality industry affects the environment positively and ultimately has an impact on more significant sustainability initiatives while improving service delivery and the future viability of the sector (Singh et al., 2024; Singh et al., 2024a; Ansari & Singh, 2024; Singh & Singh,2024a; Singh & Singh, 2024b; Sharma & Singh,2024; Bhalla et al., 2023; Singh & Bathla, 2023; Singh & Hassan, 2024). According to the Merriam-Webster Dictionary, waste is "refuse from places of human or animal habitation; material that is no longer wanted or needed after the completion of a process.

2. TYPES OF WASTE

2.1 Food Waste

Food waste is defined as edible materials disposed of or lost at any point during the production, processing, retail or consumption of foods in the hospitality and tourism sectors. It includes leftovers, trimmings and expired products and contributes to environmental and economic sustainability issues. Some chemicals involved are:

- **Methane:** Produced during anaerobic decomposition, contributing to greenhouse gas emissions and climate change.
- **Ammonia**: Released during decomposition, can contaminate air and water, causing pollution and eutrophication.
- **Nitrites and Nitrates:** From preserved foods can leach into water sources, causing algal blooms and water pollution.
- **Pesticide Residue:** Agricultural products can contaminate soil and water, harming wildlife and ecosystems.
- **Phosphates:** These issues can be mitigated through management strategies such as reduction and recycling.

2.2 Solid Waste

Solid waste is no longer of any physical or valuable use to the owners. These include commonly discarded products such as food waste, packaging papers, plastics, glass and metal. Such materials are mainly derived from the residential, commercial, industrial, and institutional sectors; hence, they must be well disposed of to help reduce their environmental effects. Some chemicals involved are:

- **Heavy Metals**: Lead, mercury, cadmium, and arsenic can leach into soil and water, causing toxicity in plants, animals, and humans.
- **Persistent Organic Pollutants (POPs)**: Chemicals like dioxins and PCBs, which are resistant to degradation, can accumulate in the environment and enter the food chain.
- **Plasticizers**: Chemicals such as phthalates from plastics can leach into soil and water, disrupting ecosystems and harming wildlife.
- **Volatile Organic Compounds (VOCs)**: Emitted from certain plastics, paints, and solvents, contributing to air pollution and respiratory problems.
- **Polybrominated Diphenyl Ethers (PBDEs)**: These are used as flame retardants in electronics and furniture. They can persist in the environment and harm wildlife and humans.

2.3 E-Waste

Stuff consists of discarded electronic devices and parts no longer helpful, broken or unwanted, such as computers, smartphones, televisions, batteries and circuit boards, which contain heavy metals such as lead, mercury and cadmium. E-waste, therefore, requires special recycling and disposal to avoid harming the ecology and our health. Some Chemicals involved are:

- **Lead:** Found in CRT monitors, lead-acid batteries, and soldering materials, which can contaminate soil and water if improperly disposed of.
- **Mercury:** Found in fluorescent lamps, thermostats, and switches, posing risks to ecosystems and human health if released into the environment.
- **Cadmium:** Present in rechargeable batteries and circuit boards, which can leach into soil and water, causing toxic effects on wildlife and aquatic organisms.
- **Brominated Flame Retardants (BFRs):** Brominated flame retardants (BFRs) are used in plastics and electronics to reduce flammability; these chemicals can persist in the environment and bioaccumulate organisms.

- **Chlorofluorocarbons (CFCs):** Found in older refrigerators and air conditioners, CFCs deplete the ozone layer if released into the atmosphere during improper disposal.

2.4 Wastewater

Water accumulates and is contaminated through shipping, residential, industrial, commercial, or agricultural activities. This water contains chemicals, organic matter, and pathogens. Severe treatment or minimisation of backflow is required to ensure safe discharge or reuse, thereby preventing environmental degradation and risks to public health. Some Chemicals involved are:

- **Pharmaceuticals and Personal Care Products (PPCPs):** Residues from medications, sunscreen, and personal care items can contaminate water sources, affecting aquatic life and potentially entering the food chain.
- **Cleaning Agents:** Harsh chemicals from cleaning products used in hotels and resorts can contribute to water pollution and harm aquatic ecosystems if not adequately treated.
- **Heavy Metals:** Heavy metals like lead, mercury, and cadmium, released into water sources through industrial activities and plumbing fixtures, can pose risks to aquatic life and human health.
- **Pesticides and Herbicides:** Used in landscaping and pest control around hospitality facilities, these chemicals can leach into water bodies, affecting water quality and aquatic organisms.
- **Nutrients:** Excess nutrients like nitrogen and phosphorus from fertilisers and food waste can cause eutrophication in water bodies, leading to algae blooms and oxygen depletion.

2.5 Hazardous Waste

Wastes that, due to their quantity, concentration or physical, chemical or infectious character, constitute or may constitute hazards to human health or the environment. – hazardous or potentially hazardous chemicals or chemical products, pharmaceuticals, pesticides, heavy metals, and radioactive substances. Appropriate handling, treatment and disposal allow for the safety and protection of public health, environmental protection and compliance with regulations. Some Chemicals involved are:

- **Cleaning Solvents:** Used for industrial cleaning, these solvents often contain volatile organic compounds (VOCs) that can contribute to air pollution and harm human health.

- **Pesticides and Herbicides:** Used for landscaping and pest control, these chemicals can contaminate soil and water, posing risks to wildlife and ecosystems.
- **Paints and Coatings** Contain heavy metals like lead and cadmium, which can leach into soil and water and cause toxicity to plants, animals, and humans.
- **Batteries**: These contain heavy metals such as lead, mercury, and cadmium, which can leach into soil and water if improperly disposed of, posing risks to wildlife and human health.
- **Medical Waste**: Includes sharps, pharmaceuticals, and infectious materials, which can spread diseases and contaminate water sources if improperly handled and disposed of.

Ultimately, proper disposal of food waste, solid waste, e-waste, wastewater, and hazardous waste within hospitality institutions is essential for sustainable business practices and environmental conservation. Suppose the hospitality industry addresses handling, recycling, and disposing of materials as anticipated. In that case, the environmental impact is reduced, improving the environment for current and upcoming generations.

3. LITERATURE REVIEW

The hospitality and tourism sector produces various waste streams, including food, solid, electronic, and hazardous waste. Effective waste management methods are necessary to prevent environmental harm. This helps to reduce the ecological footprint and establish a sustainable operation trend. Food waste management practices have gained considerable attention from international media due to environmental and economic concerns (Talukder et al., 2024; Singh et al., 2024; Singh et al., 2024a; Ansari & Singh, 2024; Singh & Singh,2024a; Singh & Singh, 2024b; Sharma & Singh,2024; Bhalla et al., 2023; Singh & Bathla, 2023; Singh & Hassan, 2024). Waste reduction can be achieved through better planning and appropriate customer education in hospitality. On the other hand, we can reduce food waste by employing systemic approaches that integrate monitoring systems, hospitality staff training, and technological innovation in the hospitality industry. It is crucial to have a comprehensive overview of food waste from an organisational perspective, including its hierarchy, processes, and tasks, to understand the food waste crisis in the context of health and safety in industry practice (Talukder et al., 2024; Singh et al., 2024; Singh et al., 2024a; Ansari & Singh, 2024; Singh & Singh,2024a; Singh

& Singh, 2024b; Sharma & Singh,2024; Bhalla et al., 2023; Singh & Bathla, 2023; Singh & Hassan, 2024).

The provision of hospitality services also poses a substantial risk: solid waste management. Insufficient waste segregation and recycling centres are common, necessitating structural advancements and effective waste management. Developing comprehensive waste management policies is crucial to safeguard the environment. Additionally, concerns regarding e-waste management arise due to the growing use of digital technology in the hospitality industry. Recycling and reusing e-waste is essential to reduce carbon emissions. Waste management policies and industry standards must be developed to address this issue comprehensively. Furthermore, wastewater can pose a significant challenge for hospitality businesses, particularly in environmentally sensitive areas, highlighting the need for effective wastewater management strategies. Hotel resorts require decentralised wastewater treatment facilities, as they can exhibit both environmental complementarities and economic efficiency. Constructed wetlands are a Nature-based solution (NBS) that can potentially address wastewater challenges in the hospitality sector and contribute to sustainable development. It is crucial to emphasise the significance of water conservation and recycling in reducing the hospitality sector's water footprint and maximising resource utility.

When improperly disposed of, hazardous waste, including chemicals and cleaning instruments, threatens the environment and individuals' health. It is imperative to examine the cases of hazardous waste management in the hospitality industry and discuss the logistics and regulations involved. This information is valuable for the hospitality sector (Singh et al., 2024; Singh et al., 2024a; Ansari & Singh, 2024; Singh & Singh, 2024a; Singh & Singh, 2024b; Sharma & Singh, 2024; Bhalla et al., 2023; Singh & Bathla, 2023; Singh & Hassan, 2024). Waste management significantly impacts the environmental outcomes of hospitality operations and influences the operational goals aimed at enhancing effective waste management practices. Innovative technologies, combined with strict guidelines from regulatory authorities and educating staff and guests on their role in maintaining environmental health, can minimise the ecological impact of hospitality facilities and maximise their contribution to the sustainability of every country.

4. DISCUSSION AND INTERPRETATION

The hospitality industry is a significant contributor to environmental waste, including food waste, solid waste, e-waste, water waste, and hazardous waste (Singh et al., 2024; Singh et al., 2024a; Ansari & Singh, 2024; Singh & Singh, 2024a; Singh & Singh, 2024b; Sharma & Singh, 2024; Bhalla et al., 2023; Singh

& Bathla, 2023; Singh & Hassan, 2024). Managing these multiple waste streams effectively is crucial to reduce their environmental impact and promote efficiency. Due to its environmental and economic implications, food waste management in the hospitality industry has received considerable attention recently. Filimonau and Delysia (2019) suggest that enhanced planning and customer awareness can be used to conserve resources in the catering industry. By analysing guests' daily, weekly, or monthly consumption patterns, hotels can predict their consumption behaviours and adjust portion sizes to reduce waste. Furthermore, educating customers about the consequences of food waste can encourage them to adopt proper consumption behaviours that minimise waste (Talukder et al., 2024; Singh et al., 2024; Singh et al., 2024a; Ansari & Singh, 2024; Singh & Singh, 2024a; Singh & Singh, 2024b; Sharma & Singh, 2024; Bhalla et al., 2023; Singh & Bathla, 2023; Singh & Hassan, 2024). Such educational campaigns can take various forms, including informative signs, electronic communications, and training sessions that may influence guest behaviour.

According to Dhir et al. (2020), it is recommended that systematic measures, including monitoring systems, staff training activities, and information technologies, be implemented to reduce the generation of food waste. Smart bins connected to the Internet of Things (IoT) and AI-based kitchen management systems can provide instant data on waste behaviours. This technology can keep records of the type and quantity of food that has been wasted, enabling management to make necessary changes in food purchasing and preparation. Additionally, AI techniques can be applied for demand forecasting, aiding organisations in planning the appropriate amount of food production. For example, AI can forecast the required amount of food using data, seasonal variation, and bookings, thereby preventing the problem of excess food. Furthermore, Lasaridi et al. (2017) emphasise that certain aspects dominant in the industry and policy frameworks are among the most significant enablers of successful food waste prevention.In terms of policy recommendations, it has been suggested that mandatory food waste audits and incentives for food donation would be beneficial in completing existing efforts to reduce food waste. Compulsory audits would help establishments to regularly examine their food waste levels and identify areas for improvement. Offering tax credits or other rewards for food donations would ensure that excess food is given to those in need rather than wasted. This would also support the community. Such policies have been recommended in various studies (Singh et al., 2024; Singh et al., 2024a; Ansari & Singh, 2024; Singh & Singh, 2024a; Singh & Singh, 2024b; Sharma & Singh, 2024; Bhalla et al., 2023; Singh & Bathla, 2023; Singh & Hassan, 2024).

Food waste buckets can be transformed into valuable resources through innovative techniques such as anaerobic digestion and composting, as reported by Papargyropoulou et al. (2014). The food waste hierarchy concept can also be used to change the public's behaviour and reduce waste through prevention, redistribution, and recycling, as Papargyropoulou et al. (2014) suggested. Additionally, collaborative platforms linking food surpluses with charities can help reduce food waste while still feeding the hungry, as Garrone et al. (2014) noted. For instance, companies such as Too Good To Go or Food Rescue US transport unsold food from hotels to shelters, where it can be consumed instead of wasted. Moreover, the hospitality industry can draw inspiration from the practices of two regions that have successfully implemented food waste management systems. For instance, campaigns such as Love Food Hate Waste in the UK have reduced food waste through awareness-raising, engagement of businesses, and provision of hints to households and businesses on how to minimise waste, as reported by Singh et al. (2024) and Ansari & Singh (2024). By establishing similar measures, hospitality units can introduce a sustainability culture within their organisational environment.

Training and empowering staff is a crucial aspect of food waste management, on par with other management systems. It is essential to train all employees, including dishwashers, preparation cooks, line cooks, commis, hosts, porters, and everyone else in the restaurant, to ensure they understand the importance of minimising waste and how to do so effectively. Organisations should conduct regular conferences and seminars to update employees on waste minimisation measures and equipment. Implementing incentive schemes that reward employees for suggesting ways and measures to reduce waste can also promote sustainability. Effective food waste management in the hospitality industry is a complex process that includes technology, policy support, staff training, and customer education. Adopting technologies such as IoT and AI can facilitate the creation of successful policy tools, increasing the chances of the industry embracing a sustainability culture. Various studies have highlighted the importance of these factors in achieving sustainability in the hospitality industry (Singh et al., 2024; Singh et al., 2024a; Ansari & Singh, 2024; Singh & Singh, 2024a; Singh & Singh, 2024b; Sharma & Singh, 2024; Bhalla et al., 2023; Singh & Bathla, 2023; Singh & Hassan, 2024).

Adopting the guidelines above can enable hotels to reduce their ecological footprint and save money by utilising water resources efficiently. Moreover, educating customers about water conservation through informative campaigns and visible signage can contribute to creating sustainable water usage patterns in line with the hotel's sustainability objectives. Effective management of hazardous waste is crucial, as improper handling can harm the environment and human health. In the hospitality industry, using various cleaning agents and chemicals can pose potential dangers, necessitating prioritising safe and efficient waste management practices to minimise adverse effects. Triantafyllou and Cherrett (2010) offer valuable insights into how the retail sector tackles hazardous waste, and their findings can also be applied to the hospitality industry. They emphasise the importance of implementing a comprehensive waste management plan that involves identifying, categorising, and separating hazardous waste to prevent mixing and ensure safe handling (Singh et al., 2024; Singh et al., 2024a; Ansari & Singh, 2024; Singh & Singh, 2024a; Singh & Singh, 2024b; Sharma & Singh, 2024; Bhalla et al., 2023; Singh & Bathla, 2023; Singh & Hassan, 2024).

Using green cleaning products and safer chemicals can help reduce the impact of hazardous substances. These products have certifications from environmental standards such as Green Seal or EcoLogo, which indicate that they are less toxic and break down quickly, making them safer for humans and the environment. International standards such as the Basel Convention can also improve hazardous waste management practices within the hospitality sector (Al-Salem et al., 2014). The Basel Convention aims to regulate the movement and disposal of hazardous waste across borders, promoting environmentally sound management practices. Compliance with these standards ensures legal adherence and demonstrates a commitment to global environmental stewardship. Combining these approaches can optimise hazardous waste management and create a more sustainable future. Effective waste management in the hospitality sector is essential for minimising environmental impact and maximising resource efficiency. Adopting integrated strategies across solid waste, food waste, water consumption, and hazardous materials ensures sustainability and regulatory compliance and enhances overall operational resilience and environmental stewardship.

5. THEORETICAL IMPLICATIONS

Theoretical implications drawn from the comprehensive study of waste management in the hospitality industry underscore several critical insights:

- Adoption of Circular Economy Principles: Adopting the principles of the circular economy can transform waste into valuable resources, promote sustainability, and minimise the environmental impact.
- Development and Execution of Policies: Effective policy frameworks are essential for establishing consistent waste management practices and encouraging sustainable behaviours in the hospitality sector. These policies provide guidelines and incentives for making environmentally conscious choices.
- Integration of Technology: Implementing advanced technologies, such as IoT-enabled systems and smart waste management solutions, can significantly enhance waste monitoring and management efficiency and accuracy. This streamlines processes and maximises results.

- Compliance with Environmental Legislation: Adhering to international environmental regulations, including the Basel Convention and EU directives, is crucial to ensuring responsible waste management. This keeps us within the bounds of the law and promotes best practices in waste management.
- Education and Behavioral Change: Raising awareness and educating stakeholders is critical for reducing waste and fostering responsible consumption. By instilling a shared understanding of the importance of waste reduction, we can cultivate a culture that actively works towards it.
- Economic Feasibility: Efficient waste management reduces operational costs and creates revenue-generating opportunities through waste recycling and resource recovery initiatives. This presents a mutually beneficial situation.
- Health and Safety Considerations: Adherence to proper waste management protocols safeguards employees and the public from the health risks associated with hazardous waste exposure. By following these protocols, we guarantee the well-being of all stakeholders involved.

6. PRACTICAL IMPLICATIONS

The practical implications of waste management practices in the hospitality industry translate theoretical insights into actionable strategies:

- Implementation of Waste Audits: Undertaking regular waste audits effectively pinpoints areas of inefficiency and devises targeted solutions for waste reduction.
- Staff Training Programs: Extensive training for employees is essential. It equips them with the necessary skills to adhere to waste management protocols, ultimately enhancing operational efficiency.
- Technology Adoption: Embracing advanced technologies can prove to be a game-changer. They aid in optimising waste collection, improving resource allocation, and reducing our environmental impact.
- Partnerships with Recycling Companies: Collaborating with certified recycling companies is mutually beneficial. It ensures that waste materials are disposed of correctly and promotes a closed-loop system.
- Consumer Engagement: Educating guests about sustainable practices is paramount. Knowledgeable individuals are more likely to act responsibly. Encouraging environmentally conscious consumption and minimising waste generation is a significant part of the solution.

- Policy Advocacy and Compliance: We must advocate for supportive policies and ensure compliance with all rules and regulations. Doing so enables us to be responsible stewards of the environment.
- Continuous Improvement: Cultivating a culture of continuous improvement is vital. Innovation in waste management strategies and practices must be nurtured to maintain progress.

7. FUTURE RESEARCH

Future research in waste management within the hospitality industry should concentrate on advancing theoretical insights into practical applications and addressing emerging challenges. Utilising cutting-edge digital technology presents opportunities to change hotel waste handling methods. To this end, investigations should consider employing IoT sensors for real-time waste monitoring and utilising data to predict and refine resource consumption. Additionally, AI and machine learning could assist in making superior waste decisions by predicting the quantity of waste that will be produced and proposing the finest methods to eliminate it.

Innovative methods for waste treatment, such as utilising tiny organisms or chemical compounds, provide hope for sustainable garbage management. However, it is vital to research how effective these methods are in various hotel environments, understand the damage or help they pose to our environment, and check if they are feasible. Knowing the impacts of these novel methods on people and the financial world is critical in gauging whether hotels should or can employ them.

Research must examine how advanced technologies like unique water filters, systems breaking down trash without oxygen, and artificial swamps are handling liquid and solid garbage. The investigation should concentrate on how these techniques benefit the surroundings and conserve our cash. Understanding how visitors perceive and act towards green waste habits is vital to altering behaviour.

8. CONCLUSION

Effective waste management practices are used in the hospitality industry to lessen the industry's impact on the environment, enhance operational efficiency, and maintain regulatory compliance. By combining theoretical insights with practical techniques, hospitality businesses can foster a culture of sustainability by reducing waste and optimising resources within their operations. Advanced technologies, such as IoT-enabled sensors and AI-driven analytics, offer promising avenues for effectively monitoring and managing waste to minimise resource wastage and lower operational costs. Furthermore, innovative technologies for waste treatment have the potential to reduce environmental impacts from waste disposal significantly. These innovations contribute to the long-term viability of sustainable waste management solutions and have significant implications for the industry's shift towards circular economy principles. However, further research is needed to understand consumer behaviours and perceptions surrounding sustainable practices and to foster a culture of environmental stewardship among guests and staff. Future studies should also focus on the scalability of decentralised waste management systems in the hospitality sector and their economic feasibility in various contexts. Additionally, robust frameworks for measuring the lifecycle impacts of waste management strategies must be developed to inform decision-making at both the business and policy levels.

REFERENCES

Al-Salem, S. M., Lettieri, P., & Baeyens, J. (2014). The Basel Convention on the Control of Transboundary Movements of Hazardous Wastes and their Disposal. *Waste Management (New York, N.Y.)*, 34(3), 293–302.

Ansari, A. I., & Singh, A. (2024). Adopting sustainable and recycling practices in the hotel industry and its factors influencing guest satisfaction. In Tyagi, P., Nadda, V., Kankaew, K., & Dube, K. (Eds.), *Examining tourist behaviors and community involvement in destination rejuvenation* (pp. 38–47). IGI Global., DOI: 10.4018/979-8-3693-6819-0.ch003

Babu, B. R., Parande, A. K., & Basha, C. A. (2007). Electrical and electronic waste: A global environmental problem. *Waste Management & Research*, 25(4), 307–318. DOI: 10.1177/0734242X07076941 PMID: 17874657

Bhalla, A., Singh, P., & Singh, A. (2023). Technological advancement and mechanization of the hotel industry. In Tailor, R. (Ed.), *Application and adoption of robotic process automation for smart cities* (pp. 57–76). IGI Global., DOI: 10.4018/978-1-6684-7193-7.ch004

Camilleri-Fenech, M., Sola, J. O., Farreny, R., & Durany, X. G. (2020). A snapshot of solid waste generation in the hospitality industry: The case of a five-star hotel on the island of Malta. *Sustainable Production and Consumption*, 21, 104–119. DOI: 10.1016/j.spc.2019.11.003

Cucchiella, F., D'Adamo, I., Koh, S. C. L., & Rosa, P. (2015). Recycling of WEEEs: An economic assessment of present and future e-waste streams. *Renewable & Sustainable Energy Reviews*, 51, 263–272. DOI: 10.1016/j.rser.2015.06.010

Dhir, A., Talwar, S., Kaur, P., & Malibari, A. (2020). Food waste in hospitality and food services: A systematic literature review and framework development approach. *Journal of Cleaner Production*, 270, 122861. DOI: 10.1016/j.jclepro.2020.122861

Diaz-Farina, E., Díaz-Hernández, J. J., & Padrón-Fumero, N. (2023). Analysis of hospitality waste generation: Impacts of services and mitigation strategies. *Annals of Tourism Research Empirical Insights*, 4(1), 100083. DOI: 10.1016/j.annale.2022.100083

Estévez, S., Feijoo, G., & Moreira, M. T. (2022). Environmental synergies in decentralized wastewater treatment at a hotel resort. *Journal of Environmental Management*, 317, 115392. DOI: 10.1016/j.jenvman.2022.115392 PMID: 35636106

Filimonau, V., & Delysia, A. (2019). Food waste management in hospitality operations: A critical review. *Tourism Management*, 71, 234–245. DOI: 10.1016/j.tourman.2018.10.009

Gautam, S., Blanchard, R. E., McDonald, T. P., & McDonald, T. (2015). Assessment of water consumption and reutilization potential in hospitality sector, a case study of Delhi. *International Journal of Environmental Engineering*, 2(2), 93–97.

Justino, S., Calheiros, C. S., Castro, P. M., & Gonçalves, D. (2023). Constructed wetlands as nature-based solutions for wastewater treatment in the hospitality industry: A review. *Hydrology*, 10(7), 153. DOI: 10.3390/hydrology10070153

Lasaridi, K., Abeliotis, K., Manios, T., Zorpas, A., Chroni, C., & Terzis, E. (2017). Food waste prevention in the hospitality sector: Prospects and challenges. Retrieved from https://www.researchgate.net/publication/356989278_Food_Waste_in_Hotel_Restaurants_Causes_Reducing_Practices_and_Barriers (accessed 28.06.2024).

Obersteiner, G., Gollnow, S., & Eriksson, M. (2021). Carbon footprint reduction potential of waste management strategies in tourism. *Environmental Development*, 39, 100617. DOI: 10.1016/j.envdev.2021.100617 PMID: 34513580

Pirani, S. I., & Arafat, H. A. (2014). Solid waste management in the hospitality industry: A review. *Journal of Environmental Management*, 146, 320–336. DOI: 10.1016/j.jenvman.2014.07.038 PMID: 25194519

Sharma, R., & Singh, A. (2024). Use of digital technology in improving quality education: Global perspectives and trends. In Nadda, V., Tyagi, P., Moniz Vieira, R., & Tyagi, P. (Eds.), *Implementing sustainable development goals in the service sector* (pp. 14–26). IGI Global., DOI: 10.4018/979-8-3693-2065-5.ch002

Singh, A., & Bathla, G. (2023). Fostering creativity and innovation: Tourism and hospitality perspective. In P. Tyagi, V. Nadda, V. Bharti, & E. Kemer (Eds.), *Embracing business sustainability through innovation and creativity in the service sector* (pp. 70–83). IGI Global. DOI: 10.4018/978-1-6684-6732-9.ch005

Singh, A., & Hassan, S. C. (2024). Service innovation through blockchain technology in the tourism and hospitality industry: Applications, trends, and benefits. In Singh, S. (Ed.), *Service innovations in tourism: Metaverse, immersive technologies, and digital twin* (pp. 205–214). IGI Global., DOI: 10.4018/979-8-3693-1103-5.ch010

Singh, A., Tyagi, P. K., & Garg, A. (Eds.). (2024). *Sustainable disposal methods of food wastes in hospitality operations*. IGI Global., DOI: 10.4018/979-8-3693-2181-2

Singh, V., Archana, T., Singh, A., & Tyagi, P. K. (2024a). Utilizing technology for food waste management in the hospitality industry hotels and restaurants. In Singh, A., Tyagi, P., & Garg, A. (Eds.), *Sustainable disposal methods of food wastes in hospitality operations* (pp. 287–295). IGI Global., DOI: 10.4018/979-8-3693-2181-2.ch019

Singh, V., & Singh, A. (2024a). Digital health revolution: Enhancing well-being through technology. In Nadda, V., Tyagi, P., Moniz Vieira, R., & Tyagi, P. (Eds.), *Implementing sustainable development goals in the service sector* (pp. 213–219). IGI Global., DOI: 10.4018/979-8-3693-2065-5.ch016

Singh, V., & Singh, A. (2024b). Revolutionizing the hospitality industry: How ChatGPT empowers future hoteliers. In Bansal, R., Ngah, A., Chakir, A., & Pruthi, N. (Eds.), *Leveraging ChatGPT and artificial intelligence for effective customer engagement* (pp. 192–203). IGI Global., DOI: 10.4018/979-8-3693-0815-8.ch011

Talukder, M. B., Kumar, S., & Das, I. R. (2024). Food wastage on the economic outcome: Evidence from the hotel industry. In Singh, A., Tyagi, P., & Garg, A. (Eds.), *Sustainable disposal methods of food wastes in hospitality operations* (pp. 68–80). IGI Global., DOI: 10.4018/979-8-3693-2181-2.ch005

Triantafyllou, M. K., & Cherrett, T. J. (2010). The logistics of managing hazardous waste: A case study analysis in the UK retail sector. *International Journal of Logistics*, 13(5), 373–394. DOI: 10.1080/13675567.2010.512150

Zaman, A. U. (2015). A comprehensive study of the environmental and economic benefits of resource recovery from global waste management systems. *Journal of Cleaner Production*, 124, 41–50. DOI: 10.1016/j.jclepro.2016.02.086

Chapter 3
Global Perspectives on Waste Management Education in Tourism and Hospitality

Pradip Kumar
https://orcid.org/0000-0002-5138-736X
Assam University, Silchar, India

Himanshu Bhusan Rout
Mizoram University, India

ABSTRACT

Waste management practices are crucial in mitigating the negative impacts and promoting sustainability in the tourism and hospitality sectors. This chapter uses relevant case studies to explore global perspectives on waste management education, regional variations, cultural considerations, innovative methodologies, obstacles, opportunities, and future recommendations. It highlights the challenges in different regions, from Asia to Europe's eco-conscious destinations and Africa's emerging markets. Challenges such as inadequate infrastructure, limited funding, and low stakeholder awareness persist, yet these also present opportunities for international cooperation and knowledge sharing. The chapter advocates leveraging technology, increasing stakeholder participation, and integrating waste management education into broader sustainability initiatives. Recommendations are provided for policy-makers, business leaders, educators, and community leaders. Emphasizing cultural sensitivity, creativity, cooperation, and long-term vision, the chapter underscores the importance

DOI: 10.4018/979-8-3693-6110-8.ch003

1. INTRODUCTION

The tourism and hospitality industry is a massive driver of the world economy, but its quick growth has created severe environmental problems, especially concerning waste production. A wide range of waste products, including food waste, plastics, paper, and hazardous materials, are produced by tourists and hospitality establishments, which adds to pollution and environmental deterioration. According to the United Nations Environment Programme, tourism contributes around 5% of the world's greenhouse gas emissions, with waste management playing a significant role in this impact (UNEP,2019). Moreover, according to the World Travel and Tourism Council (WTTC,2020), the average hotel visitor creates around 1 kilogram of waste every night, mostly in landfills or the ocean. This emphasises the critical need for sustainable waste management techniques that reduce waste output, improve recycling and reuse, and ensure efficient residual waste disposal. Sustainable waste management reduces environmental damage and has economic and social advantages by increasing resource efficiency and providing green jobs. "Development that fulfils the demands of the present without compromising the ability of future generations to meet their own needs" is how the United Nations (UN) defined sustainable development almost forty years ago (World Commission on Environment and Development (WCED), 1987, p. 43). The UN unveiled the 2030 Agenda for Sustainable Development in 2015 as a standard blueprint for a sustainable future on a global scale (United Nations, 2015). Agenda 2030 was a set of seventeen interconnected Sustainable Development Goals (SDGs) that called on all nations and sectors of the economy to form international alliances to accomplish social justice, economic expansion, and environmental preservation (Scheyvens & Cheer, 2021).

This chapter will explore worldwide perspectives on waste management education in the tourist and hospitality industries, emphasising regional differences, cultural impacts, creative training techniques, and the existing difficulties and possibilities. The chapter uses cases from different regions and methods applied to efficient waste management and preservation of the environment. Regional disparities in waste management procedures show the various issues and solutions applied worldwide, ranging from Asia's high-density tourism hotspots to Europe's ecologically aware destinations and Africa's burgeoning tourism markets. Cultural conventions, beliefs, and attitudes strongly affect waste management practises, needing instructional initiatives customised to individual cultural situations. Innovative instructional methods, such as interactive workshops, digital platforms, and gamified learning settings, have increased awareness and encouraged sustainable habits.

Despite progress, challenges such as inadequate infrastructure, limited finances, and poor stakeholder awareness remain. However, these limitations provide international cooperation, information sharing, and capacity development opportunities via global networks and cross-cultural collaborations. Emerging trends and future approaches include utilising technology, boosting stakeholder engagement, and incorporating waste management education into larger sustainability projects. This chapter makes specific recommendations to legislators, business executives, educators, and community leaders on promoting effective waste management education and developing a sustainable future for the tourist and hospitality industries. With these objectives in mind, the chapter thoroughly explains how waste management education may be effectively incorporated into the tourist and hospitality industries to promote long-term sustainability and environmental stewardship.

2. GEOGRAPHICAL VARIATIONS IN WASTE MANAGEMENT TECHNIQUES

Waste management techniques vary significantly between locations in the tourism and hospitality industry due to economic, cultural, and infrastructural factors. These variations result in unique challenges and tailored solutions catering to each destination's needs and contexts. Economic conditions influence the availability of resources for waste management infrastructure, while cultural attitudes towards waste and recycling impact the effectiveness of these practices. Additionally, the level of infrastructural development determines the capacity for efficient waste collection, processing, and disposal. Consequently, regions must develop waste management strategies that align with their local circumstances, ensuring that solutions are practical and effective in addressing their unique environmental challenges. This localised approach is essential for sustainable waste management in diverse tourism and hospitality settings.

2.1 Case Studies of the Geographical Variation of Waste Management Techniques

2.1.1 Asia: High-Density Tourism Hotspots

Asia contains some of the world's most popular and densely inhabited tourist destinations, including Bali, Thailand, and the Maldives. Waste management is a serious concern because of the many tourists in these areas. The influx of tourists in these regions significantly increases waste generation. For instance, Bali generates approximately 4,281 tons of waste per day, much of which is improperly managed and pollutes natural environments (Bali Partnership, 2020). Bali suffers from insufficient waste disposal infrastructure, substantially contaminating its beaches and rivers. Community-led efforts and government regulations to decrease plastic waste and boost recycling are being implemented to address these challenges (Hall & Gössling, 2016). Bali's efforts to reduce plastic waste have been successful. Within two years of implementing plastic bag bans and community cleanup programs, there was a 50% decrease in plastic waste on beaches (Waste Management Report, 2018).

The Maldives has undertaken several initiatives to manage waste generated through its tourism industry, recognising the significant environmental impact of this sector. A notable effort is the Waste Management Regulation (2013/R-58), which mandates resorts and hotels to establish waste management systems, including waste segregation, recycling, and proper disposal methods (Ministry of Environment and Energy, 2013). Furthermore, the Maldives are implementing the "Saafu Raajje" initiative, which aims to create a cleaner environment through community engagement and awareness programs targeting locals and tourists (Naseer, 2017). Additionally, the Waste-to-Energy (WtE) project on Thilafushi Island aims to convert waste into energy, reducing landfill usage and mitigating pollution (World Bank, 2020). Partnerships with international organisations complement these initiatives to enhance waste management infrastructure and practices, such as the UNDP-supported project to develop sustainable waste management strategies for the tourism sector (UNDP, 2018). Collectively, these efforts highlight the Maldives' commitment to addressing waste management challenges posed by its thriving tourism industry.

2.1.2 Europe: Environmentally Conscious Destinations and Advanced Waste Management Solutions

Many European locations are ecologically sensitive and have installed modern garbage management technologies. Countries like Germany, Sweden, and the Netherlands have implemented comprehensive recycling systems and waste-to-energy facilities, significantly reducing landfill use. Germany is widely recognised as an environmentally conscious through its advanced waste management practices. The nation has implemented a comprehensive recycling system known as the Green Dot program, which has significantly reduced landfill use and increased recycling rates (Schroeder et al., 2019). Germany's waste management policies emphasise the principles of waste hierarchy, prioritising waste prevention, reuse, and recycling over disposal (Mazzanti & Zoboli, 2008). Additionally, Germany has invested in waste-to-energy facilities that convert non-recyclable waste into electricity, further minimising landfill usage and reducing greenhouse gas emissions (Schüch, 2020). These efforts are supported by robust legislation and public awareness campaigns that encourage sustainable waste practices among citizens and businesses, contributing to Germany's reputation as a leader in environmental sustainability (Rogge & Reichardt, 2016).

Sweden recycles approximately half of its domestic garbage, transforming a significant percentage into electricity through incineration. These practices are backed by robust environmental legislation and public awareness initiatives encouraging sustainable behaviour among locals and visitors (GSTC, 2021). Sweden's waste-to-energy plants provide electricity and heat to 1.25 million homes. This has significantly reduced reliance on fossil fuels and improved energy security (Swedish Waste Management Association, 2020).

2.1.3 Africa: Emerging Tourism Markets and Their Unique Waste Management Issues

Emerging African tourism markets, including Kenya, Tanzania, and South Africa, confront distinct waste management challenges. Limited infrastructure and financial resources can exacerbate these difficulties. However, some novel solutions are being adopted. Kenya, for example, has implemented a ban on single-use plastics in 2017, significantly reducing plastic pollution. A survey conducted in 2019 showed a 70% reduction in plastic bag usage and a noticeable improvement in the cleanliness of urban areas and national parks (NEMA, 2019). Community-based waste management initiatives, sponsored by non-governmental organisations and foreign funding, are also critical in tackling waste concerns and encouraging sustainable tourist practices (UNWTO, 2018). Another example is Tanzania, home to iconic tourist destinations like Serengeti National Park and Zanzibar, which face substantial waste management challenges due to the rapid tourism growth. The country is developing integrated waste management systems in tourist hotspots, focusing on waste segregation, recycling, and composting. Initiatives like the Zanzibar Clean and Green Program aim to make the island a model for sustainable tourism (Tanzania Tourism Board, 2020).

2.1.4 Influence of Infrastructure, Culture, and Economics on Waste Management

The infrastructure available in a particular location significantly impacts the efficacy of waste management strategies. Developed areas with modern infrastructure may adopt complete waste management systems, including recycling and waste-to-energy technology. Developing regions, on the other hand, may have challenges with primary garbage collection and disposal due to budgetary and logistical restrictions. Cultural perspectives on waste are equally important. Public engagement in recycling programmes and waste reduction activities is higher in areas with high levels of environmental awareness, such as portions of Europe. On the other hand, there might be more littering and less recycling in areas where waste management is not a cultural priority. Economic aspects are also crucial. Wealthier regions may afford to invest in modern waste management infrastructure and technology, whilst poorer regions may rely on inefficient and ecologically harmful garbage disposal techniques. Economic incentives, such as single-use plastic fees and recycling programme subsidies, can motivate people to change their waste management habits(Di Foggia & Beccarello, 2021). By analysing these regional variations and the factors influencing them, a better understand can be undertaken on the global waste management landscape in the tourism and hospitality industries. We can also find opportunities for cross-cultural collaboration and knowledge sharing to advance sustainable practices globally.

3. CULTURAL CONSIDERATIONS IN WASTE MANAGEMENT EDUCATION

3.1 Role of Cultural Norms, Beliefs, and Attitudes in Shaping Waste Management Behaviours

Diverse cultural norms, attitudes, and beliefs significantly influence waste management practices. These cultural factors shape how people perceive waste, their willingness to adopt sustainable behaviours, and their openness to waste management education. In some cultures, waste is viewed as a valuable resource that can be recycled or reused, leading to innovative waste reduction and resource recovery practices. Conversely, in other cultures, there may be less emphasis on strict waste disposal methods, resulting in higher pollution levels and littering. For instance, in Japan, cultural norms that emphasise cleanliness and environmental responsibility have led to highly efficient recycling systems and minimal waste. In contrast, countries with less stringent cultural attitudes towards waste disposal often struggle with pollution and ineffective waste management systems (Hall & Gössling, 2016; UNEP, 2019; Mensah, 2019).

Understanding these cultural nuances is crucial for designing waste management education programs that effectively engage local communities and promote sustainable behaviours. Tailoring educational initiatives to reflect local values and practices can enhance their relevance and impact. For example, incorporating traditional waste reuse and recycling practices into modern educational programs can resonate more with local populations and encourage greater participation. Additionally, leveraging community leaders and cultural influencers to endorse and disseminate waste management education can improve acceptance and adherence to sustainable practices.

By acknowledging and integrating cultural factors into waste management strategies, we can foster more effective and sustainable waste management practices globally (Hall & Gössling, 2016; UNEP, 2019; Mensah, 2019).

3.2 Impact of Cultural Factors on Travellers, Hospitality Workers, and Local Populations

Cultural influences affect residents, visitors, and hospitality personnel, creating a complex landscape for waste management in tourism and hospitality settings. Tourists often bring their own cultural beliefs and practices regarding waste management, sometimes conflicting with local norms and practices. For instance, a tourist from a country with stringent recycling regulations may be surprised by lax waste disposal methods in another destination, leading to confusion or non-compliance with local practices. Conversely, tourists from regions with less emphasis on recycling may not follow their destination's more rigorous waste separation systems. Similarly, hospitality workers, who often come from diverse cultural backgrounds, may have varying knowledge and commitment to environmentally friendly waste practices. These disparities can lead to inconsistent waste management practices, undermining efforts to promote sustainability.

To bridge these gaps, culturally relevant education initiatives are essential. These initiatives should harmonise the diverse waste management behaviours of tourists and hospitality staff with local sustainability goals. For example, bilingual and culturally tailored waste management guidelines can effectively align tourists' practices with local norms in regions with high tourist influxes. Training programs for hospitality workers incorporating cultural sensitivity and practical waste management techniques can also enhance their understanding and commitment to sustainable practices. By implementing such culturally appropriate educational initiatives, destinations can promote a consistent and practical approach to waste management, ensuring that both visitors and staff contribute to local sustainability efforts (UNEP, 2019; WTTC,2023; Thanh & Bosch, 2020).

3.3 Effective Cultural Adaptation of Waste Management Education Programs

The "Keep Bali Clean" campaign in Indonesia is an exemplary case of effective cultural adaptation in waste management. As a prominent tourist destination, Bali faced significant challenges with waste management due to the high influx of visitors and inadequate local infrastructure. The campaign addressed these issues by educating tourists and locals through culturally resonant messaging highlighting Bali's traditional cleanliness and environmental stewardship values. Recognising the importance of these values in Balinese culture, the campaign integrated messages that aligned with local customs and beliefs, fostering a sense of shared responsibility among residents and visitors alike. The "Keep Bali Clean" campaign encouraged greater public participation in waste management efforts by tapping into Bali's cultural heritage. This approach increased awareness and inspired action, as people were likelier to engage in practices that resonated with their cultural identity. The campaign's success can be seen in the significant improvements in waste management on the island, including enhanced recycling efforts, cleaner public spaces, and reduced litter. The initiative's emphasis on cultural relevance underscores the importance of tailoring environmental campaigns to local contexts to achieve meaningful and lasting impacts (Green Hotelier, 2019).

3.4 Culturally Tailored Educational Interventions

In Japan, Kamikatsu has created a zero-waste initiative that integrates local cultural norms, including rigorous sorting and recycling. The initiative teaches locals about the necessity of garbage separation, consistent with Japanese cultural values of order and care for the environment. The effort comprises community workshops and public education campaigns that emphasise the cultural value of sustainability, resulting in high participation rates and successful waste reduction (WTTC,2023; Sustainable Hospitality Alliance, 2022). The "Plastic Bag Free Kenya" campaign in Kenya promotes waste reduction by utilising cultural customs and local community institutions. Through community gatherings and traditional storytelling—two culturally relevant modes of communication—the initiative teaches the local population about the adverse environmental effects of plastic bags. Due to the community's broad support and adherence to the plastic bag ban due to this culturally sensitive approach, the amount of plastic garbage in the area has decreased (UNWTO, 2018; GSTC, 2021; Green Hotelier, 2019). These case studies clearly show that waste management education initiatives with cultural adaptations are more successful in encouraging sustainable practices. These initiatives consider local cultural settings and employ culturally appropriate communication techniques to involve communities, tourists, and hospitality workers in waste management practices.

3.5 Innovative Educational Methodologies

Innovative teaching techniques are essential for engaging diverse audiences and fostering significant behaviour change in waste management education. These approaches leverage interactive activities, technology, and community engagement to enhance the effectiveness of education and promote sustainable habits. Interactive exercises such as waste sorting games and recycling competitions actively involve participants in learning, making it more immersive and memorable. Technology, including digital platforms and mobile apps, offers dynamic ways to deliver educational content and empower users with real-time recycling and waste disposal information. Community engagement through workshops and outreach activities fosters dialogue and collaboration, creating a sense of ownership and responsibility among participants. By combining these innovative approaches, educators can inspire meaningful behaviour change and contribute to a more sustainable future in waste management.

3.6 Case Studies of Effective Educational Programs

3.6.1 Interactive Workshops and Community Engagement Initiatives

Interactive seminars and community involvement projects provide hands-on learning opportunities, empowering participants to actively engage in waste management methods. For example, India's "Waste Warriors" program organises interactive seminars in schools and communities where participants learn about waste sorting, composting, and recycling through games and hands-on demonstrations, reaching over 100,000 individuals and significantly improving local waste management practices (Ghimire et al., 2017; Waste Warriors, 2020). Similarly, the "Clean-Up Kenya" initiative educates residents about waste management through community clean-up events and workshops, engaging thousands of volunteers and significantly reducing pollution in targeted areas (Clean-Up Kenya, 2021). In South Africa, the Eco-Schools program integrates environmental education into school curriculums, involving students in practical waste management activities and fostering sustainable habits from a young age (WESSA, 2021). Japan's "Zero Waste Academy" in Kamikatsu offers workshops and educational tours that have helped the town achieve an 80% recycling rate by teaching participants about waste sorting and recycling (Zero Waste Academy, 2020). Thailand's "Trash Hero" program combines clean-up activities with educational workshops to teach locals and tourists about the environmental impact of waste and promote sustainable behaviour (Trash Hero, 2019).

These initiatives have improved waste management practices, increased community participation, and educational impacts. Programs like Waste Warriors and Clean-Up Kenya have reduced litter and promoted recycling and composting, with Clean-Up Kenya mobilising over 5,000 volunteers (Waste Warriors, 2020; Clean-Up Kenya, 2021). The Eco-Schools program in South Africa has educated over 500,000 students on sustainable waste management, while Tanzania's community composting initiatives have provided economic benefits by producing valuable compost (WESSA, 2021). Japan's Zero Waste Academy has contributed to Kamikatsu achieving an 80% recycling rate, demonstrating the effectiveness of hands-on education and community engagement (Zero Waste Academy, 2020). These interactive workshops and community engagement initiatives foster a collective responsibility towards sustainable waste management, leading to tangible environmental and economic benefits.

3.6.2 Digital Platforms and Gamified Learning Environments

Technology is increasingly leveraged to create more engaging and accessible waste management education through digital platforms and gamified learning environments. For instance, PepsiCo Recycling's "Recycle Rally" program in the US gamifies school waste collection and recycling initiatives through a smartphone app and digital platform. This program encourages students to compete with other schools for rewards, earning points for recycling efforts. This gamified approach promotes healthy competition and sustained participation in recycling activities, effectively making waste management education more interactive and enjoyable (PepsiCo, 2020). Similarly, digital platforms like "Recyclebank" offer gamified experiences where users earn points for recycling and taking eco-friendly actions, which can be redeemed for rewards. These platforms educate users about proper waste disposal and recycling and incentivise sustainable behaviours through tangible rewards. By integrating technology and gamification, these programs foster a culture of environmental responsibility and continuous engagement, proving that digital tools can significantly enhance the effectiveness of waste management education (Recyclebank, 2021).

3.6.3 Global Best Practices Increased Awareness and Behaviour Change

In waste management education, several international best practices have emerged as effective strategies for raising awareness and promoting behaviour change. One prominent approach is the widespread adoption of the "3R" concept (Reduce, Reuse, Recycle) in education programs worldwide. This concept serves as a foundational principle, encouraging individuals to embrace more sustainable consumption and waste management habits by emphasising the importance of waste reduction at its source, material reuse, and resource recycling (UNEP, 2019). By instilling these principles in communities, waste management educators aim to cultivate a mindset shift towards more conscious consumption and responsible waste disposal practices. Moreover, comprehensive waste reduction initiatives at the city level have gained traction through movements like "Zero Waste Cities," championed by organisations such as Zero Waste Europe. These initiatives advocate for implementing policies and infrastructure focused on waste avoidance, composting, and recycling to reduce landfill waste and greenhouse gas emissions significantly (Zero Waste Europe, 2021). Promoting systemic changes at the municipal level addresses immediate waste management challenges and contributes to broader environmental goals, such as mitigating climate change and conserving natural resources. By adopting such international best practices, waste management education programs can effectively mobilise communities to adopt more sustainable waste management practices and contribute to a greener, more environmentally conscious future.

3.7 Challenges and Opportunities in Waste Management Education

3.7.1 Inadequate Infrastructure

In numerous regions, insufficient infrastructure for managing waste poses a significant challenge. This deficit encompasses various aspects, including inadequate disposal sites, recycling centres, and waste collection systems. The deficiency in infrastructure not only impedes the implementation of effective waste management initiatives but also exacerbates environmental contamination. Without proper waste disposal and recycling facilities, communities may resort to unsustainable practices such as open dumping or burning, leading to air, soil, and water pollution. Moreover, the lack of comprehensive waste collection systems results in litter accumulation in public spaces, further contributing to environmental degradation. In areas lacking recycling infrastructure, valuable resources are lost as recyclable materials are disposed of alongside general waste. Addressing these infrastructure gaps is essential for establishing sustainable waste management practices and mitigating environmental harm. Investment is required in developing disposal facilities, recycling infrastructure, and efficient waste collection systems to support comprehensive waste management initiatives and minimise environmental contamination (Ferronato & Torretta, 2019).

3.7.2 Insufficient Budget

Another major obstacle to implementing comprehensive waste management education initiatives is inadequate funding. Initiatives such as community workshops, public awareness campaigns, and infrastructure development often struggle to achieve their objectives due to insufficient financial resources. Without adequate funding, these initiatives may be limited in scope or unable to reach their target audience effectively. Community workshops, for instance, require funding for venue rental, materials, and facilitators' fees, while public awareness campaigns need funding for marketing materials, advertisements, and outreach activities. Infrastructure development, including establishing waste disposal sites, recycling centres, and waste collection systems, also requires significant investment. The lack of funding for these initiatives hampers progress in promoting sustainable waste management practices and addressing environmental concerns. To overcome this obstacle, securing adequate funding from government agencies, non-profit organisations, private sector partners, and international donors is essential to effectively support waste management education efforts (Debrah et al., 2021).

3.7.3 Low Stakeholder Awareness

The prioritisation of waste management education faces challenges from a lack of understanding among various stakeholders, including corporations, legislators, and the general public. Without a clear comprehension of the detrimental impacts of improper waste disposal on the environment and society, stakeholders may be less inclined to support or participate in waste management programs. Corporations may prioritise cost-saving measures over sustainable waste practices, legislators may allocate resources inadequately due to competing priorities, and the general public may continue with unsustainable habits due to a lack of awareness. This lack of understanding perpetuates a cycle where insufficient attention and resources are allocated to waste management education efforts, hindering progress towards more sustainable waste practices. To address this challenge, it is crucial to raise awareness among stakeholders about the importance of proper waste management and the benefits of adopting sustainable practices, thereby garnering support and facilitating meaningful engagement in waste management initiatives (Wilson et al.,2015).

4. OPPORTUNITIES FOR INTERNATIONAL COOPERATION AND INFORMATION EXCHANGE

Despite these challenges, there are opportunities for international cooperation and information exchange to address waste management issues effectively.

4.1 Cross-Cultural Collaborations

Facilitating the exchange of best practices, information, and resources in waste management education can be significantly enhanced through cooperative efforts across nations and organisations with diverse cultural backgrounds. By fostering collaboration and sharing experiences, stakeholders can learn from one another and develop more effective strategies for advancing sustainable practices. For example, a country with advanced recycling infrastructure may share its expertise with another nation struggling to establish similar systems, while a community organisation implementing successful waste reduction initiatives can offer insights to counterparts facing similar challenges. Through this exchange of knowledge and experiences, stakeholders can collectively improve their understanding of waste management issues and develop more potent plans for promoting sustainability. Additionally, cross-cultural collaboration fosters mutual respect and understanding, enriching the global discourse on waste management and strengthening collective efforts to address environmental challenges (Debrah et al., 2021; Hofstad & Torfing, 2016).

4.2 Global Networks

Stakeholders in waste management education can access a wealth of knowledge, experience, and assistance by engaging in international networks and projects. Organisations such as the International Solid Waste Association (ISWA) and the Global Sustainable Tourism Council (GSTC, 2021) provide platforms for collaboration, advocacy, and capacity building on waste management issues. Through participation in these networks, stakeholders can benefit from shared expertise, best practices, and innovative solutions worldwide. They can also engage in advocacy efforts to promote policies and practices that support sustainable waste management. Additionally, these networks offer opportunities for capacity building through training programs, workshops, and knowledge-sharing events, empowering stakeholders to enhance their skills and effectiveness in waste management education initiatives. Participation in international networks and projects enables stakeholders to leverage collective expertise and resources to address waste management challenges more effectively globally.

4.3 Case Studies of Cross-Cultural Collaborations and Global Networks

4.3.1 Waste Wise Cities Network

The World Bank's Waste Wise Cities Network is a crucial platform for promoting information exchange and capacity building among cities worldwide to enhance waste management practices. Through this network, participating cities can share insights, experiences, and best practices on waste management through various channels such as seminars, webinars, and peer-to-peer learning exchanges. By facilitating this exchange of knowledge and expertise, the network enables cities to learn from one another's successes and challenges, ultimately leading to more effective and sustainable waste management strategies. Furthermore, the network provides cities access to valuable resources, technical assistance, and funding opportunities to support the implementation of innovative solutions and policy frameworks. Overall, the Waste Wise Cities Network plays a vital role in fostering collaboration and collective action among cities to address the complex challenges associated with waste management on a global scale (World Bank, 2021).

4.3.2 Plastic Pollution Prevention Program

The Ocean Conservancy's Plastic Pollution Prevention Program is a crucial initiative to combat plastic pollution in coastal areas worldwide. By bringing together diverse stakeholders, including local communities, companies, NGOs, and governments, the program facilitates collaborative efforts to address the pervasive issue of plastic waste. The program advocates for policy reforms to minimise plastic waste and promote sustainable waste management practices through joint research, advocacy campaigns, and community engagement projects. These efforts are essential for raising awareness about the detrimental impacts of plastic pollution on marine ecosystems and coastal communities and for mobilising action to mitigate its effects. By fostering cooperation among stakeholders and advocating for systemic changes, the Plastic Pollution Prevention Program contributes to the global movement towards a more sustainable and plastic-free future (Ocean Conservancy, 2021).

5. EMERGING TRENDS AND FUTURE DIRECTIONS

5.1 Leveraging Technology for Effective Waste Management Education

Technological innovations provide encouraging opportunities to improve waste management education. Accessible and engaging methods of teaching people and communities about waste management techniques are offered by digital platforms, smartphone apps, and internet resources. Mobile applications such as "Recycle Coach" and "Waste Wizard," for instance, give users access to recycling regulations, collection schedules, and disposal choices that are customised for their area (Recycle Coach, 2022). Immersion learning experiences are made possible by virtual reality (VR) and augmented reality (AR) technology, which lets users practise sustainable habits and model waste management scenarios in a fun and secure setting (Liu et al., 2020).

5.2 Increasing Stakeholder Participation

The effectiveness of waste management education programmes depends on increasing stakeholder engagement. Involving stakeholders, such as companies, government agencies, academic institutions, and community organisations, promotes cooperation and group efforts to achieve sustainable waste management objectives. Through community-based programmes, including recycling drives, clean-up campaigns, and waste audits, individuals are empowered to tackle waste concerns in their local areas actively (UNEP, 2019). Public-private collaborations increase waste management efforts and apply creative solutions by combining the resources and knowledge of both industries (PricewaterhouseCoopers, 2020).

5.3 Incorporating Waste Management Education into Broader Sustainability Projects

The effectiveness and relevance of waste management education are increased when it is included in more extensive sustainability efforts. Stakeholders can understand how waste problems relate to other sustainability difficulties by connecting waste management to more general environmental, social, and economic sustainability goals. Waste-to-energy initiatives, for instance, emphasise the advantages of sustainable waste management techniques for the environment and economy by showcasing waste's potential as a significant resource for producing renewable energy (IEA, 2021). Initiatives for sustainable tourism include waste management education into destination management plans, encouraging eco-friendly travel behaviours and reducing the environmental impact of tourist-related activities (GSTC, 2021).

5.4 Case Studies of Innovative Approaches and Successful Implementations

5.4.1 The Wastepresso Project

The Wastepresso Project, developed by Miniwiz, represents a groundbreaking waste management and recycling innovation. This mobile recycling machine is designed to transform plastic garbage into architectural tiles, showcasing a revolutionary technology with immense potential to address plastic pollution and promote recycling, particularly in rural or underdeveloped areas. By converting plastic waste into valuable building materials, the Wastepresso Project offers a practical solution for managing plastic pollution and creates opportunities for sustainable construction practices. This innovative approach not only helps to mitigate the environmental impact of plastic waste but also contributes to the development of local economies by providing a means to repurpose waste materials into valuable resources. The Wastepresso Project exemplifies how technological innovation can be pivotal in addressing pressing environmental challenges and fostering sustainable development (Miniwiz, 2022).

5.4.2 The Circular Economy Initiative

The Circular Economy Initiative, led by the Ellen MacArthur Foundation, represents a transformative approach to waste management by promoting a regenerative model to minimise waste and maximise resource efficiency. Central to this initiative is the concept of a circular economy, which challenges traditional linear models of production and consumption by emphasising the reuse, repair, and recycling of materials throughout their lifecycle. The initiative seeks to eliminate waste and pollution by rethinking goods and systems while driving innovation and economic growth. This collaborative effort engages businesses, governments, and academics in a shared vision to transition towards a circular economy model where waste is viewed as a valuable resource rather than a disposable product. Through research, advocacy, and knowledge-sharing, the Circular Economy Initiative inspires and empowers stakeholders to adopt more sustainable practices that benefit the environment and create new opportunities for economic development and social well-being (Ellen MacArthur Foundation, 2021). By embracing emerging trends and innovative approaches, waste management education can evolve to meet the evolving needs and challenges of the 21st century, paving the way for a more sustainable and resilient future.

6. RECOMMENDATIONS FOR STAKEHOLDERS

6.1 Policymakers

Policymakers wield significant influence in shaping the regulatory landscape governing waste management practices. Governments should enact supportive regulations and policies that incentivise waste reduction, recycling, and environmentally responsible disposal methods to foster effective waste management education. Such policies might include offering incentives for businesses to adopt eco-friendly practices, implementing waste segregation and recycling standards, and imposing fines for improper waste disposal (UNEP, 2019). Additionally, policymakers should prioritise international collaboration, sharing knowledge and learning from best practices globally. Participation in international conferences and collaborations enables policymakers to gain insights into innovative approaches and strategies for addressing waste management challenges, contributing to more informed and effective policymaking (UNWTO, 2018).

6.2 Industry Leaders

Industry leaders tremendously impact waste management strategies within their respective industries. To promote sustainability and waste management education, they are encouraged to implement sustainable practices by reducing waste creation, instituting recycling and composting programs, and using environmentally friendly packaging materials (WTTC,2023). Additionally, industry leaders should invest in educational initiatives that enhance waste management knowledge and encourage sustainable practices among employees, consumers, and stakeholders. By supporting these educational efforts, industry leaders can help cultivate a culture of sustainability within their companies and communities, driving broader environmental responsibility (PricewaterhouseCoopers, 2020)

6.3 Educators

Educators play a vital role in shaping the knowledge and attitudes of future generations towards waste management. To enhance waste management education, educators are encouraged to develop culturally sensitive curricula relevant to the local context. By incorporating cultural perspectives and examples into instructional materials, educators can ensure that waste management education resonates with diverse audiences and fosters meaningful engagement (UNESCO, 2020). Additionally, educators should utilise innovative teaching methods and technologies to improve the effectiveness of waste management education. This includes interactive activities, experiential learning opportunities, and digital resources catering to various learning styles and preferences (Liu et al., 2020).

6.4 Community Leaders

Community leaders are crucial in mobilising local resources and fostering community engagement in waste management initiatives. Community leaders are encouraged to actively include residents in waste management projects through outreach programs, volunteer opportunities, and participatory decision-making processes to support waste management education at the grassroots level. By involving citizens in these initiatives, community leaders can cultivate a sense of communal responsibility and commitment to sustainability (UNEP, 2019). Additionally, they should support grassroots educational efforts by providing resources and assistance to community-based projects to raise awareness about waste management issues and promote sustainable lifestyles. This support can include funding for local projects, access to educational resources and facilities, and collaborations with local groups and institutions (UNWTO, 2018). By implementing these strategies, community leaders can contribute to building a more sustainable and resilient future by promoting effective waste management education at the local level.

7. CONCLUSION

In conclusion, tackling the environmental issues the tourism and hospitality industry raises requires a strong emphasis on waste management education. This chapter has underscored the importance of implementing sustainable waste management techniques and increasing stakeholder awareness through significant insights and global perspectives. Developing waste management education programs that engage diverse audiences necessitates cultural sensitivity, creativity, collaboration, and a long-term vision. These elements are crucial for driving meaningful change and fostering a culture of sustainability within the industry. By integrating culturally relevant content, innovative teaching methods, and collaborative efforts, stakeholders can ensure that waste management education resonates with and effectively influences various demographic groups.

To address waste management concerns and promote a sustainable future in tourism and hospitality, stakeholders must work collaboratively. Policymakers, business leaders, educators, and community leaders play vital roles in this collective effort. Policymakers need to establish supportive regulations and policies that incentivise sustainable practices and penalise improper waste disposal. Business leaders should implement eco-friendly practices and invest in educational initiatives that raise awareness among employees and customers. Educators must develop and deliver curricula highlighting the importance of waste management and sustainability. Community leaders should engage local populations through outreach programs, participatory decision-making processes, and support for grassroots initiatives. By prioritising sustainability and adopting these multifaceted approaches, stakeholders can collectively drive significant progress in waste management.

Joint efforts and a shared commitment to environmental stewardship will ultimately achieve a sustainable tourism and hospitality industry. By fostering a culture that values waste minimisation, resource preservation, and community well-being, the industry can ensure that tourism's benefits do not come at the expense of the planet's health. The collaborative efforts of all stakeholders—through regulatory frameworks, educational programs, and community engagement—will pave the way for a more sustainable model of tourism and hospitality. This holistic approach addresses immediate waste management challenges and sets a foundation for long-term environmental sustainability, allowing the industry and the planet to thrive in harmony.

REFERENCES

Bali Partnership. (2020). *Waste management report.* https://www.balipartnership.org/en_gb/

Clean-Up Kenya. (2021). *Community engagement and clean-up initiatives.* https://cleanupkenya.org/2021-in-review/

Debrah, J. K., Vidal, D. G., & Pimenta Dinis, M. A. (2021). Raising Awareness on Solid Waste Management through Formal Education for Sustainability: A Developing Countries Evidence Review. *Recycling*, 6(1), 6. DOI: 10.3390/recycling6010006

Di Foggia, G., & Beccarello, M. (2021). Designing waste management systems to meet circular economy goals: The Italian case. *Sustainable Production and Consumption*, 27, 232–240. DOI: 10.1016/j.spc.2021.01.002

Ellen MacArthur Foundation. (2021). *Circular economy initiative.* https://www.ellenmacarthurfoundation.org/topics/circular-economy-introduction/

Ferronato, N., & Torretta, V. (2019). Waste mismanagement in developing countries: A review of global issues. *International Journal of Environmental Research and Public Health*, 16(6), 1060. DOI: 10.3390/ijerph16061060 PMID: 30909625

Ghimire, P., & Jayaraman, K. (2017). Waste management education in India: A case study of the "Waste Warriors" program. *International Journal of Environmental and Science Education*, 12(8), 1801–1815.

Ghimire, S., Shrestha, R., & Shakya, S. (2017). *Waste Warriors: Promoting sustainable waste management practices.*

Global Sustainable Tourism Council (GSTC). (2021). *Sustainable tourism practices.* GSTC Guidelines. https://www.gstcouncil.org/2021/

Green Hotelier. (2019). *Waste management in the hospitality industry.* https://www.greenkey.global/stories-news-1/tag/Green+Hotelier

Hall, C. M., & Gössling, S. (2016). *Sustainable tourism: A global perspective.* Routledge.

Hofstad, H., & Torfing, J. (2016). Collaborative innovation in public service provision: New perspectives on governance. *Public Management Review*, 18(1), 82–106. DOI: 10.1080/14719037.2015.1051576

International Energy Agency (IEA). (2021). *Waste-to-energy.* IEA Publication.

Liu, C., Wang, D., & Kong, X. (2020). Application of virtual reality technology in environmental education: A review. *Frontiers in Education*, 5, 131.

Mazzanti, M., & Zoboli, R. (2008). Waste generation, waste disposal and policy effectiveness: Evidence on decoupling from the European Union. *Resources, Conservation and Recycling*, 52(10), 1221–1234. DOI: 10.1016/j.resconrec.2008.07.003

Mensah, I. (2019). Environmental education and sustainable tourism: A study of tourists and operators' perspectives in Ghana. *Journal of Sustainable Tourism*, 27(6), 770–785.

Ministry of Environment and Energy. (2013). *Waste management regulation 2013/R-58*. https://mymaldiveshome.environment.gov.mv/knowledge_articles/waste-management-regulation-2013-r-58/

Miniwiz. (2022). *Wastepresso project*. Miniwiz Projects. https://medium.com/taiwantecharena/miniwiz-the-company-changing-the-way-the-world-sees-waste-347def0f2c7

Naseer, A. (2017). Maldives' waste management efforts in the tourism sector. *Journal of Sustainable Tourism Development*, 5(3), 45–59.

National Environment Management Authority (NEMA). (2019). *Impact of plastic bag ban in Kenya*. https://www.nema.go.ke/index.php?option=com_content&view=article&id=102&Itemid=121

Ocean Conservancy. (2021). *Plastic pollution prevention program*. https://oceanconservancy.org/wp-content/uploads/2021/03/FINAL-Ocean-Conservancy-PPE-Report-March-2021.pdf

PepsiCo. (2020). *Recycle rally: Turning bottles and cans into recyclables*. PepsiCo Recycling. https://www.pepsicorecyclerally.com/

PricewaterhouseCoopers (PwC). (2020). *Public-private partnerships: A guide for local governments*. PwC Reports. https://www.pwc.com/gx/en/about-pwc/global-annual-review-2020/downloads/pwc-global-annual-review-2020.pdf

Recycle Coach. (2022). *Recycle Coach app*. https://recyclecoach.com/cities/usa-az-city-of-tucson/

Recyclebank. (2021). *How it works: Earn rewards for recycling*. https://recyclebank.com/?code=750d22e1c5e2bdbf8fb083f8920522e8&button=earn&partner=undefined

Rogge, K. S., & Reichardt, K. (2016). Policy mixes for sustainability transitions: An extended concept and framework for analysis. *Research Policy*, 45(8), 1620–1635. DOI: 10.1016/j.respol.2016.04.004

Scheyvens, R., & Cheer, J. M. (2021). Tourism, the SDGs and partnerships. *Journal of Sustainable Tourism*, 30(10), 1–11. DOI: 10.1080/09669582.2021.1982953

Schroeder, P., Anggraeni, K., & Weber, U. (2019). The relevance of circular economy practices to the sustainable development goals. *Journal of Industrial Ecology*, 23(1), 77–95. DOI: 10.1111/jiec.12732

Schüch, A. (2020). Waste-to-energy: An essential element for sustainable waste management. *Waste Management & Research*, 38(11), 1238–1245.

Sustainable Hospitality Alliance. (2022). *Best practices for waste management in hospitality*. SHA Resources. https://sustainablehospitalityalliance.org/

Swedish Waste Management Association. (2020). *Waste-to-energy statistics*. https://earth.org/sweden-waste-to-energy/

Tanzania Tourism Board. (2020). *Zanzibar's clean and green program*. https://ztite.zanzibartourism.go.tz/zct/

Thanh, N. V., & Bosch, O. J. (2020). Community participation in tourism: A case study of sustainable waste management in Vietnam. *Sustainable Tourism Journal*, 28(3), 456–472.

Trash Hero. (2019). *Engaging communities in waste management*. https://trashhero.org/category/community-waste-management/

UNDP. (2018). *Sustainable waste management strategies for the Maldives*. United Nations Development Programme. https://www.undp.org/maldives

United Nations. (2015). *Transforming our world: The 2030 agenda for sustainable development*. https://www.un.org/ga/search/view_doc.asp?symbol=A/RES/70/1&Lang=E

United Nations Environment Programme (UNEP). (2019). *Tourism's three main impact areas*. https://www.unep.org/resources/unep-annual-report-2019

United Nations World Tourism Organization (UNWTO). (2018). *Sustainable development goals and tourism*. https://www.unwto.org/global/publication/tourism-and-sustainable-development-goals-journey-2030

Waste Warriors. (2020). *Annual report: Impact and achievements*. https://www.wastewarriors.org/2020/

WESSA (Wildlife and Environment Society of South Africa). (2021). *Eco-schools program*. https://wessa.org.za/our-work/schools-programme/wessa-eco-schools/

Wilson, D. C., Velis, C. A., & Rodic, L. (2015). *Global waste management outlook*. United Nations Environment Programme (UNEP). Retrieved from https://web.unep.org/ourplanet/september-2015/unep-publications/global-waste-management-outlook

World Bank. (2020). *Maldives clean environment project: Transforming waste into energy*. https://www.worldbank.org/en/news/feature/2020/12/11/maldives-building-back-better-through-clean-energy

World Commission on Environment and Development (WCED). (1987). *Our common future*. Oxford University Press. https://digitallibrary.un.org/record/139811?v=pdf

World Travel & Tourism Council (WTTC). (2023). *Environmental impact of travel and tourism*. https://researchhub.wttc.org/product/the-environmental-impact-of-global-tourism-2023

Zero Waste Academy. (2020). *Achieving zero waste: Kamikatsu's success story*. https://academy.zerowaste.asia/

Zero Waste Europe. (2021). *Zero waste cities: Leading the change*. https://zerowasteeurope.eu/our-work/local/mission-zero-academy-miza/

Chapter 4
Waste Management in Protected Areas:
Benchmarking Recyclable Waste Emanating From Tourism

Kevin Mearns
https://orcid.org/0000-0001-5874-3542
University of South Africa, South Africa

ABSTRACT

The worldwide demand for wildlife tourism-based experiences has driven a global increase in demand for wildlife tourism experiences within protected areas. Tourism within protected areas has both positive and negative effects. The positive effects relate to the increased financial support for conservation initiatives that provide concrete substantiation for the long-term survival and conservation of protected areas. The negative effects, however, relate to the increased tourism infrastructure and the environmental impacts of accommodating guests and their hospitality staff within the protected areas. Of particular interest in this chapter is the issue of Waste management in protected areas, specifically recyclable waste. A mixed method single case study research design was employed to determine baselines and benchmarks of recyclable waste at a wildlife tourism lodge in the Greater Kruger Park area. Reflections on general waste management were also made regarding proposed principles for responsible waste management in protected areas.

DOI: 10.4018/979-8-3693-6110-8.ch004

1. INTRODUCTION

Tourism is an essential engine for economic activity and growth across the world. Over 50% of the Tourism market is driven by nature tourism and the desire to explore new environments, which is expected to grow year-on-year. The tourism sector is also the most significant contributor to the financing of protected area systems. Nature-based tourism is one of the fastest-growing subsectors of the tourism industry. Nature-based tourism focuses on travel to enjoy natural areas and biodiversity. Nature-based tourism can take many forms, including wildlife watching, birdwatching, ecotourism, walking, cycling, fishing, adventure or sun and beach tourism. Nature-based tourism includes wildlife tourism. Wildlife tourism can be broadly defined as travel to destinations primarily observing wild fauna and their associated flora. Wildlife tourism contributes millions of dollars annually to protected areas, strengthening conservation efforts and supporting sustainable development. In Africa – where nature and national parks are considered some of the most critical tourism assets – wildlife tourism generates around US$142 million in entrance fees for protected areas. However, wildlife tourism is also associated with a series of impacts in protected areas. Although tourism in protected areas can have social, economic and environmental benefits, several negative consequences may also result, such as changes in water quality, water availability, vegetation and coral reef destruction, damage to archaeological and heritage sites, pollution and damage to sensitive environments and sites as well as increased waste generation and the associated pollution resulting thereof (Eagles et al., 2002). These negative impacts should be a cause for concern and a motivation to address these issues. One of the most significant and challenging impacts associated with wildlife tourism in protected areas is waste management. Human activities within protected areas generate waste that needs to be managed carefully within protected area contexts.

Besides the growth in nature-based tourism, there is also a worldwide growing trend towards sustainable tourism (Dedeke, 2017), with water and waste management forming two crucial aspects of the sustainability of the tourism industry. Sustainable tourism consists of a balance between the environmental, economic and sociocultural aspects. Tourism companies in Africa are motivated to "go green" or become eco-efficient for a sustainable future and as a marketing ploy to keep up with their competition. A recent study by Booking.com found that more than half of global travellers want to travel more sustainably in the future (Donovan, 2021). Similarly, Morrison-Saunders et al. (2019:1) conducted a survey of visitors to the Kruger National Park (KNP), South Africa, and their expectations for responsible tourism, which recorded that "energy, water and waste management were all considered very important by all respondents". Roos et al. (2022) identified waste management as one of the most significant challenges facing many protected areas worldwide. According to Roos et al. (2023), responsible waste management must recognise that protected areas often rely on a delicate balance between nature and human intervention and that any waste created, stored, transported or disposed of should be managed in a way that respects this balance. It is also important to note that typical, generic waste management principles might not apply to protected areas because they might conflict with the conservation objectives and goals of protected areas (Roos et al., 2023).

The tourism industry plays an integral part in waste production, with a clear relationship between tourism growth and municipal solid waste production; this has, however, been the subject of little research (Arbulu et al., 2015). Inappropriate waste management can lead to adverse environmental impacts and severe ecosystem damage through water, soil and air pollution. Tourism has positive and negative effects, and the sector is vital economically for developing countries. Although tourism can be the primary source of wealth for numerous destinations, particularly developing countries in Africa, the tourism industry consumes energy and water. It produces waste with detrimental environmental effects (Del Mar Alonso-Almeida, 2012). One specific concern of increased tourism to protected areas relates to effective waste management within a protected area context (Sandham et al., 2020). This concern is echoed by Copocchi et al. (2019), which indicates that the rapid growth and increase in tourism numbers in tandem with unsustainable waste management practices does not bode well for protected areas. Municipal solid waste is the primary kind of waste generated by tourism-related activities. Tourism operators have to manage and dispose of this waste responsibly. A study by Harper et al. (2022) found that more than 70% of the tourism waste stream comprises recyclable waste. Waste management practices in protected areas by tourism operators need drastic improvement.

Tourism is a large contributor to our economy but also a threat to our wildlife and natural heritage; thus, it is vital to balance increasing tourism and sustainable tourism practices. The abundance of game reserves, national parks, and pristine coastlines significantly contribute to South African tourism. For these to remain valuable resources for our economy, they must be managed accordingly and protected to ensure these valuable attractions are conserved for many future generations to utilise and appreciate. Larger volumes of tourists equate to a larger volume of waste generated per tourist. If tourism is to remain an economic driving force, management procedures must be addressed to ensure correct waste management strategies are followed (Pirani & Arafat, 2014). Pirani and Arafat (2014) indicated that some consider waste generation the most noticeable effect of the hospitality sector on the environment, primarily because many of the establishments that make up this sector, such as hotels, use large quantities of consumer goods as part of their operations. Byer et al. (2006) conducted a study over seven days in two-, three-- and four-star hotels in Vietnam and Laos, respectively. The result showed that each guest creates between 0.43 - 0.53kg of waste per night. This includes all recyclable and landfill waste. Traditionally, waste management within the tourism industry is not seen as a high priority, and often, negligible effort and attention is given to the important management aspect.

This chapter concerns waste management in protected areas and investigates recyclable waste at wildlife tourism lodges in South Africa's Greater Kruger National Park (GKNP). This manuscript aims to determine the quantities of recyclable waste generated by the Lodge and then to determine a potential baseline and benchmark for further comparative analysis in the future. A series of general reflective observations regarding waste management practices at the Lodge are also reflected in a series of proposed principles for responsible waste management in South African protected areas as suggested by Roos et al. (2023).

2. DEVELOPING BASELINES AND BENCHMARKS

A baseline assessment is the collecting and processing of data to establish a baseline result. The purpose of establishing baselines is to provide a starting point from which future measurements and predictions can be calculated. It can aid in forecasting waste production and recycling rates on the number of expected arrivals and can assist managers in better waste production forecasting and management. Baseline assessments are not standardised, and many different formats can be used in various situations. Concerning the quantity of recycled waste in the tourism sector, the baseline recyclable waste production would be the daily volume produced per guest. The baseline can be exclusive to a specific destination, region, province, country, industry, or market segment.

Benchmarking is a comparative management tool that is widely used all over the world. Benchmarking aims to improve continuous performance and improve an organisation's best practices. The process involves measuring performance and evaluating it against the best in its class or similar organisations (Saagi et al., 2017). The four main types of benchmarking are:

a. Internal benchmarking compares a business process to a similar one inside the organisation. For example, a company such as Wilderness Safaris compares the recyclable waste production of one of their lodges to another. This method is relatively cost-efficient, and access to information is easily obtainable. However, it can be influenced by internal bias and may not yield the best-in-class comparison (Córcoles et al., 2010).

b. Competitive benchmarking compares a product, service, process, or method against direct competitors; for example, Wilderness Safaris compares its performance against Beyond or Singita Lodge Groups. Competitors may have similar regulatory issues, and competitive benchmarking could lead to partnerships. Some disadvantages of this method are that competitors might provide misleading information, exploiting opposition weaknesses, which could lead to minimum performance improvement (Córcoles et al., 2010).
c. Functional benchmarking compares similar or identical practices within the same or similar functions outside the immediate industry; for example, Beyond compares their recyclable waste production with industrial recycling waste production. Although this can result in a better improvement rate, it is more time-consuming and involves adapting best practices (Córcoles et al., 2010).
d. Generic benchmarking broadly conceptualises unrelated business processes or functions that can be practised in the same or similar ways regardless of the industry. Some advantages of this method include that it is non-competitive, innovative, and examines multiple sectors; however, it can be challenging to identify the best in its class and very time-consuming (Córcoles et al., 2010).

As mentioned earlier, baseline assessments can be used as benchmarks. By having multiple baselines, the best in a class can be set as the benchmark for the others. In this study, the author will use competitive benchmarks, establishing one wildlife lodge's baseline to facilitate benchmarking with other similar lodges. Another way of laying down a benchmark is to determine the average of multiple baselines and aim for a target improvement, in this case, a 20% reduction on the average. In this study, the author will only produce one benchmark, which will be the best-in-class benchmark. This study did not use the target-based benchmark as the author could not establish an achievable and reasonable percentage reduction from the property concerned. The purpose of monitoring recyclable waste production is to identify volumes produced to determine comparable benchmarks and to strive for continuous improvement. The most common form of management system tracking is by conducting audits (Gössling et al., 2015). Audits are performed against a predetermined set of criteria indicating whether the management system complies with company, legal or other statutory requirements. Audit results will indicate if the objectives have been met and reveal areas requiring attention and improvement. The baselines and benchmarks forthcoming from this research will guide how the auditing of recyclable waste production quantities can be compared.

3. RESPONSIBLE WASTE MANAGEMENT IN PROTECTED AREAS

Protected areas should always be viewed in collaboration with their surrounding communities. Communities often bear the brunt of disposing of the waste generated in the protected areas; as such, they should also receive any benefits associated with recycling or disposing of waste. Research by Morrison-Saunders et al. (2019), which focuses on the expectations of responsible tourism in protected areas, indicated that visitor expectations for waste management include:

- Effective waste management
- Consideration of the waste management hierarchy and the circular economy
- Improving visitors' awareness of waste management issues and waste management practices

Enacting these three crucial aspects in protected areas depends on governance structures. National parks or marine protected areas are managed by national bodies in South Africa, such as SAN parks. The corresponding provincial equivalent manages Provincial Parks. Private Parks, on the other hand, are managed by the owners or designated management authorities. Government authorities are often characterised by poor Governance, which makes effective waste management in public, protected areas more difficult (Roos et al., 2021).

Tourism-related activities in protected areas may also involve various roleplayers, contractors and service providers, including communities; this further complicates waste management. Local governments are mandated to deliver waste management services, collection, storage and disposal as specified in Section 5B of the South African Constitution (RSA, 1996). The fact that no waste may be disposed of in protected areas and should be removed from the protected area and disposed of outside the protected area, often in the community or local government land, adds to the complexity of waste management in protected areas. All these parties mentioned above have to work effectively together to ensure that waste management does not become an area of conflict or the reason for the degradation of the protected area, resulting in the loss of tourism revenue and, ultimately, the conservation of the unique natural resource.

Makgae (2011) advocates for developing principles for responsible waste management in South Africa. In line with this, Classens et al. (2022) and Roos et al. (2022) advocate for developing best practices for waste management in South African protected areas. According to Section 17 of the National Environmental Management Protected Areas At (57 of 2003) (NEMPAA), Protected areas have the following objectives:

- Provision for rehabilitation and restored ecosystems and recovery of endangered species
- Provision of a sustainable supply of environmental goods and services, and sustainable use of natural and biological resources
- Creation of nature-based tourism destinations
- Provision for human, social, cultural, spiritual and economic development

These objectives indicate the potential conflicts and effects that these objectives could have on each other if not carefully managed. For this chapter, waste emanating from primarily tourism sources is the focus of this investigation.

4. PRINCIPLES PROPOSED FOR RESPONSIBLE WASTE MANAGEMENT IN SOUTH AFRICAN PROTECTED AREAS BY ROOS ET AL. (2023)

Roos et al. (2023) propose a series of 6 principles for responsible waste management in South African protected areas. Although these principles are discussed in much greater detail in the abovementioned source, a summary of the six principles is provided next. After the baseline and benchmarking part of this chapter is completed, a brief reflection of the progress made in the case study under investigation is included at the end of this manuscript before concluding remarks are provided.

4.1 Principle 1: Protection of Ecosystems and Biodiversity

This Principle provides for the unique characteristics of protected areas. Ecosystems and biodiversity have to be conserved. No adverse impacts resulting from waste management activities in protected areas may be allowed to cause any damage to species, habitats and ecosystems that provide critical ecological services. Waste management infrastructure and collection points should be accessible for guests and visitors to facilitate litter collection, such as entrances, viewpoints, and picnic spots. These collection points should, however, be placed away from sensitive areas where adverse effects could result on ecosystems and biodiversity.

4.2 Principle 2: Prevention and Remediation of Pollution

Incorrect waste management may result in various kinds and levels of pollution. These negative impacts should be anticipated and prevented throughout the entire waste management life cycle. Recent South African studies have shown that waste management has yet to be successfully considered in Environmental Impact Assessments (EIAs) for developments within protected areas (Claasens, 2022; Wylie, 2018). Clear communication and awareness are needed to prevent illegal dumping within protected areas. Waste infrastructure has to be serviced regularly to ensure the responsible disposal of waste generated by visitors.

4.3 Principle 3: Implementation of the Waste Management Hierarchy

South African waste management regulations promote the idea of a waste management hierarchy. This hierarchy is often presented as a pyramid with the most preferred option at the top, namely, avoiding waste, followed by reduction, reuse, recycling, and recovery, with waste treatment at the bottom (DEFF, 2020). A vital element where protected areas have a significant opportunity is separating waste for recycling, which may create opportunities in the informal waste economy for community upliftment and livelihood creation through waste management activities.

4.4 Principle 4: Provision of Effective Waste Management Services and Infrastructure

Many protected areas do not receive Municipal waste services due to their remote locations. Protected areas, whether public or private, have to include waste management services in their daily management activities and budgets, even though these may be expensive and time-consuming. Waste management in protected areas must ensure that the collection, storage, transportation, and disposal are done to the highest standards and that disposal certificates and manifest documentation are continuously collected. Waste management in protected areas should strive towards cost-effectiveness and be practically implemented without compromising environmental and legal requirements. To solve this dilemma, Cointreau (2001) proposed the investigation and implementation of the Best practical environmental option (BPEO) and best available technology (BAT). These options still need to ensure minimal impact but may need additional user fee collection to ensure the best environmental options are implemented.

4.5 Principle 5: Promotion of Participation and Building of Partnerships

Building strong partnerships in the waste space may create a meaningful means of participation from various roleplayers. Sustainable partnerships are mentioned in Pillar 1 of the National Waste Management Strategy (NWMS), and Pillar 3 advocates the importance of community participation in waste management (DEFF, 2020).

Indigenous knowledge systems and processes should also be incorporated into these partnerships as communities may see alternative solutions and opportunities not identified by other role players. Meaningful participation and partnerships could play critical roles in overcoming some of the challenges experienced by protected areas in implementing effective waste management practices and services.

4.6 Principle 6: Contribution to Well-Being, Livelihood and Capacity

Waste management in protected areas primarily aims to prevent any negative impacts resulting from waste management on biodiversity and ecosystems and the well-being of the surrounding communities. Waste management in protected areas should also be about community upliftment and empowerment and contributing to meaningful livelihoods and job creation through waste management opportunities. Challenges may exist with the remote location of the protected areas and the proximity to local communities. The importance of developing skills and competencies together with education awareness and knowledge sharing may unlock potential opportunities. These opportunities could be significant in realising potential waste management-related business and economic opportunities for more responsible waste management in Protected Areas.

These proposed principles for responsible waste management in protected areas in South Africa will be reflected with specific reference to the case study lodge under investigation.

5. METHODS AND TECHNIQUES

The investigation employed a mixed method single case study design to investigate recyclable waste production at a wildlife tourism lodge in the Greater Kruger Park area. Qualitative data was collected through interviews with the property's sustainability coordinator and participant observations of the Lodge structure, operations, and activities. The observations include following the waste management processes and practices. These include collection, separation, storage, transportation and disposal on the Lodge. Secondary quantitative data was collected through a series of sustainability reports that were received from the sustainability manager of the Lodge. The relevant data concerning recyclable waste production was extracted from the sustainability reports. Further data concerning guest occupancy and number of staff overnighting at the Lodge for the seven-year period under review, namely, 2017-2023, were requested from the sustainability manager. These crucial data sets were then received from guest services and the human resource departments of the company concerned. The generic terms of Lodge and Property are used interchangeably to refer to the Wildlife tourism establishment to protect the confidentiality of the case study investigated.

Although the guests are the primary purpose of the Lodge, the guests themselves are not the main waste generators, as the operational functions and staff members themselves are the primary producers of waste. Waste is generated in the kitchens, food stores, workshops, staff, and guest accommodation areas. The operations entail servicing the chalets, housekeeping, kitchen staff, security staff and contract workers for maintenance. The generated waste is sorted for recycling into paper, tin, glass, and plastic. Service providers collect waste once a week, and the recyclable waste is sent to a reputable recycling centre. No waste may be disposed of in the protected area; the concession rules within the protected area strictly enforce this. Wet and hazardous waste are not included in this study; however, these are essential aspects that need to be considered in future investigations for improved waste management.

5.1 The Study Area

The Greater Kruger National Park (GKNP) is the largest in South Africa. It is located across two provinces, Limpopo and Mpumalanga; it further borders Mozambique to the East and Zimbabwe to the far North. In 1927, the Kruger National Park (KNP) opened its gates to tourism. Since 1927, the number of visitors to the Park has increased to 1.8 million in 2016/2017, excluding the visitors' numbers in the reserves adjacent to the park that make the GKNP, this makes it the most successful National Park in the country with regards to its local and international profile as well as the revenue generated from tourism (Brett, 2018). The Park is open to day visitors and overnight visitors who can stay in either the National Park rest camps, the private concessions within the park, or the private game reserves adjacent to the Park. The main tourist activities within the park are game drives, either self-drive or guided game drives, with walking safaris also becoming more popular. These activities, however, all focus on enjoying the natural environment in a pristine state, with the observation of wildlife forming the essential cornerstone of wildlife tourism in the GKNP.

Figure 1. Map of the Greater Kruger National Park

Source: Africa Geographic, 2020

The property under investigation is located within the Sabi Sands Game Reserve, which is adjacent to the KNP; there is no fence between the two protected areas, and animals roam freely between the two areas. The Lodge contains three sub-lodges, comprising 30 suites; it also has 128 staff rooms, bringing the total number of units to 158. These three lodges fall within a 45,000-acre concession area. The Property has a staff complement of 246 staff, with 122 staff members living onsite, however on average, 91 staff would be staying onsite at any given time, and around 31 staff would be on leave, a further 124 staff members on average would be travelling into work daily from surrounding community areas, with approximately 30 staff living offsite being on leave at any one time.

Over the study period of 2017-2023, the average number of staff members rarely fluctuated (the only exceptions being during the COVID-19 pandemic when a strict lockdown period was imposed in 2020). The property in this study provides guided game drives twice daily, walking safaris upon request, three meals per day for the guests, and pre-game drive snacks. The accommodation suites include a shower, bath, air conditioner, mini-bar, telephone, coffee machine, hair dryer, and speaker in a room; two of the sub-lodges also have a plunge pool at each suite (24 of the 30 suites). In addition, the main lodges have a television in the central area, bar, kitchen, swimming pool, boutique, gallery, spa, and fitness centre. The property has residential areas for staff to live on the property and kitchen facilities to cater for three meals a day for the staff. Most guests that frequent these lodges fly into the reserve, with a small percentage driving in. The main activities at all sub-lodges are game drives and bush walks, which are spent away from the lodge facilities during the daylight hours.

6. RESULTS

The compilation of recyclable waste data received from the sustainability reports of the wildlife tourism Lodge were collated for 7 full 12-month periods, from 2017 to 2023 (2024 data was not available yet as the year had not yet been completed). The mass (weight) of the recyclable wastes for the seven years is illustrated in Figure 2. The recyclable waste mass in kilograms for four recyclable classes of waste is indicated, namely, glass, tin, paper and plastic.

Figure 2. Recyclable waste for 2017-2023

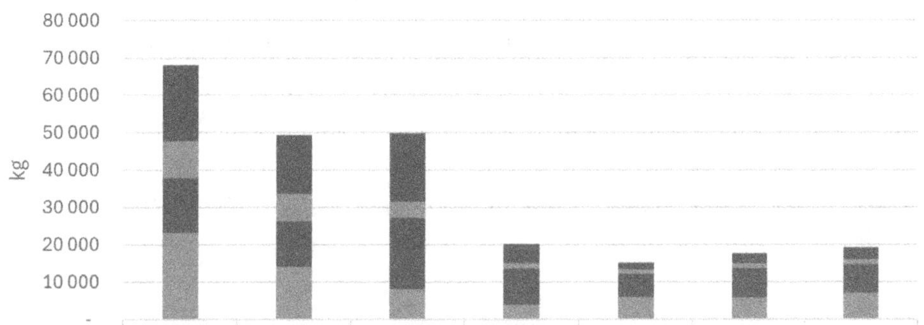

	2017	2018	2019	2020	2021	2022	2023
Glass	20 415	15 809	18 543	5 259	1 910	2 660	3305
Tin	9 808	7 196	4 104	1 254	1 051	1 309	1296
Paper	14 740	12 219	19 175	9 734	6 247	7 732	7641
Plastic	23 144	14 079	8 067	3 930	5 918	5 834	6944

Source: Author's Compilation

During consultations with staff members on the ground, it has become evident that there is a significant drive towards reducing the quantities of waste produced at the lodge in question and the company as a whole. Some interventions that have yielded results include purchasing and procuring supplies from local suppliers and buying fast-moving consumer goods in bulk, reducing the overall packaging. Some fast-moving consumer goods, such as fruit and vegetables, are also purchased from local communities where and when available. Although reducing the amount of recyclable waste produced entirely will be challenging, there has been significant progress from the highs in 2017, around 68 000kg level per year, to a new standard of around 20 000kg from 2020 to 2023. However, the question from Figure 2 is whether the significant decline in recyclable waste is not a result of the drastic COVID-19 lockdowns experienced in 2020. In response to this question, Figure 3 indicates the guest bednights and the "all bednights" which is the combination of the guest bednights and the staff bednights (as indicated previously, a large portion of staff members overnight at the site of the Lodge concerned in staff accommodation), suggests that there was a drastic decline in guest and staff accommodated in 2020. It is, however, apparent that the decline was turned around in 2021, and both the guests and the staff bednights are almost back at the levels pre-COVID-19 lockdowns. The bednights in 2023 are the highest compared to any other year during the study period; this increase in bednights only had a marginal increase in the quantity of recyclable waste. When comparing the results of Figure 2 with the bednight numbers in Figure 3, it is apparent that there will be a slight increase in the quantity of recyclable waste from 2021 to 2023. A new norm in the levels of recyclable waste produced by the Lodge is evident. There is also a strong correlation between the recyclable waste produced between 2021 and 2023 and the associated all bednights for the same period (Figure 2 and Figure 3). This reduction proves that the interventions mentioned before are effective and have led to a new baseline level in the recyclable waste produced at the Lodge under investigation.

Figure 3. Guest bednights and All bednights (guests and staff) 2017-2023

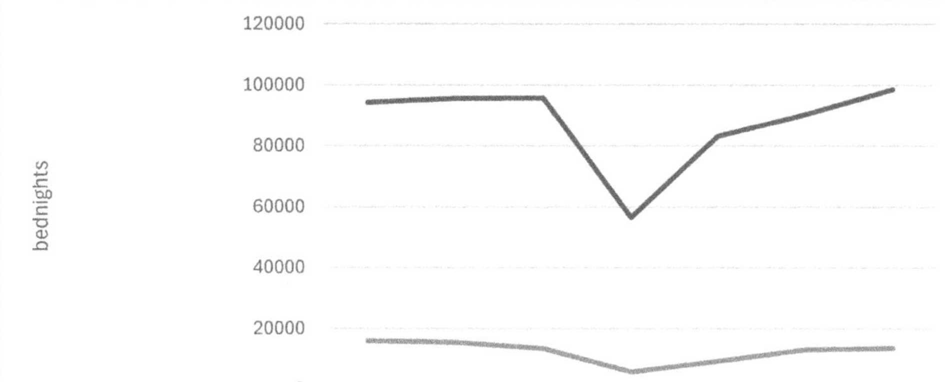

	2017	2018	2019	2020	2021	2022	2023
All bednights (guests+staff)	94 245	95 684	95 722	56 555	83 081	90 275	98 566
Guest bednights	15 981	15 481	13 437	5 889	9 258	13 101	13 624

Source: Author's Compilation

Figure 3 indicates that the staff nights are significantly higher than the guest nights. The international universal unit for benchmarking for the tourism industry is guest nights for the tourism industry. This benchmark unit is unsuitable for the wildlife tourism sector within a southern African context as the staff members also reside at the tourism ventures, contributing towards the waste production of the tourism venture as they work and live on the tourism premises. The recommendation for establishing wildlife tourism benchmarks for the tourism industry that includes guest nights and staff nights is supported by Mearns and Grobler (2016) and Baker and Mearns (2017).

Benchmarks and baselines are essential when comparing similar wildlife tourism-based lodges with significant staff living on the tourism premises. Figure 4 provides this information. The quantity of recyclable waste generated per guest night declined over the study period from 4,26kg/guest bednight in 2017 to 1,34 kg/guest bednight in 2022. Then it increases slightly in 2023 to around 1,41kg of recyclable waste per guest bednight. The results indicate continual improvement in the waste reduction efforts with recyclable waste in the case study under investigation. However, no benchmarks exist for the quantity of recyclable waste produced by wildlife tourism lodges where most staff are accommodated on the premises.

Figure 4. Recyclable waste per bednight and per all bednights (guests and staff) for 2017-2023

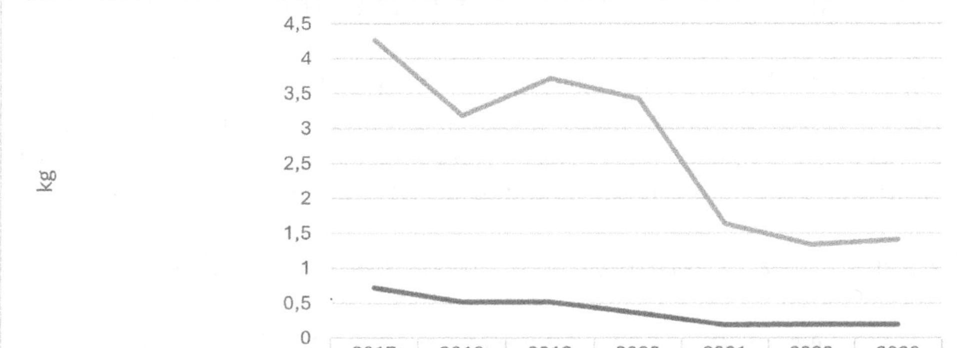

	2017	2018	2019	2020	2021	2022	2023
Recycled waste in kg per guest bednight	4,26	3,18	3,71	3,43	1,63	1,34	1,41
Recycled waste kg per all bednights (guests+staff)	0,72	0,52	0,52	0,36	0,18	0,19	0,19

Source: Author's Compilation

The International Tourism Partnership (ITP) (2008) proposes a total waste per guest night of <0,6kg/guest night as excellent, <1,2kg/guest night as satisfactory, and <2,0kg/guest night as high. In the case study being investigated it is clear that the the quantity of recyclable waste per guest bednight is not comparable with this standard proposed. The minimum quantity of recyclable waste during the study period per Guest bednight is 1,34 kg, which is considered high according to the ITP (2008) benchmark. These standards were developed for the hotel industry, where all the employed staff live offsite and can, as a result, not be used as a benchmark. The proposed ITP (2008) benchmark includes all waste, recyclable, wet, hazardous and sanitary waste. This further indicates that the proposed benchmark does not apply to the wildlife tourism Lodge industry, where staff are accommodated on-site.

Therefore, this investigation proposes a benchmark and a baseline for the wildlife tourism Lodge industry that includes Guest bednights and staff bednights. The baseline for comparative reasons can therefore be set at 0,72 kg/bednight (which includes both staff and guest bednights) in 2017. And a best-in-class Benchmark which can be used for comparative reasons is 0,18kg/bednight (staff and guest bednights) set in 2021.

7. CASE STUDY REFLECTION ON THE PROPOSED PRINCIPLES TOWARDS RESPONSIBLE WASTE MANAGEMENT IN SOUTH AFRICAN PROTECTED AREAS (ROOS ET AL., 2023)

7.1 Principle 1: Protection of Ecosystems and Biodiversity

The case study under investigation contributed to Principle 1 through the responsible collection, separation and disposal of waste, thereby minimising the potential negative impacts on ecosystems and biodiversity. The waste management infrastructure on site is well-placed and managed to ensure minimal impact.

7.2 Principle 2: Prevention and Remediation of Pollution

Due to the effective and efficient waste management at the Lodge under investigation, pollution prevention and remediation are unnecessary at this stage. It is important to the Lodge that additional measures should be implemented to prevent potential pollution if it occurs accidentally on site in the future.

7.3 Principle 3: Implementation of the Waste Management Hierarchy

The results obtained from the investigation of the recycled waste and communication with the sustainability manager have made it evident that the Lodge has already implemented several measures to avoid waste. Some measures for reducing waste have been through bulk buying from suppliers and purchasing certain fast-moving consumer goods from local communities, such as fruit and vegetables, where no packaging results. The core part of this investigation into recyclable waste shows that elements of the waste management hierarchy have been implemented already. The characterisation of recycled waste assists with understanding the waste stream and the potential opportunities communities may utilise.

7.4 Principle 4: Provision of Effective Waste Management Services and Infrastructure

Upon site visits, it has become evident that the lodge's waste services are effective and that necessary infrastructure has been implemented for well-managed waste management. Reducing recycled waste indicates a continued drive towards better, more responsible waste management.

7.5 Principle 5: Promotion of Participation and Building of Partnerships

Although the Tourism Lodge concerned was responsible for managing waste from the Lodge, the necessary measures have been taken to do this well. The participation and collaboration of other lodges and the Management of the Sabie Sands Game Reserve have been established to develop a combined Waste Management Plan for the Reserve for all tourism establishments and the reserve. Although this plan has not been finalised, the early signs of building successful partnerships within the Protected Areas are developing. More collaborative participation with local communities for empowerment and livelihood development has started. The protected area, tourism, and local community stakeholders will have to work more closely with one another in the future to ensure responsible waste management.

7.6 Principle 6: Contribution to Well-Being, Livelihood and Capacity

This recycling investigation has established that there are definite opportunities to contribute to the well-being of local communities. The fact that recycled waste has been characterised and classified also allows for better planning for local community involvement and capacity building in the future. Additional composting, craft production, waste collection, and disposal opportunities could still be explored.

These principles provide an excellent framework for evaluating waste management in Protected Areas. Additional development of these principles and a series of usable tools for advancing and measuring responsible waste management in protected areas should be encouraged.

8. CONCLUSION

Waste management is one of the most critical aspects that must be managed carefully within a protected area - wildlife tourism nexus to ensure that tourism within protected areas does not destroy the resources upon which the long-term success and sustainability of tourism in protected areas are based. This investigation has successfully established a benchmark for the quantities of recyclable waste produced at wildlife tourism lodges where most staff live on the accommodation premises. The benchmark is also unique based on the combined guest and staff bednights. Further research into the refinement of this and other waste, water, energy, and other essential indicators of the sustainability of wildlife tourism in protected areas is necessary. Waste management within the context of wildlife tourism in protected areas should always be contextualised and studied within the context of the surrounding communities, as waste disposal and recycling may not take place within protected areas. These aspects create additional pressure on surrounding communities to dispose of, recycle and process waste that originates within the protected area. Upon reflection of the six proposed principles for more responsible waste management by Roos et al. (2023), it is clear that they provide an excellent framework for evaluating waste management in protected areas. Additional tools for evaluating responsible waste management in protected areas should also be developed to facilitate better measurement and management of waste management in protected areas. Potential future research opportunities exist for the fine-tuning of the waste management processes and the investigation of the Best practical environmental option (BPEO) and best available technology (BAT) for specific protected areas to ensure that waste management services are provided as required and that plays an active role in the conservation of the protected areas assets.

REFERENCES

Africa Geographic. (2020). The real economic value of Greater Kruger National Park. Available at: https://africageographic.com/stories/compensation-for-damage-causing-animals-near-kruger-np/kruger-national-park-and-greater-kruger-map/ [Retrieved August 12 2024].

Arbulu, I., Lozano, J., & Rey-Maquieira, J. (2015). Tourism and solid waste generation in Europe: A panel data assessment of the Environmental Kuznets Curve. *Waste Management (New York, N.Y.)*, 46, 628–636. DOI: 10.1016/j.wasman.2015.04.014 PMID: 26294012

Baker, M. S. P., & Mearns, K. (2017). Applying sustainable tourism indicators to measure the sustainability performance of two tourism lodges in the Namib Desert. *African Journal of Hospitality, Tourism and Leisure*, 6(2), 1–22.

Brett, M. R. (2018). Tourism in the Kruger National Park: Past development, present determinants and future constraints. *African Journal of Hospitality, Tourism and Leisure*, 7(2), 1–28.

Byer, P. H., Hoang, C. P., Nguyen, T. T. T., Chopra, S., Maclaren, V., & Haight, M. (2006). Household, hotel and market waste audits for composting in Vietnam and Laos. *Waste Management & Research*, 24(5), 465–472. DOI: 10.1177/0734242X06068067 PMID: 17121118

Capocchi, A., Vallone, C., Pierotti, M., & Amaduzzi, A. (2019). Overtourism: A literature review to assess implications and future perspectives. *Sustainability (Basel)*, 11(12), 1–18. DOI: 10.3390/su11123303

Claassens, C. E., Cilliers, D. P., Retief, F. P., Roos, C., & Alberts, R. C. (2022). The consideration of waste management in Environmental Impact Assessment (EIA) for developments in protected areas. *Impact Assessment and Project Appraisal*, 40(4), 320–330. DOI: 10.1080/14615517.2022.2080491

Cointreau, S. (2001). *Declaration of principles for sustainable and integrated waste management*. World Bank.

Córcoles, J. I., de Juan, J. A., Ortega, J. F., Tarjuelo, J. M., & Moreno, M. A. (2010). Management evaluation of Water Users Associations using benchmarking techniques. *Agricultural Water Management*, 98(1), 1–11. DOI: 10.1016/j.agwat.2010.07.018

Dedeke, A. N. (2017). Creating sustainable tourism ventures in protected areas: An actor-network theory analysis. *Tourism Management*, 61, 161–172. DOI: 10.1016/j.tourman.2017.02.006

Del Mar Alonso-Almeida, M. (2012). Water and waste management in the Moroccan tourism industry: The case of three women entrepreneurs. *Women's Studies International Forum*, 35(5), 343–353. DOI: 10.1016/j.wsif.2012.06.002

Department of Environmental, Forestry and Fisheries (DEFF). (2020). National waste management strategy, GN 56 in Government Gazette 44116 of 28 January 2021. *Government Gazette*.

Donovan, N. (2021). New research reveals an increased desire to travel more sustainably. Available at https://partner.booking.com/en-gb/click-magazine/new-research-reveals-increased-desire-travel-more-sustainably [Retrieved December 12, 2021].

Eagles, P. F., McCool, S. F., & Haynes, C. D. (2002). *Sustainable tourism in protected areas: Guidelines for planning and management*. IUCN.

Gössling, S., Hall, C. M., & Scott, D. (2015). *Tourism and water*. Channel View Publications. DOI: 10.21832/9781845415006

Harper, C., Roos, C., Retief, F. P., Alberts, R. C., & Cilliers, D. P. (2022). Applying the theory of planned behaviour to tourism-related waste behaviour in Marine Protected Areas: The Aliwal Shoal case study. *Journal of Environmental Management and Tourism*, 3(59), 853–876. DOI: 10.14505/jemt.v13.3(59).23

International Tourism Partnership (ITP). (2008). *Environmental management for hotels: The industry guide to sustainable operation* (3rd ed.). ITP.

Makgae, M. (2011). Key areas in waste management: A South African perspective. In Kumar, S. (Ed.), *Integrated waste management* (Vol. 2, pp. 1–20). IntechOpen.

Mearns, K., & Grobler, H. (2016). Managing water for tourism in the Phinda Private Nature Reserve, KwaZulu-Natal, South Africa. *Athens: ATINER'S Conference Paper Series, No: WAT2016-2022*, 4th Annual International Forum on Water, Athens, Greece, 11 July-14 July 2016, ISSN 2241-2891, 1-23.

Morrison-Saunders, A., Hughes, M., Pope, J., Douglas, A., & Wessels, J. A. (2019). Understanding visitor expectations for responsible tourism in an iconic national park: Differences between local and international visitors. *Journal of Ecotourism*, 18(3), 284–294. DOI: 10.1080/14724049.2019.1567740

Pirani, S. I., & Arafat, H. A. (2014). Solid waste management in the hospitality industry: A review. *Journal of Environmental Management*, 146, 320–336. DOI: 10.1016/j.jenvman.2014.07.038 PMID: 25194519

Republic of South Africa (RSA). (1996). *Constitution of the Republic of South Africa, Act 108 of 1996*. Republic of South Africa.

Republic of South Africa (RSA). (2003). *National Environmental Management: Protected Areas Act 57 of 2003, GN181 in Government Gazette 26025 of 18 February 2004*. Republic of South Africa.

Roos, C., Alberts, R. C., Retief, F. P., Cilliers, D. P., & Bond, A. J. (2023). Proposing principles towards responsible waste management in South African protected areas. *Koedoe*, 65(1), a1753. DOI: 10.4102/koedoe.v65i1.1753

Roos, C., Retief, F. P., Alberts, R. C., Cilliers, D. P., Hodgson, W., & Olivier, I. (2021). Understanding waste management behaviour in private nature reserves through the theory of planned behaviour. *Parks*, 27(2), 47–56. DOI: 10.2305/IUCN.CH.2021.PARKS-27-2CR.en

Saagi, R., Kroll, S., Flores-Alsina, X., & Jeppson, U. (2017). A model library for simulation and benchmarking of integrated urban wastewater systems. *Environmental Modelling & Software*, 93, 282–295. DOI: 10.1016/j.envsoft.2017.03.026

Sandham, L. A., Huysamen, C., Retief, F. P., Morrison-Saunders, A., Bond, A. J., Pope, J., & Alberts, R. C. (2020). Evaluating Environmental Impact Assessments report quality in South African National Parks. *Koedoe*, 62(1), a1631. DOI: 10.4102/koedoe.v62i1.1631

Wylie, D. M., Bhattacharjee, S., & Rampedi, I. T. (2018). Evaluating the quality of environmental impact reporting for proposed tourism-related infrastructure in the protected areas of South Africa: A case study on selected EIA reports. *African Journal of Hospitality, Tourism and Leisure*, 7(3), 1–14.

Chapter 5
Sustainable Waste Management Practices in the Hospitality Industry:
Towards Environmental Responsibility and Economic Viability

Sarah Hussain
https://orcid.org/0000-0003-0161-4699
Jamia Millia Islamia, India

Gagandeep Soni
Banarsidas Chandiwala Institute of Hotel Management and Catering Technology, India

ABSTRACT

This chapter examines sustainable waste management practices within the hospitality industry, a crucial segment of the global tourist sector that significantly impacts environmental health through waste production (Oluwole et al., 2020). Initially, it outlines the industry's structure and its role in waste generation, emphasizing the necessity for sustainable practices to mitigate environmental harm. The chapter highlights the environmental repercussions of poor waste management, including pollution and biodiversity loss, stressing the urgency for proactive measures. Additionally, it reviews the legal frameworks governing waste management and their importance in ensuring sustainability. Strategies for reducing, reusing, recycling, recovering, and disposing of waste are discussed, along with the role of technology in enhancing efficiency. Case studies illustrate successful initiatives, underscoring

DOI: 10.4018/979-8-3693-6110-8.ch005

sustainable waste management's economic, social, and corporate benefits. These benefits include cost savings, improved public perception, and enhanced corporate responsibility.

1. INTRODUCTION

The hospitality industry, an essential component of the global tourist sector, is widely recognised for delivering lodging, dining, and recreational amenities to millions worldwide (Nikolskaya et al., 2019). Although the hospitality business excels at offering its customers comfort, elegance, and pleasure, it also encounters a significant obstacle when generating waste. Hospitality facilities generate waste daily because their operations contribute to ecological deterioration (Oluwole et al., 2020). The variety of waste streams is vast, encompassing food waste, packing materials, linens, amenities, and disposable products. As a result, the accumulation of waste creates environmental issues and brings about logistical and financial difficulties for waste management and disposal. Over 45 percent of luxury hotels in India have incorporated advanced technology into their food management systems, resulting in a significant 30 percent decrease in food waste during the previous year (Trends in food waste reduction, 2024). This chapter discusses sustainable waste management in the hospitality sector, focusing on waste generation, environmental impact, legal frameworks, and sustainable solutions. It highlights the negative environmental consequences of inadequate waste management, discusses legislative frameworks, and discusses sustainable waste management methods. The chapter emphasises practical benefits, social responsibility, and future trends in waste management techniques.

The hospitality industry increasingly prioritises sustainable waste management practices due to the environmental consequences and societal demands. Sustainable waste management involves implementing measures prioritising waste reduction, resource efficiency, and environmental stewardship at every stage of hospitality operations. By incorporating sustainability concepts into waste management processes, businesses can reduce their ecological impact, improve process efficiency, and cultivate a favourable brand image among environmentally conscious consumers. Adopting sustainable waste management strategies can provide numerous benefits to hospitality organisations. Firstly, it reduces the amount of waste sent to landfills, relieving pressure on waste disposal facilities and minimising environmental contamination. Additionally, it promotes resource preservation through measures such as recycling, composting, and waste-to-energy conversion, thereby saving valuable resources and reducing the release of greenhouse gases. Furthermore, implementing sustainable waste management strategies can lead to financial benefits such as decreased waste disposal costs, improved operational efficiency, and the potential to generate income from recycling materials or recovering energy. Furthermore, sustainability in waste management corresponds to the changing needs of customers, investors, and regulatory authorities. Amidst a period characterised by heightened environmental awareness, customers progressively prefer businesses strongly dedicated to sustainability and social responsibility. Similarly, investors closely examine companies' environmental performance as a component of their investment criteria, while governments are enacting rules and providing incentives to promote sustainable business practices.

2. UNDERSTANDING WASTE GENERATION IN THE HOSPITALITY INDUSTRY

2.1 Types and Sources of Waste Generated in the Hospitality Sector

Restaurants and food outlets contribute significantly more biodegradable waste than hotels, according to Juvan et al. (2023). Each hotel guest reportedly produces 2.5 kilogrammes of waste daily, with 58.5% of this waste capable of decomposing naturally and 25.8% recyclable. Zein et al. (2008) categorise hotel waste as non-hazardous and hazardous. Non-hazardous categories include household waste such as food or hotel kitchen waste, glass, paper and packaging, plastic bags, and other items generated by various departments. The purchase section and other departments generate cardboard and wood packaging waste. Paper waste comes from printed documents in administration, front desk, guest rooms, and food and beverage outlets. Metal waste is generated from tin containers and jars, food containers, and aluminium from the kitchen, food and beverage outlets, and guest rooms. Cloth waste consists of restaurant and bed linen, cleaning rags and clothes from the kitchen, food and beverage outlets, public toilets, and guest rooms. Flowers, leaves, and grass are generated as waste from food and beverage outlets, guest rooms, and gardens.

2.2 Factors Contributing to Waste Generation in Hospitality Industry

While essential for accommodating travellers and offering recreational experiences, the hospitality industry generates substantial waste due to various factors. These include consumption patterns, operational procedures, and industry-specific underlying structural elements.

First and foremost, consumption habits play a crucial role. Guests frequently expect amenities such as disposable toiletries, single-use cutlery, and individually packaged items, resulting in significant waste (Legrand et al., 2022). Moreover, the prevalent culture of indulgence in the hospitality industry encourages excessive consumption, leading to food waste, underutilised amenities, and abandoned promotional materials. Hotel operational practices also contribute substantially (Jawabreh et al., 2021). Everyday housekeeping procedures often involve excessive cleaning chemicals and frequent linen changes, generating chemical waste and increasing water usage for laundry. Additionally, inefficient energy practices, such as leaving lights and appliances on when unnecessary, exacerbate the environmental damage (Ali et al., 2008).

In addition, the transient nature of hospitality establishments creates obstacles for waste management. The frequent turnover of guests makes it challenging to implement sustainable practices consistently. Furthermore, fluctuations in occupancy levels due to seasonal variations can lead to unpredictable waste generation patterns, making it challenging to efficiently allocate resources and implement waste reduction initiatives (Guerra-Lombardi et al., 2024). To effectively address waste creation in the hotel industry, a comprehensive strategy must consider customer behaviour, operational efficiency, infrastructure improvements, and industry-wide collaboration. To minimise the environmental impact of hospitality operations, it is recommended to implement measures such as promoting reusable alternatives, optimising supply chains, investing in energy-efficient equipment, and educating workers and visitors about waste reduction.

2.3 Waste Generation in Different Sectors of Hospitality

Hotel waste encompasses the food waste produced by dining establishments and room service, using disposable hygiene products and packaging in guest rooms, and disposing of items such as paper and plastics from administrative activities. Evaluating waste in hotels involves assessing the rates of daily consumption, levels of guest occupancy, and the composition of waste using methods such as waste audits and tracking systems integrated into daily operations.

Restaurants generate significant waste through food preparation, dining services, and kitchen operations. Addressing food waste, including pre-consumer waste, trimmings and expired ingredients, and post-consumer waste, such as leftover food from plates (Principato et al., 2021) is crucial. Moreover, using disposable packaging, single-use utensils, and paper items adds to the overall volume of waste in the environment (Graulich et al., 2021). Quantifying restaurant waste involves measuring food deterioration, portion sizes, plate waste, and packaging utilisation, typically done using waste sorting and weighing techniques.

Resorts, which function similarly to hotels, generate additional waste from recreational activities and amenities, including spa services, golf courses, and entertainment facilities (Okumus et al., 2020). To assess waste in resorts, it is essential to consider the specific offerings and services provided, the participation rates of guests and the waste they produce. Furthermore, other hospitality industry segments, such as event venues, theme parks, and cruise ships, also contribute to waste production (Burlakovs et al., 2020). Nonetheless, measuring waste in these sectors necessitates customised techniques due to their distinct operational characteristics.

3. POOR WASTE MANAGEMENT AND THE ENVIRONMENT

3.1 Pollution of Air, Water, and Soil

Ineffective waste management has significant environmental consequences by polluting the air, water, and soil (Sharma et al., 2018). Improper disposal and waste handling cause air pollution, producing harmful gas and particulate emissions. The incineration of waste releases pollutants like dioxins, furans, and heavy metals, leading to health issues such as respiratory problems, cardiovascular disorders, and cancer (de Titto & Savino, 2019; Domingo et al., 2020). Landfill sites also produce methane, a greenhouse gas that contributes to climate change (Wang et al., 2024).

Water contamination is caused by leachate, a liquid that forms when rainfall comes into contact with waste in landfills (Mhammedsharif & Kolo, 2023). Leachate can infiltrate groundwater, rivers, and lakes, spreading harmful elements such as heavy metals, pathogens, and organic contaminants (Jagaba et al., 2024). Contaminated water sources harm aquatic ecosystems, resulting in a decline in biodiversity and disruptions to food webs. Additionally, contaminated water sources can harm human communities by spreading waterborne diseases and rendering water unsuitable for drinking or agricultural use.

Soil pollution, a consequence of inadequate waste management, arises from the direct disposal of industrial and domestic waste. This results in the accumulation of toxic substances like lead, cadmium, and mercury in the soil, rendering it unsuitable for plant growth and posing potential dangers to plants (Sharma et al., 2018). These toxins can contaminate the food chain by contaminating crops, posing health risks to animals and humans. Additionally, as per Salam et al. (2023), plastic waste degrades into microplastics, which impair soil structure and disrupt the microbial communities responsible for nutrient cycling.

3.2 Ecological Consequences of Improper Waste Disposal

Poor waste management has significant ecological consequences affecting terrestrial and aquatic ecosystems. On land, improper waste disposal can result in habitat destruction and soil pollution. According to Madhav et al. (2024), harmful toxic compounds such as heavy metals and persistent organic pollutants in electronic and industrial waste seep into the soil, negatively impacting plant growth and soil health. These contaminants can accumulate in living organisms and become part of the food chain, posing dangers to wildlife and humans. Inadequate waste removal in aquatic habitats is highly susceptible. Marine debris, mainly plastic waste, poses a significant threat to marine organisms. Nelms et al. (2023) reported that plastic ingestion or entanglement often occurs in animals such as fish, seabirds, and marine mammals, leading to harm, malnutrition, and death and additionally, breaking down plastics into microplastics results in extensive pollution, affecting even the smallest marine species and potentially entering the human food chain. Improper waste disposal disrupts ecological balance by promoting the spread of exotic species. Khattak et al. (2024) argued that waste sites can act as habitats for non-native species, enabling them to outperform and displace indigenous plants and animals, resulting in a decline in biodiversity. Furthermore, waste in natural habitats can alter nutrient cycles, deteriorate water quality, and disrupt ecosystem services, ultimately weakening ecosystem resilience and health. Adopting effective waste management practices is vital to reduce these detrimental ecological effects and preserve biodiversity.

3.3 Impact on Greenhouse Gas Emissions and Climate Change

The hospitality business significantly impacts greenhouse gas (GHG) emissions and global warming due to its various activities and services (Laeis & Legrand, 2023). The main consequence of this impact is seen in the energy consumed, waste generated, and supply chain operations related to accommodation, food and beverage services, and event management. Energy usage in hotels and resorts is a primary source of greenhouse gas (GHG) emissions. These businesses require significant energy for heating, cooling, lighting, and operating equipment. A large number of hotels depend on non-renewable energy sources, such as natural gas, coal, and oil, which release substantial amounts of carbon dioxide (CO_2) during combustion (Nisa et al. (2024). The constant demand for energy-intensive facilities, such as swimming pools, spas, and gyms, exacerbates the industry's carbon emissions.

The solid waste the hotel industry generates is also a significant contributor to greenhouse gas emissions. The anaerobic decomposition of food waste in landfills, which occurs without oxygen, is a significant concern due to the production of methane, a potent greenhouse gas (Merouani et al. (2024). The manufacturing and disposal of disposable plastics and other non-recyclable materials frequently used in hotels substantially contribute to carbon emissions and environmental pollution. Supply chain activities, which procure food, beverages, and other items (Aditya & Kurniawati, 2023), also contribute to the industry's greenhouse gas (GHG) emissions. The transportation of these goods, often over long distances, requires substantial fuel use and the release of CO_2 emissions. Moreover, the manufacturing processes of numerous hotel products, such as textiles, electronics, and furniture, require significant energy and result in high carbon emissions.

The hospitality industry must adopt sustainable practices to minimise its impact on the environment, particularly concerning climate change. To this end, resources should be directed towards developing renewable energy sources, enhancing energy efficiency, reducing food and material waste, and sourcing locally produced products. In addition, embracing green construction standards and advocating environmentally responsible practices can significantly decrease the industry's carbon dioxide emissions and contribute to broader efforts to combat climate change.

4. REGULATORY FRAMEWORK AND STANDARDS

4.1 Overview of Relevant Regulations and Policies

Various regulations and policies govern the hospitality industry to minimise environmental impact and promote sustainability. Key international frameworks include the United Nations' Sustainable Development Goals (SDGs), particularly SDG 12, emphasising responsible consumption and production. National and regional policies also shape waste management practices, often mandating reduction, reuse, and recycling measures. The European Union's Waste Framework Directive sets the legislative framework for waste management, emphasising waste hierarchy and extended producer responsibility. The U.S. Environmental Protection Agency (EPA) regulates waste under the Resource Conservation and Recovery Act (RCRA), encouraging waste minimisation and resource recovery. Local regulations, such as San Francisco's Mandatory Recycling and Composting Ordinance, enforce waste segregation and recycling ordinances. Voluntary certifications and standards, like the Green Key Eco-Rating Program and ISO 14001, provide guidelines for the hospitality sector to implement effective waste management systems. These frameworks encourage best practices and continuous improvement, ensuring compliance with legal requirements and contributing positively to environmental sustainability. In the following section, we will look at these in depth.

4.2 Compliance Requirements for Waste Management

The hospitality business must adhere to strict compliance regulations to ensure efficient and sustainable waste management. These standards are determined by a blend of global treaties, domestic laws, and municipal regulations to reduce environmental harm and encourage sustainable behaviours.

4.2.1 Global and Domestic Regulations

Internationally, agreements such as the United Nations' Sustainable Development Goals (SDGs) set out broad aims for sustainability. Goal 12 emphasises responsible consumption and production, urging corporations, including those in the hospitality industry, to implement efficient waste management policies.

The Waste Framework Directive (2008/98/EC) of the European Union provides a comprehensive regulatory framework for waste management. The waste hierarchy is established, prioritising prevention, reuse, and recycling over disposal. Member states are required to incorporate these principles into their national legislation. Moreover, the directive mandates expanded producer responsibility, which compels businesses, particularly hospitality operators, to take responsibility for the waste generated by their products.

In the United States, hazardous and non-hazardous waste management is governed by the Resource Conservation and Recovery Act (RCRA) (1976). The Environmental Protection Agency (EPA) enforces this legislation, requiring firms to accurately classify, manage, store, and dispose of waste. Hospitality establishments, including hotels and restaurants, are mandated by the Resource Conservation and Recovery Act (RCRA) to keep thorough records of how they dispose of waste and to verify that any hazardous waste is managed in compliance with federal regulations.

4.2.2 Regulations at the Local Level

The Circular Economy principle advocates for a system of production and utilisation that emphasises distribution, letting, reusing, repairing, restoring, and recycling materials and products for as long as feasible. The management of waste for hotels in major world cities varies, reflecting the local environmental policies and priorities of these areas.

a. New York City

The New York City Department of Sanitation (DSNY) (2020) enforces stringent waste management regulations to promote recycling and reduce landfill waste. Hotels are required to:

- **Recycle**: Separate recyclable materials (paper, metal, glass, and plastics) from regular waste. The city mandates commercial establishments, including hotels, to comply with specific recycling rules to minimise environmental impact.
- **Composting**: Participate in organic waste separation and composting programs. This includes food waste from hotel kitchens and dining areas.
- **Waste Reduction Plans**: Create and execute strategies to minimise waste as a component of broader sustainability efforts.

b. Berlin

Berlin's waste management system, guided by the Circular Economy Act (Kreislaufwirtschaftsgesetz - KrWG), 2012), emphasises recycling and circular economy principles. Hotels have to adhere to the following requirements:

- **Mandatory Recycling**: Hotels must sort and recycle waste, including paper, plastics, metals, and organic waste.
- **Waste Prevention**: The city encourages hotels to adopt waste prevention measures, such as reducing packaging waste and promoting the reuse of materials.
- **Biowaste Collection**: Hotels must participate in the city's bio-waste collection program, separating organic waste for composting.

c. European Union (EU) Context

Under the EU Waste Framework Directive, cities like Berlin follow specific regulations that impact waste management practices (European Commission, 2023):

- **Waste Hierarchy**: Prioritize waste prevention, reuse, recycling, and recovery over disposal.
- **Separate Collection**: By 2025, mandate the separate collection of key waste streams, including hazardous waste and biowaste.
- **Extended Producer Responsibility**: Implement economic tools like 'pay-as-you-throw' schemes and deposit-refund systems to encourage waste reduction and recycling.

These regulations ensure that hotels in major cities like New York, Tokyo, and Berlin actively contribute to sustainable waste management practices, aligning with local and international environmental goals.

4.2.3 International Standards and Certifications

International standards and certifications guide the hospitality industry towards a circular economy. These frameworks provide comprehensive guidelines, promote best practices, and help businesses enhance their environmental performance.

a. ISO 14001: Environmental Management Systems

The International Organization for Standardization (2015), established in 1996 and last revised in 2021, is one of the most widely recognised international standards for environmental management systems. It specifies a framework for businesses, including those in the hospitality industry, to develop, implement, and continuously improve their environmental management strategies. ISO 14001 emphasises waste reduction, resource efficiency, and regulatory compliance, enabling establishments to effectively manage and reduce their environmental impacts. Certification requires regular audits and compliance with stringent environmental performance criteria.

b. Green Key Eco-Rating Program

The Green Key Eco-Rating Program, established in 1994 and last updated in 2021, is an international certification tailored for the hospitality industry (Green Key—*Green Key Programme*, 2021). It assesses hotels, resorts, and other hospitality establishments' environmental and social sustainability performance. The program includes guidelines for waste management, such as waste reduction, recycling, and composting initiatives. Green Key certification encourages continuous improvement and recognises establishments that demonstrate leadership in sustainability.

c. EarthCheck Certification

EarthCheck is a globally recognised environmental certification and benchmarking program for the travel and tourism industry (EarthCheck, n.d.). It provides a scientific approach to measuring and managing waste, energy, water, and social impacts. EarthCheck certification requires businesses to meet rigorous environmental standards and undergo annual assessments to ensure ongoing compliance and improvement.

d. LEED Certification

LEED (Leadership in Energy and Environmental Design) Certification, developed by the U.S. Green Building Council (USGBC), is a globally recognised green building and sustainability certification system. It is a green building certification from the United States of America that focuses on sustainable design, construction, and waste management and is applicable internationally and used worldwide (U.S. Green Building Council, n.d.). Hospitality facilities with LEED certification demonstrate sustainability through efficient resource utilisation and waste reduction strategies.

Adopting international standards like ISO 14001, Green Key, EarthCheck, and LEED helps hospitality businesses implement effective waste management practices. These certifications enhance the industry's reputation for environmental stewardship and attract eco-conscious travellers. They also set a high standard for operational excellence.

5. SUSTAINABLE WASTE MANAGEMENT PRACTICES

The hospitality industry generates significant unnecessary waste annually, which challenges the environment. The need for sustainable waste management is increasingly becoming a priority for stakeholders, as there is growing pressure from ecological groups and other interested parties. The hospitality industry is slowly embracing sustainability, recognising the benefits of reducing waste and protecting natural resources while meeting visitors' expectations.

An effective waste management system must consider the entire process, including generation, collection, and disposal. Adopting advanced waste management methods prioritising waste diversion from disposal and utilising feedback loops is necessary to achieve a sustainable society. The current reductionist approach is unsustainable due to its narrow focus and lack of adaptability. A sustainable waste management system should prioritise processes, incorporate adaptability, and focus on waste diversion. Shifting to a sustainable waste management system requires identifying and utilising leverage points that drive change. The waste management hierarchy model, shown in Figure 1, depicts the strategies for managing waste, including prevention, reduction, reuse, recycling, recovery, and disposal (Waste Online: After it has been binned, 2006). The model is represented visually in the shape of a pyramid, demonstrating the hierarchy of recommended actions for minimising and handling waste (Mensah, 2020).

Figure 1. Waste management hierarchy

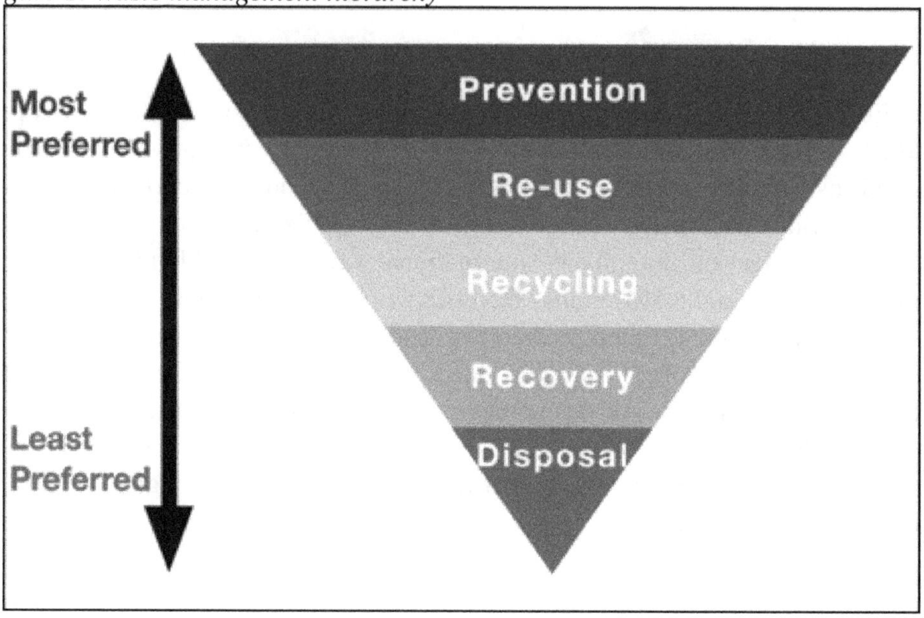

Source: Waste Online (2006)

5.1 Reduce: Minimizing Waste Generation

Minimising waste generation encompasses less waste generation through reduction at the source level before it contaminates the environment. Reduction in waste generation can be achieved by incorporating alterations in the procedure and product simultaneously with continuous censoring and recycling (Babu et al., 2009).

5.1.1 Source Reduction Techniques

Waste reduction at the source level diminishes the waste-borne hazards. The assembly and shipping of waste are more hazardous and expensive than the process related to waste minimisation. Waste treatment at the source lowers the transportation cost and the risks associated with collection and transportation. Cleaner technologies can be adopted for source-level reduction as they are beneficial and help uphold long-standing sustainability in the waste sector (Zsigraiová et al., 2009).

5.1.2 Menu Planning and Portion Control

The food industry faces a formidable challenge in reducing food waste. Principato et al. (2021) proposed a food waste map that illustrates the generation of food waste in three stages: food preparation, food service, and customer consumption. Large portion sizes significantly contribute to customer leftovers. Tansel et al. (2021) found that excessively piled plates by guests from buffet counters are a substantial reason for food waste, as guests cannot consume so much food. Agyeiwaah (2020) recommends that providing guests with appropriately portioned dishes can help to minimise food waste. Training staff in resource conservation can also be beneficial. Chefs can develop menus based on customer requirements and serve reasonably sized portions (Principato et al., 2021). Menu redesigning and utilising food delivery platforms can effectively decrease food waste. Pirani and Arafat (2014) recommend substituting single-use plastics with reusable containers. Betz et al. (2015) suggest offering clients a range of portion sizes to select according to their needs.

5.2 Reuse: Maximizing the Reuse of Materials

Implementing the " Reuse " concept in the hospitality sector is directly associated with the concept of ``Reduce". The concept of reuse involves transforming items after their initial use to prevent additional waste by all possible means. Reusing also benefits society by minimising the usage of natural resources.

5.2.1 Creative Reuse Strategies

Reusing in the hotels is directed towards the three major elements: water and energy. Hotels can save water by effectively treating the wastewater for reuse. Water and energy consumption can be lowered by encouraging guests to reuse their linen and towels in the guestroom. Reuse practices also need to be implemented to treat generated waste. Hotels can reuse the furniture and crockery, and products made and packaged using recycled materials can be chosen from the suppliers to reduce the carbon footprint. Capital expenditure can be reduced by using equipment and machinery for other purposes (Ioannidis et al., 2021).

Innovative methods for product development by collaborating with different users and devising clear sustainable strategies for communication with individual users by the governments, businesses, and local authorities to increase awareness of a contribution towards the future generation are a must. Rising environmental degradation necessitates that consumers alter their behaviour by following the use-transform-reuse concept over the single-use attitude. This can be achieved by spreading responsiveness amongst the guests. Holding open innovation workshops to educate individuals about prolonging product use for different purposes after their initial usage will also inspire creative reuse of products (Tarabashkina et al., 2022).

5.3 Recycle: Implementing Effective Recycling Programs

As global waste generation rates rise, recycling is becoming increasingly crucial. Strategists must choose the most effective recycling practices from the various options available to avert waste from landfills.

5.3.1 Sorting and Separation Techniques

Pham Phu et al. (2019) mentioned that a waste reduction of 30-70 percent could be achieved through recycling in the hotel industry. This results in positive alterations in waste characterisation for incineration and composting processes while generating commercial gains for the hotels. Successful recycling depends on effective waste sorting, which requires consumers to categorise waste into paper, plastics, and organics. Effective waste sorting is essential for recycling, increasing the quantity and quality of recyclable materials, thereby decreasing landfill waste. It is necessary to install individual containers for plastics, paper, and other types of waste in hotel bedrooms and common areas such as receptions, restaurants, kitchens, and conference halls (Scholz et al., 2020). A few of the latest techniques for sorting and separating waste are mentioned below:

- Robot-assisted systems for sorting and separating recyclable materials in waste processing for households and municipal waste that is already presorted (Aschenbrenner et al., 2023).
- Material Recovery Facilities (MRFs) enable mechanical waste segregation based on the physical characteristics of every material. For example, electrical conductivity is used for segregating aluminium waste with the help of specialised equipment (Ip et al., 2018).
- Near-infrared (NIR) spectroscopy is a speedy, non-destructive analysis method for detecting polymers. It can also safely segregate plastics (Wu et al., 2020).
- Automated solutions powered by Artificial Intelligence (AI) are being used to segregate medical waste using image-texture-dependent features (Kumar et al.,2021).

- IoT-based sensors can be installed in waste bins to enable waste measurement. The data can then be sent to the server for further processing. This measure can synchronously plan waste collection routines for the workers (Shyam et al., 2017).

5.3.2 Collaborating With Recycling Facilities

Due to the limited financial and technical resources available to small and medium-sized hotels, collaboration between these establishments, government organisations, and other stakeholders is necessary to facilitate recycling efforts (Mensah, 2014). Local authorities can focus on fostering associations among small hotel organisations to discuss ideas and information about solid waste management. Working together between small hotels, local authorities, and other stakeholders successfully involves hotels in sustainable practices. Hotel operators can coordinate with their outsourced waste haulers to incorporate additional eco-friendly alternatives within the waste management framework. The public sector needs to support hotels in effectively managing their solid waste through policy guidelines, incentives, tools, facilities, training for hotel staff, and network-building efforts for environmental improvement (Radwan et al., 2012).

5.4 Recovery: Extracting Value From Waste

From a sustainability standpoint, enhancing waste management practices within the hospitality industry is a crucial component of its broader environment-friendly approach.

5.4.1 Composting Organic Waste

Hotels must implement the practice of composting in order to decrease organic waste in landfills and mitigate environmental impact. The critical factor is properly segregating waste at its source to ensure effective composting. If all hotels adopt food waste composting, a 50% reduction in food waste could be achieved within the hotel industry (Pirani & Arafat, 2016). Food waste from hotels originates from food items and can be highly advantageous when combined with horticultural waste. This mixture can produce high-quality compost and soil conditioner. For long-term soil sustainability, this type of compost offers numerous benefits and can be used by hotels for consistent fruit and vegetable production (Singh et al., 2014). Wang (2012) emphasised that composting is gaining increasing recognition worldwide, mainly for hotels with extensive landscaping, such as golf resorts. These hotels can utilise the compost for their gardening requirements. The Environmental Protection

Agency in the United States reported that the Eastern Band of Cherokee Indians of North Carolina casinos and restaurants compost almost 544 kilograms of food waste generated daily (Estes, 2000).

5.4.2 Energy Recovery From Waste

Solid waste can also recover energy and reduce greenhouse gas emissions. A study in Central Macedonia revealed that the food waste generated in hotels can be utilised for energy recovery. The research presented that if food waste from hotels is transported to a centralised anaerobic digestion facility, it can generate 2,740 MWh of energy during the six peak tourist months each year. This energy can be sold to the nationwide grid, generating revenue. The amount of recoverable energy will depend on the technologies used, which can vary significantly in design, manufacturing, and performance (Perkoulidis & Moussiopoulos, 2017). Another study conducted in Poland stated that installing drain water heat recovery systems is a significant feature for energy conservation in hotels that also decreases CO2 emissions. This grey water heat retrieval technology enables the capture of the waste heat from the drain line network of the washbasin or shower and is used for preheating the cold shower water (Sayegh et al., 2021).

5.5 Responsible Disposal: Managing Non-Recyclable Waste

Sustainable waste disposal requires formulating and developing efficient policies and procedures related to the composition and magnitude of waste generated. Technological integration through recycling and resource recovery incorporates understanding the composition and quantification of the solid waste generated (Senzige et al.,2014). Non-recyclable waste comprises composite polymers in landfills and oceans as they are challenging to recycle. Many developed countries make it mandatory to significantly reduce the amount of non-recyclable waste generated by hotels before sending it to landfills for disposal (Emery et al., 2006).

5.5.1 Hazardous Waste Management and Proper Disposal Methods

Specific guidelines govern different categories of hazardous waste and have specific methods of treatment and disposal (Orloff & Falk, 2003). Visvanathan (2006) detailed methods for safely disposing of hazardous waste in his study. He mentioned that for the safe disposal of hazardous waste, proper pre-treatment encompasses the hazardous waste management system. Different factors, such as the availability of capital, maintenance, operational costs, and the knowledge of technically handling hazardous waste, help decide the disposal method. A combination of more than one method is generally opted for. When the hazardous waste cannot be recycled, reduced, or averted to a landfill, incineration (high or low temperatures) is the best disposal option where, through a process of thermal oxidation, the hazardous waste is transformed with the help of oxygen into gases and non-combustible compact deposit. This results in the reduction of volume and weight along with the detoxification of the hazardous waste processed. Carpenter et al. (1996) highlighted the importance of separating hazardous waste from non-hazardous waste, which is necessary to prevent the disposal of the same into a landfill designated for non-hazardous waste. Another process, immobilisation, relies on mixing the toxic wastes with specific materials that render them impermeable and transfer them into a solid matrix form, causing the waste material to be fixed in this structure. Solidification of hazardous waste is a technique where this waste is converted into transportable-sized structures of solid mass that are structurally stable and do not require any secondary containers for transportation. Landfilling, offshore, and deep well disposal are other hazardous waste methods (Visvanathan, 2006).

5.6 Implementation of Technology for Waste Tracking and Management

Information and communication technology (ICT) have become vital in the design and planning of solid waste tracking and management, encompassing the assembly, transport, handling, and monitoring of solid waste. ICT can be broadly classified into four categories: geospatial technology, data capture and communication technology, and identification technology. Geospatial and global information systems are utilised to identify waste bin locations and the schedules of waste collection by vehicles. Radio frequency identification tags can be installed in trucks for waste collection to manage the fleet effectively. This technology aids in the cost-effective and efficient management of employees and collection vehicles. All data is consistently transmitted to a central hub using general packet radio systems (Karadimas et al., 2005). Data capture technology facilitates weight tracking, sorting, and combustion. The filling levels of waste bins and waste sorting can also be monitored using imaging technology. Dynamic routing enhances waste collection efficiency by providing real-time status updates on waste bins (Hannan et al., 2015). However, the authors noted several challenges associated with implementing ICT in solid waste management, such as the high cost of the network infrastructure, inadequate data availability, and the lack of real-time data.

6. CASE STUDIES AND BEST PRACTICES

Solid waste (SW) management is increasingly recognised as a critical issue in the hospitality industry. According to Bohdanowicz (2006), Swedish hotels have successfully implemented effective practices for waste sorting and segregation. These hoteliers tend to purchase products from suppliers that have minimal environmental impact. In Nepal, a hotel in Kathmandu has made significant efforts towards solid waste management (SWM) by collaborating with a private organisation. Waste segregation is carried out in every department using a colour-coded bin system, and the amount of waste generated and recovered is monitored. The waste is then transported to a material recovery facility, while the food waste is sent to animal farms as feed (Giri, 2021). In Langkawi, Malaysia, hotels employ the "first in, first out" principle to reduce food waste generation. Ingredients are appropriately labelled, showing the date of expiry and storage instructions (Kasavan et al., 2017).

In Czechia, hotels practice various green measures to promote waste sorting, recycling, and reuse of recyclable materials. They also provide colour-coded bins and engage in composting (Scholz et al., 2020). In Vietnam, hotels have implemented several good practices to limit waste generation, such as refilling shampoo bottles, bulk purchasing certain products, and using eco-friendly non-CFC refrigerators. Some hotels have also replaced their old minibars. Encouraging guest participation and involvement through washroom environmental protection messages may also be beneficial (Kumar, 2005). The ITC hotels in India operate the largest wind energy farms for captive consumption, utilising wind energy for power generation.

The hospitality industry faces various obstacles in its pursuit of sustainable strategies. The segregation of waste can be costly, time-consuming, and logistically challenging. Furthermore, some countries lack the policy framework to support eco-friendly initiatives. Hotels may also face financial and marketing difficulties in implementing these sustainable strategies. The managers of small hotel properties often lack the knowledge, time, and resources to prioritise solid waste management (SWM) practices. They also feel that they do not receive enough support from external stakeholders and that local authorities should provide them with the necessary tools to facilitate SWM practices (Pirani & Arafat, 2014).

7. ECONOMIC BENEFITS OF SUSTAINABLE WASTE MANAGEMENT

The waste management process is a collaborative initiative that can be realised when all stakeholders participate together and jointly work towards fulfilling environmental, social and economic benefits (Ribić et al., 2017). A study by Sorin &

Sivrajah (2021) evaluated the economic benefits realised by deploying the circular economy concept in the hotels of Scandinavia covering Denmark, Finland, Sweden and Norway. The authors revealed that the circular acceptability framework brings about a reduction in the cost of waste management.

7.1 Cost Savings Associated with Waste Reduction and Recycling/ Potential for Revenue Generation Through Waste Diversion Programs

A comprehensive framework for solid waste management (SWM) offers both environmental benefits and economic and financial advantages. Bohdanowicz (2006) emphasised that standard management techniques in hotels include composting organic and food waste, partnering with recycling firms to process reduced waste, and providing bins for collecting paper, cans, and food waste. Hotel managers in Macau have embraced eco-friendly practices that lead to cost savings. Wan et al. (2017) highlighted that green practices can result in economic benefits, reduce negative environmental impacts, and enhance employee satisfaction. LEED-certified buildings may increase construction costs by 1-2%, but these costs can be recovered within two years due to energy and water savings and reduced solid waste expenses (Butler, 2008).

Research conducted in Hong Kong hotels revealed that decreasing water consumption can result in lower water and sewage charges and reduced greenhouse gas emissions from energy-intensive water transport, heating, and sewage treatment processes (Chan et al., 2009). Recycling solid waste to generate energy offers economic benefits with unit costs. Waste economic efficiency is achieved when the cost of reducing one unit of waste equals the financial and environmental benefits of not generating one less unit of waste (Pizzol et al., 2013). Diversion-based activities provide both environmental and economic advantages for communities. These activities offer immediate economic benefits by mitigating the environmental hazards of bioactive landfill sites. Additionally, diversion activities create jobs, generate revenue, and stimulate economic sectors (Ogishi& Goldman, 2001).

Effective and efficient waste management brings immediate financial gains through reduced waste, lower recycling expenses, and indirect advantages such as an improved brand image, government support, and favourable guest feedback. Consequently, many establishments implement eco-friendly hotel guidelines (Singh et al., 2014). Additionally, local authorities benefit from waste recycling as disposal costs are decreased and landfill lifespan is prolonged, ultimately reducing transportation expenses. Implementing sustainable practices for food waste management in the hospitality industry can result in cost savings, enhanced reputation, food waste upcycling, energy conversion, tax incentives, and efficient resource allocation. By optimising portion sizes, managing inventory effectively, and minimising overproduction, establishments can reduce costs, attract more customers, generate additional revenue, and potentially qualify for tax incentives or grants from the government. This approach promotes sustainability and enhances revenue (Addressing Food Waste in the Hospitality Industry, 2024).

8. SOCIAL AND CORPORATE RESPONSIBILITY

According to recent findings, the hotel and lodging sector is responsible for preventing approximately 45% of municipal waste from being disposed of in landfills. This waste, including wet and dry components, significantly contributes to environmental pollution. Eco-friendly initiatives undertaken by companies can positively impact guests' and customers' perceptions of the company and its products and services. Such initiatives, such as water conservation and waste reduction, demonstrate a hotel's corporate and social responsibility (Han et al., 2018). Understanding customer interest and participation in green activities is critical for organisations to develop effective and efficient green programs. Hotel managers must communicate the relevance of these activities to guests to ensure their participation and promote awareness and education (Baker et al., 2014).

8.1 Engaging Employees and Guests in Waste Reduction Efforts

Encouraging guests to participate in pro-environmental efforts made by hotels is essential. It is necessary to explore novel approaches rather than mere message communication, as Brodie et al. (2013) suggested. Securing commitment through gaining trust and empathy and introducing a savings program to foster customer reciprocation is essential to encourage guest participation. Warren et al. (2019) recommend offering eco-feedback and appreciating guests' efforts as practical steps towards encouraging guest participation. Promotional campaigns and "cascade training" for spreading moral awareness amongst guests are also effective measures towards food waste reduction. Alsuwaidi et al. (2022) highlight that guest needs and values, personal norms, and behaviour are important factors to consider while designing and devising efficient waste reduction and eco-friendly programs.

Green management practices in Zimbabwean hotels include waste management, water conservation, and wastewater reuse, which help save resources and ultimately lead to cost reduction, as Mbasera et al. (2016) noted. Luu (2020) emphasised that front-line hotel employees connect the hotel organisation and guests, making it crucial for hotel managers to integrate and promote green marketing strategies and service interactions among employees and external stakeholders. Effective communication mechanisms, such as posters, speeches, webpages, reports, and employee training, can serve as catalytic agents to magnify the employee's intention and behaviour to reduce food waste, as effective communication suggests.

8.2 Building Partnerships with Local Communities

Implementing environmentally friendly initiatives and green programs can enhance the hotel's brand image in the eyes of guests and local communities while also benefiting the hotel by attracting more potential customers. The hotel industry in Malaysia recognises the significant role that customer satisfaction and community growth play in the overall success of tourism (Kasimu et al., 2012). In Ghana, homestay accommodation owners and managers relied on local communities to interact with guests and establish connections, which led them to support the local community and environment through their actions (Agyeiwaah, 2020).

Corporate sustainability requires modifying and developing internal activities, organisation, and management. Collaboration between stakeholders empowers participants and facilitates smooth decision-making for integrated solid waste management (SWM) practices. The principles of sustainable development can be further enhanced through a participatory approach (Boluk, 2017). Community-based organisations, led by respected and reputable leaders, can act as intermediaries for society. These organisations are crucial in efficiently disposing of solid waste from local areas regarding environmental protection and sustainability, as well as safeguarding health and community service (Joseph, 2006). A study in the US hotel industry found that implementing effective waste management programs is one of the most critical environmental activities conducted under corporate social responsibility (CSR) initiatives. The authors suggested that hotels can improve their sustainable measures by training employees, reusing grey water for gardening, strengthening employee relationships with local communities, and continuously evaluating and monitoring their practices (Levy & Park, 2011).

9. FUTURE TRENDS AND CHALLENGES

The transformation of urban settlements and environments into Smart areas requires a seamless integration of technology, people, and data. The pursuit of sustainable development has led to the emergence and growth of smart cities. Implementing technology-driven tools in waste management programs can help achieve the circular economy model. Artificial Intelligence, the Internet of Things, Radio Frequency Identification tags, cloud computing, and wireless communication technologies are set to play a crucial role in waste management and recycling programs by facilitating real-time information collection and transport management (Shukla & Hait, 2022). It is essential to align technological integration with the needs of local communities through a gradual implementation process encompassing the 4-Rs of waste management—reduce, reuse, recycle, and recover—along with proper evaluation and monitoring of the financial investments required for the process. The active involvement of all stakeholders, including guests, is crucial for overcoming challenges related to sustainable waste management practices. Clearly defining hotel employees' administrative, regulatory, and monitoring roles and responsibilities is also necessary (Khatib, 2011).

Addressing the challenge of plastic waste requires the hotel industry to focus on sustainable innovation. Replacing disposable plastics with biodegradable or reusable alternatives, as well as investing in environmentally friendly products, are essential measures that hospitality managers must adopt globally in the context of waste management in the hospitality sector after the COVID-19 pandemic (Filimonau, 2021). Proactivity in environmentally responsible hotel behaviour is linked to innovative and knowledge-based capabilities. Competitive advantages can only be achieved by effectively using an organisation's capabilities. Employees having access to appropriate information for critically evaluating past decisions and accepting novel ideas on a common ground facilitates the process and development of proactive environmental actions undertaken by hotels. The hospitality industry is also implementing advanced food technology to reduce waste. AI-driven predictive analytics, intelligent food labels, automated inventory management, mobile applications, circular economy projects, and zero-waste certificates are implemented to enhance efficiency and sustainability (Addressing Food Waste in the Hospitality Industry, 2024).

10. CONCLUSION

Due to its large-scale operations, the hospitality industry generates significant amounts of waste daily, which causes environmental issues and logistical and financial difficulties. Sustainable waste management practices are becoming increasingly important to reduce environmental impact, improve operational efficiency, and cultivate a favourable brand image among environmentally aware consumers.

The culture of extravagance, excessive utilisation of cleaning chemicals, frequent changes of linen, and inefficient energy practices contribute to waste generation. Inadequate waste management significantly impacts the environment by polluting the air, water, and soil. Improper disposal can lead to habitat destruction, soil pollution, and damage to land and water ecosystems. The hospitality industry is a noteworthy contributor to greenhouse gas emissions and climate change through its activities and services. Regulations and policies in the hospitality sector aim to minimise environmental impact and promote sustainability. International standards and certifications like ISO 14001, Green Key Eco-Rating Program, EarthCheck Certification, and LEED Certification guide the hospitality industry towards sustainable waste management practices. A sustainable approach involves reducing waste generation, using source reduction techniques, menu planning and portion control, maximising the reuse of materials, and implementing creative reuse strategies. The hotel industry can achieve a 30-70 percent waste reduction through effective waste sorting and separation techniques, such as robot-assisted systems, Material Recovery Facilities (MRFs), near-infrared spectroscopy, automated solutions powered by AI, and IoT-based sensors. An alliance between hotels, government organisations, and other trades is necessary to facilitate recycling efforts. Enhancing waste management practices is crucial for a broader green approach, including composting organic waste, energy recovery from waste, and efficient policies and procedures related to waste composition and quantity. Information and communication technology (ICT) is essential for solid waste tracking and management. Smart cities require technology integration, including AI, IoT, and wireless communication, for efficient waste management. Local communities need involvement, and green innovation is crucial for plastic waste reduction. Hotels must adopt environmentally friendly practices involving employee access, risk-taking, and innovative techniques.

Stakeholders ought to revolutionise the hospitality sector by adopting sustainable waste management practices. Recycling programs, reduction of single-use plastics, and composting can significantly minimise the industry's environmental impact. Our dedication to sustainability enhances the sector's reputation, attracts environmentally conscious guests, and reduces expenses. Collectively, we can lead the way in fostering a greener future. Immediate action is necessary to implement and promote eco-friendly waste management solutions. Let us establish a new standard for environmental responsibility in the hospitality sector and leave a lasting impression on the planet.

REFERENCES

Addressing food waste in the hospitality industry. (2024, January 19). *Shapiroe*. https://shapiroe.com/blog/food-waste-hospitality-industry/

Aditya, A., & Kurniawati, K. (2023). Food waste management challenges and strategies in the hotel industry in Jakarta. *Journal of Social Research*, 2(9), 2983–2989. DOI: 10.55324/josr.v2i9.1295

Agyeiwaah, E. (2020). The contribution of small accommodation enterprises to sustainable solid waste management. *Journal of Hospitality and Tourism Management*, 44, 1–9. DOI: 10.1016/j.jhtm.2020.04.013

Al-Aomar, R., & Hussain, M. (2017). An assessment of green practices in a hotel supply chain: A study of UAE hotels. *Journal of Hospitality and Tourism Management*, 32, 71–81. DOI: 10.1016/j.jhtm.2017.04.002

Ali, Y., Mustafa, M., Al-Mashaqbah, S., Mashal, K., & Mohsen, M. (2008). Potential of energy savings in the hotel sector in Jordan. *Energy Conversion and Management*, 49(11), 3391–3397. DOI: 10.1016/j.enconman.2007.09.036

Alsuwaidi, M., Eid, R., & Agag, G. (2022). Tackling the complexity of guests' food waste reduction behaviour in the hospitality industry. *Tourism Management Perspectives*, 42, 100963. DOI: 10.1016/j.tmp.2022.100963

Aschenbrenner, D., Colloseus, C., Khoury, R., & Fangerow, N. (2023). Robot-assisted automated sorting techniques for plastic recycling. *Procedia CIRP*, 120, 1232–1237. DOI: 10.1016/j.procir.2023.09.154

Babu, B. R., Bhanu, S. U., & Meera, K. S. (2009). Waste minimization in electroplating industries: A review. *Journal of Environmental Science and Health. Part C, Environmental Carcinogenesis & Ecotoxicology Reviews*, 27(3), 155–177. DOI: 10.1080/10590500903124158 PMID: 19657919

Baker, M. A., Davis, E. A., & Weaver, P. A. (2014). Eco-friendly attitudes, barriers to participation, and differences in behavior at green hotels. *Cornell Hospitality Quarterly*, 55(1), 89–99. DOI: 10.1177/1938965513504483

Betz, A., Buchli, J., Göbel, C., & Müller, C. (2015). Food waste in the Swiss food service industry—Magnitude and potential for reduction. *Waste Management (New York, N.Y.)*, 35, 218–226. DOI: 10.1016/j.wasman.2014.09.015 PMID: 25305683

Bohdanowicz, P. (2006). Environmental awareness and initiatives in the Swedish and Polish hotel industries—Survey results. *International Journal of Hospitality Management*, 25(4), 662–682. DOI: 10.1016/j.ijhm.2005.06.006

Boluk, K. A. R. L. A. (2017). Corporate sustainability as an opportunity for tourism partnerships: A case study on Fairmont Hotels & Resorts. In *Linking Urban and Rural Tourism* (pp. 216–232). Strategies in Sustainability. DOI: 10.1079/9781786390141.0216

Brodie, R., Hollebeek, L., Juric, B., & Ilic, A. (2011). Customer engagement: Conceptual domain, fundamental propositions, and implications for research. *Journal of Service Research*, 14(3), 252–271. DOI: 10.1177/1094670511411703

Burlakovs, J., Jani, Y., Kriipsalu, M., Grinfelde, I., Pilecka, J., & Hogland, W. (2020, March). Implementation of new concepts in waste management in tourist metropolitan areas. In *IOP Conference Series: Earth and Environmental Science* (Vol. 471, No. 1, p. 012017). IOP Publishing. DOI: 10.1088/1755-1315/471/1/012017

Butler, J. (2008). The compelling "hard case" for "green" hotel development. *Cornell Hospitality Quarterly*, 49(3), 234–244. DOI: 10.1177/1938965508322174

Carpenter, D. O., Suk, W. A., Blaha, K., & Cikrt, M. (1996). Hazardous wastes in Eastern and Central Europe [Meeting report]. *Environmental Health Perspectives*, 104(3), 244–248. DOI: 10.2307/3432874 PMID: 8919756

Chan, W., Wong, K., & Lo, J. (2009). Hong Kong hotels' sewage: Environmental cost and saving technique. *Journal of Hospitality & Tourism Research (Washington, D.C.)*, 33(3), 329–346. DOI: 10.1177/1096348009338525

Circular Economy Act (Kreislaufwirtschaftsgesetz - KrWG). (2012). *Climate Laws*. https://climate-laws.org/document/circular-economy-act-kreislaufwirtschaftsgesetz-krwg_7e83

de Titto, E., & Savino, A. (2019). Environmental and health risks related to waste incineration. *Waste Management & Research*, 37(10), 976–986. DOI: 10.1177/0734242X19859700 PMID: 31319775

Domingo, J. L., Marquès, M., Mari, M., & Schuhmacher, M. (2020). Adverse health effects for populations living near waste incinerators with special attention to hazardous waste incinerators: A review of the scientific literature. *Environmental Research*, 187, 109631. DOI: 10.1016/j.envres.2020.109631 PMID: 32460091

EarthCheck. (n.d.). *EarthCheck Certification: Environmental Certification and Benchmarking for the Travel and Tourism Industry*. https://earthcheck.org/

Emery, A., Davies, A., Griffiths, A., & Williams, K. (2006). *Environmental and economic modelling: A case study of municipal solid waste management scenarios in Wales*. Elsevier.

Estes, L. (2000, May 15). Waste reduction tips for hotels and casinos in Indian Country. *Tribal P2*. https://tribalp2.org/waste-reduction-tips-for-hotels-and-casinos-in-indiancountry

European Commission. (2023, May 24). Waste Framework Directive (WFD) - Directive (EU) 2018/851. *Circular Cities and Regions*. https://circular-cities-and-regions.ec.europa.eu/support-materials/eu-regulations-legislation/waste-framework-directive-wfd-directive-eu-2018851-0

Filimonau, V. (2021). The prospects of waste management in the hospitality sector post COVID-19. *Resources, Conservation and Recycling*, 168, 105272. DOI: 10.1016/j.resconrec.2020.105272 PMID: 36568327

Fraj, E., Matute, J., & Melero, I. (2015). Environmental strategies and organizational competitiveness in the hotel industry: The role of learning and innovation as determinants of environmental success. *Tourism Management*, 46, 30–42. DOI: 10.1016/j.tourman.2014.05.009

Giri, S. (2021). Integrate solid waste management: A case study of a hotel in Kathmandu, Nepal. *EPRA International Journal of Multidisciplinary Research*, 7(5), 264–268.

Goal 12: Ensure sustainable consumption and production patterns. (2015). *United Nations*. https://www.un.org/sustainabledevelopment/sustainable-consumption-production/

Graulich, K., Köhler, A., Löw, C., Sutter, J., Watson, D., Mehlhart, G., & Lippert Tange, I. (2021). *Impact of COVID-19 on single-use plastics and the environment in Europe*. European Environment Agency.

Green Key - Green Key Programme. (2021). *Green Key*. https://www.greenkey.global/our-programme

Guerra-Lombardi, V., Hernández-Martín, R., & Padrón-Fumero, N. (2024). Drivers, barriers, and key practices of corporate sustainability strategy implementation in hotels. *International Journal of Hospitality Management*, 120, 103791. DOI: 10.1016/j.ijhm.2024.103791

Han, H., Lee, J. S., Trang, H. L. T., & Kim, W. (2018). Water conservation and waste reduction management for increasing guest loyalty and green hotel practices. *International Journal of Hospitality Management*, 75, 58–66. DOI: 10.1016/j.ijhm.2018.03.012

Hannan, M. A., Al Mamun, M. A., Hussain, A., Basri, H., & Begum, R. A. (2015). A review on technologies and their usage in solid waste monitoring and management systems: Issues and challenges. *Waste Management (New York, N.Y.)*, 43, 509–523. DOI: 10.1016/j.wasman.2015.05.033 PMID: 26072186

International Organization for Standardization. (2015). *ISO 14001:2015 Environmental management systems – Requirements with guidance for use*. ISO. https://www.iso.org/standard/60857.html

Ioannidis, A., Chalvatzis, K. J., Leonidou, L. C., & Feng, Z. (2021). Applying the reduce, reuse, and recycle principle in the hospitality sector: Its antecedents and performance implications. *Business Strategy and the Environment*, 30(7), 3394–3410. DOI: 10.1002/bse.2809

Ip, K., Testa, M., Raymond, A., Graves, S. C., & Gutowski, T. (2018). Performance evaluation of material separation in a material recovery facility using a network flow model. *Resources, Conservation and Recycling*, 131, 192–205. DOI: 10.1016/j.resconrec.2017.11.021

ISO - International Organization for Standardisation. (2021). *ISO 18513:2021 – Tourism and related services – Hotels and other types of tourism accommodation – Terminology*. ISO. Retrieved from https://www.iso.org/obp/ui/en/#iso:std:iso:18513:ed-2:v1:en

Jagaba, A. H., Lawal, I. M., Birniwa, A. H., Affam, A. C., Usman, A. K., Soja, U. B., & Yaro, N. S. A. (2024). Sources of water contamination by heavy metals. In *Membrane Technologies for Heavy Metal Removal from Water* (pp. 3–27). CRC Press., DOI: 10.1201/9781003326281-2

Jawabreh, O., Abdelrazaq, H., & Jahmani, A. (2021). Business sustainability practice and operational management in the hotel industry in Aqaba Special Authority Economic Zone Authority (ASEZA). *Geo Journal of Tourism and Geosites*, 38(4), 1089–1097. DOI: 10.30892/gtg.38414-748

Joseph, K. (2006). Stakeholder participation for sustainable waste management. *Habitat International*, 30(4), 863–871. DOI: 10.1016/j.habitatint.2005.09.009

Juvan, E., Grün, B., & Dolnicar, S. (2023). Waste production patterns in hotels and restaurants: An intra-sectoral segmentation approach. *Annals of Tourism Research Empirical Insights*, 4(1), 100090. DOI: 10.1016/j.annale.2023.100090

Karadimas, N. V., Mavrantza, O. D., & Loumos, V. G. (2005, November). GIS integrated waste production modeling. In *EUROCON 2005-The International Conference on "Computer as a Tool"* (Vol. 2, pp. 1279-1282). IEEE. DOI: 10.1109/EURCON.2005.1630190

Kasavan, S., Mohamed, A. F., & Halim, S. A. (2017). Sustainable food waste management in hotels: Case study Langkawi UNESCO Global Geopark. *Planning Malaysia, 15.*

Kasimu, A. B., Zaiton, S., & Hassan, H. (2012). Hotels' involvement in sustainable tourism practices in Klang Valley, Malaysia. *International Journal of Economics and Management*, 6(1), 21–34.

Khatib, I. A. (2011). Municipal solid waste management in developing countries: Future challenges and possible opportunities. *Integrated Waste Management*, 2, 35–48.

Khattak, W. A., Sun, J., Hameed, R., Zaman, F., Abbas, A., Khan, K. A., Elboughdiri, N., Akbar, R., He, F., Ullah, M. W., Al-Andal, A., & Du, D. (2024). Unveiling the resistance of native weed communities: Insights for managing invasive weed species in disturbed environments. *Biological Reviews of the Cambridge Philosophical Society*, 99(3), 753–777. DOI: 10.1111/brv.13043 PMID: 38174626

Kumar, N. M., Mohammed, M. A., Abdulkareem, K. H., Damasevicius, R., Mostafa, S. A., Maashi, M. S., & Chopra, S. S. (2021). Artificial intelligence-based solution for sorting COVID-related medical waste streams and supporting data-driven decisions for smart circular economy practice. *Process Safety and Environmental Protection*, 152, 482–494. DOI: 10.1016/j.psep.2021.06.026

Kumar, S. (2005). Resource use and waste management in Vietnam hotel industry. *Journal of Cleaner Production*, 13(2), 109–116. DOI: 10.1016/j.jclepro.2003.12.014

Laeis, G. C., & Legrand, W. (2023). Climate change: Leading question: A vicious circle? How does the hospitality industry impact the climate and how is it impacted by climate change? In *Critical Questions in Sustainability and Hospitality* (pp. 3–12). Routledge.

Legrand, W., Chen, J. S., & Laeis, G. C. (2022). *Sustainability in the hospitality industry: Principles of sustainable operations*. Routledge.

Levy, S. E., & Park, S. Y. (2011). An analysis of CSR activities in the lodging industry. *Journal of Hospitality and Tourism Management*, 18(1), 147–154. DOI: 10.1375/jhtm.18.1.147

Luu, T. T. (2020). Reducing food waste behavior among hospitality employees through communication: Dual mediation paths. *International Journal of Contemporary Hospitality Management*, 32(5), 1881–1904. DOI: 10.1108/IJCHM-09-2019-0779

Madhav, S., Mishra, R., Kumari, A., Srivastav, A. L., Ahamad, A., Singh, P., Ahmed, S., Mishra, P. K., & Sillanpää, M. (2024). A review on sources identification of heavy metals in soil and remediation measures by phytoremediation-induced methods. *International Journal of Environmental Science and Technology*, 21(1), 1099–1120. DOI: 10.1007/s13762-023-04950-5

Mbasera, M., Du Plessis, E., Saayman, M., & Kruger, M. (2016). Environmentally-friendly practices in hotels. *Acta Commercii*, 16(1), 1–8. DOI: 10.4102/ac.v16i1.362

Mensah, I. (2014). Different shades of green: Environmental management in hotels in Accra. *International Journal of Tourism Research*, 16(5), 450–461. DOI: 10.1002/jtr.1939

Mensah, I. (2020). Waste management practices of small hotels in Accra: An application of the waste management hierarchy model. *Journal of Global Business Insights*, 5(1), 33–46. DOI: 10.5038/2640-6489.5.1.1081

Merouani, E. F. O., Ferdowsi, M., Buelna, G., Jones, J. P., Benyoussef, E. H., Malhautier, L., & Heitz, M. (2024). Exploring the potential of biofiltration for mitigating harmful gaseous emissions from small or old landfills: A review. *Biodegradation*, 35(5), 1–23. DOI: 10.1007/s10532-024-10082-5 PMID: 38748305

Mhammedsharif, R. M., & Kolo, K. Y. (2023). A case study of environmental pollution by pathogenic bacteria and metal(oid)s at Soran Landfill Site, Erbil, Iraqi Kurdistan Region. *Environmental Monitoring and Assessment*, 195(7), 811. DOI: 10.1007/s10661-023-11352-0 PMID: 37284886

Nelms, S. E., Clark, B. L., Duncan, E. M., Germanov, E., Godley, B. J., Parton, K. J., . . . Rodríguez, Y. (2023). Plastic pollution and marine megafauna: Recent advances and future directions. In *Plastic Pollution in the Global Ocean* (pp. 97-138). DOI: 10.1142/9789811259111_0005

New York City Department of Sanitation. (2020, July 31). Commercial organics requirements. Retrieved from https://www.nyc.gov/site/dsny/businesses/materials-handling/commercial-organics-requirements.page

Nikolskaya, E. Y., Anzorova, S. P., Potapov, S. V., Dekhtyar, G. M., & Lebedev, K. A. E. (2019). Methodical approaches to estimate hotel facilities' efficiency. *Journal of Environmental Management and Tourism*, 9(8), 1664–1669. DOI: 10.14505//jemt.v9.8(32).04

Nisa, R. K., Haryono, E., & Santosa, L. W. (2024). Assessing the environmental footprint: A life cycle assessment of greenhouse gas emissions from energy consumption in 3 and 4-star hotels in Pontianak, West Kalimantan. *Jurnal Teknologi Lingkungan, 25*(1), 001-009. DOI: 10.55981/jtl.2024.964

Ogishi, A., & Goldman, G. E. (2001). The economic impacts of solid waste disposal and diversion in California.

Okumus, B., Taheri, B., Giritlioglu, I., & Gannon, M. J. (2020). Tackling food waste in all-inclusive resort hotels. *International Journal of Hospitality Management*, 88, 102543. DOI: 10.1016/j.ijhm.2020.102543

Oluwole, E. A., Attahiru, H., Oyediran, O. B., Omotugba, S. K., Meduna, P. N., & Kolawole, G. T. (2020). Assessment of waste management system among hotels and guest houses in Minna. *Journal of Environmental Management and Tourism*, 11(7), 1609–1616. DOI: 10.14505//jemt.v11.7(47).02

Orloff, K., & Falk, H. (2003). An international perspective on hazardous waste practices. *International Journal of Hygiene and Environmental Health*, 206(4-5), 291–302. DOI: 10.1078/1438-4639-00225 PMID: 12971684

Perkoulidis, G., & Moussiopoulos, N. (2017). Perspectives of energy recovery from food waste generated from hotels.

Pham Phu, S. T., Fujiwara, T., Hoang, M. G., Pham, V. D., & Tran, M. T. (2019). Waste separation at source and recycling potential of the hotel industry in Hoi An city, Vietnam. *Journal of Material Cycles and Waste Management*, 21(1), 23–34. DOI: 10.1007/s10163-018-0807-5

Pirani, S. I., & Arafat, H. A. (2014). Solid waste management in the hospitality industry: A review. *Journal of Environmental Management*, 146, 320–336. DOI: 10.1016/j.jenvman.2014.07.038 PMID: 25194519

Pirani, S. I., & Arafat, H. A. (2016). Reduction of food waste generation in the hospitality industry. *Journal of Cleaner Production*, 132, 129–145. DOI: 10.1016/j.jclepro.2015.07.146

Pizzol, M., Møller, F., & Thomsen, M. (2013). External costs of atmospheric lead emissions from a waste-to-energy plant: A follow-up assessment of indirect exposure via topsoil ingestion. *Journal of Environmental Management*, 121, 170–178. DOI: 10.1016/j.jenvman.2013.02.043 PMID: 23542215

Prakash, S., Sharma, V. P., Singh, R., Vijayvargy, L., & Nilaish, . (2023). Adopting green and sustainable practices in the hotel industry operations-an analysis of critical performance indicators for improved environmental quality. *Management of Environmental Quality*, 34(4), 1057–1076. DOI: 10.1108/MEQ-03-2022-0090

Principato, L., Di Leo, A., Mattia, G., & Pratesi, C. A. (2021). The next step in sustainable dining: The restaurant food waste map for the management of food waste. *Italian Journal of Marketing*, 189-207. https://doi.org/DOI: 10.1007/s10203-020-00445-3

Radwan, H. R., Jones, E., & Minoli, D. (2012). Solid waste management in small hotels: A comparison of green and non-green small hotels in Wales. *Journal of Sustainable Tourism*, 20(4), 533–550. DOI: 10.1080/09669582.2011.621539

Ramirez, A., & George, B. (2019). Plastic recycling and waste reduction in the hospitality industry: Current challenges and some potential solutions. *Economics. Management and Sustainability*, 4(1), 6–20. DOI: 10.14254/jems.2019.4-1.1

Resource Conservation and Recovery Act (RCRA). (1976). Retrieved from https://www.epa.gov/rcra

Ribić, B., Voća, N., & Ilakovac, B. (2017). Concept of sustainable waste management in the city of Zagreb: Towards the implementation of circular economy approach. *Journal of the Air & Waste Management Association*, 67(2), 241–259. DOI: 10.1080/10962247.2016.1229700 PMID: 27650014

Salam, M., Zheng, H., Liu, Y., Zaib, A., Rehman, S. A. U., Riaz, N., Eliw, M., Hayat, F., Li, H., & Wang, F. (2023). Effects of micro (nano) plastics on soil nutrient cycling: State of the knowledge. *Journal of Environmental Management*, 344, 118437. DOI: 10.1016/j.jenvman.2023.118437 PMID: 37343476

Sayegh, M. A., Ludwińska, A., Rajski, K., & Dudkiewicz, E. (2021). Environmental and energy saving potential from greywater in hotels. *The Science of the Total Environment*, 761, 143220. DOI: 10.1016/j.scitotenv.2020.143220 PMID: 33131865

Scholz, P., Linderová, I., & Konečná, K. (2020). Green management tools as a way to sustainable behaviour in the hotel industry: Case study from Czechia. *Sustainability (Basel)*, 12(23), 10027. DOI: 10.3390/su122310027

Senzige, J. P., Makinde, D. O., Njau, K., & Nkansah-Gyeke, Y. (2014). Factors influencing solid waste generation and composition in urban areas of Tanzania: The case of Dar-es-Salaam.

Sharma, B., Vaish, B., Srivastava, V., Singh, S., Singh, P., & Singh, R. P. (2018). An insight to atmospheric pollution-improper waste management and climate change nexus. In *Modern Age Environmental Problems and Their Remediation* (pp. 23-47). DOI: 10.1007/978-3-319-64501-8_2

Shukla, S., & Hait, S. (2022). Smart waste management practices in smart cities: Current trends and future perspectives. In *Advanced Organic Waste Management* (pp. 407-424). DOI: 10.1016/B978-0-323-85792-5.00011-3

Shyam, G. K., Manvi, S. S., & Bharti, P. (2017, February). Smart waste management using Internet-of-Things (IoT). In *2017 2nd International Conference on Computing and Communications Technologies (ICCCT)* (pp. 199-203). IEEE. DOI: 10.1109/ICCCT2.2017.7972282

Singh, N., Cranage, D., & Lee, S. (2014). Green strategies for hotels: Estimation of recycling benefits. *International Journal of Hospitality Management*, 43, 13–22. DOI: 10.1016/j.ijhm.2014.07.006

Soni, G., Prakash, S., Kumar, H., Singh, S. P., Jain, V., & Dhami, S. S. (2020). An interpretive structural modeling of drivers and barriers of sustainable supply chain management: A case of stone industry. *Management of Environmental Quality*, 31(5), 1071–1090. DOI: 10.1108/MEQ-09-2019-0202

Sorin, F., & Sivarajah, U. (2021). Exploring circular economy in the hospitality industry: Empirical evidence from Scandinavian hotel operators. *Scandinavian Journal of Hospitality and Tourism*, 21(3), 265–285. DOI: 10.1080/15022250.2021.1921021

Tansel, T., Yeshenkulova, G., & Nurmanova, U. (2021). Analyzing waste management and recycling practices for the hotel industry. *Journal of Environmental Management and Tourism*, 12(2), 382–391. DOI: 10.14505//jemt.v12.2(50).07

Tarabashkina, L., Devine, A., & Quester, P. G. (2022). Encouraging product reuse and upcycling via creativity priming, imagination and inspiration. *European Journal of Marketing*, 56(7), 1956–1984. DOI: 10.1108/EJM-06-2020-0442

Trends in food waste reduction. (2024, June 7). Retrieved from https://www.hotelierindia.com/fb/trends-in-food-waste-reduction#:~:text=According%20to%20a%202024%20report,waste%20over%20the%20past%20year

Visvanathan, C. (2006). Solid waste management in Asian perspectives. *Environmental Management Tools, 131*.

Wan, Y. K. P., Chan, S. H. J., & Huang, H. L. W. (2017). Environmental awareness, initiatives and performance in the hotel industry of Macau. *Tourism Review*, 72(1), 87–103. DOI: 10.1108/TR-06-2016-0016

Wang, R. (2012). Investigations of important and effective effects of green practices in restaurants. *Procedia: Social and Behavioral Sciences*, 40, 94–98. DOI: 10.1016/j.sbspro.2012.03.166

Wang, Y., Fang, M., Lou, Z., He, H., Guo, Y., Pi, X., Wang, Y., Yin, K., & Fei, X. (2024). Methane emissions from landfills differentially underestimated worldwide. *Nature Sustainability*, 7(4), 1–12. DOI: 10.1038/s41893-024-01307-9

Warren, C., Becken, S., & Coghlan, A. (2019). Using persuasive communication to co-create behavioural change–Engaging with guests to save resources at tourist accommodation facilities. In *Marketing for Sustainable Tourism* (pp. 67–86). Routledge., DOI: 10.4324/9780203711668-5

Waste Framework Directive. (2008/98/EC). (2008, November 19). Retrieved from https://www.eea.europa.eu/: https://www.eea.europa.eu/policy-documents/waste-framework-directive-2008-98-ec

Waste Online. After it's been binned. (2006). Retrieved July 15, 2024, from http://www.wasteonline.org.uk

Wu, X., Li, J., Yao, L., & Xu, Z. (2020). Auto-sorting commonly recovered plastics from waste household appliances and electronics using near-infrared spectroscopy. *Journal of Cleaner Production*, 246, 118732. DOI: 10.1016/j.jclepro.2019.118732

Zein, K., Wazner, M., & Meylan, G. (2008). Best environmental practices for the hotel industry. [online]. Sustainable Business Associates (SBA).

Zsigraiová, Z., Tavares, G., Semiao, V., & de Graça Carvalho, M. (2009). Integrated waste-to-energy conversion and waste transportation within island communities. *Energy*, 34(5), 623–635. DOI: 10.1016/j.energy.2008.10.015

Chapter 6
Minimizing Waste in the Hospitality Industry:
The Opportunities and Challenges of Lean Tourism

Abdul Waaje
https://orcid.org/0009-0004-7207-0750
Michigan Technological University, USA

Rejaul Karim
https://orcid.org/0000-0001-6175-8825
Varendra University, Bangladesh

Md. Mustaqim Roshid
https://orcid.org/0000-0001-8517-6336
University of Rajshahi, Bangladesh

Tamanna Nusrat Meem
https://orcid.org/0009-0006-6612-7946
Jahangirnagar University, Bangladesh

ABSTRACT

Despite economic difficulties, global warming, and the ongoing recovery from the pandemic, the hospitality industry, a key player in the global economy, is under more urgency to adopt sustainable practices. This chapter examines 'Lean Hospitality,' an approach to the hospitality industry that borrows lean concepts from manufacturing operations to strengthen efficiency, reduce waste, and satisfy customers more. It examines how these concepts have changed and how they manage different kinds of waste, such as solids, liquids, gases, food choices, and potentially hazardous materials. The 5S Lean Philosophy, Value Stream Mapping, Standardized Work,

DOI: 10.4018/979-8-3693-6110-8.ch006

and Poka-Yoke are some of the methodologies studied to see how they might promote operational excellence and continuous improvement, offering a promising future for the industry. Examples of real-world advantages and disadvantages are provided through case studies, including the Ritz-Carlton in Riyadh's effective lean implementation.

1. INTRODUCTION

Despite the increasing mechanisation of modern society, people continue to maintain a strong desire for travel as they seek opportunities to discover new places and rejuvenate themselves through international travel. Various factors, such as seasonal shifts, weather conditions, family incomes, and the political landscapes of nations, all impact the complex hospitality and tourism sector (Legrand et al., 2023). The significance of this industry is undeniable, considering the substantial contributions it makes to the global economy and the role it plays in promoting sustainable development, particularly in emerging and underdeveloped nations that are in the process of transitioning towards development (Salem et al., 2023, Karim 2020, Gazi et al., 2024). Following the pandemic, the hotel sector has experienced remarkable growth, which has marked a worldwide rebound. According to the United States Travel Association, the travel sector in the United States made a substantial contribution to the economy by accounting for half a trillion dollars in direct expenditure (Abdelmoety et al., 2022). This resulted in a total monetary impact of 2.6 trillion dollars, equivalent to the levels seen in 2019. The industry significantly contributed to the labour market in 2022, supporting approximately 15 million jobs in the United States, with immediate workforce estimates exceeding 8 million. With a 22% growth from the previous year, the travel and tourism industry had its best year for employment in 2022, supporting one out of every eleven jobs globally and contributing 7.6% to the global GDP, according to the World Travel and Tourism Council's 2022 Economic Report (Colmekcioglu et al., 2022; Khan et al., 2023). A global interaction trend with other cultures and customs has driven this expansion.

The hotel business has encountered substantial obstacles in establishing a sustainable and distinctive approach to operations, management, and guest services due to economic hardship, climate change, and the COVID-19 epidemic. Over time, the need for change in the tourist industry has become more and more pressing. Although the manufacturing sector has had significant advancements in transitioning from handicrafts to Industry 4.0 (Roshid et al., 2024), the hospitality business has yet to undergo such revolutionary transformations during the industrial age. The advent of information technology and the digital era has brought about the introduction of new opportunities. These new opportunities substantially impact our day-to-day activities, such as consuming goods and services, travelling, and staying (Carvalho et al., 2023). Within the hotel industry, many different tactics have been implemented to reach a condition of operations that is both seamless and sustainable. On the other hand, there is a growing school of thought among academics that advocates for the sole implementation of the "Lean" methodology in the hospitality industry. It is currently being advocated as the prevailing technique that this method, which has shown significant success in assuring sustainability, profitability, and customer happiness within manufacturing businesses, notably in the automobile manufacturing industry, must be utilised (Al-Aomar & Hussain, 2018).

In response to rising expenses for disposal and growing worries about the environment, the hotel industry is coming under increasing pressure to implement waste management procedures that are more environmentally friendly (Sucheran, 2021). The demand for efficient operations that reduce waste is growing as the tourism industry expands worldwide. Revenues are anticipated to reach around 8 trillion US dollars by 2028. There is a high level of competition in the hospitality industry, and sustainable methods not only coincide with environmental aims but also improve operational efficiency, potentially increasing revenue. Over the past ten years, many studies have concentrated on the effects of the growing tourism industry on the environment. While this increased scrutiny is taking place, the hotel business is making significant contributions to the worsening of the global climate through its vast use of fossil fuels, water, and non-resilient items (Tanchangya, T., et al., 2024). Therefore, it is of the utmost importance for hotels to commit to the implementation of management strategies that are both responsible and environmentally friendly (Sucheran, 2017).

Eliminating waste from the production floor is the primary goal of the Lean approach to manufacturing. By eliminating waste and reconfiguring the whole operational structure, from management to staff hierarchies, adopting this ideology can provide hotel owners with an efficient way to cut operating expenses. This approach might be a sustainable waste management strategy that helps advance sustainability and the growth of the tourist and hospitality sectors. This would be accomplished by clearly outlining roles and responsibilities. Service quality is an essential element of the hospitality industry, and its crucial role in the sector's success cannot be emphasised enough (Ramphal, 2017). Serviceability can be significantly improved by applying Lean concepts to meet guests' expectations. With this method, you can restructure the operational frameworks of different hotel sub-departments and simultaneously fulfil modern needs. Lean Tourism is a sustainable waste management technique that will be explained in this chapter. It is used in the hotel business. The following topics will be covered: the history of Lean, why it is necessary to apply Lean principles in the tourism industry, the basics of Lean Tourism that are in line with Lean principles, the strategies and terminology of Lean Tourism, the difficulties in implementing Lean, real-life examples of hotels implementing Lean, and where the field is headed in the future. This will cover how Lean can improve customer satisfaction by streamlining hotel procedures, dealing with serviceability problems, and eliminating waste through continuous improvement.

"What are the opportunities and challenges associated with implementing Lean Tourism practices to minimise waste in the hospitality industry?" is the main research question that directs this study. This inquiry seeks to identify the possible advantages and challenges that companies in the hospitality industry may face when using lean approaches. Through this investigation, the study offers a thorough grasp of how lean concepts can be customised to the environment of the hospitality sector, ultimately leading to more effective and sustainable operations. The hospitality sector, which plays a significant role in the world economy, is under increasing pressure to improve its sustainable standards. This challenge includes the crucial element of minimising waste, which has ramifications for the environment and the economy. Waste generation increases with industry growth, raising serious issues with resource consumption, pollution, and cost control. This study investigates how implementing Lean Tourism principles might address waste minimisation in the hospitality industry. Lean Tourism emphasises efficiency and sustainability by maximising value and minimising waste, drawing inspiration from lean manufacturing.

2. THE CONCEPT OF LEAN

The early Japanese literature offered a more concise outline of the Toyota Production System (TPS) than the following academic books. It gave a more in-depth description of the critical facets that comprised it. Recent studies have concentrated more on elaborating on aspects of the system rather than confronting the system holistically. The philosophical definition of lean is guided ideas and overarching aims (Spear & Bowen, 1999). Lean is based on the principle of minimising waste while maximising customer value. It is a methodical strategy for finding and doing away with things that don't add value while constantly improving. This strategy includes methods such as the 5S for organising the workplace, standardising tasks, and continuously monitoring and changing processes to obtain the best results (Bubber et al., 2022; Ferreira et al., 2023a; Garcia-Garcia et al., 2022; Goyal et al., 2022). In contrast, the pragmatic definition is a set of management techniques, devices, and protocols that can be witnessed firsthand (Mastos et al., 2022). To comprehend the Lean idea, it is crucial to initially establish the definition of waste, as the primary aim of Lean is to eradicate waste. Waste generally encompasses any component that incurs expenses without contributing value to a product or service. This definition is universally relevant in all sectors; however, the precise forms of trash may differ from one business to another. Academics recognise various common types of waste, which include flaws that do not meet the required standards, idle time that causes delays in production, service, or delivery due to insufficient resources, equipment, or information, and unnecessary movement that does not contribute value, such as transferring materials between subprocesses or changing operators or service providers between stages. Some other types of waste are overproduction or over-servicing, which happens when supply exceeds demand or when services exceed what customers want; excess inventory related to raw materials, work-in-progress, services in process, or finished goods; and inefficiencies caused by a lack of skill, effort, or the wrong use of machine capacity. These sorts of trash are widely acknowledged as standard in both the production and service sectors.

Researchers claim that Lean is a methodical approach to waste management that achieves the commercialisation of merchandise and services while lowering buffers overhead (Hopp & Spearman, 2004). It uses half the labour in the manufacturing facility, half the space utilised in making things, half the money spent on equipment, and half the number of skilled labour hours needed to develop a fresh product in fifty percent of the time required to maintain merchandise (Womack et al., 2007). It also produces fewer errors and defects and offers a broader range of items and amenities. Since then, lean manufacturing ideas have been implemented in various industries, including developing software, medical treatment, and business services. A few industries, including the hospitality industry, the banking industry, and aircraft operations, are also implementing lean principles to increase their profitability and capacity to provide timely service to their clients.

3. HISTORICAL BACKGROUND OF LEAN

Many of the fundamental ideas that underpin Lean may be traced back to the Japanese business known as Toyota. The Toyota Production System (TPS), which is seen to be the embodiment of these ideas, can trace its roots back to the early part of the twentieth century. Henry Ford's methodology greatly impacted the development of what would eventually be known as Lean, which can be traced back to a time before TPS was even conceived. Ford's methods of producing their products impressed the Japanese businessman Kiichiro Toyoda, who later founded Toyota when he visited the factories where Ford produced automobiles. These discoveries, along with the concepts found in Ford's 1926 book "*Today and Tomorrow*," served as the foundation for the creation of TPS. This book outlines Ford's groundbreaking manufacturing processes, including assembly line production and waste elimination. Ultimately, this method improved and built upon these concepts, contributing to the lean manufacturing processes being disseminated worldwide (Ferreira et al., 2023b; Lin et al., 2022; Vanichchinchai, 2023).

It was in 1991 that James P. Womack, Daniel T. Jones, and Daniel Roos from the Massachusetts Institute of Technology were the ones who first introduced the concept of Lean to the academic community. At the time, it was the Toyota Production System (TPS). Their landmark article, "The Machine That Changed the World," presented a comparative examination of Japanese and American businesses, promoting Toyota Motor Company and the TPS as the most efficient model. Following the principles of Lean, this system was acknowledged as the first framework of its kind. In 2001, Womack and Jones published "Lean Thinking: Banish Waste and Create Wealth in Your Corporation," laying out the fundamental principles underpinning Lean Philosophy (Dekier, 2012). They further refined these notions in this book. As is now recognised, Lean Manufacturing is considered a direct descendant of TPS. It uses the methodology and technologies that Toyota initially established. The first step in defining a lean idea is to trace its development over time and classify the viewpoints often used to describe it. Figure 1 shows the significant progressions that have led to our present knowledge of lean production.

Figure 1. Evolution of lean

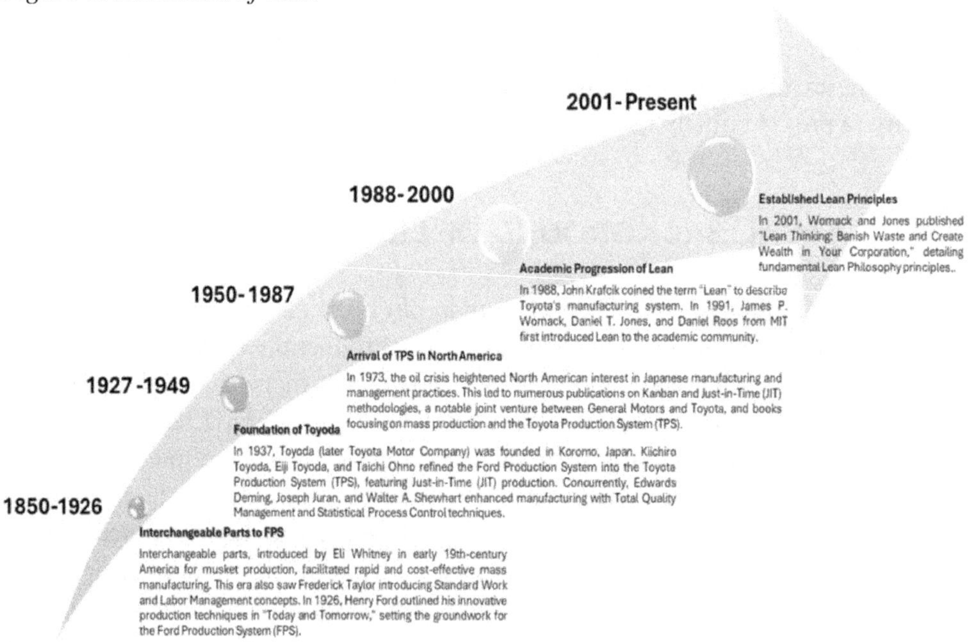

Source: Authors' Compilation

4. WASTE MANAGEMENT IN THE TOURISM SECTOR

Waste in the tourist industry refers to all kinds of waste produced by tourism-related activities, including gaseous waste (e.g., emissions from transportation) and liquid waste (e.g., wastewater from hotels and resorts). Minimising the environmental effect, protecting natural resources, and keeping tourist locations appealing depends on efficient industry waste management (Diaz-Farina et al., 2023). The many tasks included in waste management plans are recycling, appropriate disposal, and waste production reduction.

Solid waste in the tourism sector primarily includes materials such as packaging, disposable items, and general refuse generated by guests and staff. Proper waste management involves effective collection, segregation, recycling, and disposal practices. Implementing waste reduction strategies such as using biodegradable or reusable materials and promoting recycling can significantly minimise the environmental impact of solid waste.

Liquid waste management is crucial in tourism, particularly in hotels and resorts where large quantities of wastewater are generated from kitchens, bathrooms, and laundry services. Effective treatment and disposal of liquid waste are essential to prevent pollution of local water bodies. Advanced treatment systems and water conservation measures can help reduce the volume of liquid waste and mitigate its environmental impact.

Gaseous waste includes emissions from heating, cooling, and cooking systems and transportation services associated with tourism activities. Managing gaseous waste involves implementing energy-efficient technologies, using cleaner fuel options, and promoting alternative transportation methods such as cycling and walking. Reducing carbon footprints through these measures contributes to the overall sustainability of the tourism sector.

Food waste is a significant challenge in the hospitality industry due to the large volumes of food prepared and consumed. Effective food waste management includes portion control, proper inventory management, and donating surplus food to local communities. Composting and other organic waste recycling methods can also divert food waste from landfills and create valuable by-products.

Hazardous waste in the tourism sector includes chemicals from cleaning products, batteries, electronic waste, and other materials that pose risks to human health and the environment. Proper handling, storage, and disposal of hazardous waste are critical to prevent contamination and ensure safety. Implementing environmentally friendly alternatives and educating staff and guests about safe waste management practices can help reduce the generation of hazardous waste.

Table 1. Key aspects of waste and waste management in tourism

Type of Waste	Sources	Management Strategies
Solid Waste	Restaurants, lodging, events	Use of biodegradable materials, reduce, reuse, recycle, and effective waste sorting systems.
Liquid Waste	Hotel waste and cruise ship discharges	Treatment facilities, recycling of greywater, and appropriate disposal procedures
Gaseous Waste	Transport (cars, buses, and flights), facility energy consumption	Technologies for reducing emissions, substitute fuels, and energy efficiency programs
Food Waste	Buffets, catering, restaurants	Donation schemes, composting, and portion control
Hazardous Waste	Electronics, batteries, cleaning products	Ecological substitutes, personnel training, safe handling and disposal

Source: Authors' Compilation

Businesses in the tourism industry can take many steps to manage their trash better. First and foremost, we must implement waste reduction programs. These programs should include measures to decrease single-use plastics, promote waste minimisation among guests, and prioritise digital communication over paper-based ones. Another critical factor is the establishment of recycling and composting activities, including the placement of recycling stations, the composting of organic waste, and collaboration with nearby recycling centres (Camilleri, 2021).

It is also crucial to prioritise suppliers with sustainable practices and engage in sustainable procurement, purchasing things that may be recycled, reused, or composted. Teaching employees and visitors how to dispose of trash properly is critical to effective waste management. Lastly, collaborative efforts with other businesses, communities, and local governments can achieve better waste management infrastructure and practices. In addition to assisting in environmental protection, improving local communities, quality of life, and tourists' overall experience can be achieved through effective waste management in the tourism industry. (Obersteiner et al., 2021). Implementing thorough waste management systems can preserve popular tourist spots and lower the industry's environmental footprint.

5. LEAN TOURISM AND HOSPITALITY AND SUSTAINABLE WASTE MANAGEMENT

The manufacturing sector has been primarily responsible for developing the Lean concept throughout its history. Before the beginning of the 20th century, there needed to be more written material and actual implementation of Lean principles within the service industry. Not only has the hospitality industry been more active in adopting, implementing, and transitioning from conventional functional and service structures to those influenced by Lean methodology, but it has also been prolonged. Furthermore, particular critiques have been levelled against lean manufacturing practices. Many academics have concluded that many businesses wrongly consider Lean a collection of tools and practices rather than a holistic philosophy for enhancing processes (Spear, 2014). Some have argued that Lean is not the only effective technique for increasing customer happiness, but it could be more effective when used independently.

As an essential service industry component, the hotel sector follows the conventional organisational, operational, service-related, and financial practices commonly observed in this field. Lean services are regarded as a socio-technological system consisting of two separate sets of practices (Shah & Ward, 2007). We will provide a more detailed explanation of the Lean Hospitality idea by expanding upon these distinctive methodologies. Lean Hospitality is an approach for neutralising waste in the hospitality industry by reducing non-value-added procedures, intending to optimise client satisfaction while improving effectiveness and customer care quality by combining lean production concepts with the separate practices of lean services to consumers. Lean Hospitality is intrinsically multidimensional, illustrating the internal interconnections among different in-house divisions in the hospitality industry and the principles of Lean Services. From the perspective of Lean Hospitality, these three elements significantly contribute to waste reduction, exemplifying a genuine commitment to sustainable waste management within the industry.

Figure 2. Structure of lean hospitality

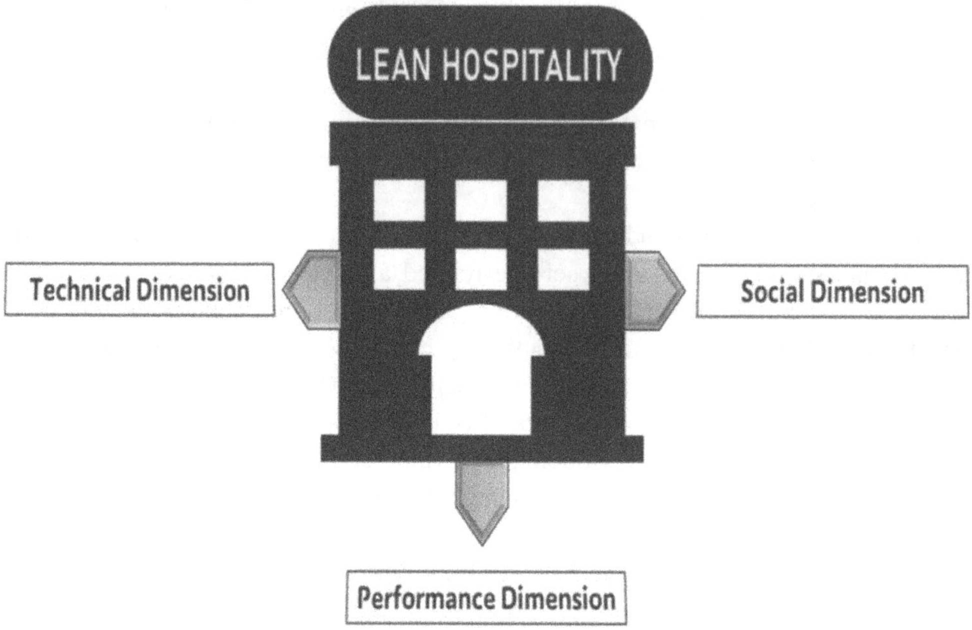

Source: Authors' Compilation

Lean Tourism includes components while sticking to the fundamental principles of the *lean production concept* to establish itself as a recognised and sustainable waste management method within the hospitality industry. Removing operations that do not add value and cutting down on visible waste across all operational areas, including upfront office housekeeping operations, food and drink, and service and technology, is intended to be recognised as a comprehensive and sustainable waste management system. One of the foremost objectives of putting Lean concepts into practice is to fulfil the requirements of customers by catering to the requirements of internal customers that pop up during operations and service delivery. Achieving profitability, cultivating an ecologically sound brand image, ensuring the safety of employees, and making a significant contribution to creating a healthy planet for the future are all goals attained with this strategy. When viewed through the lens of Lean Hospitality, the principle can be broken down into three core principles. Each of these fundamental principles incorporates various tools and methods to facilitate the successful realisation of Lean tactics and the transition from the current waste management and control solutions.

6. LEAN HOSPITALITY FROM A TECHNO-SOCIO-PERFORMANCE PERSPECTIVE

The first dimension of Lean Hospitality is concerned with the technical concepts, and it is part of the larger framework that consists of five dimensions. These concepts are linked with a comprehensive collection of tools and methods that are derived from the idea of lean manufacturing management. For departments within the hotel sector to efficiently achieve their desired goals and establish competency, these tools play a significant role in shaping essential practices. The purpose of this part is to investigate the development and use of these technological principles to improve operational efficiency and correspond with the overarching goals of Lean Hospitality.

6.1 5S Lean Philosophy

The 5S housekeeping technique, derived from five Japanese terms, has achieved widespread adoption across various enterprises. Several physical and intangible benefits have been realised due to implementing the 5S Lean Philosophy (Chandrayan et al., 2019) and its continuous execution. Some examples of these areas are efficiency, productivity, and quality service delivery enhancements. It is possible to make such changes by converting the workplace into a cleaner, more pleasant, and more professional environment. This idea, which has its roots in Japan, serves as the basis for the concepts of the Visual Workplace, articulated by Hiroyuki Hirono, and the five keys to a comprehensive quality environment developed by Takashi Osada (Moulding, 2010). These concepts center on socio-dimensions such as training, employee involvement, empowerment, and safety. The five S's derived from individual Japanese words beginning with the letter "S" are presented in Table 2 below.

Table 2. 5S Lean Philosophy

Japanese Word	English Word (Translated)
Seiri	Sort/Organization
Seiton	Set in Order/Straighten/Stabilize/Orderliness/Neatness
Seiso	Shine/Scrub/Cleanliness
Seiketsu	Standardise/Systematize/Adherence
Shitsuke	Sustain/Discipline/Training

Source: Authors' Compilation

6.1.1 Sort/Organization (Seiri)

The first step of the 5S process is Seiri, also called Sort, and it entails clearing the workspace of anything that is not essential for daily operations. At this point, you figure out what is important and what is not. A red document, called a red tag, identifies and labels unnecessary, out-of-place, or redundant items. After that, anything with a red tag is sent to a unique temporary storage facility, often called the Seiri area. To make sure nobody gets confused, this spot is appropriately designated. The goal is to eliminate everything that does not have a place, clean the area, and make things easier to use going forward. In a hotel, this may entail clearing out storage facilities, guest rooms, and workstations to guarantee that only necessary objects are retained. This simplifies activities such as cleaning and maintenance and decreases the possibility of losing or misplacing expensive objects.

6.1.2 Set in Order/Straighten/Stabilize/ Orderliness/Neatness (Seiton)

The second concept of the 5S system is called Seiton, which translates to "Set in Order." This principle highlights the importance of effectively arranging tools and equipment to maximise workflow. By placing the tools in the order in which they are used, the objective is to limit the time wasted searching for items and prevent errors. This will be accomplished by reducing unnecessary movement. Designating where goods belong entails using labels, colour-coding, and visual cues such as floor markings. This guarantees that every item has a specific location, which improves orderliness and makes it possible to retrieve it quickly and easily. As a result, Seiton can streamline processes and increase productivity in the workplace. As an illustration, chefs working in a hotel kitchen should arrange their cooking implements and ingredients consistent with their culinary procedures. This will allow them to expedite service and reduce the required movement.

Figure 3. Set in order/straighten/stabilize/orderliness/neatness

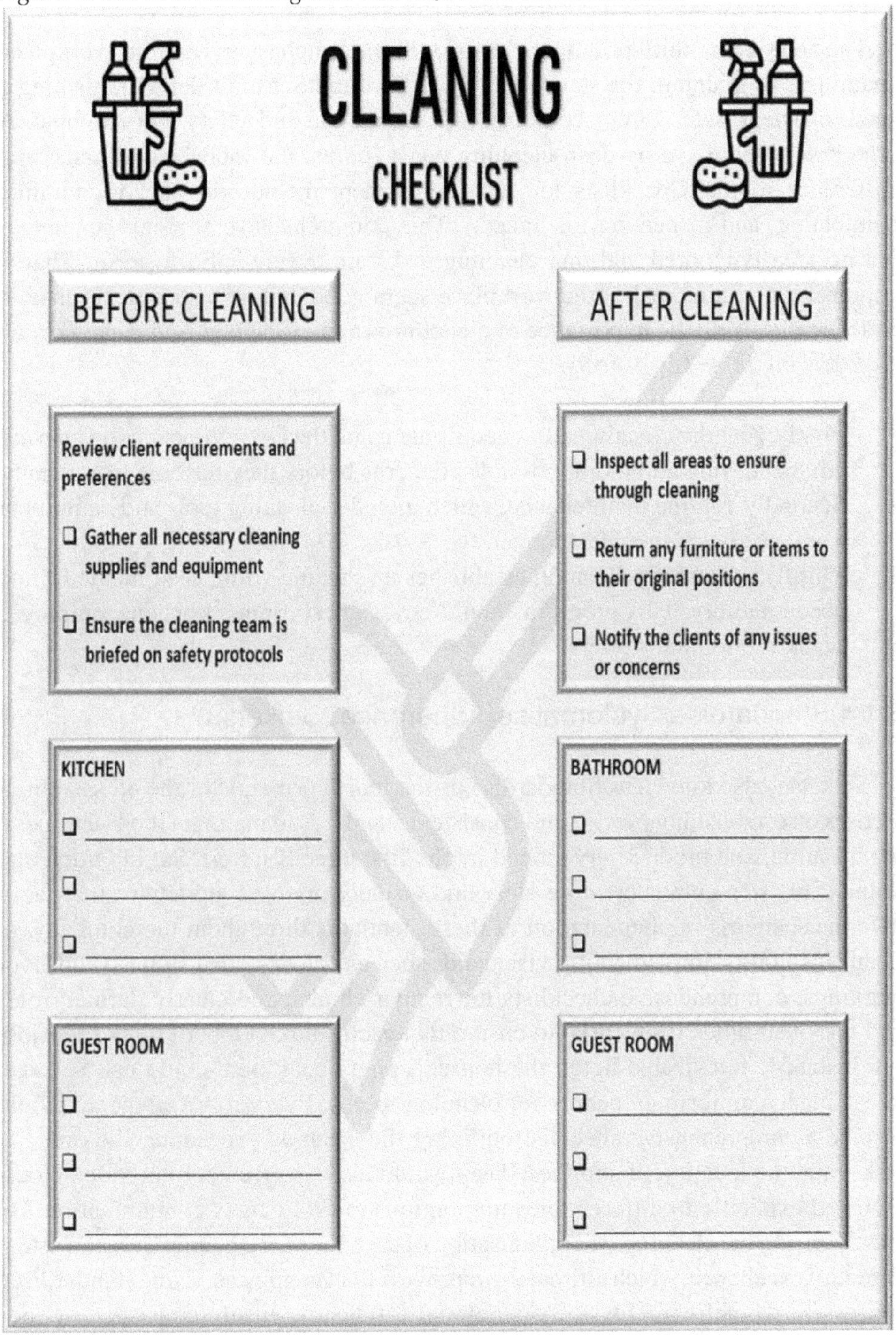

Source: Authors' Compilation

6.1.3 Shine/Scrub/Cleanliness (Seiso)

The 5S system's third principle, Seiso, or Shine, emphasises rigorous workplace cleanliness to maintain the standards set by Sort and Set in Order. Establishing a clean, efficient, safe workplace boosts worker morale and safety. For example, A large hotel uses Seiso by deep-cleaning guest rooms, the lobby, restaurants, and conference rooms. Checklists for each department include dusting, vacuuming, disinfecting, and inspecting for upkeep. This comprehensive strategy guarantees that no area is ignored and that cleaning tasks are evenly split, fostering shared responsibility. Shine makes the workplace seem good and gives people a sense of pride, emphasising the importance of cleanliness and efficiency. *Shine emphasises the following three key aspects.*

- Firstly, Regular cleaning allows equipment and the environment to be inspected, identifying and fixing possible concerns before they become significant.
- Secondly, routine maintenance, which includes cleaning tools and equipment routinely, helps them last longer.
- Thirdly, systematic cleaning establishes a schedule with tasks, methods, and accountability. This program should cover guest rooms, kitchens, employee break rooms, and storage.

6.1.4 Standardise/Systematise/Adherence (Seiketsu)

Seiketsu, also known as Standardise, is the fourth principle of the 5S system. It focuses on establishing precise and consistent standards to maintain the cleanliness, organisation, and productivity gained by the first three S's: Sort, Set in Order, and Shine. This step entails creating clear and visually oriented guidelines that facilitate the seamless implementation of these standards throughout the company. In Lean Hospitality, implementing visual aids such as colour-coded signage and floor markings, comprehensive checklists for regular chores, and clearly defined roles and responsibilities is essential to ensure that each team member knows their job. For instance, at a sizable hotel, the housekeeping department could use Seiketsu to establish a uniform procedure for cleaning rooms. Every room attendant would receive a comprehensive checklist outlining the cleaning procedure, the order of tasks, and the location of supplies. They would also be given colour-coded cloths assigned explicitly to different cleaning regions to avoid cross-contamination. By standardising the cleaning and organisation of each room, we guarantee a consistent degree of excellence, which ultimately improves client happiness. Using standardised processes, a hotel can enhance operational efficiency, ensure consistent service quality, and create a foundation for ongoing improvement. Consistently updating and evaluating these standards, considering staff input and new operational knowledge, guarantees that the hotel adjusts to evolving requirements while upholding exceptional standards. This method maintains robust operating processes and promotes a culture of discipline and ongoing improvement, which is crucial in the competitive hospitality business.

6.1.5 Sustain/Discipline/Training (Shitsuke)

The 5S system's fifth and final principle, Shitsuke, emphasises the significance of sustaining and practising the requirements to embed them into an organisation's culture. Ongoing guidance, surveillance, and punishment guarantee that sorting, placing in order, and shining processes are followed, and that the workplace stays efficient, safe, and organised. Maintaining excellent standards requires regular audits, continual observations, recognition, and awards. Hotel housekeepers' regular habits might illustrate Shitsuke. Staff might renew their training each morning before inspecting each room for cleanliness and preparedness. This regimen promotes discipline and worker dedication to high standards. Monthly reviews when teams discuss best practices and areas for development may help sustain the 5S principles across the enterprise, ensuring that cleanliness, order, and efficiency are part of the hotel's operational culture. This dedication minimises operational interruptions and improves guest pleasure, demonstrating how Shitsuke promotes long-term hotel business success.

Figure 4. 5S - Foundation of continuous improvement

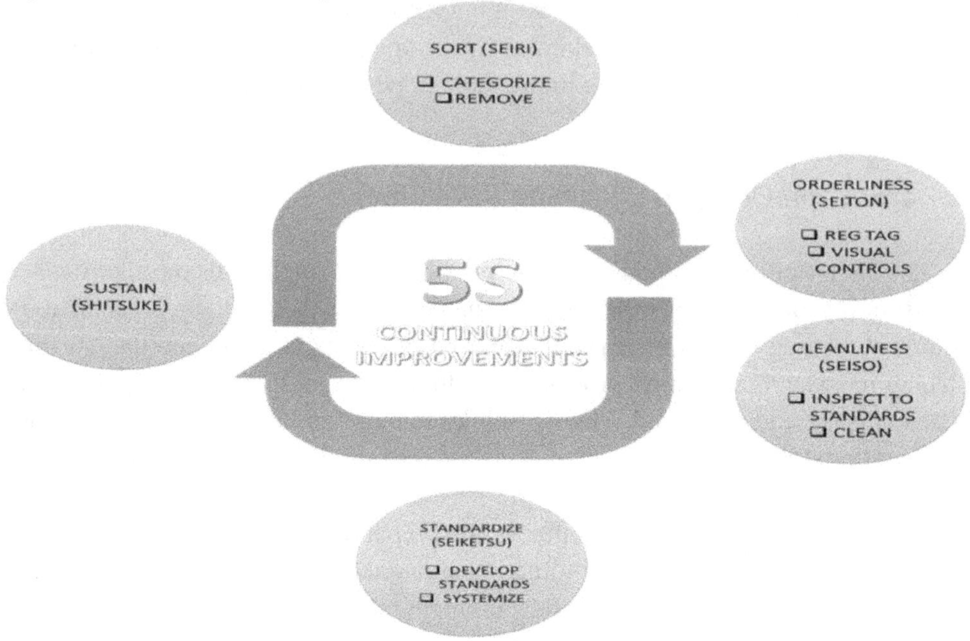

Source: Authors' Compilation

When applied to Lean Hospitality, the 5S Lean Philosophy is a durable waste management strategy promoting ongoing enhancement. Hospitality firms may improve their operations, decrease waste, and increase customer delivery by diligently adopting the concepts of Sort, Set in Order, Shine, Standardize, and Sustain (Martindale, 2010). This systematic approach enhances environmental sustainability and fosters a culture of efficiency and ongoing improvement, guaranteeing that hospitality operations can adjust and flourish in a dynamic market context.

6.2 Value Stream Mapping

It is becoming increasingly common for the services industry, incredibly Lean Hospitality, to use value stream mapping, a vital tool for locating waste and optimising processes. This method can be applied to both product delivery flows and customer delivery flows. Enhancing operational efficiency is the primary focus of the product flow component, which aims to optimise the procedures involved in delivering and finishing a product or service. Regarding the hospitality business, the customer flow component focuses on the processes necessary to fulfil the needs and expectations of end-users, hence creating a seamless experience for guests. The

VSM method impacts the increase in Value-Added time and the reduction in the proportion of non-value-added but necessary activities and waste. (Shou et al., 2020).

In the hospitality industry, Value Stream Mapping (VSM) can be used in different operations, such as preparing rooms for guests, preparing food for guests, and managing the hotel's total inventory by providing material and information flow. Meanwhile, developing a current state map for the hotel's shop floor is a prerequisite to drawing the future state map. As an illustration of VSM in the context of Lean Hospitality, consider the operations at Hotel Purbani Inn, which is completely booked for a conference that will last for three days. The operations manager's goal is to shorten the service cycle times to improve the level of satisfaction experienced by guests and to ensure repeat bookings. An approach that is more streamlined to how VSM could be utilised is as follows:

6.2.1 Current State Map for the Hotel Purani

a. **Preparation of the Room**
   - *Tasks:* Cleaning, upkeep, and arrangement of amenities.
   - *Key Performance Indicators:* Duration spent per room, allocation of personnel, guest satisfaction with room preparedness.
   - *Waste Discovered:* The period between check-out and cleaning and the surplus inventory of amenities.

b. **Catering**
   - *Tasks:* Conducting inventory, creating meal plans, preparing food, and delivering services.
   - *Key Performance Indicators:* Guest Wait times, ingredient utilisation, and meal preparation.
   - *Waste Discovered:* Overcooking of food and service interruptions during busy hours.

c. **Management of Inventory**
   - *Tasks:* Storing food supplies, toiletries, and linens.
   - *Key Performance Indicators:* Frequency of restocking, stock quantities, and waste rates.
   - *Waste Discovered:* Stock expiration from overstocking and underuse of storage space are the results.

6.2.2 Future State Map for the Hotel Porbany

a. **Preparation of the Room**

- *Optimized Tasks:* Efficient cleaning schedules synchronised with up-to-date check-out information.
- *Aim:* By dynamically reallocating resources, reduce the time spent on preparation by 20% and enhance the personnel's effectiveness.

b. **Catering**
- *Optimized Tasks:* Batch cooking will be introduced to optimise and simplify food provision, and designated mealtimes for conference participants will be established.
- *Aim:* Enhance efficiency by decreasing meal preparation and waiting durations by 30% and enhance precision in ingredient consumption.

c. **Management of Inventory**
- *Optimized Tasks:* Implement a just-in-time inventory system for non-perishable commodities and essential everyday goods.
- *Aim:* Optimize inventory levels to achieve a 25% reduction in waste and storage expenses.

d. **Execution and Surveillance**
- Recommended actions: Train personnel on updated procedures, integrate scheduling software to streamline room preparation and modify inventory ordering methods.
- Conduct periodic evaluations of service cycle times and guest satisfaction ratings to verify the effectiveness of improvements and make any necessary adjustments.

This VSM example for Hotel Purbani Inn, given at the Monico LLC conference, shows how targeted improvements in the preparation of the room, food service, and inventory management may streamline operations and improve visitor experiences. This Hotel Purbani Inn Value Stream Mapping (VSM) analysis reveals how lean principles might transform the hospitality sector. The hotel discovered and fixed inefficiencies in food service, inventory management, and room preparation, improving guest happiness and service. Methodical usage of VSM streamlines operations and establishes excellence and customer loyalty, demonstrating how purposeful lean interventions may increase operational efficiency and the visitor experience. This case study is an excellent resource for hospitality companies wishing to apply Lean Hospitality to enhance operations and compete.

6.2.3 Standardized Work

Standardised work is one of the core components of the Lean Concept to guarantee the objectives of service optimisation, quality, and worker safety in the hospitality sector. A collection of work procedures (standard routines) intended to establish the best-repeated sequence of operations, given to a single employee at the operation floor, balanced to takt time, were satisfied by standardised work (Bragança & Costa, 2015), which met the three dimensions of Lean Hospitality. When Standard Work is appropriately implemented, the process becomes the most efficient and productive means for the business to guarantee quality for its guests staying at the hotel and the safest and most accessible for the authorised hospitality employee. Using standard work to create uniformity is a primary defence against irregularities such as unclean rooms, improper floor cleaning, and food workers adhering to health department-mandated protocols.

"Where there is no standard, there can be no improvement."

-Ohno Taiichi, the father of the Toyota production system

Throughout history, specialisation has frequently resulted in uniformity in physical labour and specific types of intellectual activity. The need for uniform work has become evident in the hospitality business. Standardisation is necessary for physical labourers and all individuals in a specific department. They should be included in a system of standardised work processes. This guarantees consistency and effectiveness at all levels of operation. Standardised work comprises three essential components described below to ensure a continuous flow of uniformity and efficiency.

a. **Takt Time**

Takt time establishes the frequency of operator processes and gauges the speed of client requirements. Following Takt time promotes a steady and uninterrupted flow of goods by helping to create a steady operational pace. It is usually calculated, not observed. The following method should be used to calculate Takt time.

Takt Time = Total Available Production or Service Processing Time/Average Customer Demand or Volumes

In the tourism industry, for instance, wait time is dictated by flight schedules and the demand for seats for each shuttle journey for an airport shuttle service. Rescheduling the shuttle service so that it runs often enough to meet customers' needs would result in reduced wait times and better coordination with arriving and leaving flights.

b. **Sequential Workflow**

Sequential workflow is a sequence of tasks an operator performs to finish a process securely. This procedure values consistency, making each step repetitive and reliable throughout units of measurement. Sequential workflow emphasises uniformity, yet stabilising work phases improves proficiency and reveals improvement opportunities. The work sequence, which orders manual tasks, may differ from the process sequence. Emphasising sequence importance helps identify waste and stabilise the process, optimising routine operations. One example from the tourism industry is the expected order of duties for housekeeping in a hotel. For example, restrooms may be cleaned after stripping beds, floors vacuumed, and fresh linens placed. Following this order guarantees that all rooms are cleaned to a high standard, reducing the risk of cross-contamination and making guests happy.

c. **Work/Service in Process**

Work/Service in Process (W/SIP) refers to the suitable minimum merchandise demanded to sustain manufacturing or service without interruptions, guaranteeing a smooth and uninterrupted flow of activities for service or production itself. In tourism, a tour operator's W/SIP typically consists of a fixed quantity of instructional pamphlets, audio guides, and first aid kits made accessible for every tour group. This preparation guarantees that guides are well-prepared to offer tourists a smooth, educational experience without interruptions or deficiencies. These tools, part of the Lean Production Paradigm, were initially utilised almost solely by Ohno's team. It is sophisticated and requires the creation of three necessary diagrams: 1) The Percentage Load Chart, 2) The Standard Work Combination Sheet, and 3) The Standard Work Sheet. Continuous improvement is at the heart of the Kaizen process, based on documenting the work and establishing a uniform approach to operations. It takes a lot of work to implement and keep up with standard work.

6.2.4 Error/Mistake- Proofing (Poka Yoke)

Poka-Yoke, a Japanese word for "mistake-proofing," is a systematic lean methodology that detects root causes and eliminates and corrects errors before they reach the customer by inspecting 100% of the pieces independently on the operator's attention span. (Grout & Toussaint, 2010; Middleton, 2001)The notion of poka-yoke has existed in many iterations for a significant duration. Still, Shigeo Shingo, a Japanese industrial engineer, enhanced it into a tool with the specific objective of attaining a state of zero flaws. While Zero Quality Control may not consistently achieve the goal of zero faults, it can nevertheless be aligned with the critical concepts of eliminating waste and anomalies in service operations, therefore keeping the ethical norms it aims to promote.

Implementing error-proofing strategies requires inspection. Three primary inspection methods—Source Inspection, Judgement Inspection, and Informative Inspection—help eliminate faults and achieve Zero Quality Control. Source Inspection first finds and fixes the sources of service issues rather than just detecting them. Second, Judgement Inspection evaluates error-prone processes before the service process begins to prevent problems. Finally, Informative Inspection monitors service flow to report issues so providers can fix and reduce mistakes. Based on inspection methods, the error-proofing methodology can be divided into two primary methods, each with its distinct sub-methods.

Figure 5. Hotel check-in procedure

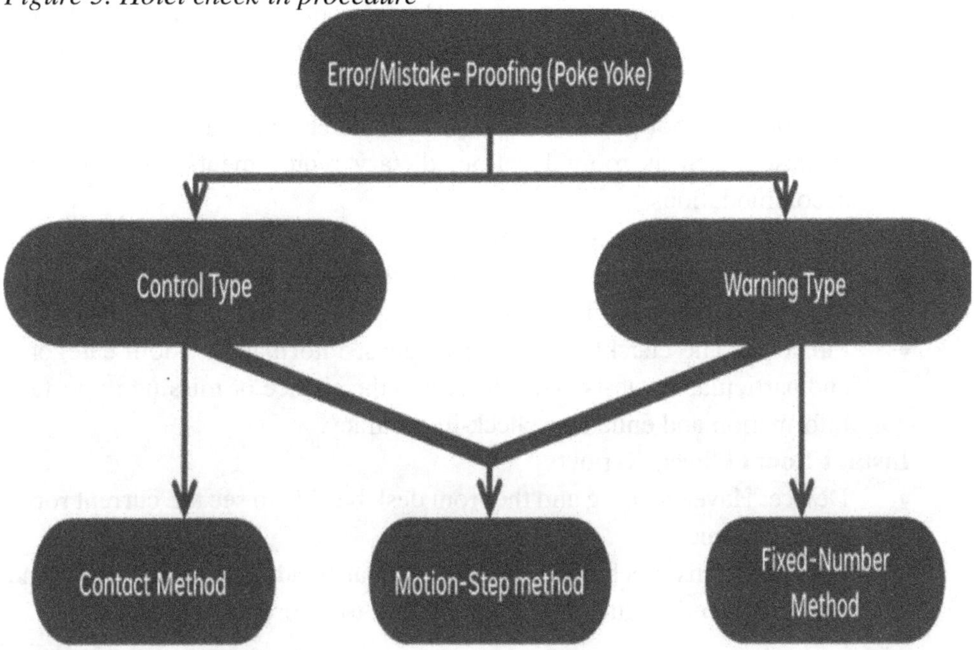

Source: Authors' Compilation

The first approach, called Control Poka-Yoke, is to halt the process to fix problems as they occur temporarily. This can be done by locking clamps or turning off a machine. The second approach, Warning Poka-Yoke, uses signals such as buzzers or lights to notify employees of possible faults; nevertheless, if the warnings are ignored, the process may proceed. There are also specialised sub-methods for finding flaws, such as the Contact method, which looks for evidence of physical contact between a product's features and devices to spot abnormalities; the Fixed-number method, which checks that a set number of movements or actions have been executed, and the Motion-step method, which checks that a process's necessary steps have been executed correctly.

Implementing Error/Mistake Proofing (Poka-Yoke)

Automatic Room Allocation

- Device: Utilise a computerised system that seamlessly combines reservations with real-time lodging availability.
- Function: This mechanism guarantees the automatic allocation of guests to the appropriate room type based on their booking, hence avoiding the likelihood of manual input mistakes.

Keeping Tabs on Digital Preferences

- Device: Incorporate a computerised system that notifies front desk employees of guest preferences upon check-in and records them during the scheduling process.
- Function: Without staff memory or manual checks, this meets all guest requests, such as room location, dietary requirements, or accessible accommodations.

Verification Procedures for Errors

- Device: Establish a sequential digital checklist that the front desk agent must complete throughout check-in.
- Function: The checklist will verify guest information, room category, and particular requests. This decreases the chance of missing important information and enhances check-in accuracy.

Instant Room Check Reports

- Device: Have cleaning and the front desk be able to see the current room status on time.
- Function: This mechanism guarantees that 'ready' rooms are clean and ready, preventing guests from being sent to ill-prepared guestrooms.

By implementing these Poka-Yoke strategies, the hotel may optimise its check-in process and significantly improve visitor happiness by reducing disruptions and errors. These enhancements enhance operations efficiency, optimise resource utilisation, and improve service quality, reflecting the core concepts of Lean Hospitality.

6.2.5 Just in Time

Just-In-Time (JIT) production control reduces raw materials and work-in-process/work-in-service inventories, controls errors, stabilises output, continuously simplifies the production process, and creates a flexible, multi-skilled workforce. (Fullerton & McWatters, 2001). JIT emphasises simplicity in production and service efficiency and excellence through enhancements to productivity and minimise waste (Schonberger & Gilbert, 1983). This attitude of development goes beyond production practices and helps U.S. firms regain and retain global competitiveness. The use of just-in-time (JIT) can considerably improve the delivery of services, shorten costs, and enhance client loyalty in the hospitality business. This is accomplished by minimising the waste associated with excessive output, delays, and additional inventory.

Using Just-in-Time (JIT) for Hotel Housekeeping

Scenario: The housekeeping operations of a large hotel are being streamlined to improve both efficiency and the level of satisfaction its guests feel. JIT principles are going to be implemented by the hotel to manage its linen inventory and the distribution of its housekeeping workers.

Implementation involves:

1. *Need-Based Linen Supplies*

The hotel uses a JIT system to distribute linens to floor storage containers based on capacity and guest requests. This method saves storage space and capital from extensive, underutilised linen stocks.

2. *Scheduling Housekeeping Services in Real-Time*

The hotel management system's real-time guest departure and arrival times data determine cleaning staff timetables. This optimises labor consumption and prepares rooms for future guests by cleaning rooms precisely in time.

3. *Adaptive Staff Re-Positioning*

Housekeeping supervisors use a smartphone application to track hotel service demand during actual time and reassign staff to high-demand areas. This method reduces staff downtime and allocates services as required.

4. *Vendor Coordinating for Resources*

Suppliers work with the hotel to provide toiletries and minibar products on the schedule based on consumption trends and expected demand. JIT prevents oversupply and keeps facilities fresh.

By demonstrating how JIT principles may be successfully applied in the housekeeping department of a hotel, this example exhibits how the operation can become more efficient, responsive, and cost-effective. These Just-In-Time (JIT) tactics can also be expanded to other areas of hospitality management, such as establishment maintenance and dining services, to improve service delivery across the industry further.

6.2.6 Continuous Improvement (Kaizen)

Kaizen is a Japanese term that encompasses the notion of ongoing enhancement in all facets of life, encompassing personal, domestic, social, and professional domains. The term “Kaizen” is derived from the combination of two Japanese words: “Kai” (meaning change) and “Zen” (meaning good) (Brunet & New, 2003). Together, these terms represent the concept of improvement or positive change. Historically, as described by Masaaki Imai, Kaizen encompasses the active participation of all individuals within an organisation, including managers and workers, in a continuous endeavour to improve procedures and minimise inefficiencies without necessarily requiring significant financial investments. The approach emphasises gradual and incremental modifications rather than large-scale transformations, fostering employee engagement and creativity (Suárez-Barraza et al., 2011)This strategy seeks to enhance operational efficiencies and quality and promotes a work climate prioritizes continual learning and problem-solving.

Great strides can be made in the hotel sector by applying kaizen principles and prioritising customer happiness and operational efficiency. This methodology promotes teamwork amongst all employees, from upper management to frontline workers, to pinpoint inefficiencies and discuss potential remedies, promoting a proactive culture of continual improvement.

Lean Hospitality: HR Service Process Kaizen Case Study (Suárez-Barraza & Ramis-Pujol, 2010)

To improve the efficiency of its Human Resources (HR) processes and enhance service delivery, MexiServe, a public service organisation in Mexico, embarked on a journey to streamline its HR procedures using Lean Kaizen methodologies. Lean Kaizen, a blend of principles from Lean focusing on waste reduction and Kaizen's philosophy of continuous improvement, was chosen to address bureaucratic inefficiencies and improve overall operational effectiveness within the HR department. One Mexican public service organisation implemented Lean Kaizen in its HR procedures. The aim was to optimise HR screening and hiring cycle times to address the dynamic demands of public sector roles and the need for efficient staffing solutions.

Background

MexiServe, a government entity responsible for providing various public services, faced significant challenges in its HR department. The issues included:

- Prolonged recruitment and onboarding processes.
- Inefficient handling of employee records.

- High levels of employee dissatisfaction due to administrative delays.
- Redundant paperwork and lengthy approval workflows.

These challenges hindered the department's productivity and impacted on the employees' morale, leading to increased turnover and decreased public trust in the organisation.

Objectives

The primary objectives of implementing Lean Kaizen in MexiServe's HR department were:

- Reduce process cycle times for recruitment, onboarding, and employee record management.
- Eliminate wasteful activities and redundant paperwork.
- Enhance employee satisfaction by streamlining HR services.
- Foster a culture of continuous improvement within the department.

Implementation Process

The Lean Kaizen implementation was carried out in several phases:

Preparation and Training

- The HR team underwent comprehensive training on Lean Kaizen principles.
- A Kaizen team, comprising HR staff and Lean experts, was formed to lead the initiative.

Mapping Current Processes

- The Kaizen team conducted workshops to map out existing HR processes.
- Value Stream Mapping (VSM) was used to visualise the entire flow of HR activities and identify non-value-added steps.

Identifying Areas of Improvement

- Through VSM and employee feedback, key pain points and bottlenecks were identified.
- Common issues included excessive approval layers, data entry duplication, and document processing delays.

Implementing Improvements

- Standardizing Forms and Processes: Standardized forms were introduced to

reduce variability and errors.

- Streamlining Approval Workflows: Approval steps were consolidated, and electronic signatures were implemented to expedite processes.
- Digitizing Records: A shift from paper-based to digital record-keeping was initiated, reducing physical storage needs and retrieval times.
- Cross-training employees: Staff members were cross-trained to handle multiple HR tasks, improving flexibility and resource utilisation.

Monitoring and Continuous Improvement

- Performance metrics, such as process cycle time, error rates, and employee satisfaction scores, were established to monitor improvements.
- Regular Kaizen events and feedback sessions were held to sustain momentum and address new issues promptly.

Results

The implementation of Lean Kaizen in MexiServe's HR department led to significant improvements:

- **Recruitment Cycle Time Reduction:** The time taken to recruit new employees was reduced by 40%, from an average of 60 days to 36 days.
- **Onboarding Efficiency:** Onboarding processes saw a 50% reduction in completion time, allowing new hires to start contributing sooner.
- **Employee Record Management:** Digitization reduced the time required to update and retrieve employee records by 60%.
- **Increased Employee Satisfaction:** Employee satisfaction surveys indicated a 30% increase in positive responses regarding HR services.
- **Waste Reduction:** Paper usage was cut by 70%, and physical storage requirements were reduced by 50%.

Summary

Implementing Lean Kaizen methodologies within MexiServe's HR department showcased the potential for continuous improvement in enhancing public service operations. The department improved operational efficiency, employee satisfaction, and service delivery by eliminating waste and optimising processes. The success of this initiative has set an example for other departments within MexiServe to adopt similar Lean Kaizen strategies and foster a culture of continuous improvement across the organisation.

On a technological, social, and performance-oriented level, applying Lean principles in Hospitality combines technical expertise with social and performance goals. Value Stream Mapping and the 5S Lean Philosophy have played a crucial role in streamlining processes and promoting continual improvement and efficiency. This approach creates a safer, more organised, and more engaging workplace that enhances operations and employee satisfaction. The focus on technical and socio-performance factors in Lean Hospitality meets and exceeds the unique needs of the hospitality industry. Various case studies and techniques demonstrate how Lean approaches can revolutionise hospitality service delivery, employee empowerment, and customer satisfaction.

This chapter's section emphasises that the Lean Tourism Principles for Sustainable Waste Management have significantly transformed the hospitality industry by focusing on waste reduction, operational efficiency enhancement, and customer satisfaction improvement. By incorporating Lean approaches, the sector has successfully adopted sustainable practices that promote environmental stewardship and service quality. These practices have been fully implemented across various sectors, including the Front Office, Housekeeping, Food and Beverage, Service, and Technology. The fundamental concepts of Lean, such as minimising waste and reducing non-value-added operations, are crucial for encouraging sustainability and defining the hotel industry's role as a responsible contributor to a healthy planet. The industry's dedication to economic success and ecological responsibility is demonstrated by all employees actively participating in continuous improvement efforts. The hospitality industry is exceptionally well-positioned to serve as a model for sustainable operations, significantly contributing to global environmental objectives and enhancing the overall visitor experience. This is because the hospitality industry is best suited through the strict application of Lean principles.

7. PROMOTING LEAN HOSPITALITY FOR SUSTAINABLE WASTE MANAGEMENT

Lean approaches are widely accepted as a paradigm for industrial management and an efficient waste management strategy in the service sectors. Lean, when applied in the hotel industry as a garbage management system, can improve working environments while contributing to environmental initiatives by reducing the amount of garbage that is produced that is not essential (Al-Aomar & Hussain, 2018). Consequently, the costs of detecting, collecting, handling, monitoring, and disposing of trash are decreased. The hospitality business stands out among the many other kinds of organisations, such as those in the healthcare and finance sectors (Karim et al., 2023; Karim et al., 2024). Notably, the Grand Palace Hotel & Spa is the first hotel in India to implement Lean Management effectively, using Japanese work culture techniques in its operations. This achievement is particularly noteworthy. Platinum Award from ABK AOTS - Dosakai was bestowed upon the hotel in 2017 as a token of appreciation for its many accomplishments. Since this is the case, Lean Hospitality offers tremendous prospects for long-term sustainability, mainly through the efficient waste management procedures that it implements.

7.1 Reduced Food Waste

An amalgamation of wastes designed explicitly for the service industry, such as waiting times for clients, replication, pointless movement, hazy communication, inaccurate inventory, lost (not pulled in or preserved) consumers, faults in the delivery of services (errors), and flaws in the standard of service. In the hospitality business, poor inventory management, incorrect storage of fresh foods and supplies, and lack of kitchen and serving staff training can lead to severe food waste. Restaurants, pubs, bistros, fast food chains, and catering businesses cause food waste. According to the Natural Resources Defence Council (NRDC), US restaurants created 22–33 billion pounds of food waste in 2015, with schools, hotels, and hospitals contributing 7–11 billion pounds (Gunders, 2017).

Implementing lean concepts can significantly decrease food waste by enhancing stock handling, streamlining prepared meals, and enforcing portion control. Strategies such as just-in-time inventory and analysing client consumption trends guarantee that food preparation is tightly synchronised with actual demand, reducing surplus and spoiling. Within the framework of Lean Hospitality, which focuses on minimising food waste, five impactful measures may be adopted throughout the hospitality sector.

- **First, Measurements:** To determine the amount and sources of waste, create a "food wastage registry." This registry will help managers pinpoint the primary sources of waste production, rank them, and monitor continuous improvements.
- **Second, Engage Staff:** Define waste reduction duties for kitchen and customer service employees and provide management direction and assistance.
- **Third, Minimize Surplus Production:** Batch cooking, buffet sets, and substantial casserole trays often overproduce. Food waste can be reduced by encouraging cook-to-order.
- **Fourth, Reassess Inventories and Procurement:** Align inventory management and purchasing with demand to reduce over-ordering and food deterioration.
- **Fifth, Recycle Leftover Food:** Recognize that client demand prediction is difficult and plan to employ surplus ingredients. Food can be used for other menu items, discounted, or donated to charities.

7.2 Optimization of the Use of Resources

When individual resource efficiency is maximised without considering its integration into the system, the efficiency paradox can result in system-wide inefficiencies. Resource efficiency requires precise coordination to improve organisational performance by minimising input and maximising output for individual aspects. It has been proved that the adaptation of lean is done in such a silo manner within a specific service function that it would be difficult to achieve when lean initiatives are focused on enhancing the flow of subjects and their experience across all the connected processes in the service delivery chain, this will lead to desired and sustained operational outcomes. Although there are obstacles, it is possible to achieve resource efficiency through Lean Adoption (Tay et al., 2017). Applying Lean principles to improve flow and experience throughout the service delivery chain can produce improved and long-lasting operational results. Lean approaches aim to maximise the efficiency and effectiveness of all resources, encompassing energy, water, and materials (Abreu et al., 2017). This efficiency is attained by implementing streamlined procedures, including improved scheduling and machinery servicing, reducing energy consumption and water usage. This reduces expenses and mitigates hospitality services' ecological footprint.

7.3 Sustainable Purchasing Practices

Lean Hospitality adopts a revolutionary approach to sustainable operations. This approach involves streamlining processes, reducing waste, and improving efficiency across service delivery. This approach identifies and refines value streams within hospitality settings to decrease activities that do not provide value. This ultimately results in practices that are more environmentally friendly and cost-effective. (Wilson & Roy, 2009). Lean Hospitality is tightly aligned with sustainability goals so that it may provide both immediate benefits and long-term resilience in the hospitality business. This is accomplished by continuously developing and adjusting these tactics. Implementing lean principles acts as a catalyst to encourage the purchase of environmentally friendly products, such as biodegradable consumables and energy-efficient appliances (Karmakar et al., 2023; Waaje et al., 2018). Selecting suppliers that adhere to sustainable practices is another step that contributes to establishing a green supply chain (Hossain, 2023; Karmakar et al., 2022). Value Strem Mapping is one of the best tools for sustainable purchasing practices in Lean Hospitality.

7.4 Separation of Waste and Reuse and Recycling

The Lean Hospitality methodology encourages implementing stringent recycling and sorting procedures for essential waste disposal components. Hotels and restaurants can dramatically reduce the amount of garbage sent to landfills if they develop clear guidelines for separating recyclable waste. Lean systems minimise waste disposal costs and promote sustainability by segregating and recycling garbage. (Franchetti & Barnala, 2013). Waste is precisely sorted for handling and recycling, reducing administrative expenses and harmful environmental effects. Waste can be recycled into valuable resources, supporting a circular economy within and outside the organisation (Karim et al., 2024). Lean practices boost operational efficiency by integrating waste-reducing techniques into daily tasks. Technological advances, especially in Industry 4.0, can increase manufacturing processes, reduce waste, and provide high-quality goods and services. Integrating real-time data monitoring with machinery can improve operations and choice-making. This synergy streamlines operations and reduces resource use and waste, benefiting the planet and economy.

7.5 Highly Optimised Operative Procedures

Lean practices boost operational efficiency by integrating waste-reducing techniques into daily tasks. Technological advances, especially in Industry 4.0, can also help. Lean approaches can increase manufacturing processes, reduce waste, and provide high-quality goods and services. (Tortorella et al., 2019). Integrating real-time data monitoring with machinery can improve operations and choice-making. This synergy streamlines operations and reduces resource use and waste, benefiting the planet and economy. As part of their commitment to continuous development, lean techniques routinely assess and tweak processes to minimise waste. This could necessitate the implementation of new, more efficient technologies or rethinking existing operations to reduce resource use.

7.6 Enhancing Employee Flexibility and Conformity to Standards

Lean and sustainable training gives employees a stake in waste reduction, promoting ongoing development. This empowerment is essential for trash management, environmental awareness, and regulatory compliance. Since modern consumers value sustainability, efficient waste management helps firms meet legal requirements and improves their public image (Subramanian & Suresh, 2024). This comprehensive Lean Hospitality method cuts costs, improves client happiness, and makes businesses sustainable. Training in sustainable practices and lean concepts is essential for employees to actively participate in waste reduction efforts—a critical component of waste management initiatives. Knowing how their actions affect environmental goals helps employees comply with ecological standards and improves the organisation's reputation. Demonstrating efficient waste management techniques may significantly enhance a brand's reputation, help it draw more eco-conscious customers, and assist it avoid potential regulatory penalties as sustainability becomes a more critical component in consumer choice.

Lean Hospitality promotes sustainable waste management using lean methods to improve yield and ethical behavior. Lean Hospitality practices, including waste segregation, recycling, and resource optimisation, reduce waste and improve service delivery, cost reduction, and environmental safety. Lean Hospitality boosts firms' ecological credibility and competitiveness by training personnel and using sustainable materials. This strategy encourages ongoing improvement, ensuring hospitality industry sustainability and resilience.

8. CHALLENGES IN THE IMPLEMENTATION OF LEAN METHODS IN THE HOSPITALITY INDUSTRY

It is possible that the successful adoption of lean techniques and their capacity to be maintained over time could be hampered by many issues associated with adopting lean management principles in the hotel industry. As a result of research that was carried out on lean management in a wide range of different sectors, the following five fundamental difficulties that are specific to the hospitality industry have been identified by several scholars:

8.1 Cultural Restrictions and Employee Reluctance

An employee's aversion to change is expected in the hospitality industry, mainly when the change requires them to learn new procedures or alter their jobs. Traditional processes are firmly embedded inside the organisation, and there needs to be a better awareness of lean principles (Govender & Mara, 2018). This further complicates the inherent resistance that is already there within the organisation.

8.2 The Tricky Structure of Services Delivery

Instead of dealing with tangible outputs and processes that are frequently repeated, the hospitality business works with intangible and extremely varied interactions with clients. Manufacturing, on the other hand, produces tangible products (Rauch et al.,

2020). The complexity and variety of service delivery make it difficult to detect and reduce waste while maintaining service quality. This challenge must be overcome.

8.3 Insufficient Competency and Understanding of Lean Methods

The hotel industry needs to gain more knowledge of lean management. Many firms have a significant shortage of workers with sufficient expertise and experience in lean processes (Fercoq et al., 2016). This makes it challenging to lead and maintain lean efforts without external help.

8.4 A Wide Range of Services to Meet the Needs of Clients

Customer demands, which can vary significantly from one guest to the next, are a significant factor in the hospitality industry, characterised by a wide range of services. Although lean leadership frequently relies on standardisation to decrease waste and improve efficiency, the diversity of the processes makes it difficult to standardise them.

8.5 Prioritizing the Short-Term against the Long-Term Gains

The hotel industry's tendency to prioritise immediate monetary objectives and instant client happiness may cause it to overlook the long-term benefits of implementing lean principles. Companies may be encouraged to commit to lean management's start-up and continual improvement segments because they concentrate on the short term. However, these phases are essential for achieving large profits in the long run. These problems highlight the challenges of implementing lean management within the hospitality industry. They also highlight the importance of developing strategies considering the sector's distinctive characteristics and focusing on clients. These challenges require a strategic approach to change management, thorough training, and dedication to ambitious goals. Only then will it be possible to fully realise the benefits of lean principles in improving operational effectiveness and the level of contentment experienced by guests.

9. CASE STUDY: RIYADH RITZ-CARLTON LEAN MANAGEMENT

The Ritz-Carlton Hotel in Riyadh, Saudi Arabia, symbolises the successful utilisation of manufacturing efficiency concepts in the travel and tourism sector. The hotel's dedication to furnishing outstanding hospitality is demonstrated by its triumphant reception of two Malcolm Baldrige National Quality Awards. It reflects adopting lean principles in the hotel's daily operations, particularly cleaning and quality assurance.

9.1 Essential Lean Concepts Used

9.1.1 Human dignity

The Ritz-Carlton promotes the slogan, "We are Ladies and Gentlemen Serving Ladies and Gentlemen," which exemplifies a fundamental tenet of appreciating and recognising visitors and employees. This is achieved by consistently communicating stakeholder goals, engaging people in transformation planning, and providing frequent retraining.

9.1.2 Continuous Improvement

Professionals are incentivised to recognise and notify flaws, frequently called "Mr. BIV" (Mistakes, Rework, Breakdowns, Inefficiencies, Variation). These concerns are being addressed using a problem-solving procedure consisting of six steps, guaranteeing enhancements will be maintained.

9.1.3 Standardized Processes

Housekeeping procedures are thoroughly standardised for consistency and quality. From cleaning the area to restocking supplies, everything is done methodically. As one illustration, the A.B.C. (Away trash and linens, Bed making, Clean the bathroom) method of cleaning the bedroom guarantees uniformity in performance.

9.1.4 Empowering Employees

Professionals are given the authority to satisfy guests' demands on time and document their favourite foods in a system to manage knowledge for further review. The result is a service that is both personalised and prompt.

9.2 Success in Employing Lean Methodology

Implementing lean concepts has dramatically improved operations, eliminated waste, and improved the Ritz-Carlton Hotel's service level. Standards and ongoing efforts to enhance the routine housekeeping process have made it more efficient and uniform. Client satisfaction and commitment levels may be mainly attributed to the empowered and engaged personnel urged to respond to guests' requirements and provide excellent service quickly. Lean concepts have given The Ritz-Carlton Hotel an operational advantage in tourism by combining them into everyday operations. This case study shows that Lean Hospitality's business approaches may increase customer satisfaction and managerial effectiveness.

10. FUTURE RESEARCH DIRECTION

Lean approaches in hospitality provide several research possibilities. Focusing on efficiency, environmental sustainability, and satisfied consumers, additional research may investigate the unique problems and complexities of adopting lean concepts in this business. Several significant aspects need additional investigation; research should examine how hospitality firms may reconcile the necessity for standardisations in lean methods to provide personalised guest experiences. This includes investigating flexible, lean, and efficient adaptive techniques. Second, evaluate how digital technology and business 4.0 might improve lean implementation in the hotel business. IoT, AI, and big data analytics might be studied to enhance operations, waste, and execution of services. Future studies should align lean concepts with sustainable practices in hospitality. This involves assessing lean's environmental effect and finding hospitality operations' greenest practices. Efficient solutions for overcoming cultural resistance and improving staff engagement with lean concepts are essential for employee training and cultural shift. Training, leadership, and change management options that ease lean implementation might be studied. Research on lean approaches in various hospitality settings, including luxury, budget, resort, and boutique hotels, might reveal how to adapt to individual contexts and operational needs. Understanding the impact of lean deployment on customer perception and satisfaction is crucial. Lean techniques should be studied for their direct and indirect effects on visitor satisfaction and loyalty. Longitudinal research on the economic impact of lean adoption in hospitality can highlight cost reductions, profitability, and ROI. Consider short- and long-term financial gains. Research should explore the legislative and governing consequences of lean adoption in hospitality. This includes how government rules and industry standards affect lean adoption.

11. CONCLUSION

Based on Lean management principles, "Lean Tourism" provides a solid foundation for the hotel sector to improve operational efficiency, reduce waste, and balance sustainability and profitability. The study's main conclusions stress that Lean

Tourism improves service quality while cutting costs by eliminating unnecessary steps and using available resources better. Improved client experiences, staff buy-in for constant growth, cutting-edge tech like the Internet of Things (IoT), and data analytics to monitor resources are all benefits. On the other hand, there are many obstacles to overcome, such as the high cost of entry, the necessity of changing company culture, the importance of constantly striving to improve, and the need for extensive training and development programs. Amidst these obstacles, Lean Tourism promises to accomplish economic and environmental objectives in the hospitality industry. Its advantages include less waste, improved operational efficiency, and increased customer pleasure.

REFERENCES

Abdelmoety, Z. H., Alamoudi, H., Alharthi, M., Sharkasi, N., & Agag, G. (2022). Ensuring a Sustainable Hospitality and Tourism Industry in the COVID-19 Era: Using an Open Market Valuation Technique. *Sustainability 2022, Vol. 14, Page 13190, 14*(20), 13190. https://doi.org/DOI: 10.3390/su142013190

Abreu, M. F., Alves, A. C., & Moreira, F. (2017). Lean-Green models for eco-efficient and sustainable production. *Energy*, 137, 846–853. DOI: 10.1016/j.energy.2017.04.016

Al-Aomar, R., & Hussain, M. (2018). An assessment of adopting lean techniques in the construct of hotel supply chain. *Tourism Management*, 69, 553–565. DOI: 10.1016/j.tourman.2018.06.030

Bragança, S., & Costa, E. (2015). An application of the Lean Production tool Standard Work. [Sciences and Engineering]. *Jurnal Teknologi*, 76(1), 47–53. DOI: 10.11113/jt.v76.3659

Brunet, A. P., & New, S. (2003). Kaizen in Japan: An empirical study. *International Journal of Operations & Production Management*, 23(11–12), 1426–1446. DOI: 10.1108/01443570310506704

Bubber, D., Jain, R. K., Babber, G., & Shashi, . (2022). Transforming product development and production to be lean for improving business performance. *Benchmarking*, 30(9), 3021–3039. DOI: 10.1108/BIJ-01-2022-0004

Camilleri, M. A. (2021). Sustainable Production and Consumption of Food. Mise-en-Place Circular Economy Policies and Waste Management Practices in Tourism Cities. *Sustainability (Basel)*, 13(17), 9986. DOI: 10.3390/su13179986

Carvalho, A. B., Carvalho, M., Mota, M., Fonseca, S., & Martins, S. (2023). Lean Thinking and Tourism Management—An Airbnb Case Study in Douro. *Smart Innovation. Systems and Technologies*, 345, 263–279. DOI: 10.1007/978-981-99-0337-5_23

Colmekcioglu, N., Dineva, D., & Lu, X. (2022). "Building back better": The impact of the COVID-19 pandemic on the resilience of the hospitality and tourism industries. *International Journal of Contemporary Hospitality Management*, 34(11), 4103–4122. DOI: 10.1108/IJCHM-12-2021-1509

Dekier, L. (2012). The Origins and Evolution of Lean Management System. *Journal of International Students*, 5(1), 46–51. DOI: 10.14254/2071-8330.2012/5-1/6

Diaz-Farina, E., Díaz-Hernández, J. J., & Padrón-Fumero, N. (2023). Analysis of hospitality waste generation: Impacts of services and mitigation strategies. *Annals of Tourism Research Empirical Insights*, 4(1), 100083. DOI: 10.1016/j.annale.2022.100083

Fercoq, A., Lamouri, S., & Carbone, V. (2016). Lean/Green integration focused on waste reduction techniques. *Journal of Cleaner Production*, 137, 567–578. DOI: 10.1016/j.jclepro.2016.07.107

Ferreira, L. M. D. F., Moreira, A. C., & Silva, P. (2023a). Lean implementation in product development processes: A framework proposal. *Production Planning and Control*, 1–17. DOI: 10.1080/09537287.2023.2217429

Ferreira, L. M. D. F., Moreira, A. C., & Silva, P. (2023b). Lean implementation in product development processes: A framework proposal. *Production Planning and Control*, 1–17. DOI: 10.1080/09537287.2023.2217429

Franchetti, M., & Barnala, P. (2013). Lean six sigma at a material recovery facility: A case study. *International Journal of Lean Six Sigma*, 4(3), 251–264. DOI: 10.1108/IJLSS-05-2013-0026

Fullerton, R. R., & McWatters, C. S. (2001). Production performance benefits from JIT implementation. *Journal of Operations Management*, 19(1), 81–96. DOI: 10.1016/S0272-6963(00)00051-6

Garcia-Garcia, G., Singh, Y., & Jagtap, S. (2022). Optimising Changeover through Lean-Manufacturing Principles: A Case Study in a Food Factory. *Sustainability (Basel)*, 14(14), 8279–8279. DOI: 10.3390/su14148279

Gazi, M. A. I., Islam, H., Islam, M. A., Karim, R., Momo, S. M., & Senathirajah, A. R. B. S. (2024). Unlocking sustainable development in East Asia Pacific and South Asia : An econometric exploration of ESG initiatives. *Sustainable Environment*, 10(1), 1–15. DOI: 10.1080/27658511.2024.2366558

Govender, C., & Mara, C. (2018). Lean Practices for Efficiency in the South African Hospitality Sector. *African Journal of Hospitality, Tourism and Leisure*, 7, 1–14. https://www.researchgate.net/publication/328305365

Goyal, A., Vaish, D. C., Agrawal, R., Choudhary, S., & Nayak, R. (2022). Sustainable Manufacturing through Systematic Reduction in Cycle Time. *Sustainability (Basel)*, 14(24), 16473–16473. DOI: 10.3390/su142416473

Grout, J. R., & Toussaint, J. S. (2010). Mistake-proofing healthcare: Why stopping processes may be a good start. *Business Horizons*, 53(2), 149–156. DOI: 10.1016/j.bushor.2009.10.007

Gunders, D. (2017). *Wasted: How America Is Losing Up to 40 Percent of Its Food from Farm to Fork to Landfill*. Natural Resources Defense Council. https://www.nrdc.org/resources/wasted-how-america-losing-40-percent-its-food-farm-fork-landfill

Hopp, W. J., & Spearman, M. L. (2004). To Pull or Not to Pull: What Is the Question? *Manufacturing & Service Operations Management*, 6(2), 133–148. DOI: 10.1287/msom.1030.0028

Hossain, M. S. (2023). Attributes Influence Customers Towards Ice-Cream Purchase With Brand Preference And Income Effect On purchasing Ice Cream: An Empirical Study on Dhaka City. *Global Journal of Management and Business Research*, 23(B5), 1–11. https://gjmbr.com/index.php/gjmbr/article/view/1968

Karim, R. (2020). Green Banking Practices in Bangladesh: Present Scenario and Future Challenges. *Journal of Business Studies*, (January), 34–44.

Karim, R., Hasan, M. M., Waaje, A., Yesmin, M. S. H., & Roshid, M. M. (2023). Financial Performance of Bangladeshi Listed Commercial Banks: A Generation-based Analysis. International Journal of Finance & Banking Studies (2147-4486), 12(4), 46-56.

Karim, R., Roshid, M., & Waaje, A. (2024). Circular Economy and Supply Chain Sustainability. In *Strategic Innovations for Dynamic Supply Chains* (pp. 1–30). https://doi.org/DOI: 10.4018/979-8-3693-3575-8.ch001

Karim, R., Roshid, M. M., Shamme, F. B., & Hasan, M. M. (2022). Non-performing loans and bank profitability: Evidence from Bangladesh. International Journal of Finance & Banking Studies (2147-4486), 11(4), 95-102.

Karmakar, A., Khan, Z. H., & Roshid, M. (2022, November). M., Shamme, F. B., & Rahman, Md. A. (2022). A Study of Farmers Direct Marketing: Assessing the Factors for the Home to Land Approach. *European Journal of Business and Management*. Advance online publication. DOI: 10.7176/EJBM/14-22-05

Karmakar, A., Hossain Khan, Z., Roshid, M., Hoshain Yesmin, S., & Bashar Shamme, F. (2023). Consumer Learning and Split-Brain Theory: Potential Usage in an Advertisement. *International Journal of Economics. Finance and Management Sciences*, 11(3), 168–176. DOI: 10.11648/j.ijefm.20231103.21

Khan, A., Bibi, S., Li, H., Fubing, X., Jiang, S., & Hussain, S. (2023). Does the tourism and travel industry really matter to economic growth and environmental degradation in the US: A sustainable policy development approach. *Frontiers in Environmental Science*, 11, 1147504. DOI: 10.3389/fenvs.2023.1147504

Legrand, W., Kuokkanen, H., Marucco, F., Hazenberg, S., & Fischer, F. (2023). Survival of the Fittest? A Call for Hospitality to Incorporate Ecology Into Business Practice and Education. Https://Doi.Org/10.1177/19389655231182083, *65*(1), 68–87. https://doi.org/DOI: 10.1177/19389655231182083

Lin, C. C., Chueh, J. W., Chen, H. M., Chiu, Y. H., & Chu, D. (2022). Applying the Toyota production system to decrease the time required to transport patients undergoing surgery from the general ward to the operating room and reviewing the essence of lean thinking. *Frontiers in Medicine*, 9, 1054583. Advance online publication. DOI: 10.3389/fmed.2022.1054583 PMID: 36569138

Martindale, W. (2010). Waste: Uncovering the global food scandal. *International Journal of Sustainable Engineering*, 3(2), 144–145. DOI: 10.1080/19397030903573317

Mastos, T., Gkotzamani, K., & Kafetzopoulos, D. (2022). Development and Validation of a Measurement Instrument for Sustainability in Food Supply Chains. *Sustainability (Basel)*, 14(9), 5203. DOI: 10.3390/su14095203

Middleton, P. (2001). Lean Software Development: Two Case Studies. *Software Quality Journal*, 9(4), 241–252. DOI: 10.1023/A:1013754402981

Obersteiner, G., Gollnow, S., & Eriksson, M. (2021). Carbon footprint reduction potential of waste management strategies in tourism. *Environmental Development*, 39, 100617. DOI: 10.1016/j.envdev.2021.100617 PMID: 34513580

Ramphal, R. R. (2017). Lean Six Sigma Framework for the hospitality industry. *African Journal of Hospitality, Tourism and Leisure*, 6(4). http//:www.ajhtl.com

Rauch, E., Matt, D. T., & Linder, C. (2020). Lean Management in Hospitality: Methods, Applications and Future Directions. *International Journal of Services and Operations Management*, 36(3), 303–326. DOI: 10.1504/IJSOM.2020.108115

Roshid, M., Waaje, A., Meem, T. N., & Sarkar, A. (2024). Logistics 4.0: A Comprehensive Literature Review of Technological Integration, Challenges, and Future Prospects of Implementation of Industry 4.0 Technologies. *International Journal of Technology, Knowledge and Society*, 20(1), 65–85. DOI: 10.18848/1832-3669/CGP/v20i01/65-85

Salem, I. E., Abbas, H., Akram, H. W., & Elbaz, A. M. (2023). Contemporary sustainability themes in tourism and hospitality during and post COVID-19: Critical review and a step strategies forward for post the pandemic. *Sustainable Development (Bradford)*, 31(5), 3946–3964. DOI: 10.1002/sd.2636

Schonberger, R. J., & Gilbert, J. P. (1983). Just-in-Time Purchasing: A Challenge for U.S. Industry. *California Management Review*, 26(1), 54–68. DOI: 10.2307/41165050

Shah, R., & Ward, P. T. (2007). Defining and developing measures of lean production. *Journal of Operations Management*, 25(4), 785–805. DOI: 10.1016/j.jom.2007.01.019

Shou, W., Wang, J., Wu, P., & Wang, X. (2020). Value adding and non-value adding activities in turnaround maintenance process: Classification, validation, and benefits. *Production Planning and Control*, 31(1), 60–77. DOI: 10.1080/09537287.2019.1629038

Spear, S., & Bowen, H. K. (1999). *Decoding the DNA of the Toyota Production System*. Harvard Business Review. https://hbr.org/1999/09/decoding-the-dna-of-the-toyota-production-system

Spear, S. J. (2014). Learning to Lead at Toyota. *Harvard Business Review*. https://www.researchgate.net/publication/8558897

Suárez-Barraza, M. F., & Ramis-Pujol, J. (2010). Implementing Lean-Kaizen in the human resource service process: A case study in a Mexican public service organisation. *Journal of Manufacturing Technology Management*, 21(3), 388–410. DOI: 10.1108/17410381011024359

Suárez-Barraza, M. F., Ramis-Pujol, J., & Kerbache, L. (2011). Thoughts on kaizen and its evolution: Three different perspectives and guiding principles. *International Journal of Lean Six Sigma*, 2(4), 288–308. DOI: 10.1108/20401461111189407

Subramanian, N., & Suresh, M. (2024). Lean HRM practices in manufacturing SMEs: exploring the interplay among the influencing factors. *International Journal of Organizational Analysis, ahead-of-print*(ahead-of-print). https://doi.org/DOI: 10.1108/IJOA-08-2023-3916/FULL/XML

Sucheran, D. R. (2017). Barriers to environmental management in hotels in Kwazulu-Natal, South Africa. *PEOPLE: International Journal of Social Sciences*, 3(3), 1352–1364. DOI: 10.20319/pijss.2018.33.13521364

Sucheran, R. (2021). Preliminary economic impacts of the COVID-19 pandemic on the hotel sector in South Africa. *African Journal of Hospitality, Tourism and Leisure*, 10(1), 115–130. DOI: 10.46222/ajhtl.19770720-90

Tanchangya, T., Mia, M. R., Roshid, M. M., Chakma, J., Islam, F., & Rahman, M. H. (2024). Role of green financing and energy production to mitigate climate change in emerging economy like Bangladesh. *International Journal of Sustainable Energy and Environmental Research*, 13(2), 114–139. DOI: 10.18488/13.v13i2.3935

Tay, H. L., Singh, P. J., Bhakoo, V., & Al-Balushi, S. (2017). Contextual factors: Assessing their influence on flow or resource efficiency orientations in healthcare lean projects. *Operations Management Research: Advancing Practice Through Research*, 10(3–4), 118–136. DOI: 10.1007/s12063-017-0126-3

Tortorella, G. L., Giglio, R., & van Dun, D. H. (2019). Industry 4.0 adoption as a moderator of the impact of lean production practices on operational performance improvement. *International Journal of Operations and Production Management, 39*(6/7/8), 860–886. https://doi.org/DOI: 10.1108/IJOPM-01-2019-0005/FULL/XML

Vanichchinchai, A. (2023). Contextual factors on Toyota Way and Agile Manufacturing: An empirical investigation. *Operations Management Research : Advancing Practice Through Research*, 16(3), 1–12. DOI: 10.1007/s12063-023-00352-5

Waaje, A., Islam, J. M. S., & Akter, T. (2018). Study on Gas Sensor and Its Application as a IOT Device. *EWU Institutional Repository*. http://dspace.ewubd.edu:8080/xmlui/handle/2525/3056

Wilson, M. M. J., & Roy, R. N. (2009). Enabling lean procurement: A consolidation model for small-and medium-sized enterprises. *Journal of Manufacturing Technology Management*, 20(6), 817–833. DOI: 10.1108/17410380910975096

Womack, J. P., Jones, D. T., & Roos, D. (2007). *The Machine That Changed the World: The Story of Lean Production-- Toyota's Secret Weapon in the Global Car Wars That Is Now Revolutionizing World Industry*. 352. https://books.google.com/books/about/The_Machine_That_Changed_the_World.html?id=9NHmNCmDUUoC

Chapter 7
Beyond Bin Basics: Confronting Waste in Tourism and Hospitality

Bhaskar Sailesh
https://orcid.org/0000-0002-5666-3597
PES University, Bangalore

ABSTRACT

The tourism and hospitality sectors face significant challenges in managing waste, which poses environmental, economic, and social risks. This chapter explores waste management challenges within these sectors, examining the current state of waste management, environmental impacts of poor waste management, economic and social implications, innovative solutions and best practices, case studies of successful initiatives, and future outlooks. Key findings highlight the urgency for sustainable waste management practices, emphasising the need for collaboration between businesses, governments, communities, and consumers. The chapter concludes with a call to action for integrating sustainability into operations, investing in infrastructure and technology, and advocating for supportive policies. Future outlooks focus on embracing innovation, increasing awareness, and fostering collaboration to build a more sustainable future for the tourism and hospitality sectors.

1. INTRODUCTION

The tourism and hospitality sectors play a vital role in the global economy, providing employment opportunities, fostering cultural exchange, and driving economic growth. According to the World Travel & Tourism Council (WTTC), the travel and tourism sector contributed 10.4% to global GDP and supported over 319 million jobs in 2019 (WTTC, 2020). With the increasing globalisation and travel accessi-

DOI: 10.4018/979-8-3693-6110-8.ch007

bility, these sectors have experienced unprecedented growth, leading to positive and negative environmental impacts. As tourism and hospitality activities expand, so does the volume of waste generated by these industries. From disposable amenities in hotels to single-use plastics in restaurants and attractions, the environmental footprint of tourism-related waste is substantial (European Environment Agency, 2019). In addition to physical waste, tourism can also contribute to pollution, habitat degradation, and biodiversity loss in fragile ecosystems (Gössling & Scott, 2019). Therefore, effective waste management practices are crucial for mitigating these environmental impacts and promoting sustainable tourism development.

The tourism and hospitality sectors encompass diverse businesses and services to meet travellers' needs and desires. This includes accommodation providers such as hotels, resorts, vacation rentals and food and beverage establishments such as restaurants, cafes, and bars (World Tourism Organization, 2019). Additionally, transportation services such as airlines, cruise lines, and tour operators significantly facilitate travel experiences for tourists worldwide (World Travel & Tourism Council, 2020). These sectors operate within a complex ecosystem of interconnected businesses, suppliers, and destination communities, each contributing to the tourism experience (Oxford Economics, 2019). From luxury resorts in tropical destinations to budget hostels in urban centres, the diversity of tourism offerings caters to a wide range of travellers, from leisure tourists to business travellers and adventure seekers (Dolnicar & Ring, 2014; Page, 2019).

Waste management is paramount in the tourism and hospitality sectors due to the significant environmental, economic, and social impacts of waste generation (Tang, 2020). In addition to reducing the visual pollution and environmental degradation caused by waste, effective waste management practices can lead to cost savings, resource efficiency, and enhanced brand reputation for businesses (Gössling & Higham, 2021). Furthermore, sustainable waste management contributes to the preservation of natural resources, protection of wildlife habitats, and maintenance of ecosystem services, which are essential for the long-term viability of tourism destinations (Scheyvens & Cheer, 2022).

This chapter explores the waste management challenges facing the tourism and hospitality sectors and proposes sustainable solutions. By examining the current state of waste management practices, environmental impacts, economic implications, and regulatory frameworks within these industries, this chapter seeks to raise awareness of the importance of waste management in achieving sustainable tourism development. Furthermore, this chapter aims to identify critical challenges and barriers hindering effective waste management in tourism and hospitality businesses, such as inadequate infrastructure, limited resources, and lack of awareness. By understanding these challenges, stakeholders can develop tailored strategies and initiatives to overcome them and promote a more circular approach to waste management. This chapter will explore innovative solutions, best practices, and case studies worldwide to highlight the potential for collaboration and knowledge sharing among stakeholders to drive positive change. This chapter seeks to inspire collective action towards a more sustainable and resilient future for the tourism and hospitality sectors by engaging policymakers, businesses, local communities, and travellers in the dialogue on sustainable waste management.

2. CURRENT STATE OF WASTE MANAGEMENT IN TOURISM AND HOSPITALITY

Waste management in the tourism and hospitality sectors is a critical issue, influenced by various factors, including the type of establishment, destination characteristics, and local regulations. Statistics and data on waste generation provide insights into the scale and scope of the waste management challenge in tourism and hospitality. According to a report by the World Tourism Organization (UNWTO), the tourism sector generates a significant amount of waste annually, estimated at over 5.2 million tons of solid waste per year globally (UNWTO, 2019). This waste includes various types, such as organic waste from food and beverage operations, packaging materials, plastics, paper, and other non-biodegradable items.

Furthermore, waste generation rates vary depending on the type and size of tourism establishments and the level of tourist activity. For example, large-scale resorts and cruise ships may produce substantial waste due to their high occupancy rates and diverse services. In 2018, it was reported that cruise ships generate an average of 3.5 kg of waste per person per day (Butt, 2007). Similarly, large hotels can produce up to 1 kg of waste per guest per day (Radwan et al., 2012). In contrast, smaller accommodations and eco-lodges may have lower waste generation rates, often averaging around 0.5 kg of waste per guest per day. However, they still face challenges in managing their waste effectively (Mensah, 2006).

2.1 Types of Waste Commonly Generated

Several types of waste are commonly generated in the tourism and hospitality sectors, each presenting unique challenges for waste management. Food waste is one of the most significant contributors to kitchen operations, buffets, and guest consumption. Recent studies have shown that hotels generate approximately 1.4 kg of food waste per guest per day (Papargyropoulou et al., 2021). Packaging waste, including single-use plastics, cardboard, and glass containers, is another prevalent type of waste, particularly in the hospitality industry, where products are often packaged for guest convenience. Gössling et al. (2023) found that single-use plastics contribute significantly to environmental waste, with hotels generating around 8 million tons of plastic waste annually. Moreover, using disposable amenities such as toiletries, towels, and linens in hotels contributes to waste generation, with reports indicating that the hospitality industry produces around 0.7 kg of textile waste per guest per day (Tzeng & Lin, 2023). Electronic waste from outdated equipment and appliances is also a challenge, with hotels frequently disposing of substantial electronic devices (Zhou et al., 2023). Additionally, recreational tourism, events, and excursions can result in waste, including litter, debris, and hazardous materials, which pose environmental and health risks if not managed properly (Baker et al., 2023).

Table 1 outlines various types of waste generated by the tourism and hospitality sector and their descriptions, challenges, and sources. This table highlights the diverse waste streams associated with this sector and underscores the significant environmental and health challenges posed by each type of waste.

Table 1. Types of waste generated by the tourism and hospitality sector

Type of waste	**Description**	**Challenges**	**Sources**
Plastic Waste	Includes bottles, bags, straws, and packaging materials that are commonly discarded.	Long degradation time, harmful to marine life, microplastic pollution.	Jambeck et al. (2015), "Plastic waste inputs from land into the ocean", *Science*, 347(6223), 768-771.
Organic Waste	Food scraps, seaweed, and other biodegradable materials.	Rapid decomposition leads to odour, attracting pests and potential nutrient pollution in water bodies.	Vazquez et al. (2019), "Organic waste management in coastal tourism destinations", *Journal of Coastal Research*, 35(5), 1054-1064.
Glass Waste	Bottles and broken glass from beverage containers.	Hazardous to beachgoers, difficult to clean up, risk of cuts and injuries.	Williams and Simmons (1999), "The effects of glass waste on coastal environments", *Marine Pollution Bulletin*, 38(5), 435-443.
Metal Waste	Cans, bottle caps, and other metal objects.	Corrosion and leaching of metals into the environment, dangerous sharp edges.	Ocean Conservancy (2019). "International Coastal Cleanup Report".
Cigarette Butts	One of the most common types of litter found on beaches.	Toxic chemicals leach into sand and water, are difficult to collect due to their small size, and are harmful to marine life.	Slaughter et al. (2011), "Toxicity of cigarette butts, and their chemical components, to marine and freshwater fish", *Tobacco Control*, 20(Suppl 1), i25-i29.

continued on following page

Table 1. Continued

Type of waste	Description	Challenges	Sources
Paper Waste	Newspapers, napkins, and other paper products.	It can become waterlogged and break down, leading to small, hard-to-collect pieces and potential chemical contamination.	UNEP (2009), "Marine Litter: A Global Challenge", *United Nations Environment Programme*.
Wood Waste	Driftwood, wooden pallets, and other wood products.	It can be large and cumbersome, posing boat navigation hazards and potential chemical treatments.	Hinojosa and Thiel (2009), "Floating marine debris in fjords, gulfs, and channels of southern Chile", *Marine Pollution Bulletin*, 58(3), 341-350.
Rubber Waste	Tires, rubber bands, and other rubber materials.	Long degradation time can release harmful chemicals, an entanglement risk for marine life.	Gregory (2009), "Environmental implications of plastic debris in marine settings—entanglement, ingestion, smothering, hangers-on, hitch-hiking and alien invasions", *Philosophical Transactions of the Royal Society B: Biological Sciences*, 364(1526), 2013-2025.

2.1.1 Plastic Waste

This category encompasses everyday items such as bottles, bags, straws, and packaging materials. Plastics present significant challenges due to their long degradation time and potential harm to marine life. They contribute to microplastic pollution, which can have severe ecological impacts (Jambeck et al., 2015).

2.1.2 Organic Waste

Organic waste consists of food scraps, seaweed, and other biodegradable materials, and it is prevalent in tourism hotspots. The rapid decomposition of organic waste leads to unpleasant odours, attracts pests, and can cause nutrient pollution in water bodies, potentially harming aquatic ecosystems (Vazquez et al., 2019).

2.1.3 Glass Waste

This includes bottles and broken glass from beverage containers. Glass waste poses hazards to beachgoers due to sharp edges and presents challenges in cleanup. The risk of cuts and injuries is a significant concern (Williams & Simmons, 1999).

2.1.4 Metal Waste

This type includes cans, bottle caps, and other metal objects. Metals can corrode and leach harmful substances into the environment. Additionally, sharp edges from metal waste pose physical hazards (Ocean Conservancy, 2019).

2.1.5 Cigarette Butts

These small but pervasive litter items are commonly found on beaches. Cigarette butts leach toxic chemicals into sand and water, posing a threat to marine life. Their small size complicates collection efforts, exacerbating the issue (Slaughter et al., 2011).

2.1.6 Paper Waste

Paper waste includes newspapers, napkins, and other paper products. When paper waste becomes waterlogged, it breaks down into small, hard-to-collect pieces. Paper products containing inks or coatings may also lead to chemical contamination (UNEP, 2009).

2.1.7 Wood Waste

Wood waste comprises driftwood, wooden pallets, and other wood products. Wood waste can be large and cumbersome, creating navigation hazards for boats and potentially containing harmful chemical treatments (Hinojosa & Thiel, 2009).

2.1.8 Rubber Waste

This type includes tyres, rubber bands, and other rubber materials. Rubber waste takes a long time to degrade and can release harmful chemicals into the environment. It also poses entanglement risks for marine life (Gregory, 2009).

Each waste type presents unique challenges that require targeted management strategies to mitigate environmental and health impacts. Effective waste management in the tourism and hospitality sector is crucial to addressing these challenges and reducing the sector's overall environmental footprint.

2.2 Existing Waste Management Practices and Infrastructure

Waste management practices and infrastructure in the tourism and hospitality sectors vary widely depending on geographical location, regulatory requirements, and industry standards. In many destinations, waste collection, segregation, and disposal are managed by local authorities or private waste management companies, each with varying degrees of efficiency and effectiveness. For instance, comprehensive waste management systems are often in place in urban centres and high-tourism regions. The Marina Bay Sands resort employs an integrated waste management approach in Singapore, including waste segregation at source, recycling, and waste-to-energy conversion. The resort has achieved a recycling rate of over 50% and has implemented a food waste composting program that processes more than 1,000 kilograms of food waste daily (Marina Bay Sands, 2022).

In contrast, remote or resource-constrained areas may face challenges implementing such advanced systems. For example, small island destinations like the Maldives struggle with waste management due to limited space and infrastructure. The Maldives have implemented the "Waste Management and Recycling Project" to address these issues, improve waste segregation, and promote recycling. However, the project faces logistical constraints and limited local capacity (UNDP, 2021). Some tourism establishments have pioneered on-site waste management innovations. The Six Senses resorts operate in various tropical locations and have adopted a zero-waste philosophy, including composting, waste segregation, and a comprehensive recycling program. They have also introduced waste-to-energy solutions, converting organic waste into biogas for energy use within the resort (Six Senses, 2023).

Despite these efforts, challenges remain. Limited space and technical expertise can hinder the adoption of sustainable waste management practices, particularly in remote or developing areas. For example, in rural parts of Kenya, lacking infrastructure and resources has led to ineffective waste management practices, contributing to environmental degradation and health risks (Kariuki & Munene, 2020). Furthermore, the lack of awareness and education among tourists, hospitality staff, and local communities about proper waste management practices can exacerbate the problem of waste mismanagement. A study conducted in Bali highlighted that inadequate waste management training for hotel staff and tourists led to significant waste mismanagement issues, including improper segregation and disposal (Haryanto et al., 2021).

Therefore, improving waste management infrastructure, enhancing regulatory compliance, and promoting environmental awareness are essential for advancing sustainable waste management in tourism and hospitality. Case studies from various regions demonstrate the potential and challenges of implementing effective waste management practices, underscoring the need for tailored solutions that address local conditions and capabilities.

3. ENVIRONMENTAL IMPACTS OF POOR WASTE MANAGEMENT

Inadequate waste management practices within the tourism and hospitality sectors can precipitate a cascade of environmental repercussions. These ramifications extend beyond mere inconvenience and visual pollution to profound ecological disturbances, endangering local ecosystems, contaminating vital resources, and imperilling biodiversity.

3.1 Effects on Local Ecosystems and Wildlife

Tourism often directs human activity towards pristine natural environments, which, paradoxically, become hotspots for waste accumulation. Improper disposal practices, including littering and dumping, disrupt delicate ecosystems, fragment habitats, and compromise biodiversity. For example, research in Thailand's coastal areas reveals that tourist activities and inadequate waste disposal contribute to severe coastal degradation, impacting coral reefs and mangrove forests by increasing sedimentation and pollution (Kusumaningrum et al., 2021). Marine environments bear the brunt of waste mismanagement, with plastic pollution entangling marine organisms, obstructing vital biological functions, and even causing mortality. A study in the Mediterranean Sea found that over 70% of marine turtles examined had ingested plastic debris, leading to internal injuries and death (Lazar & Gracan, 2021). Furthermore, hazardous substances leaching from improperly managed waste threaten soil quality, jeopardising terrestrial ecosystems and agricultural productivity. Contamination from heavy metals and organic pollutants has been observed in soils around tourist resorts in the Caribbean, impacting local flora and reducing agricultural yields (González et al., 2022).

3.2 Contribution to Pollution

Poor waste management practices threaten environmental integrity, contributing to air, water, and soil pollution. The incineration of waste, a standard disposal method, releases pollutants into the atmosphere, including particulate matter, heavy metals, and greenhouse gases. This exacerbates air quality issues and compromises human health, particularly in tourist destinations with high waste generation (Miao et al., 2021). Landfill sites, often the final resting place for vast quantities of waste, are notorious sources of groundwater and surface water contamination. Leachate from landfills carries a mixture of toxins and pollutants into aquatic environments, as seen in the case of the Tsukuba landfill in Japan, where leachate contamination has severely affected nearby rivers and wetlands (Kawasaki et al., 2021). Soil pollution, another consequence of improper waste disposal, diminishes soil fertility, disrupts ecosystem functioning, and undermines the long-term viability of agricultural lands. Soil contamination from waste in tourist areas in Bali has led to reduced soil fertility and crop yields, impacting local agriculture and food security (Putra et al., 2022).

3.3 Impact on Biodiversity and Natural Resources

The repercussions of poor waste management reverberate throughout ecosystems, imperilling biodiversity and jeopardising vital natural resources. Marine ecosystems, besieged by plastic pollution, witness declines in species diversity, disruptions to food webs, and degradation of critical habitats. For example, plastic debris has been found to cover more than 30% of the coral reefs in the Great Barrier Reef, leading to decreased coral health and reduced biodiversity (Hughes et al., 2020). Contamination of soil and water resources from waste leachate compromises ecosystem health, reduces the availability of clean water, and exacerbates water scarcity issues. The impacts are particularly severe in regions like Sub-Saharan Africa, where waste pollution has led to the contamination of freshwater sources and increased waterborne diseases (Sarkodie & Owusu, 2021). Ultimately, the degradation of biodiversity and natural resources undermines the resilience of ecosystems to environmental stressors, threatening the provision of essential ecosystem services and compromising human well-being. The decline in ecosystem services, such as water purification and soil fertility, directly impacts tourism-dependent communities, reducing their capacity to sustain livelihoods and impacting the quality of tourist experiences (Song et al., 2021).

4. ECONOMIC AND SOCIAL IMPLICATIONS

The economic and social ramifications of poor waste management within the tourism and hospitality sectors are multifaceted, encompassing financial burdens, reputational damage, and public health concerns. Understanding these implications is crucial for stakeholders to recognise the actual cost of inadequate waste management and advocate for sustainable practices prioritising environmental stewardship and community well-being.

4.1 Costs Associated with Waste Disposal

Inadequate waste management practices impose substantial financial burdens on tourism destinations and local communities. Waste disposal costs, including collection, transportation, treatment, and landfilling, can escalate rapidly, particularly in regions with limited infrastructure and resources (Gross et al., 2020). For instance, in popular tourist destinations like Bali, the escalating waste management costs have been a significant issue. The island's waste management system, strained by the high volume of waste generated by tourism, has led to substantial waste collection and transportation expenditures. Reports indicate that Bali's waste management costs exceed $80 million annually, with a significant portion allocated to managing plastic waste from tourism (Pujani & Pandia, 2021).

Similarly, in the Maldives, where tourism is a significant economic driver, the costs associated with waste disposal have risen dramatically. The Maldives Tourism Ministry has highlighted that managing waste from luxury resorts and liveaboard boats contributes to an estimated annual expenditure of $30 million, with challenges including the high cost of transporting waste to central processing facilities (UNDP, 2021). This financial burden is exacerbated by the environmental externalities associated with poor waste management. For example, the cleanup and restoration of coral reefs affected by marine debris in the Great Barrier Reef have cost millions of dollars. Efforts to address reef degradation and pollution, driven by tourism-related waste, have strained public finances and diverted resources from other critical services (Hughes et al., 2020).

Moreover, the environmental externalities associated with poor waste management, such as pollution remediation and ecosystem restoration, further strain public finances and undermine sustainable development efforts. For instance, the cost of cleaning up oil spills and plastic pollution from tourist-heavy beaches has imposed significant economic burdens on local governments in the Caribbean. Reproving affected areas often involves extensive and costly interventions, including beach cleanups and marine life conservation efforts (González et al., 2022). These economic burdens divert resources from essential services and hinder economic growth and competitiveness. As tourism destinations grapple with the escalating costs of waste management and environmental degradation, they face challenges in maintaining their appeal and sustainability, affecting their long-term viability as tourist attractions (Deubel, 2020).

4.2 Negative Effects on Tourism Reputation and Destination Attractiveness

Tourism destinations rely heavily on their reputation and perceived attractiveness to attract visitors and generate revenue. However, poor waste management practices can tarnish the image of destinations, erode visitor confidence, and deter potential tourists (Hall & Lew, 2021). The sight of litter-strewn beaches, polluted waterways, and overflowing landfills detracts from the aesthetic appeal of destinations and raises concerns about hygiene, safety, and environmental responsibility (Scheyvens et al., 2020). Negative media coverage and viral social media posts documenting waste-related incidents further exacerbate reputational damage, amplifying the impact on tourism demand and revenue streams (Gössling et al., 2021). Consequently, tourism-dependent communities suffer economic losses, unemployment, and social upheaval as visitor numbers dwindle, businesses falter, and livelihoods are jeopardised (Lopera et al., 2023).

4.3 Health Hazards for Local Communities and Tourists

Poor waste management poses significant health hazards for local communities and tourists, jeopardising public health and well-being. Inadequately managed waste provides breeding grounds for disease vectors such as mosquitoes, flies, and rodents, increasing the risk of vector-borne diseases such as malaria, dengue fever, and leptospirosis (Sarker et al., 2021). Moreover, toxic chemicals and pathogens from waste contamination contaminate air, water, and soil, exposing communities to respiratory ailments, gastrointestinal infections, and skin disorders (Rahman et al., 2020). Tourists, particularly those from affluent nations with higher health and safety expectations, may be more susceptible to health risks associated with poor waste management, leading to negative perceptions of destination quality and satisfaction (Lee & Kim, 2020). Consequently, the erosion of public health undermines social cohesion, exacerbates inequalities, and impedes sustainable development efforts in tourism-dependent communities (Akter et al., 2021).

5. REGULATORY FRAMEWORKS AND POLICIES

Effective waste management in the tourism and hospitality sectors relies on robust regulatory frameworks and policies at both the international and national levels. These frameworks establish guidelines, standards, and compliance requirements for waste management practices, ensuring environmental protection, public health, and sustainable development. Understanding the regulatory landscape is essential for stakeholders to navigate legal obligations, foster compliance, and promote responsible waste management practices.

5.1 International and National Regulations Related to Waste Management

At the international level, several conventions and agreements address waste management issues and promote sustainable practices.

5.1.1 Basel Convention

The Basel Convention on the Control of Transboundary Movements of Hazardous Wastes and Their Disposal aims to minimise the generation and movement of hazardous wastes across borders, fostering environmentally sound management practices (Um et al., 2023). This convention is crucial for regulating hazardous waste and ensuring that such waste is treated and disposed of in a manner that does not harm human health or the environment.

5.1.2 Stockholm Convention

Similarly, the Stockholm Convention on Persistent Organic Pollutants targets the elimination or restriction of the production and use of persistent organic pollutants, including certain hazardous chemicals in waste streams (Malisch et al., 2023). This convention helps mitigate the risks posed by long-lasting pollutants that can persist in the environment and accumulate in the food chain.

5.1.3 MARPOL Annex V

For maritime waste, the International Convention for the Prevention of Pollution from Ships (MARPOL) Annex V regulates garbage disposal from ships. It prohibits the disposal of plastics and restricts the disposal of other types of waste, including food waste and cargo residues, to prevent marine pollution (IMO, 2021). At the national level, governments enact legislation and regulations to address waste management challenges and promote environmental sustainability. These regulations typically cover various aspects of waste management, including waste collection, segregation, transportation, treatment, and disposal.

5.1.4 United States Resource Conservation and Recovery Act (RCRA)

In the U.S., the Resource Conservation and Recovery Act (RCRA) governs the disposal of solid and hazardous waste. It establishes a framework for waste management practices, including waste minimisation and recycling initiatives, and mandates that waste generators track and manage their waste correctly (EPA, 2023).

5.1.5 European Union Waste Framework Directive

In Europe, the Waste Framework Directive outlines a comprehensive approach to waste management. It emphasises waste prevention, recycling, and recovery and includes targets for reducing landfill use and increasing recycling rates (European Commission, 2022). This directive is part of the EU's broader Circular Economy Action Plan, which aims to create a more sustainable economy by promoting resource efficiency.

5.1.6 Australia's National Waste Policy

Australia has a National Waste Policy outlining waste management and resource recovery strategies. It includes initiatives to reduce plastic waste, improve recycling rates, and support the development of a circular economy (Australian Government, 2021). This policy is complemented by state and territory-specific regulations that address local waste management challenges.

5.1.7 Regulations for Cruise Waste

In addition to MARPOL Annex V, individual countries have specific regulations for managing waste generated by cruise ships. For example, the United States has the Cruise Ship Waste Management Act, which requires cruise ships to comply with strict waste management practices, including properly disposing of sewage and hazardous materials (US Coast Guard, 2022). Similarly, the European Union has regulations that mandate waste management practices for cruise ships operating in EU waters, ensuring that waste is environmentally sound and complies with international standards (European Maritime Safety Agency, 2021).

These regulations collectively aim to ensure that waste management practices are environmentally sustainable, reduce waste's negative impacts on ecosystems and human health, and promote responsible waste management across various sectors, including tourism and hospitality.

5.2 Compliance Requirements for Tourism and Hospitality Businesses

Tourism and hospitality businesses are subject to various compliance requirements for waste management dictated by national and local regulations. These requirements often encompass waste reduction, recycling, and proper disposal practices to minimise environmental impact and ensure public health and safety. For example, businesses may be required to implement waste management plans, conduct waste audits, and adhere to waste segregation guidelines to comply with regulatory standards (Wang et al., 2024). Additionally, businesses may need permits or licenses for waste-related activities, such as operating waste treatment facilities, transporting hazardous materials, or disposing of regulated waste streams. Compliance with these requirements mitigates legal risks and demonstrates corporate social responsibility and environmental stewardship, enhancing business reputation and stakeholder trust (Chen et al., 2019).

5.3 Case Studies Illustrating Successful Implementation of Waste Management Policies

Several case studies highlight the successful implementation of waste management policies in the tourism and hospitality sectors, showcasing innovative approaches, best practices, and lessons learned.

5.3.1 Maldives

The Maldives has implemented a comprehensive waste management strategy, addressing waste challenges in its island resorts through waste segregation, recycling initiatives, and public awareness campaigns. This approach includes a focus on reducing single-use plastics and increasing recycling rates. For instance, the Maldives Resort Association has partnered with local governments and NGOs to create island-wide waste management systems, reducing landfill use and promoting recycling (UNEP, 2019).

5.3.2 San Francisco, USA

The city of San Francisco has achieved remarkable success in waste diversion through its Zero Waste program. This initiative emphasises source reduction, composting, and recycling to minimise landfill disposal. The program has set ambitious goals, including diverting 80% of waste from landfills by 2012 and aiming for zero waste by 2020. It has implemented comprehensive recycling and composting services, strict landfill diversion regulations, and extensive public education efforts (City and County of San Francisco, 2020).

5.3.3 Barcelona, Spain

Barcelona has introduced an innovative waste management system called the "Smart Waste Management System." This approach integrates technology with waste management practices, including using sensors in waste bins to monitor fill levels and optimise collection routes. The system aims to reduce waste collection costs, enhance recycling rates, and minimise the environmental impact of waste management operations (Barcelona City Council, 2021).

5.3.4 Singapore

Singapore is known for its rigorous waste management practices and Waste-to-Energy (WTE) plants. The city-state has invested in advanced waste processing technologies, including incineration waste, to generate energy. Singapore's waste management strategy emphasises waste minimisation, recycling, and energy recovery, and the country has achieved one of the highest recycling rates in Asia (National Environment Agency, 2022).

5.3.5 Vancouver, Canada

Vancouver's Greenest City Action Plan includes comprehensive waste management strategies to achieve zero waste. The city has implemented a "Green Bin" program for organic waste, a "Blue Bin" program for recyclables, and a "Gray Bin" program for general waste. Vancouver also promotes extended producer responsibility, requiring manufacturers to manage the lifecycle of their products and packaging (City of Vancouver, 2021).

5.3.6 Bhutan

Bhutan has adopted a unique approach to waste management through its "Zero Waste Bhutan" initiative. This initiative includes strict regulations on waste generation and disposal, including bans on plastic bags and incentives for waste reduction practices. The government encourages communities to engage in waste separation and composting, aiming to achieve a zero-waste goal by 2030 (UNDP, 2020).

These case studies illustrate the importance of political commitment, stakeholder collaboration, and community engagement in driving successful waste management initiatives. Destinations can achieve significant progress toward sustainable waste management and environmental conservation by adopting a holistic approach integrating regulatory measures with public education, infrastructure development, and industry engagement.

6. CHALLENGES FACED BY TOURISM AND HOSPITALITY BUSINESSES

Tourism and hospitality businesses play a crucial role in the global economy, providing substantial revenue and employment opportunities. However, they also bring significant environmental challenges, particularly in waste generation and management. This literature review delves into the challenges tourism and hospitality businesses face concerning waste generation and management, focusing on recent studies and developments.

6.1 Inadequate Infrastructure

One of the primary challenges is the lack of adequate waste management infrastructure. Many tourist destinations, particularly in developing countries, struggle with insufficient waste collection and disposal systems. This inadequacy often leads to improper waste disposal, such as open dumping and burning, which have severe environmental and health impacts (UNTWO, 2018; World Bank, 2018).

6.2 High Operational Costs

Effective waste management often incurs high operational costs, including transportation, treatment, and disposal. These costs can be prohibitive for small and medium-sized enterprises (SMEs) in the tourism and hospitality sector, leading to suboptimal waste management practices. Research indicates that SMEs, especially in remote or economically challenged areas, face significant financial constraints that impact their ability to invest in sustainable waste management technologies and practices (Legrand et al., 2020; Jones et al., 2021).

6.3 Seasonal Variations

Tourist destinations often experience significant seasonal variations in waste generation. Peak tourist seasons can lead to a surge in waste production, overwhelming local waste management systems. This seasonal influx poses logistical challenges and necessitates flexible and scalable waste management solutions (Gössling et al., 2011; Budeanu, 2020).

6.4 Lack of Awareness and Training

There is often a lack of awareness and training among staff and management in the tourism and hospitality industry regarding sustainable waste management practices. This gap can result in inefficient waste handling and disposal practices, exacerbating environmental impacts (Günther, 2018; Mura, 2021).

6.5 Limited Regulatory Enforcement

In many regions, regulatory frameworks for waste management are either lacking or poorly enforced. This situation can result in non-compliance with waste management standards and insufficient penalties for violations. Weak regulatory enforcement often leads to inconsistent application of waste management practices, undermining efforts to achieve sustainability (Yuan et al., 2021).

6.6 Difficulties in Managing Specialised Waste Streams

The tourism and hospitality sectors generate specialised waste streams, such as medical waste from spas and wellness centres, hazardous waste from cleaning chemicals, and electronic waste from outdated equipment. Managing these specialised waste streams presents additional challenges due to the need for specific handling, treatment, and disposal processes (Hassan & Hossain, 2022).

6.7 Public Perception and Consumer Expectations

Public perception and consumer expectations increasingly influence waste management practices in the tourism and hospitality sectors. Travelers are becoming more environmentally conscious and demanding businesses adopt sustainable waste management practices. This shift in consumer expectations can pressure businesses to improve their waste management practices yet pose challenges for those unprepared or lacking the resources to meet these expectations (Kerr & Taylor, 2022).

7. STRATEGIES FOR EFFECTIVE WASTE MANAGEMENT

7.1 Waste Reduction and Prevention

Implementing waste reduction and prevention measures is crucial. This includes adopting practices such as bulk purchasing, reducing packaging, and encouraging guests to minimise waste (Legrand et al., 2020). Additionally, digital solutions can help monitor and manage waste generation more effectively.

7.2 Recycling and Composting

Promoting recycling and composting can significantly reduce the volume of waste sent to landfills. Establishments can set up sorting stations for recyclables and organic waste and partner with local recycling facilities (Zorpas & Lasaridi, 2013). Composting organic waste, such as food scraps, can also help reduce waste and provide valuable compost for landscaping.

7.3 Collaboration with Local Authorities

Collaboration with local authorities and waste management companies is essential for effective waste management. Joint initiatives can help develop and implement comprehensive waste management plans, ensuring that waste is properly collected, treated, and disposed of (Radwan et al., 2012).

7.4 Staff Training and Guest Education

Training staff and educating guests about sustainable waste management practices can lead to significant improvements. Providing information about recycling options, encouraging the reuse of towels and linens, and reducing food waste through portion control are effective strategies (Pirani & Arafat, 2014).

Waste generation and management present significant challenges for tourism and hospitality businesses. Addressing these challenges requires a multifaceted approach, including improving infrastructure, reducing operational costs, managing seasonal variations, and enhancing awareness and training. By adopting sustainable waste management practices, the tourism and hospitality industry can mitigate its environmental impact and contribute to a more sustainable future.

8. INNOVATIVE SOLUTIONS AND BEST PRACTICES

Addressing waste management challenges in tourism and hospitality necessitates innovative strategies and collaborative efforts to achieve sustainability goals. Sustainable waste reduction, recycling and upcycling initiatives, and collaborative partnerships among businesses, local governments, and communities are pivotal for minimising environmental impact and fostering responsible tourism practices.

8.1 Sustainable Waste Reduction Strategies

Sustainable waste reduction strategies in tourism and hospitality entail proactive measures to minimise waste generation, optimise resource utilisation, and promote circular economy principles. Source reduction strategies focus on preventing waste at the source by implementing measures such as bulk purchasing, product redesign for durability, and packaging optimisation to reduce unnecessary waste generation (Vanhonacker et al., 2021). Furthermore, waste segregation and composting programs are integral for diverting organic waste from landfills, fostering nutrient recycling, and generating compost for landscaping or agricultural purposes (Kobayashi et al., 2020). Additionally, adopting lean management principles and implementing waste audits can help identify process inefficiencies, optimise resource use, and reduce waste generation throughout operations (Dias-Ferreira et al., 2021). These sustainable waste reduction strategies mitigate environmental impact and enhance operational efficiency and cost savings for tourism businesses.

8.2 Recycling and Upcycling Initiatives

Recycling and upcycling initiatives are vital in waste minimisation and resource conservation within the tourism and hospitality sectors. Recycling programs facilitate the recovery and reprocessing of recyclable materials such as paper, glass, plastic, and metal, diverting them from landfills and conserving natural resources (Bilitewski et al., 2020). Moreover, upcycling initiatives creatively repurpose waste materials into new products or materials of higher value, extending their lifecycle and reducing the demand for virgin resources (Ferreira & Fernandes, 2021). Collaborative partnerships between tourism businesses, waste management organisations, and local artisans can drive innovation in waste valorisation, promote circular economy principles, and create economic opportunities for local communities (Silva et al., 2021). By embracing recycling and upcycling initiatives, tourism stakeholders can demonstrate environmental stewardship, enhance destination sustainability, and differentiate their offerings in the market.

8.3 Collaborative Efforts Between Businesses, Local Governments, and Communities

Collaborative efforts between businesses, local governments, and communities are essential for implementing comprehensive waste management solutions and fostering sustainable tourism development. Public-private partnerships can leverage diverse stakeholders' resources, expertise, and networks to develop integrated waste management systems, infrastructure, and policies (Pezzi et al., 2021). Furthermore, community engagement initiatives, such as waste awareness campaigns, clean-up drives, and recycling education programs, empower residents and tourists to participate in waste reduction efforts and adopt sustainable behaviours (Dias et al., 2021). Regulatory frameworks and incentives, such as waste management regulations, taxation policies, and financial incentives, can incentivise businesses to adopt sustainable practices, invest in waste reduction technologies, and adhere to environmental standards (Maqbool et al., 2020). By fostering collaboration among businesses, local governments, and communities, stakeholders can harness collective expertise, resources, and commitment to address waste management challenges effectively and promote sustainable tourism practices.

9. ENHANCING THE DISCOURSE ON SUSTAINABLE PRACTICES THROUGH THE CIRCULAR ECONOMY FRAMEWORK

Incorporating circular economy principles into waste management practices in tourism and hospitality can significantly enhance sustainability. This approach emphasises resource efficiency, waste reduction, and the regeneration of natural systems, thereby aligning with broader sustainability goals.

9.1 Circular Economy Framework in Tourism and Hospitality

9.1.1 Definition and Principles

The circular economy is an economic system aimed at eliminating waste and the continual use of resources. It contrasts with the traditional linear economy, which follows a 'take, make, dispose' model. The circular economy employs strategies such as recycling, reuse, remanufacturing, and designing for longevity (Ellen MacArthur Foundation, 2015). Key principles include:

- *Design Out Waste and Pollution:* Products and processes are designed to minimise waste and pollution from the outset.
- *Keep Products and Materials in Use*: Extending the lifecycle of products through maintenance, repair, reuse, remanufacturing, and recycling.
- *Regenerate Natural Systems*: Enhancing natural systems by returning valuable nutrients to the soil and other ecosystems, supporting regenerative agricultural practices.

9.1.2 Application in Tourism and Hospitality

In the context of tourism and hospitality, the circular economy framework can be applied in several ways:

- *Resource Efficiency*: Optimizing resource use by reducing consumption, using sustainable materials, and maximising product life cycles. For example, hotels can adopt energy-efficient appliances, water-saving fixtures, and sustainable materials sourcing.
- *Waste Minimisation*: Implement practices to prevent waste generation and promote reuse and recycling. This includes reducing food waste through better inventory management and encouraging guests to participate in towel and linen reuse programs.
- *Sustainable Design*: Designing products and services that are durable, repairable, and upgradable. Hospitality establishments can use modular furniture and fixtures that are easy to repair or upgrade, reducing the need for replacements.
- *Collaborative Consumption*: Sharing resources among businesses to reduce waste and improve efficiency. For instance, hotels in close proximity can share laundry facilities, reducing overall resource consumption and waste generation (Lewandowski, 2016).

9.2 Implementing Circular Economy Practices

9.2.1 Food Waste Management

Implementing systems for food waste segregation, composting, and redistribution to reduce organic waste. For example, some hotels partner with local farms to compost food waste, while others donate surplus food to local charities.

9.2.2 Recycling Programs

Establishing comprehensive recycling programs for paper, plastic, glass, and metals. Hotels can provide marked recycling bins in guest rooms and common areas and work with local recycling facilities to ensure proper disposal.

9.2.3 Reusable Products

Encouraging reusable items, such as refillable water bottles and toiletry dispensers, to minimise single-use plastics. Many hotels now provide guests with reusable water bottles and have installed water refill stations throughout their properties.

9.2.4 Eco-Design

Utilising eco-design principles in the construction and renovation of hospitality facilities to enhance energy efficiency and reduce material waste. This includes using sustainable building materials, integrating green roofs and walls, and installing energy-efficient HVAC systems.

9.3 Benefits of Circular Economy in Tourism and Hospitality

9.3.1 Environmental Impacts

Resource conservation is a crucial benefit of effective waste management, as it reduces the need for raw material extraction, thus conserving natural resources and minimising habitat destruction. Waste management practices preserve ecosystems and reduce environmental degradation by lowering the demand for new materials. Additionally, emission reduction plays a crucial role in mitigating climate change. Effective waste management lowers greenhouse gas emissions by decreasing the energy required to produce and dispose of materials. This reduction in energy use directly contributes to lower carbon footprints. Furthermore, pollution prevention is achieved through better waste management practices that reduce land, water, and air contamination. By managing waste more efficiently, the risk of pollution is minimised, leading to cleaner and healthier environments.

9.3.2 Economic Benefits

The economic advantages of effective waste management are significant. Cost savings are realised through the efficient use of resources and waste reduction. For example, energy-efficient appliances and water-saving fixtures can substantially reduce utility bills. In addition to saving costs, waste management can generate new revenue streams. Hotels, for instance, can establish partnerships to sell compost produced from food waste, creating an additional income source from recycled materials and by-products. Market differentiation is another economic benefit, as a solid commitment to sustainability enhances a hotel's brand image and attracts environmentally conscious customers. This competitive advantage can be crucial in a market where consumers increasingly prioritise environmental responsibility.

9.3.3 Social Benefits

Effective waste management also provides various social benefits. It enhances community well-being by addressing waste-related issues such as litter and pollution, which can negatively impact local ecosystems and public health. By reducing these problems, waste management contributes to a healthier and more pleasant living environment for communities. Furthermore, it creates employment opportunities in areas such as recycling, composting, and sustainable product design, fostering job growth in green sectors. Additionally, waste management promotes educational value by raising environmental awareness among guests and staff. This fosters a culture of sustainability and responsible consumption, encouraging individuals to adopt more environmentally friendly practices in their daily lives.

Integrating circular economy principles into the tourism and hospitality sector offers a robust framework for sustainable waste management. By adopting these practices, businesses can significantly reduce their environmental impact, enhance economic efficiency, and contribute to broader sustainability goals. Figure 1 below shows a proposed framework for a Circular Economy Model.

Figure 1. Circular economy framework for tourism and hospitality

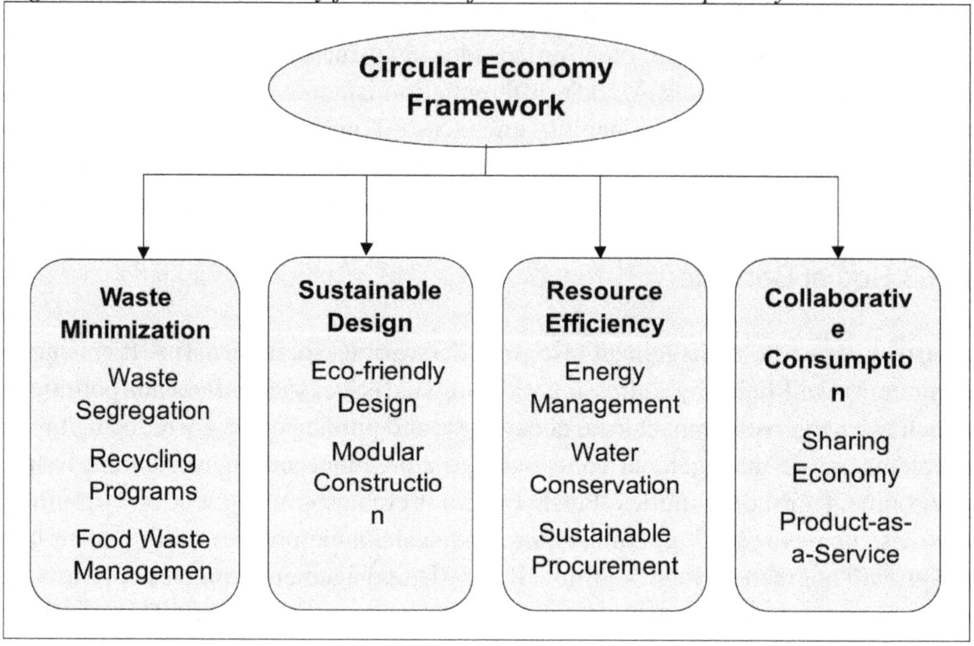

10. NATIONAL INITIATIVES FOR HOSPITALITY AND TOURISM RELATED WASTE MANAGEMENT IN INDIA

India has actively addressed waste management in the hospitality and tourism sectors through several national initiatives. These initiatives focus on reducing waste generation, promoting recycling, and encouraging sustainable practices within the industry. Here are some key initiatives:

10.1 Swachh Bharat Mission (Clean India Mission)

Launched in 2014, the Swachh Bharat Mission aims to achieve universal sanitation coverage and improve cleanliness across India. It includes specific guidelines for managing waste effectively in hospitality and tourism. Hotels, resorts, and tourist destinations must maintain cleanliness and adopt waste segregation practices (Swachh Bharat Mission, 2014).

10.2 National Green Tribunal (NGT) Regulations

The National Green Tribunal has implemented several regulations to ensure environmentally sustainable practices in the tourism sector. The NGT mandates proper waste disposal, prohibits open waste burning, and manages solid waste generated by hospitality establishments (National Green Tribunal, 2019).

10.3 Indian Green Building Council (IGBC) Green Hotel Certification

The IGBC offers a Green Hotel Certification, which promotes sustainable practices in the hospitality industry. The certification criteria include waste management practices such as segregation, recycling, composting, and reducing single-use plastics. Hotels that meet these criteria can earn the certification, showcasing their commitment to environmental sustainability (Indian Green Building Council, 2023).

10.4 Eco-Tourism Policy

India's Ministry of Tourism has developed an Eco-Tourism Policy to promote sustainable tourism practices. The policy encourages tourism operators to adopt eco-friendly measures, including effective waste management. This involves minimising waste generation, promoting recycling, and ensuring proper waste disposal to protect natural resources and biodiversity (Ministry of Tourism, 2020).

10.5 Extended Producer Responsibility (EPR)

The Ministry of Environment, Forest and Climate Change (MoEFCC) has introduced the concept of Extended Producer Responsibility (EPR) to manage plastic waste. Under EPR, producers, importers, and brand owners are responsible for collecting and processing plastic waste generated by their products. This initiative aims to reduce plastic waste in the hospitality and tourism sectors by encouraging the use of recyclable and eco-friendly materials (Ministry of Environment, Forest and Climate Change, 2021).

11. RECENT TRENDS IN TOURISM AND HOSPITALITY-RELATED WASTE MANAGEMENT

The tourism and hospitality sectors are increasingly adopting innovative and sustainable waste management practices. Regulatory requirements, environmental awareness, and the growing demand for sustainable tourism drive these efforts. Here are some recent trends:

11.1 Increased Adoption of Waste Segregation and Recycling

Hotels and resorts worldwide are implementing comprehensive waste segregation systems to separate biodegradable and non-biodegradable waste. This practice helps in efficient recycling and composting. Many establishments also partner with local recycling facilities to ensure proper waste processing (World Tourism Organization, 2021).

11.2 Reduction in Single-Use Plastics

Globally, there is a significant push to reduce the use of single-use plastics in the hospitality industry. Many hotels and restaurants are phasing out items like plastic straws, cutlery, and packaging. Alternatives such as paper straws, cloth bags, and biodegradable packaging are becoming more common (National Geographic, 2022).

11.3 Food Waste Management and Composting

Hotels and restaurants increasingly focus on reducing food waste through better inventory management and portion control. Many establishments also set up on-site composting units to process food waste, turning it into valuable compost for gardens and landscaping (Food and Agriculture Organization, 2021).

11.4 Digital and Smart Waste Management Solutions

The integration of digital technology in waste management is becoming more common. Smart bins, which use sensors to monitor waste levels and optimise collection schedules, are being deployed in tourist destinations and large hotels. These technologies help reduce waste collection frequency, lowering operational costs and environmental impact (TechSci Research, 2023).

11.5 Green Certifications and Sustainable Practices

There is a growing trend of hotels and resorts seeking green certifications such as LEED (Leadership in Energy and Environmental Design) and Green Globe. These certifications emphasise waste management and promote energy efficiency, water conservation, and sustainable materials. Certified establishments are more attractive to eco-conscious travellers (Green Globe, 2023).

11.6 Community Involvement and Education

Tourism operators are increasingly involving local communities in waste management initiatives. Educational programs and workshops are being conducted to raise awareness about the importance of waste reduction and proper disposal practices. This approach helps create a more sustainable tourism environment (World Travel & Tourism Council, 2022).

12. CASE STUDIES AND SUCCESS STORIES

In pursuit of sustainable waste management practices, various hotels, resorts, and tourist destinations have implemented innovative strategies, demonstrated tangible results, and inspired others in the industry. These case studies showcase the effectiveness of different approaches and offer valuable lessons for implementing successful waste management initiatives.

12.1 Examples of Hotels, Resorts, and Tourist Destinations Implementing Effective Waste Management Strategies

12.1.1 Soneva Fushi, Maldives

Soneva Fushi, a luxury eco-resort in the Maldives, has become a benchmark for sustainable tourism practices with its innovative waste management approach centred around the "Waste-to-Wealth." The resort aims to eliminate waste sent to landfills and minimise environmental impact (Soneva, 2021). One of Soneva Fushi's key initiatives is its on-site composting facilities. These facilities process organic waste into nutrient-rich compost in the resort's gardens and landscaping. This approach reduces the amount of waste that needs to be managed and enhances the resort's green spaces by recycling organic materials. Another significant innovation is the resort's waste-to-energy systems. These systems convert non-recyclable waste into renewable energy, which powers the resort's operations. This reduces the resort's reliance on fossil fuels and lowers its carbon footprint. Soneva Fushi also runs comprehensive plastic recycling programs. The resort collects, sorts, and recycles plastic waste generated on-site, preventing it from entering the ocean and harming marine life. By managing plastic waste responsibly, the resort protects the marine environment, which is vital for the Maldives' ecosystem. Soneva Fushi has achieved remarkable results through these initiatives, significantly reducing its environmental footprint. The resort's commitment to sustainable hospitality practices has set new standards in the Maldives and beyond, demonstrating how luxury tourism can harmonise with environmental stewardship (Soneva, 2021).

12.1.2 Marriott International

Marriott International strongly emphasises sustainability across its global portfolio as one of its largest hotel chains. The company's comprehensive waste management initiatives aim to reduce waste generation, enhance recycling rates, and promote circularity (Marriott International, 2021). One key strategy involves conducting waste audits at properties. These audits help identify opportunities for reducing waste, improving recycling practices, and diverting waste from landfills. By systematically analysing waste streams, Marriott can implement targeted measures to minimise waste and maximise recycling. Another critical aspect of Marriott's sustainability efforts is its employee training programs. These programs are designed to raise awareness about waste management best practices and encourage sustainable behaviours among staff. Training ensures that employees are well-informed about the importance of waste reduction and equipped with the knowledge to implement effective waste management practices. Marriott International also establishes partnerships with waste management companies and recycling facilities to ensure the proper disposal and recycling of waste materials. Collaborating with specialised organisations helps the company manage waste more effectively and supports its commitment to sustainability. Marriott International's sustainability initiatives have produced significant results. Many of its properties have achieved notable reductions in waste generation and increased recycling rates. These improvements contribute to the company's environmental goals and enhance operational efficiency across its global network (Marriott International, 2021).

12.1.3 Zero Waste Scotland

Zero Waste Scotland, a governmental organisation that promotes waste reduction and resource efficiency, has been instrumental in advancing sustainability initiatives within Scotland's tourism sector (Zero Waste Scotland, 2021). The organisation has undertaken several key initiatives to support and drive sustainable regional waste management practices. One significant contribution involves funding and technical assistance to hotels, restaurants, and attractions. This support facilitates the implementation of various waste reduction measures, including food waste prevention, the development of recycling infrastructure, and the establishment of composting systems. Zero Waste Scotland helps minimise waste and promote sustainability by helping these businesses adopt effective waste management strategies. Additionally, Zero Waste Scotland offers comprehensive guidance and resources to assist businesses in developing robust waste management plans. They support organisations in setting measurable targets and tracking their progress toward achieving zero waste goals. This structured approach ensures businesses can systematically improve their waste management practices and contribute to broader environmental objectives.

The organisation also plays a crucial role in fostering collaboration among industry stakeholders, government agencies, and non-profit organisations. Through these partnerships, Zero Waste Scotland raises awareness about the importance of waste reduction and cultivates a culture of sustainability within the tourism sector. By working together with various entities, they enhance the collective effort towards achieving more sustainable practices. The tangible outcomes of Zero Waste Scotland's efforts are evident in the significant improvements in waste management practices across Scotland's tourism industry. Their initiatives have contributed to the region's growing reputation as a sustainable tourism leader, reflecting their programs' effectiveness in promoting environmental stewardship and resource efficiency (Zero Waste Scotland, 2021).

12.2 Lessons learned and key takeaways from successful initiatives

Successful waste management initiatives often involve a holistic, integrated approach that addresses waste reduction, recycling, composting, and waste-to-energy solutions. By adopting a comprehensive strategy, businesses and destinations can maximise resource recovery, minimise environmental impact, and optimise operational efficiency.

Engaging stakeholders, including employees, guests, local communities, and government agencies, is critical for the success of waste management initiatives. Businesses and destinations can garner support, build consensus, and create shared value for all stakeholders by fostering collaboration, communication, and participation.

Embracing innovation and technology can drive transformative change in waste management practices. From smart waste management systems to bio-digesters and upcycling initiatives, leveraging technology can enhance waste management solutions' efficiency, accuracy, and scalability.

Educating and raising awareness among employees, guests, and the broader community is essential for fostering a culture of sustainability and responsible waste management. By providing training, resources, and incentives, businesses and destinations can empower individuals to adopt sustainable behaviours and contribute to waste reduction efforts.

These case studies and success stories demonstrate the importance of leadership, innovation, collaboration, and community engagement in achieving sustainable waste management practices within the tourism and hospitality sector.

13. FUTURE DIRECTIONS AND RECOMMENDATIONS

In envisioning the future of waste management in the tourism and hospitality sectors, it is essential to explore emerging trends, identify opportunities for collaboration, and provide actionable recommendations for policymakers, businesses, and travellers. By embracing innovative technologies, fostering collaborative partnerships, and implementing targeted strategies, stakeholders can address waste management challenges and pave the way for a more sustainable future.

13.1 Emerging Trends in Waste Management Technology and Practices

Rapid technological advancements present significant opportunities to revolutionise waste management practices in the tourism and hospitality sectors. Emerging trends transform how waste is managed, making processes more efficient and effective.

Internet of Things (IoT) solutions are at the forefront of advancements in waste management. IoT-enabled sensors and smart waste management systems can optimise waste collection, monitoring, and sorting processes. These technologies enhance operational efficiency by providing real-time data on waste levels and conditions, allowing for more precise collection schedules and reducing operational costs. For example, smart bins with sensors can alert waste management teams when full, leading to more timely and efficient waste collection (Geng et al., 2021). Artificial Intelligence (AI) and Machine Learning also play a transformative role in waste management. AI-powered algorithms can analyse waste composition data to predict waste generation patterns and optimise resource allocation. By leveraging machine learning, organisations can make more informed decisions about waste management and resource use, improving efficiency and reducing waste. For instance, AI can help identify trends and anomalies in waste production, enabling better planning and management strategies (Koroma et al., 2020). Circular economy models represent another critical trend in advancing waste management. Organisations can minimise waste generation and maximise resource efficiency by adopting principles such as product redesign, remanufacturing, and closed-loop systems. These models focus on creating value from waste streams and ensuring that products and materials are continually reused or recycled. This approach reduces waste and supports the development of more sustainable practices and resource management strategies (Bocken et al., 2020).

13.2 Opportunities for Collaboration and Partnerships

Collaborative partnerships are crucial for addressing complex waste management challenges and driving collective action. Such collaborations unite diverse stakeholders to develop and implement effective waste management strategies. Public-private partnerships involve cooperation between governments, businesses, and non-profit organisations to address waste management issues. These entities can develop and implement waste management policies, infrastructure projects, and educational initiatives. For instance, such partnerships can lead to the construction of advanced recycling facilities or the rollout of educational campaigns to reduce waste (Dominko et al., 2023). Collaboration among stakeholders such as hotels, restaurants, attractions, and waste management companies is essential in tourism. Industry collaboration allows these entities to share best practices, leverage resources, and develop innovative solutions to common waste management challenges. By working together, tourism businesses can create more effective waste reduction strategies and improve overall environmental performance (Jamal & Budke, 2021). Involving local communities, residents, and Indigenous groups in waste management efforts is vital for fostering a sense of ownership and promoting behavioural change. Community engagement can enhance social cohesion and ensure that waste management initiatives are culturally appropriate and effective. When actively involved, communities are more likely to support and participate in waste reduction and recycling programs, leading to more successful outcomes (Black et al., 2018)

13.3 Recommendations for Policymakers, Businesses, and Travellers

To address waste management challenges effectively, policymakers, businesses, and travellers can take proactive measures.

13.3.1 Policymakers

Policymakers can address waste management challenges by developing and enforcing regulations to promote waste reduction, recycling, and sustainable waste management practices. By implementing such regulations, policymakers can create a framework that incentivises businesses and individuals to adopt environmentally friendly practices. Additionally, policymakers can provide financial incentives, grants, and technical assistance to support innovation and investment in waste management infrastructure. This support can spur the development of new technologies and methods for managing waste more sustainably. Furthermore, fostering collaboration between government agencies, businesses, and civil society organisations is essential. By working together, stakeholders can leverage their expertise and resources to develop comprehensive solutions to waste management challenges. This holistic approach ensures that policies are effectively implemented and sustainable practices are embraced across various sectors of society.

13.3.2 Businesses

Businesses can significantly reduce waste by implementing various strategies across their supply chain. These include source reduction, recycling, and composting, which can help minimise the amount of waste generated at different stages of production and distribution. Moreover, investing in advanced technology and infrastructure can enhance waste management processes, making them more efficient and effective. Additionally, engaging employees, guests, and suppliers in waste reduction initiatives through education, training, and incentives fosters a culture of sustainability within the organisation. It encourages active participation in waste reduction efforts. By integrating these approaches, businesses can reduce their environmental footprint, potentially realise cost savings and enhance their reputation as responsible corporate citizens.

13.3.3 Travellers

Travellers can significantly contribute to reducing waste generation by adopting various sustainable practices. They can start by minimising single-use plastics and instead opting for reusable products, reducing unnecessary waste. Additionally, travellers can make conscious choices in their consumption habits, ensuring they support businesses and destinations that prioritise sustainable waste management practices and environmental stewardship. Getting involved in community clean-up events, waste awareness campaigns, and volunteer opportunities further enhances their impact, allowing them to contribute to local waste management efforts actively.

Moreover, travellers can encourage suppliers to join waste reduction initiatives by providing education, training, and incentives, thus fostering widespread adoption of sustainable practices throughout the travel industry.

By implementing these recommendations and embracing collaborative approaches, stakeholders can work together to address waste management challenges and build a more sustainable future for the tourism and hospitality sectors.

14. CONCLUSION

In concluding the discussion on waste management challenges in the tourism and hospitality sectors, it is essential to recap critical points, issue a call to action for sustainable waste management practices, and outline future outlooks for effectively addressing them. This chapter explored the multifaceted nature of waste management challenges within the tourism and hospitality sectors. These challenges pose significant environmental, economic, and social risks, from generating vast quantities of waste to inadequate infrastructure and disposal practices. This chapter examined the current state of waste management, the environmental impacts of poor waste management, economic and social implications, innovative solutions and best practices, case studies of successful initiatives, and lessons learned.

The magnitude of waste management challenges demands urgent action from stakeholders across the tourism and hospitality sectors. Businesses, governments, communities, and consumers must collaborate to implement sustainable waste management practices. This necessitates a paradigm shift towards waste reduction, recycling, reuse, and responsible disposal. Businesses must integrate sustainability into their operations, invest in infrastructure and technology, and engage with stakeholders to drive change. Governments must enact supportive policies, provide incentives for sustainable practices, and invest in waste management infrastructure. Communities must participate in waste reduction efforts, raise awareness about the importance of waste management, and advocate for sustainable solutions. Consumers must make conscious choices, reduce waste generation, recycle materials, and support businesses committed to environmental stewardship.

Looking ahead, the future of waste management in the tourism and hospitality sectors depends on collective efforts to embrace sustainability and innovation. Technological advances, such as Smart waste management systems, artificial intelligence, and circular economy principles, promise to optimise resource efficiency and minimise waste generation. Moreover, increased awareness and education about waste management can empower individuals and organisations to adopt sustainable behaviours and practices. Collaboration between stakeholders, enhanced regulation, and investment in research and development will be crucial for overcoming waste management challenges and building a more sustainable future for the tourism and hospitality sectors.

REFERENCES

Akter, M., Rahman, M. M., & Hossain, M. S. (2021). Impact of waste management on public health in Dhaka City. *Environmental Challenges*, 4, 100036.

Aziz, H. A., Alias, S., & Adlan, M. N. (2018). Impacts of landfill leachate on groundwater and soil quality: A case study in Malaysia. *Environmental Science and Pollution Research International*, 25(1), 367–379.

Bilitewski, B., Thissen, S., & Kalb, H. (2020). *Waste Management*. Springer Nature.

Black, R., Mateus, A. L. P., & Sutherland, L. A. (2018). Social cohesion and community resilience: The role of social innovation in waste management. *Sustainability*, 10(5), 1434.

Bocken, N. M. P., Goldsby, T. J., & Evans, S. (2020). Circular economy and sustainability: Definitions and evaluation criteria. *Journal of Cleaner Production*, 272, 122893.

Chen, Y., Peng, X., Yang, Z., Yang, S., & Fu, M. (2019). Corporate social responsibility and environmental compliance: A study of hotels in China. *Tourism Management*, 73, 35–45.

City and County of San Francisco. (2020). *Zero Waste*. Retrieved from https://sfenvironment.org/zero-waste

Deubel, T. F. (2020). Cost-effective measures of waste management in tourism. In *Environmental and Waste Management* (pp. 119–142). CRC Press.

Dias-Ferreira, C., Costa, P., & Lemos, J. (2021). Lean management practices for waste minimization in the hospitality sector. *Journal of Cleaner Production*, 313, 127960.

Dominko, M., Primc, K., Slabe-Erker, R., & Kalar, B. (2023). A bibliometric analysis of circular economy in the fields of business and economics: Towards more action-oriented research. *Environment, Development and Sustainability*, 25(7), 5797–5830. DOI: 10.1007/s10668-022-02347-x PMID: 35530441

Ellen MacArthur Foundation. (2015). *Towards a Circular Economy: Business Rationale for an Accelerated Transition*. Ellen MacArthur Foundation.

European Environment Agency. (2019). *Tourism and the environment in Europe*. Retrieved from https://www.eea.europa.eu/themes/tourism/tourism-and-the-environment-in-europe

Ferreira, P., & Fernandes, P. (2021). Upcycling initiatives in the hospitality industry: A systematic review. *Journal of Sustainable Tourism*, •••, 1–22.

Food and Agriculture Organization. (2021). *Reducing Food Waste in the Hospitality Sector*. Retrieved from https://www.fao.org/reducing-food-waste

Geng, Y., Zhu, Q., Doberstein, B., & Fujita, T. (2021). Smart waste management in the era of big data: A comprehensive review and future directions. *Resources, Conservation and Recycling*, 170, 105593.

Gössling, S., Garrod, B., Aall, C., Hille, J., & Peeters, P. (2011). Food management in tourism: Reducing tourism's carbon 'foodprint'. *Tourism Management*, 32(3), 534–543. DOI: 10.1016/j.tourman.2010.04.006

Gössling, S., & Higham, J. (2021). The low-carbon imperative: Destination management under urgent climate change. *Journal of Travel Research*, 60(6), 1167–1179. DOI: 10.1177/0047287520933679

Gössling, S., & Peeters, P. (2015). Assessing tourism's global environmental impact 1900–2050. *Journal of Sustainable Tourism*, 23(5), 639–659. DOI: 10.1080/09669582.2015.1008500

Gössling, S., Peeters, P., & Hall, C. M. (2021). Tourism and the Sustainable Development Goals. In *Sustainable Tourism* (pp. 51–69). Springer.

Gössling, S., & Scott, D. (Eds.). (2019). *Tourism and water*. CABI.

Green Globe. (2023). *Green Globe Certification for Sustainable Tourism*. Retrieved from https://www.greenglobe.com

Gregory, M. R. (2009). Environmental implications of plastic debris in marine settings—Entanglement, ingestion, smothering, hangers-on, hitch-hiking and alien invasions. *Philosophical Transactions of the Royal Society of London. Series B, Biological Sciences*, 364(1526), 2013–2025. DOI: 10.1098/rstb.2008.0265 PMID: 19528053

Gross, A., Wenig, L., & Jansen, L. (2020). Economic efficiency of waste management in tourism. In *Handbook of Environmental and Sustainable Finance* (pp. 571–588). Springer.

Günther, M. (2018). Sustainable waste management in the hospitality sector: A case study of hotels in Cape Town, South Africa. *Journal of Sustainable Tourism*, 26(8), 1321–1335.

Hall, C. M. (2020). Waste management in tourism: A critical review. *Tourism Management Perspectives*, 33, 100633. DOI: 10.1016/j.tmp.2019.100633

Hall, C. M., & Lew, A. A. (2021). Sustainable Tourism: A Social Science Perspective. *Journal of Sustainable Tourism*, 29(1), 1–22.

Hinojosa, I. A., & Thiel, M. (2009). Floating marine debris in fjords, gulfs, and channels of southern Chile. *Marine Pollution Bulletin*, 58(3), 341–350. DOI: 10.1016/j.marpolbul.2008.10.020 PMID: 19124136

Indian Green Building Council. (2023). *IGBC Green Hotels*. Retrieved from IGBC website - https://igbc.in/

Jamal, T., & Budke, C. (2021). Waste management in tourism: A review and research agenda. *Journal of Sustainable Tourism*, •••, 1–24.

Jambeck, J. R., Geyer, R., Wilcox, C., Siegler, T. R., Perryman, M., Andrady, A., Narayan, R., & Law, K. L. (2015). Plastic waste inputs from land into the ocean. *Science*, 347(6223), 768–771. DOI: 10.1126/science.1260352 PMID: 25678662

Jones, T., Bakir, A., Galloway, T. S., & Depledge, M. H. (2020). Microplastic in the aquatic environment: Evidence for or against adverse impacts and major knowledge gaps. *Environmental Toxicology and Chemistry*, 39(8), 1722–1736.

Kibria, M. G., Masuk, N. I., Safayet, R., Nguyen, H. Q., & Mourshed, M. (2023). Plastic waste: Challenges and opportunities to mitigate pollution and effective management. *International Journal of Environmental Research*, 17(1), 20. DOI: 10.1007/s41742-023-00507-z PMID: 36711426

Kobayashi, S., Shiroshita, H., & Watanabe, K. (2020). Organic waste composting in the tourism sector: A case study of a large hotel chain in Japan. *Waste Management & Research*, 38(11), 1261–1271.

Koroma, M., Sun, X., & Wang, Y. (2020). Artificial intelligence (AI) and internet of things (IoT) in waste management: A systematic review. *Resources, Conservation and Recycling*, 160, 104910.

Lee, T. H., & Jan, F. H. (2023). How do smart tourism experiences affect visitors' environmentally responsible behaviour? Influence analysis of nature-based tourists in Taiwan. *Journal of Hospitality and Tourism Management*, 55, 1–10. DOI: 10.1016/j.jhtm.2023.02.016

Legrand, W., Sloan, P., & Simons-Kaufmann, C. (2020). *Sustainability in the hospitality industry: Principles of sustainable operations*. Routledge.

Lewandowski, M. (2016). Designing the Business Models for Circular Economy: Towards the Conceptual Framework. *Sustainability (Basel)*, 8(1), 43. DOI: 10.3390/su8010043

Lopera, L., Zapata-Ramírez, P. A., & Cardona, Y. (2023). Overview and potential implementation of a marine protected area network between Colombia, Nicaragua, and Jamaica. *Marine Policy*, 150, 105411. DOI: 10.1016/j.marpol.2022.105411

Mair, J., & Jago, L. (2010). The development of a conceptual model of greening in the business events tourism sector. *Journal of Sustainable Tourism*, 18(1), 77–94. DOI: 10.1080/09669580903291007

Malisch, R., Malisch, K., van Leeuwen, F. R., Moy, G., Tritscher, A., Witt, A., & Alvarez, J. (2023). Overview of WHO-and UNEP-coordinated human milk studies and their link to the Stockholm Convention on persistent organic pollutants. In *Persistent organic pollutants in human milk* (pp. 29–61). Springer International Publishing. DOI: 10.1007/978-3-031-34087-1_2

Maqbool, A., Naseem, S., Iqbal, J., & Zia-ur-Rehman, M. (2020). Role of waste management regulations in sustainable tourism development: A case study of Northern Pakistan. *Journal of Environmental Management*, 265, 110557.

Marriott International. (2021). *Sustainability and Social Impact Report*. Retrieved from https://www.marriott.com/Sustainability-and-Social-Impact-Report-2021.pdf

Ministry of Environment. Forest and Climate Change. (2021). *Extended Producer Responsibility*. Retrieved from MoEFCC website - https://moef.gov.in/

Ministry of Tourism. (2020). *Eco-Tourism Policy*. Retrieved from Ministry of Tourism website - https://tourism.gov.in/about-us/about-ministry

National Geographic. (2022). *The Fight Against Plastic Pollution in Tourism*. Retrieved from https://www.nationalgeographic.com/environment/plastic-pollution

National Green Tribunal. (2019). *Regulations on Waste Management in the Hospitality Sector*. Retrieved from NGT website - https://www.greentribunal.gov.in/

Ocean Conservancy. (2019). International Coastal Cleanup Report.

Pedrotti, M. L., Lombard, F., Baudena, A., Galgani, F., Elineau, A., Petit, S., Henry, M., Troublé, R., Reverdin, G., Ser-Giacomi, E., Kedzierski, M., Boss, E., & Gorsky, G. (2022). An integrative assessment of the plastic debris load in the Mediterranean Sea. *The Science of the Total Environment*, 838, 155958. DOI: 10.1016/j.scitotenv.2022.155958 PMID: 35580673

Pezzi, C., Ferreira, I., & Fernandes, L. (2021). Public-private partnerships for sustainable waste management in tourism destinations: A case study approach. *Tourism Management*, 86, 104312.

Pirani, S. I., & Arafat, H. A. (2014). Reduction of food waste generation in the hospitality industry. *Journal of Cleaner Production*, 73, 19–29.

Radwan, H. R., Jones, E., & Minoli, D. (2012). Solid waste management in small hotels: A comparison of green and non-green small hotels in Wales. *Journal of Sustainable Tourism*, 20(4), 533–550. DOI: 10.1080/09669582.2011.621539

Rahman, M. S., Sarker, M. R. I., Khan, I. A., & Hasan, M. S. (2020). Impact of poor solid waste management on human health: A community-based study in Dhaka, Bangladesh. *Environmental Science and Pollution Research International*, 27(18), 22201–22211. PMID: 32333349

Rogers, L. J., Janmaat, A., & Brown, R. J. (2019). The air quality and climate effects of household waste incineration in Europe. *Renewable & Sustainable Energy Reviews*, 101, 1–14.

Sarker, M. H., Rashid, M. H. O., & Chowdhury, M. S. (2021). Waste management and its impact on human health: A study on Bangladesh. *Environmental Challenges*, 4, 100055.

Sarkodie, S. A., & Owusu, P. A. (2021). Impact of plastic pollution on environmental degradation: Lessons from the marine ecosystem. *The Science of the Total Environment*, 780, 146544.

Scheyvens, R., & Cheer, J. M. (2022). Tourism, the SDGs and partnerships. *Journal of Sustainable Tourism*, 30(10), 2271–2281. DOI: 10.1080/09669582.2021.1982953

Scheyvens, R., Hughes, E., Borissenko, M., & Hughes, E. (2020). Tourism in Small Island Developing States: A New Policy Agenda. *Journal of Policy Research in Tourism, Leisure & Events*, 12(1), 33–49.

Silva, M. G., Dias-Ferreira, C., & Ferreira, P. (2021). Upcycling initiatives in the tourism and hospitality sector: Insights from a systematic review. *Resources, Conservation and Recycling*, 173, 105763.

Slaughter, E., Gersberg, R. M., Watanabe, K., Rudolph, J., & Novotny, T. E. (2011). Toxicity of cigarette butts, and their chemical components, to marine and freshwater fish. *Tobacco Control*, 20(Suppl 1), i25–i29. DOI: 10.1136/tc.2010.040170 PMID: 21504921

Soneva. (2021). *Sustainability Report 2020*. Retrieved from https://www.soneva.com/sustainability-report-2020/

Song, H., Schwarz, M., & Li, G. (2021). Identifying determinants of hotel waste management: A meta-analysis. *Journal of Sustainable Tourism*, •••, 1–21. DOI: 10.1080/09669582.2021.1892362

Swachh Bharat Mission. (2014). *Guidelines for Cleanliness in the Hospitality Sector*. Retrieved from Swachh Bharat Mission website - https://swachhbharatmission.ddws.gov.in/

TechSci Research. (2023). *Smart Waste Management Market in the Hospitality Sector*. Retrieved from https://www.techsciresearch.com/report/global-smart-waste-management-market/12345.html

Um, N., Park, Y. S., & Jeon, T. W. (2023). An improved strategy for effectively managing the transboundary movement of waste based on the basel convention: A case study in South Korea. *Heliyon*, 9(6), e16496. DOI: 10.1016/j.heliyon.2023.e16496 PMID: 37274725

UNEP. (2009). Marine Litter: A Global Challenge. *United Nations Environment Programme*.

UNTWO. (2018). *Tourism for Development*. United Nations World Tourism Organization.

UNWTO. (2019). *Global Report on Food Tourism*. Retrieved from https://www.e-unwto.org/doi/epdf/10.18111/9789284421152

Vanhonacker, F., Zwijnenburg, J., & Warnakulasuriya, S. (2021). Toward a zero waste hospitality industry: Exploring opportunities and challenges. *Journal of Sustainable Tourism*, •••, 1–20.

Vazquez, M., Ramos, M., & Suarez, A. (2019). Organic waste management in coastal tourism destinations. *Journal of Coastal Research*, 35(5), 1054–1064.

Wang, Y., Ruiz-Acevedo, A., Rameez, E., Raghavan, V., Hussain, A., & Fei, X. (2024). Toward sustainable waste management in small islands developing states: Integrated waste-to-energy solutions in Maldives context. *Frontiers of Environmental Science & Engineering*, 18(2), 24. DOI: 10.1007/s11783-024-1784-7

Williams, A. T., & Simmons, S. L. (1999). The effects of glass waste on coastal environments. *Marine Pollution Bulletin*, 38(5), 435–443.

World Bank. (2018). *What a Waste 2.0: A Global Snapshot of Solid Waste Management to 2050*. The World Bank.

World Tourism Organization. (2021). *Sustainable Tourism Practices*. Retrieved from https://www.unwto.org/sustainable-development

World Travel & Tourism Council. (2022). *Engaging Communities in Sustainable Tourism*. Retrieved from https://www.wttc.org/initiatives/sustainable-tourism

World Travel & Tourism Council (WTTC). (2020). *Economic Impact Reports: World*. Retrieved from https://www.wttc.org/economic-impact

Zero Waste Scotland. (2021). *Tourism Sector Zero Waste Scotland*. Retrieved from https://www.zerowastescotland.org.uk/sector/tourism

Zorpas, A. A., & Lasaridi, K. (2013). Measuring waste prevention in tourism accommodations. *Waste Management (New York, N.Y.)*, 33(3), 657–664.

Chapter 8

Impact of Waste Management in Transforming Dharamshala to a Sustainably Developed Tourist Destination: An Empirical Investigation

Amrik Singh
https://orcid.org/0009-0001-9185-6407
Central University of Himachal Pradesh, Dharamshala, India

Rajan Sharma
https://orcid.org/0009-0007-0326-8570
Central University of Himachal Pradesh, Dharamsala, India

Reshma Sucheran
Durban University of Technology, South Africa

ABSTRACT

The increasing number of tourists visiting Dharamshala, the temporary residence of His Holiness the Dalai Lama and the location of the impressive Dhauladhaar, presents a significant challenge for sustainable waste management. This requires a comprehensive understanding of the effects on stakeholders. This study investigates the consequences of sustainable waste management in the context of increasing tourists. It examines the intricate dynamics between stakeholders and waste man-

DOI: 10.4018/979-8-3693-6110-8.ch008

agement practices to identify the benefits and challenges of this paradigm shift. The perspectives of various stakeholders, including local communities, businesses, government agencies, non-governmental organisations, and tourists, are analysed to illuminate their diverse interests, concerns, and aspirations. With the growth of tourism, the amount of waste generated also increases, necessitating sustainable waste management practices. The need for collaboration, innovation, and stakeholder engagement is also highlighted to develop a sustainable path.

1. INTRODUCTION

Dharamshala, a renowned tourist destination in the Himalayas, is a cultural melting pot with diverse landscapes. Nevertheless, the city confronts the challenge of effectively managing waste to preserve its natural beauty and cultural identity. Waste management is essential for sustainable tourism development as it mitigates environmental degradation, enhances the visitor experience, and fosters community well-being. This chapter examines the influence of waste management practices in transforming Dharamshala into a sustainably developed tourist destination. The surge in waste generation, including solid waste, wastewater, and litter, due to the influx of tourists can pose significant environmental and health hazards. Inadequate waste disposal can tarnish the aesthetic appeal of tourist sites, diminishing the overall visitor experience and Dharamshala's attractiveness as a tourist destination. This empirical investigation aims to illuminate the impact of waste management practices in transforming Dharamshala into a sustainably developed tourist destination.

Dharamshala, situated amidst the tranquil and beautiful Dhauladhar range of the Himalayas, Dharamshala has become a well-known tourist attraction, drawing people from all over the world who come to experience the breathtaking natural beauty, spiritual solitude, and cultural immersion. Dharamshala, the exiled Tibetan government's capital and home to His Holiness the Dalai Lama, is a cosmopolitan city that combines many different traditions, cultures, and topographies. Dharamshala's charm is found in its diverse array of attractions, which spans from the serene surroundings of McLeod Ganj to the historic temples of Bhagsu Nag, and from the picturesque pathways of Triund to the bustling Tibetan marketplaces. To maintain its pristine environment and rich cultural legacy, Dharamshala must successfully manage its waste in addition to the city's explosive increase in tourists (Diederich,1993; Bodo, 2019; Jasrotia et al, 2020).

In 2017, Dharamshala received 2,823,289 visitors; however, this number surged to 676,836 following the COVID-19 pandemic in 2023 (https://himachaltourism.gov.in). The proposed international airport and four-lane highways will boost tourism and attract more visitors by improving transportation options. Consequently, the significant growth in the tourism industry has resulted in a substantial increase in waste generation. Therefore, there is an urgent need for effective waste management solutions to deal with the escalating volume of waste and to ensure environmental sustainability. Implementing efficient waste management practices is crucial to preserve Dharamshala's natural beauty and provide a satisfying experience for both locals and visitors.

. Tourist arrivals at Kangra from 2012-2023

Year	Domestic Tourists	Foreigners
2023	660614	16222
2022	404305	6538
2021	234251	2701
2020	217827	9921
2019	2226897	115990
2018	2117675	107451
2017	2684948	138341
2016	2528837	123895
2015	2396970	112843
2014	2230888	102479
2013	2098028	102595
2012	2190302	115109

Source: https://himachaltourism.gov.in/wp-content/uploads/2024/02/Tourist-Statistics.pdf

Waste control is essential to ensure the sustainable development of Dharamshala's tourism industry. Increased waste produced from tourist inflow includes waste, wastewater, and solid waste, all of which pose serious risks to the environment and public health if allowed unchecked. According to Thakur (n.d.), poor waste management can damage the visual appeal of tourist attractions, which lowers Dharamshala's appeal as a travel destination and detracts from the entire visiting experience. Various parties, including local government agencies, community organisations, and tourism enterprises, have launched initiatives to solve this issue, recognising the vital role that waste management plays in maintaining the growth of the tourism industry. These programs cover a variety of waste management techniques, such as public awareness campaigns, composting facilities, recycling programs, and waste segregation. Furthermore, partnerships between waste management and non-governmental organisations (NGOs) have improved Dharamshala's waste management infrastructure and capabilities (Gupta et al., 2023).

This chapter's empirical study thoroughly analyses empirical data, including waste creation rates, waste composition studies, and stakeholder interviews, to clarify the success of current waste management programs and pinpoint areas for improvement. By comprehending the complex relationship between waste management and sustainable tourism development, the authors aim to offer valuable insights and recommendations to policymakers, practitioners, and other stakeholders dedicated to promoting sustainable tourism in Dharamshala and beyond.

2. WASTE MANAGEMENT UNIT

According to Table 2, between 1 March 2023 and 30 April 2024, the Dharamshala waste processing unit gathered 416.3 metric tons (416,268 kg) of rubbish from various wards and municipal staff; 7,451 truckloads were received. This data highlights the city's continuous efforts to manage waste effectively, essential for developing and optimising waste management plans to preserve Dharamshala's clean and sustainable environment.

Table 2. Monthly trend of total waste

Month	(Waste in kg)	Month	(Waste in kg)
Apr-23	15398	Nov-23	24222
May-23	22528	Dec-23	35520
Jun-23	33859	Jan-24	44943
Jul-23	31472	Feb-24	37981
Aug-23	19309	Mar-24	44366
Sep-23	35357	Apr-24	42071
Oct-23	29241		

Source: DHM Urban MRF Waste Monitoring (May 18, 2024)

The waste processing unit received 287.1 metric tons of waste between 1 October 2023 and 30 April 2024; 214.08 metric tons were successfully processed. Furthermore, 261.15 metric tons of waste were transferred to recycling facilities after being thoroughly separated (Figure 1). This initiative demonstrates how effectively the unit manages and diverts waste into environmentally friendly recycling methods, thereby improving the community's waste management results.

Figure 1. Waste inward and outward

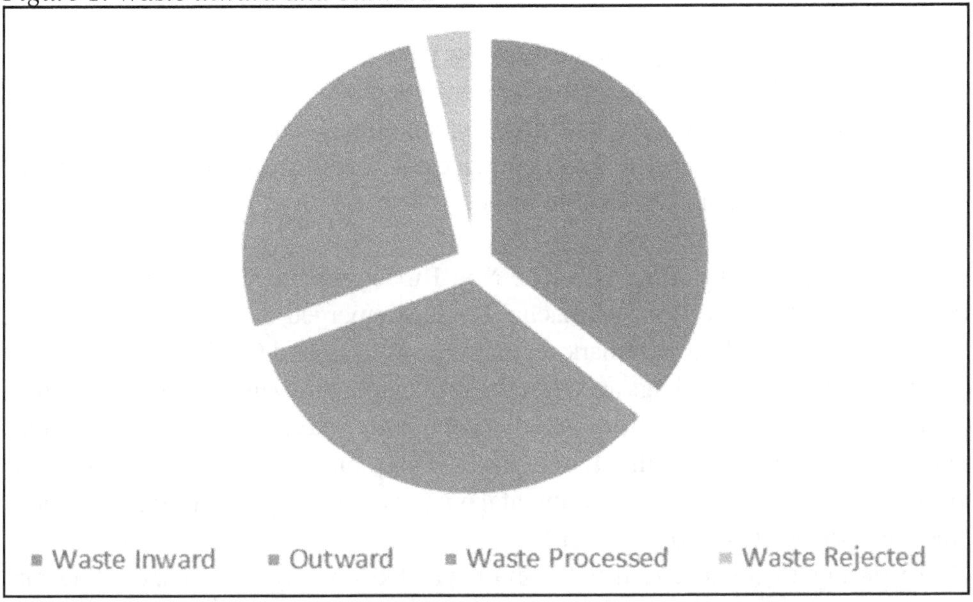

Source: DHM Urban MRF Waste Monitoring (2024)

The pivotal role of waste management in community capacity-building focuses on healthcare component care to ensure a holistic approach to public health. According to Van den Berg (2005), the components of healthcare are as follows:

a. **Clean Environment:** Waste management initiatives are essential for maintaining a clean environment free from harmful pollutants and contaminants. Proper waste disposal practices prevent the accumulation of waste in public spaces, thereby reducing the risk of environmental degradation and its adverse effects on community health.
b. **Clean Air:** Effective waste management helps minimise air pollution by reducing the release of harmful emissions from waste decomposition. Strategies, such as recycling and waste-to-energy programs, contribute to cleaner air quality, safeguarding respiratory health and overall well-being.
c. **Clean Water:** Contaminated waste can leach harmful substances into water sources, posing significant health risks to communities. By implementing proper waste collection and treatment methods, communities can safeguard their water supply and ensure access to clean and safe drinking water.
d. **Clean Food:** Waste management practices influence food safety and security by preventing the contamination of agricultural lands and food production areas. Through responsible waste disposal and recycling, communities can mitigate

the risk of foodborne illness and ensure the availability of clean and nutritious food options.

e. **Sunlight:** Incorporating sustainable waste management practices can harness the power of sunlight for energy generation and therapeutic purposes. By promoting renewable energy sources and green technologies, communities can reduce their reliance on fossil fuels and embrace eco-friendly alternatives, leading to a healthier and more sustainable future.

With the enactment of the H.P. The Non-Biodegradable Waste Control Act of 1995 and the Rules of 1996, Himachal Pradesh emerged as a national leader in waste management. This landmark regulation was designed to address the issue of coloured recycled plastic bags and effectively manage solid waste. By implementing these rules, Himachal Pradesh promoted sustainability and environmental protection, setting a high standard for other Indian states. This pioneering project is a testament to the state's commitment to environmental preservation and promoting eco-friendly practices (www.https://hppcb.nic.in).

According to the Dharamshala Municipal Corporation (n.d.) and (Sharma, 2020), the waste management practices in Dharamshala, Himachal Pradesh, and Municipality Area, as well as the segregation of waste collected through different channels, are as follows:

a. Municipal Corporation administrative 17 wards (cover civilian populations, government offices, including hospital waste, business and semi-business populations, hotels, resorts, and guesthouses, and commercial tourist vehicles parking door-to-door waste/waste material collection agents and staff.
b. Waste Warriors, a non-governmental group, manages waste collecting facilities that separate waste into four categories: paper/cardboard, plastic, e-waste, and biodegradable items. Next, the waste is put to new use. Rough paper is used as a raw material instead of being used to tar roads, biodegradable materials are used in biogas plants, and plastic and paper are used as fuels in cement factories. Waste is efficiently recycled and reused through their efforts, promoting environmental sustainability. Dry waste is divided into more than 17 categories. One effective waste management program is collecting and recycling tetra packs into rigid sheets for indoor furniture (Images 1, 2, 3).

Figure 2. Types Of dry waste

Source: Authors Photography at Processing Centre

Figure 3. Tetra packs compressed for recycling

Source: Authors Photography at Processing Centre

Figure 4. Hard board made out of tetra packs

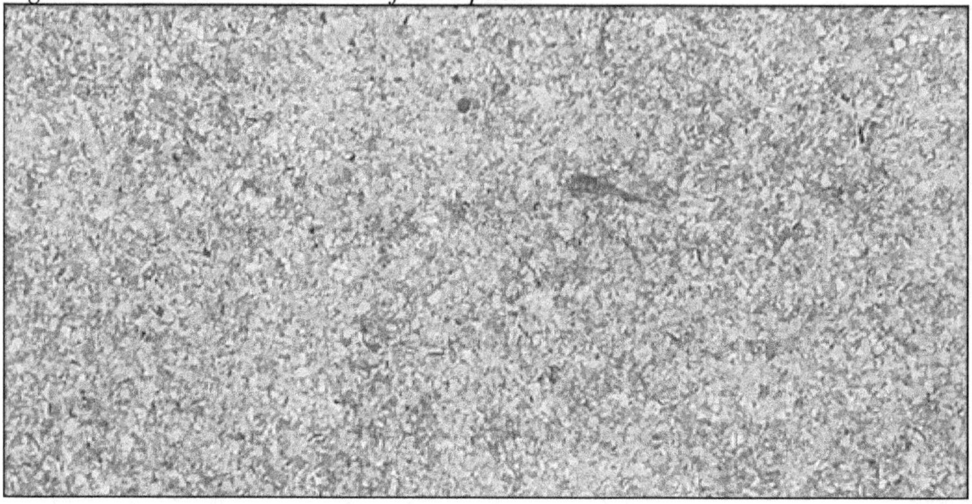

Source: Authors Photography at Processing Centre

c. Dhauladhar Cleaner is a non-governmental organisation (NGO) engaged in waste/waste cleanliness drives in public places such as tourist tracking sites and nearby river bed areas. It also runs used clothes collection vehicles in most Municipal Corporation wards. It distributes them, using those collected cloths as raw materials to produce usable materials like rugs and mats. It is a group of volunteers based in Dharamshala who, through community efforts, make a green process of collaboration and sharing to proactively solve environmental issues and make a greener and cleaner tomorrow. Their mission was to provide a solution to mountain pollution.
d. Educational Institutions, including government and private academic institutes/ecosystems, schools, colleges, and vocational institutes, are moving forward in the direction of a vision led by various ministries whereby they undertake onsite/offsite cleaning activities and their continuous effort ongoing under the SWACHATA DRIVES (cleanliness drives) through various clubs, such as the National Cadet Corps, National Service Scheme, Nehru Yuva Kendra, and Yuva Tourism Club in the academic world during national calls from the Honourable Prime Minister of India as well as the Ministry of Education.
e. Stakeholders, such as civilian populations, government offices, hospitals, businesses, semi-businesses, hotels, resorts, guest homes, and commercial tourist vehicle parking facilities, have all actively participated, supported, and contributed to this effort. Together, these efforts are directing projects in the right direction and promoting advancement and community development.

3. LITERATURE REVIEW

Waste management is a critical component of sustainable tourism development, significantly impacting the environmental, economic, and social sustainability of tourist destinations (Sharma, 2020; Tsai et al., 2021; Perkumienė et al., 2023; Burlakov et al., 2020; Ezeah et al., 2015). Studies have identified various challenges and opportunities in waste management within tourism, focusing on factors like waste generation rates, waste composition, collection and disposal methods, infrastructure, stakeholder engagement, and policy frameworks. Research emphasises the importance of tailored waste management strategies that address each destination's unique needs. Ineffective waste management can lead to pollution of natural ecosystems, degradation of cultural heritage sites, public health risks, aesthetic degradation, and negative visitor perceptions. In contrast, effective practices enhance destination attractiveness, visitor satisfaction, community well-being, and long-term sustainability (Pham et al., 2019). Dharamshala, a prominent hill station in India, exemplifies the challenges of waste management amid rising tourist numbers. Research by Gupta et al. (2023) reveals that these challenges have led to environmental degradation and loss of natural beauty, prompting efforts to transform Dharamshala into a sustainably developed tourist destination through effective waste management strategies.

3.1 Challenges and Strategies in India's Waste Management

India's waste management system faces significant challenges, including insufficient leadership, planning, infrastructure, and resources (Brar, 2018). The inadequacy of infrastructure and maintenance, particularly in municipal solid waste management, results in poor rubbish collection. For instance, in Himachal Pradesh, the daily waste production exceeds 350 tons. However, the collection efficiency is only 60-70% in areas such as Sunder Nagar and Solan. The inefficiencies are attributed to the lack of disposal vehicles, insufficient bins, inadequate equipment and a shortage of labour (Sharma et al., 2018). To improve India's waste management, it is recommended that waste reduction and sustainability be prioritised through recycling, composting, biogas plants, public awareness, and regular waste collection. Effective waste segregation at the source, provision of community waste bins, composting, sanitary landfills, and anaerobic digestion is crucial for enhanced waste management (Joshua, 2015; Joshi et al., 2016). Long-term strategies should focus on infrastructure improvements, public awareness campaigns, and the adoption of innovative techniques (Kumar et al., 2017).

3.2 Global Perspectives on Waste Management

Waste management challenges are not unique to India; they are globally linked to resource consumption and economic growth. High-income nations face issues with high consumption, while middle- and low-income nations struggle with inadequate waste management infrastructure and rising waste rates (Singh et al., 2014). Intricate product compositions and comprehensive strategy deficiencies have impeded waste management technology advances. Successful strategies worldwide emphasise coordinated actions, strict regulations, public-private partnerships, and investments in infrastructure.

In Corfu, Greece, poor waste management during peak tourism damages the island's reputation, highlighting the need for source separation and infrastructure investments (Koliotasi et al., 2023). Similar issues are evident in Lapland's tourist areas, where inadequate funding, oversight, and infrastructure exacerbate waste problems. Despite awareness, the persistence of single-use goods remains challenging, emphasising the need for sustainable practices and collaboration (Benkherouf, 2023).

Tourists seek clean settings, so sustaining tourism requires proper solid waste management (SWM). National and local governments manage waste collection and disposal in Tunisia, with assistance from the populace and the travel and tourism sector. Sustainable practices, integrated methods, and group activities are necessary for effective SWM. The study underscores the significance of Extended Producer Responsibility (EPR) in tackling waste issues, specifically in tourist locations. It also emphasises the necessity of unambiguous law, stakeholder cooperation, and sustained funding. The study also notes that more investigation is required into the financial and technical aspects of SWM solutions in Tunisia's tourist areas (Chaabane et al., 2019).

Similarly, Perkumienė et al. (2023) investigated waste management difficulties in the tourism and recreation industries by interviewing experts from Morocco, Lithuania, and Turkey. The high volume of garbage produced, the requirement for more stringent rules, the significance of public-private partnerships for raising awareness, and the lack of funding for waste management are some of the major conclusions. Sustainable tourism, public health, and the environment depend on efficient trash management. In order to guarantee long-term benefits for tourism and leisure in these nations, the study emphasises the necessity of further research and sustainable methods.

3.3 Technological and Policy Innovations

Technological innovations like artificial intelligence (AI) can revolutionise tourism by optimising pricing strategies, reducing waste, increasing efficiency, and promoting sustainable practices. AI can analyse data on energy use, waste management, carbon emissions, and environmental impacts, helping implement sustainable tourism practices and attract eco-aware tourists (Singh, 2024). Additionally, an IoT-enabled architecture for waste management promotes intelligent infrastructure, innovative business models, and interdisciplinary cooperation while emphasising waste avoidance, upstream separation, timely collection, and value recovery (Esmaeilian et al., 2018).

Policy innovations are equally critical in waste management. The Basel Convention and PACE II highlight zero-waste programs in Australia, Brazil, China, Lebanon, and New Zealand, focusing on recycling, community engagement, and legislative support. These initiatives aim to cut emissions, conserve energy, and promote social inclusion through sustainable waste management practices. The Basel Convention also requires countries to adjust their national laws by January 2021 to implement new waste minimisation and management procedures (B3011, 2021).

The improper disposal of electronic waste poses severe risks to human health and the environment, making it a significant source of pollution (Kandel, 2022). The "e-Market for Returned Deposit" system in the United States aims to address e-waste challenges by increasing low-income communities' access to technology, decreasing environmental impacts, improving material recovery, and promoting ethical disposal practices (Kahhat et al., 2008). This approach leverages market principles and economic incentives to provide sustainable solutions, ensuring proper handling of e-waste.

3.4 Environmental and Health Impacts

Tourism in the Himalayan ecosystem contributes significantly to carbon emissions, waste generation, pollution, and soil, water, and air quality degradation. Despite the economic benefits of tourism, these environmental impacts have led to growing criticism. Sustainable tourism practices are urgently needed, especially in West Bengal, Uttarakhand, and Himachal Pradesh (NGT, 2022). Effective policies, such as visitor restrictions, vehicle laws in Himachal Pradesh, and plastic prohibitions in Sikkim, are crucial for mitigating these impacts. The risks associated with improper e-waste disposal are considerable, as it is a significant source of environmental pollution and a danger to public health. Recycling electrical waste at certified facilities with the required equipment ensures safe disposal procedures, reduces environmental pollution, and safeguards public health (Kandel, 2022). Similarly, multinational corporations, particularly in fast-moving consumer goods sectors, contribute substantially to plastic pollution, with Coca-Cola, PepsiCo, Nestle, Danone, and Philip Morris International being major offenders (Ujala, 2024). The Indian Supreme Court has also highlighted the link between climate change and the constitutional right to a stable environment, stressing the need for sustainable practices to protect vulnerable populations from the adverse effects of climate change (Ujala, 2024). Himachal Pradesh faces severe geo-environmental risks, including the degradation of Dal Lake, landfills, and slate mining operations due to poor waste management. Regular monitoring of illegal mining and open waste disposal, along with the beautification and maintenance of Dal Lake, is necessary. Sustainable geo-environmental practices are essential for building ecologically stable communities and protecting life-support systems for future generations (Thakur et al., 2013).

3.5 Community Engagement and Sustainability Initiatives

Community involvement is essential for addressing waste management and sanitation challenges. The Human-Centred Design (HCD) approach emphasises comprehensive strategies that align with sustainable development goals (SDGs) 3, 5, 6, 13, and 17, promoting the use of innovative materials like plastic bricks for construction to enhance sanitation, empower women, and encourage self-sufficiency within communities (Balu et al., 2020). In India, targeted levies, decentralized composting, waste categorisation, public education on source segregation, and partnerships with educational institutions are recommended for effective waste management (Joshi et al., 2016). The Himachal Pradesh government's waste management initiatives, as highlighted in a report by the Comptroller and Auditor General of India, underscore the importance of collaboration among various organisations to ensure effective waste management. The Department of Urban Development develops state-level plans, while local urban bodies oversee waste collection, processing, and disposal. Environmental standards are enforced by the State Pollution Control Board, with waste generators required to pay user fees and separate waste at the source (India, C. A. A. G., 2022).

Waste management is crucial for achieving environmental sustainability, particularly in tourism and urban development. A transition to a circular economy, where waste is viewed as a resource rather than a problem, is necessary. This requires comprehensive waste management strategies, technological advancements, public-private partnerships, and strong governance. By fostering community engagement and implementing innovative policies, destinations can enhance their sustainability, mitigate environmental impacts, and ensure long-term viability for tourism and broader urban development goals. The global community can effectively address waste management challenges by engaging producers, consumers, and authorities through a systems approach that integrates design, manufacturing, consumption, and waste management. The need for large-scale adjustments in corporate governance, consumer behaviour, and corporate structures is imperative, with the United Nations and the European Union playing crucial roles in global policymaking (Bohdanowicz, 2006; Ujala, 2024). Sustainable development of the tourism industry, including hotel business practices, requires a coordinated effort involving education, awareness, and collaboration among academia, industry, and governments.

4. METHODOLOGY

This study employed a mixed methods approach to investigate the effects of waste management on transforming Dharamshala into a sustainably developed tourist destination. It analysed waste management practices and their impact on tourism development through primary and secondary data sources. In-depth interviews were conducted with 92 key informants, including nine government officials, three waste management experts, seven tourism industry representatives, 15 visitors (both national and international), 47 residents (a few from each ward), two taxi unions, and nine community leaders (ward chancellors of the Municipal Corporation, Dharamshala). The interviews and questionnaires explored stakeholders' perspectives on Dharamshala's waste management challenges, strategies, and outcomes. The study sample included a diverse range of participants to ensure the representativeness and validity of the findings. Both domestic and international tourists visiting Dharamshala were interviewed to capture their perceptions and experiences of waste management and tourism. Residents of Dharamshala were also interviewed to assess their waste disposal habits, attitudes towards waste management, and perceived impacts of tourism on waste generation. Waste management authorities and officials from municipal agencies and waste management organisations were interviewed to understand the policies, practices, and challenges of waste management in Dharamshala. Tourism industry representatives, including hoteliers, tour operators, and cultural heritage site managers, were interviewed to explore their roles and perspectives in waste management and sustainable tourism development. By employing a multi-method approach and engaging diverse stakeholders, this study aims to provide comprehensive insights into the complex dynamics of waste management and its implications for sustainable tourism in Dharamshala. The researchers visited all four sites to observe Dharamshala's waste management infrastructure, facilities, and practices. Site visits provided firsthand insights into implementing waste management initiatives and their impacts on the local environment and tourism sites.

5. RESULTS

5.1 Impact of Waste Management on Environmental Sustainability

This study reveals various waste management practices in Dharamshala, including waste segregation at source, collection, transportation, and disposal methods. A representative NGO stated the following, which identifies the key stakeholders involved in waste management, such as municipal authorities, NGOs, and community-based organisations:

Our NGO raised waste management and segregation awareness by launching a dry waste collection program in six panchayats under the Dharamshala Block. These challenges include low public awareness, little funding, poor enforcement of regulations, and inadequate infrastructure. Progress requires more financing and state-level initiatives.

One hotelier indicated:

Waste is managed through segregation at the source and disposal via municipal services. As part of our CSR, we prioritise sustainability practices. Maintaining hygiene is crucial for business, especially with foreign tourism bodies. We also organise clean-up drives every 15 days to keep our surroundings clean.

The analysis indicates the effectiveness of the waste management infrastructure, including the availability of waste bins, recycling facilities, composting units, and recycling processes, as shown in images 1 to 8.

Figure 5. Compressed tetra packs being loaded and transported to recycling facility

Source: Authors Photography

Figure 6. Segregation of waste

Source: Authors Photography

Figure 7. Stacking of segregated and compressed waste

Source: Authors Photography

Figure 8. Segregation of waste and sending to shredder for using as cement factory fuel

Source: Authors Photography

Figure 9. Weighing of daily waste

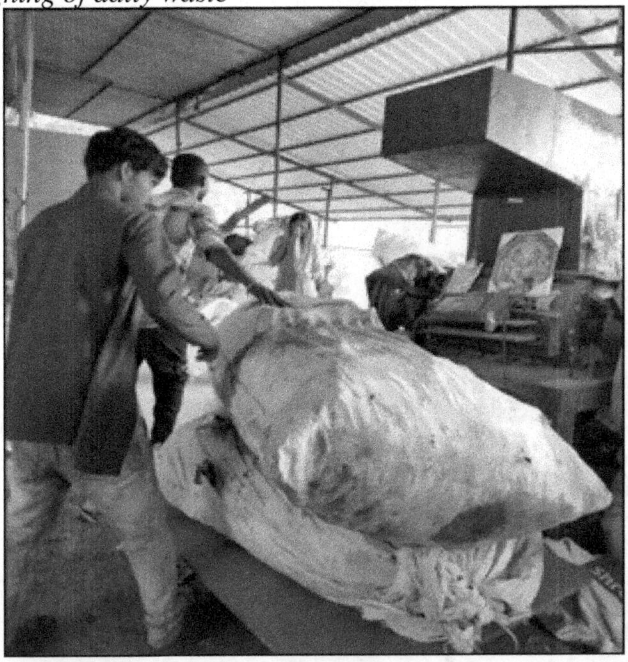

Source: Authors Photography

The research indicates a beneficial influence of waste management schemes on environmental sustainability in Dharamshala. The primary data unequivocally shows that nearly 92% of the population is cognizant of waste management practices and actively engages in them. Furthermore, this same proportion of individuals believe that efficient waste management is indispensable for the sustainability of Dharamshala as a tourist destination, thereby highlighting the extensive recognition of its significance in the area's future, as depicted in Figures 2 and 3.

Figure 10. Awareness of waste management practices

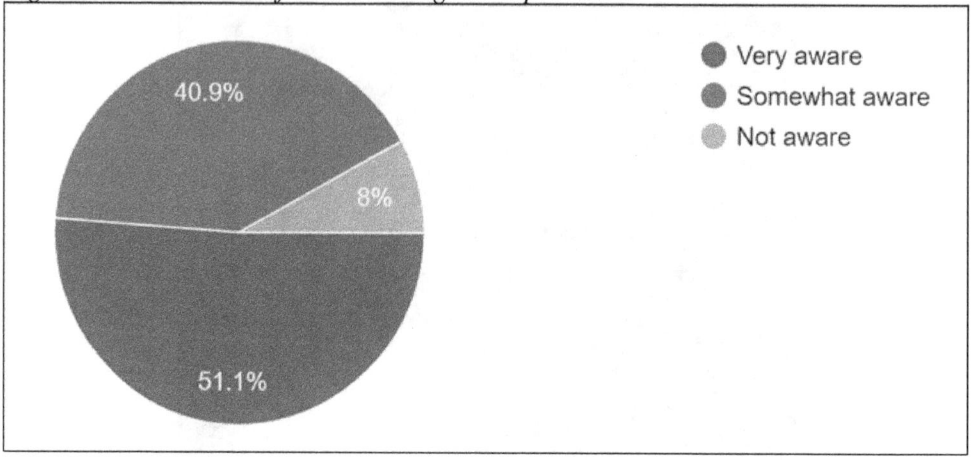

Note: (n=88)

Figure 11. Perception of the importance of waste management in the sustainability of Dharamshala as a tourist destination

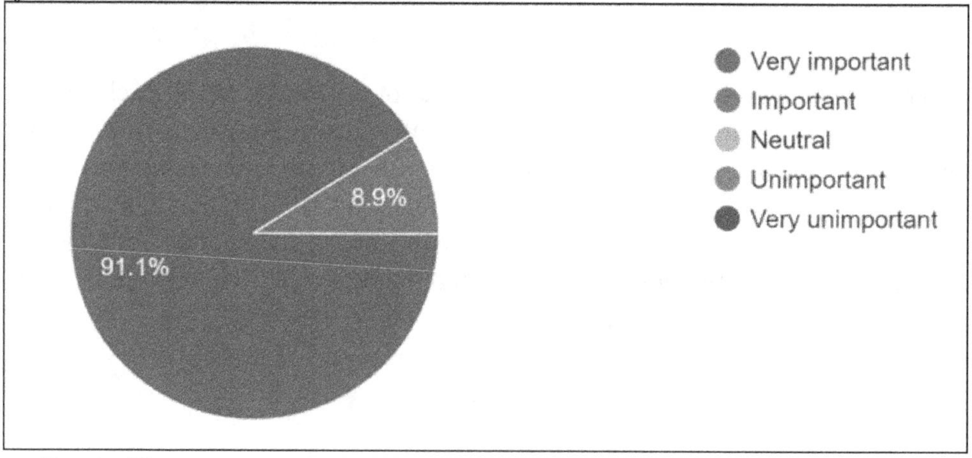

Note: (n=90)

Effective waste management practices contribute to the reduction of littering, pollution, and ecological degradation in tourist destinations and the surrounding areas. Community leaders stated:

> *Waste management has also been incorporated into the curriculum through awareness campaigns, educational videos, and activities. Schools prioritise cleanliness, particularly the management of e-waste, and encourage the 5Rs. Dustbins*

positioned thoughtfully, and frequent instructions on how to dispose of waste serve to reinforce these lessons, encouraging sustainability and responsibility.

5.2 Tourist Perceptions and Satisfaction with Waste Management Initiatives

Tourist behaviour significantly influences waste generation and management in tourist destinations such as Dharamshala. An international visitor positively stated:

Cleanliness is aesthetically attractive, and I rarely witness such efficient waste management techniques in India. I would most definitely love to return to Dharamshala if advancements keep up.

Waste management is seen by visitors as “very important” in keeping Dharamshala a popular travel destination. They believe that to maintain cleanliness and manage waste, there should be closer cooperation between the municipality and the locals. Even though waste output has increased due to population growth and tourism, waste sorting, transportation, and disposal efficiency are challenging. Even though Dharamshala has improved and is thought to be cleaner than other hill stations, more work has to be done, like increasing the number of waste cans in public spaces. Increasing waste management awareness among tourists through signage, information boards, and interactive programs can encourage responsible behaviour, such as proper waste disposal and recycling. Tourists' decisions to return are influenced by proper waste management, which includes recycling and converting waste into renewable energy. Visitors stated that better practices would increase the area's cleanliness and appeal. Tourists express satisfaction with visible waste bins, clean streets, and well-maintained public spaces, all contributing to a positive destination image. Overall, the empirical findings underscore the critical role of effective waste management in promoting environmental sustainability and enhancing the tourist experience in Dharamshala. By addressing existing challenges and building on successful initiatives, Dharamshala can leverage waste management as a tool for sustainable tourism development.

Implementing sustainable waste management strategies helps preserve natural resources, protect biodiversity, and mitigate adverse environmental impacts. Improved Tourist Perceptions and Satisfaction with Waste Management Initiatives. These findings indicate that tourists perceive waste management as essential to their overall tourism experience in Dharamshala.

5.3 Stakeholder Perspectives

To understand and address stakeholder perspectives, Dharamshala can develop comprehensive waste management strategies that are inclusive, participatory and sustainable. Effective collaboration between local authorities, communities, and tourists is key to lasting waste management and environmental conservation improvements.

5.3.1 Role of Local Authorities in Waste Management

Government offices' waste management procedures involve sorting waste into designated containers, which sanitation workers transport to appropriate disposal sites using public transportation. Nevertheless, these offices have recommended banning plastic from the outset. This change would aid in environmental restoration and prevent the addition of new plastics to the ecosystem.

Although there have been improvements in waste management in Dharamshala, some lingering issues remain. While inadequate infrastructure and low public awareness persist, initiatives such as source separation, door-to-door collection of materials, and local waste disposal have progressed. The cooperative efforts of stakeholders, including sanitation drives, waste hotspot beautification initiatives, and awareness campaigns conducted by self-help groups (SHGs), have contributed to improving waste management in the city. Implementing visible helplines, weekly hotspot cleaning plans, fines, and penalties have also enhanced waste management. Dharamshala's cleanliness compared to other cities demonstrates the effectiveness of these efforts. Educational initiatives and outward signage, such as banners and posters, have helped raise awareness about appropriate recycling and waste segregation. Moreover, the city's commitment to maintaining a cleaner environment is evident in the provision of dustbins in residential and commercial areas and the prohibition of open dumping. These actions positively impact tourism and enhance the city's reputation. The citizens' involvement in keeping the city clean and the noticeable changes in the city demonstrate the progress made in waste management.

5.3.1 Community Participation and Engagement

Divergent opinions regarding Dharamshala's waste management are expressed in the community. Many notable advancements have been made, particularly in recycling, municipal services, and source segregation. However, problems exist, such as poor finances, public awareness, and inadequate infrastructure. Social media campaigns, community gatherings, and announcements by local governments are examples of successful projects. These have significantly increased awareness and enhanced procedures. Community members suggest more waste cans, law enforcement, and increased public participation in waste management. Although some believe Dharamshala is cleaner than other locations, everyone agrees that there is always room for development. Further progress can be made by implementing waste management procedures with more rigour, improving recycling initiatives, and raising awareness through educational initiatives. Cooperation between local governments and citizens and observable advancements in waste management could increase the city's sustainability and tourism appeal.

5.3.2 Industry Representatives

Hoteliers and members of the tourist taxi union have suggested that car manufacturers should prioritise the inclusion of small dustbins in their vehicle designs to facilitate waste segregation and ensure proper disposal at designated sites. Although the existing doorstep rubbish collection services have been satisfactory, there is a desire for additional incentives to encourage greater participation in waste management programs. Suggested benefits include receiving manure from deposited organic waste, biogas for cooking meals for staff, and customised recycled brown paper products such as paper bags and other handcrafted items. Implementing these measures could significantly enhance the city's sustainability efforts and waste management practices.

5.3.3 Educational Institutions

The findings from educational institution interviews indicated that teachers proactively promoted the correct disposal of waste. However, they acknowledged that additional effective models are necessary to bolster these efforts. To ensure that children gain an early understanding of the principles of reduction, reuse, and recycling, it is recommended that industrial visits to waste-recycling facilities be organised. These practical experiences are instrumental in helping them grasp these critical environmental concepts at the outset of their academic journey. The government could look to the recycling plants set up in Dharamshala for such initiatives.

5.4 Challenges and Opportunities

The primary data in Figure 4 reveals the barriers to effective waste management in Dharamshala.

5.4.1 Lack of Infrastructure

Like many growing cities, Dharamshala struggles with insufficient waste management infrastructure, which significantly hampers the effectiveness of waste collection, transportation, and disposal. Respondents indicated that the absence of adequate facilities means that waste often accumulates in public areas, leading to environmental degradation and health hazards. Moreover, without proper waste segregation systems and disposal sites, managing the various types of waste efficiently, particularly hazardous and non-biodegradable materials, becomes difficult. Enhancing infrastructure is crucial to managing the increasing volume of waste in Dharamshala, especially with the city's growing population and tourism industry. Views of a community member are quoted for better insight:

Lack of community involvement, inadequate money, poor enforcement of laws, inadequate infrastructure, and low public knowledge are the main issues facing Dharamshala's waste management efforts. Inadequate dustbins and the requirement for stronger public-private partnerships to effectively handle waste management exacerbate these problems.

5.4.2 Limited Resources

Respondents stated that financial constraints are a significant bottleneck in improving waste management practices. Limited budgets restrict the ability of local governments to invest in essential infrastructure, such as waste treatment plants, recycling facilities, and efficient transportation systems. Additionally, financial limitations hinder the adoption of modern technologies that can optimise waste management processes, such as automated waste sorting or energy recovery from waste. The reliance on outdated systems and manual labour due to budgetary constraints also leads to inefficiencies and higher long-term costs. Finding sustainable financing models through public-private partnerships or international aid could be vital in addressing these limitations. Views of a community member is:

They are merely attempting to manage the situation; there is no particular plan. It appears that they lack infrastructure, resources and manpower.

5.4.3 Behavioural Change

According to the respondents, changing public attitudes and behaviours towards waste disposal is a significant challenge in Dharamshala. Ingrained habits, such as littering and illegal dumping, are deeply rooted in societal norms and require extensive education and awareness campaigns to alter. Raising awareness about the environmental and health impacts of improper waste disposal, combined with providing accessible and convenient waste disposal options, is essential. Public participation is critical, and fostering a sense of community responsibility can lead to more effective waste management. Introducing penalties for non-compliance, alongside positive reinforcement for proper waste disposal practices, can also support behavioural change. Additionally. a community member's view on the habits of people is stated:

> *"Organise community clean-ups, recycling drives, and workshops to raise awareness about the importance of waste management. Encourage the public to follow waste management rules, which can effectively change behaviour and lead to a cleaner environment."*

5.4.4 The Informal Waste Sector

Study respondents believe the informal waste sector plays a dual role in Dharamshala's waste management system. While informal waste pickers and recyclers contribute to waste reduction and recycling, they complicate formal waste management systems. This sector's lack of regulation and integration poses challenges, such as uncoordinated waste collection efforts and unsafe working conditions for waste pickers. Informal workers often operate outside the formal waste management framework, leading to inefficiencies and difficulties in tracking waste flows. Integrating the informal sector into the formal system through training, social security benefits, and creating formal employment opportunities can help mitigate these challenges while acknowledging this sector's valuable role in recycling and waste recovery.

Figure 12. Major challenges facing waste management in Dharamshala

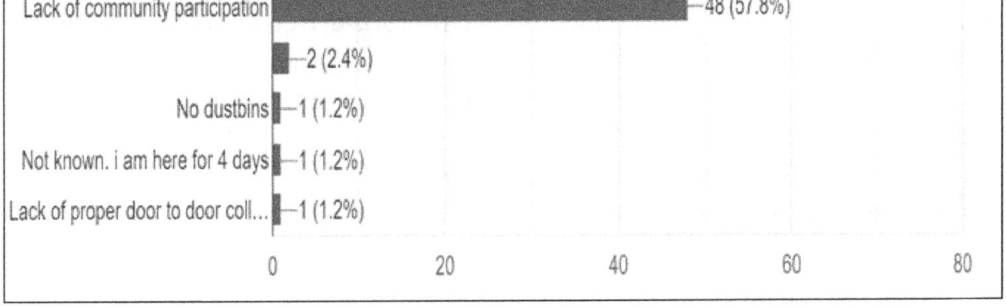

5.6 Opportunities for Improvement and Innovation

The findings from the stakeholder interviews suggest various promising avenues for improving waste management in Dharamshala. However, while these opportunities have the potential to transform the waste management system, each also comes with its own set of complexities that must be critically examined.

5.6.1 Technology Adoption

Respondents indicated that advanced technologies like smart bins, GPS tracking, and waste-to-energy solutions offer significant potential to optimise waste management. Smart bins can alert authorities when they are full, improving collection efficiency. GPS tracking can ensure better route management for waste collection vehicles, reducing fuel consumption and operational costs. Waste-to-energy technologies can also help convert waste into energy, reducing landfill usage and contributing to energy needs. The deployment of these technologies requires substantial initial investment, which may be a challenge given Dharamshala's budgetary constraints. Furthermore, the success of such technologies depends on the availability of skilled personnel for maintenance and operation and public cooperation. There is also the risk of technology being underutilised or failing to integrate with existing systems, leading to inefficiencies rather than improvements. Additionally, waste-to-energy solutions must be carefully managed to avoid environmental issues, such as air pollution from incineration processes.

5.6.2 Public-Private Partnerships (PPPs)

Respondents believe PPPs can leverage private sector expertise and investment to enhance waste management services. This collaboration can lead to more innovative and cost-effective solutions, with private companies bringing in new technologies, management practices, and financial resources that might be inaccessible to the public sector alone. PPPs can also help scale up waste management infrastructure more quickly. While PPPs can provide significant benefits, they also come with the risk of prioritising profit over public interest. If profit-driven companies dominate the partnership, consumers may face increased costs. Additionally, these partnerships require robust legal frameworks and oversight to ensure that public interests are safeguarded and that there is accountability and transparency. The negotiation and management of PPPs can also be complex, requiring strong governance capacity that may be lacking in smaller municipalities like Dharamshala.

5.6.3 Community Engagement

According to the study respondents, engaging the local community is crucial for the success of any waste management initiative. Awareness campaigns, training programs, and involving the community in decision-making processes can lead to more effective waste segregation at the source, increased recycling rates, and better compliance with waste management protocols. Empowered communities are more likely to take ownership of local waste management issues, leading to sustainable behaviour change. However, achieving meaningful community engagement is often challenging, especially in areas with low awareness and education levels about waste management. It requires sustained efforts and resources to build trust and motivate communities to participate actively. There is also the challenge of ensuring inclusivity, where marginalised or less vocal groups are given a platform to contribute. Inconsistent participation or resistance to change can hinder the effectiveness of community engagement efforts.

5.6.4 Circular Economy Initiatives

Stakeholder interviews indicate that promoting a circular economy can drastically reduce waste generation by encouraging practices like reducing, reusing, and recycling materials. This shift conserves resources and creates economic opportunities by developing new industries, such as recycling businesses or upcycling initiatives. By closing the loop, Dharamshala could reduce its reliance on landfills and decrease environmental pollution. Transitioning to a circular economy requires a systemic change that can be difficult to implement. It demands changes in consumer behaviour, business practices, and government policies, all of which must be aligned to make circular initiatives successful. Moreover, the infrastructure for supporting circular economy activities, such as efficient recycling centres or markets for reused products, may not be fully developed. There may also be resistance from industries relying on the traditional linear economy and consumers accustomed to disposable products.

5.6.5 Policy Reforms

According to the study respondents, strengthening waste management policies can provide the regulatory backbone to enforce sustainable practices. Effective policy reforms can include stricter regulations on waste disposal, incentives for recycling and waste reduction, and penalties for non-compliance. Policy reforms can also facilitate the integration of informal waste workers into the formal economy, ensuring better working conditions and contributing to overall waste management efficiency. However, policy reforms often face resistance from various stakeholders, particularly if perceived as restrictive or economically burdensome. Implementing new policies requires political will, administrative capacity, and enforcement resources. There is also the risk of unintended consequences, such as increased illegal dumping, if waste disposal becomes too costly or inconvenient for businesses and households. Ensuring policies are well-designed, effectively communicated, and enforced is critical to their success.

6. RECOMMENDATIONS AND FUTURE RESEARCH DIRECTIONS

The recommendations for effective waste management in Dharamshala offer an approach that integrates policy-level changes and practical strategies to address the city's waste challenges. Strengthening waste management regulations is a crucial first step. Enforcing existing regulations and introducing new policies will promote sustainable practices such as waste segregation, recycling, and resource recovery. However, this approach hinges on local authorities' capacity to monitor and enforce compliance, necessitating robust governance and stakeholder consultation to ensure the policies are practical and effective.

Incentivising compliance through financial incentives, tax breaks, or subsidies can motivate businesses and households to adopt sustainable waste management practices. These incentives can lower cost barriers and encourage widespread adoption of recycling and composting. Alongside these measures, investment in waste management infrastructure is fundamental. Allocating sufficient funds for developing efficient waste collection and transportation systems and establishing recycling facilities and composting plants will significantly enhance Dharamshala's waste management capacity. The challenge, however, lies in securing the necessary funds, which may require exploring diverse funding sources, including public-private partnerships.

Public-private partnerships (PPPs) present a valuable opportunity to leverage additional resources, expertise, and innovation. Dharamshala can introduce more efficient and sustainable waste management practices by facilitating collaborations between government agencies, private sector companies, and NGOs. These partnerships, however, require clear agreements and regulatory frameworks to protect public interests and ensure long-term sustainability. Raising public awareness through education campaigns is essential for instilling the importance of proper waste disposal and recycling. Effective campaigns can lead to behaviour change and greater public participation, although their success depends on consistent engagement and the ability to reach diverse audiences.

Another critical strategy is implementing mandatory waste segregation at the source, supported by separate bins for different waste types. Proper segregation facilitates recycling and resource recovery but requires substantial public education and infrastructure to ensure compliance. Local authorities must implement responsive waste management strategies to address the significant increase in waste during peak tourist months. These could include increased waste collection, deploying additional resources, and enhancing public awareness about responsible waste disposal. Addressing these seasonal surges is essential to protecting Dharamshala's natural beauty and public health, which are central to its appeal as a tourist destination.

Community participation is another vital component of effective waste management. Engaging local communities through volunteer programs, citizen science initiatives, and clean-up drives fosters a sense of ownership and responsibility. However, maintaining consistent community involvement requires ongoing support and incentives. Supporting the informal waste sector by integrating informal waste pickers and recyclers into the formal system can also enhance efficiency and provide social benefits.

Adopting innovative technologies, such as smart bins, sensor-based systems, and mobile applications, can optimise waste collection and improve efficiency. These technologies provide significant benefits; however, their adoption necessitates investment, technical expertise, and public acceptance. Ensuring that these technologies are user-friendly and accessible is crucial for their successful implementation.

7. FUTURE RESEARCH DIRECTIONS

Future research directions in waste management for Dharamshala are crucial for ensuring interventions' long-term sustainability and effectiveness. These research areas can provide valuable insights and guide the continuous improvement of waste management practices in the region.

- **Long-term impact assessment** is essential for understanding the sustainability of waste management interventions. Longitudinal studies can evaluate how these interventions affect environmental quality, public health, and tourism development. Such studies can help identify any unintended consequences or areas where improvements are needed, ensuring that waste management practices remain effective and adaptable to changing conditions. By tracking these impacts, researchers can also provide evidence-based recommendations for policy adjustments and resource allocation, contributing to more resilient waste management systems.

- **Comparative analysis** offers another valuable research direction by examining waste management practices across tourist destinations. By comparing Dharamshala's practices with those of other locations, researchers can identify best practices, lessons learned, and strategies that can be adapted or transferred to Dharamshala. This comparative approach allows for a broader understanding of what works and what does not in similar contexts, potentially leading to more effective and innovative solutions tailored to the specific challenges faced by Dharamshala.
- **Socioeconomic analysis** is vital for understanding the broader implications of sustainable waste management initiatives. Research in this area can explore how these initiatives impact local communities, businesses, and the tourism industry, focusing on job creation, income generation, and community development. Understanding the economic and social benefits, as well as any potential drawbacks, can help in designing waste management programs that not only address environmental concerns but also contribute positively to the local economy and social fabric. This approach ensures that waste management is integrated into the broader sustainable development goals.
- **Stakeholder engagement** is another critical area for future research, as successful waste management relies heavily on the active participation of all stakeholders. Exploring novel approaches for engaging residents, tourists, businesses, and government agencies in the co-design and co-management of waste management solutions can lead to more inclusive and effective strategies. Research could focus on identifying the most effective methods for fostering collaboration, trust, and shared responsibility among stakeholders, ensuring that waste management initiatives are more widely supported and sustainably implemented.

8. CONCLUSION

In conclusion, this study sheds light on the critical role of waste management in Dharamshala's journey towards sustainable tourism development. Key findings highlight the significance of effective waste management practices for enhancing environmental sustainability, preserving natural resources, and improving the tourist experience. Dharamshala has recycled 430,731 kg of waste over the last 14 months, with January 2024 having the most extensive monthly collection of 44,322 kg. According to this data, the city generates between 1.5 and 1.7 metric tons of waste daily. Even though NGOs such as Waste Warriors and Dhauladhar Cleaners play a significant role in waste management, more stakeholder participation is needed to make these efforts even more effective. The district government must consider creating small-scale businesses that use waste as a raw material to turn non-biodegradable waste into priceless objects such as awards and exhibits. By assisting Dharamshala in reaching its zero-waste city objective, these programs would help the city transition from a developing destination to a developed destination.

Through the analysis of waste management practices, it is evident that Dharamshala's current approach has both strengths and challenges. Although efforts have been made to implement waste management initiatives, there is still room for improvement, particularly in areas such as waste segregation, recycling infrastructure, and community participation. These findings have profound implications for Dharamshala's sustainable tourism development. Dharamshala can strengthen its position as a sustainable tourist destination by addressing identified challenges and leveraging opportunities for improvement. This can lead to several benefits, including enhanced environmental quality, quality of life, increased visitor satisfaction, and long-term economic prosperity. Moving forward, stakeholders, including local authorities, community members, and tourism organisations, must collaborate effectively and implement targeted strategies for improving waste management practices. By doing so, Dharamshala can realise its potential as a leading example of sustainable tourism development, ensuring the preservation of its natural beauty for future generations.

REFERENCES

Amasuomo, E., & Baird, J. (2016). The concept of waste and waste management. *Journal of Management and Sustainability*, 6(4), 88. DOI: 10.5539/jms.v6n4p88

Ayeleru, O. O., Okonta, F. N., & Ntuli, F. (2018). Municipal solid waste generation and characterization in the City of Johannesburg: A pathway for implementing zero waste. *Waste Management (New York, N.Y.)*, 79, 87–97. DOI: 10.1016/j.wasman.2018.07.026 PMID: 30343815

Balu, S., Sampath, K. S., Hari, L. R., Ramprasath, M., Aishwarya, N. K., Aiswarya, M. S., & Nandanan, K. (2020, December). *Assessment of waste management practices and its impact on the village ecosystem in Himachal Pradesh, India. In 2020 IEEE 8th R10 Humanitarian Technology Conference (R10-HTC)*. IEEE., DOI: 10.1109/R10-HTC49770.2020.9357043

Basel Convention. (n.d.-a). EPR Forum. *Basel Convention.* https://www.basel.int/Implementation/Plasticwaste/PlasticWastePartnership/EPRForum/tabid/9658/Default.aspx

Basel Convention. (n.d.-b). Zero waste strategies. *Basel Convention.* https://www.basel.int/Implementation/PublicAwareness/NewsFeatures/Zerowastestrategies/tabid/9866/Default.aspx

Benkherouf, M. (2023). Waste management practices in tourism destinations of Finnish Lapland. *Tourism and Hospitality*, 4(1), 1–13. DOI: 10.3390/tourhosp4010001

Bodo, T. (2019). Garden city or waste city: The case of Port Harcourt metropolis, Rivers State, Nigeria. *Asian Journal of Advanced Research and Reports*, 3(3), 1–11. DOI: 10.9734/ajarr/2019/v3i330091

Bohdanowicz, P. (2006). Environmental awareness and initiatives in the Swedish and Polish hotel industries—Survey results. *International Journal of Hospitality Management*, 25(4), 662–682. DOI: 10.1016/j.ijhm.2005.06.006

Brar, Z. (2018). Management of urban solid waste in Shimla—A case study. In *Water Quality Management: Select Proceedings of ICWEES-2016* (pp. 329–337). Springer Singapore. DOI: 10.1007/978-981-10-5801-1_34

Burlakovs, J., Jani, Y., Kriipsalu, M., Grinfelde, I., Pilecka, J., & Hogland, W. (2020, March). Implementation of new concepts in waste management in tourist metropolitan areas. In IOP Conference Series: Earth and Environmental Science (Vol. 471, No. 1, p. 012017). IOP Publishing. DOI: 10.1088/1755-1315/471/1/012017

Chaabane, W., Nassour, A., Bartnik, S., Bünemann, A., & Nelles, M. (2019). Shifting towards sustainable tourism: Organizational and financial scenarios for solid waste management in tourism destinations in Tunisia. *Sustainability (Basel)*, 11(13), 3591. DOI: 10.3390/su11133591

Diederich, M. D.Jr. (1993). Does waste have standing: Democracy, flow control and a principled constitutional approach to municipal solid waste management. *Pace Environmental Law Review*, 11(1), 157–182. DOI: 10.58948/0738-6206.1455

Esmaeilian, B., Wang, B., Lewis, K., Duarte, F., Ratti, C., & Behdad, S. (2018). The future of waste management in smart and sustainable cities: A review and concept paper. *Waste Management (New York, N.Y.)*, 81, 177–195. DOI: 10.1016/j.wasman.2018.09.047 PMID: 30527034

Ezeah, C., Fazakerley, J., & Byrne, T. (2015). Tourism waste management in the European Union: Lessons learned from four popular EU tourist destinations. *American Journal of Climate Change*, 4(5), 431–445. DOI: 10.4236/ajcc.2015.45035

Forti, V., Balde, C. P., Kuehr, R., & Bel, G. (2020). The Global E-waste Monitor 2020: Quantities, flows and the circular economy potential. *United Nations University (UNU), United Nations Institute for Training and Research (UNITAR)–co-hosted SCYCLE Programme, International Telecommunication Union (ITU), and International Solid Waste Association (ISWA)*. https://doi.org/DOI: 10.25634/ewm2020

Gupta, P., Chanjta, A., & Mehta, Y. (2023). Reflections on ecological sustainability: Assessing development through sustainable initiatives of government and non-government organizations in Himachal Pradesh. *International Journal of Environmental Sciences*, 12(4), 53–65. DOI: 10.1007/s40899-023-00147-4

Gupta, S., Mohan, K., Prasad, R., Gupta, S., & Kansal, A. (1998). Solid waste management in India: Options and opportunities. *Resources, Conservation and Recycling*, 24(2), 137–154. DOI: 10.1016/S0921-3449(98)00033-0

India Comptroller and Auditor General. (2022). A compendium report International Centre for Environment Audit and Sustainable Development (iCED), Jaipur, India [Report, Government of Himachal Pradesh Report No. 2 of the Year 2021].

India Comptroller and Auditor General. (2022). *Social, general and economic sectors*. Non-Public Sector Undertakings. [Report, Government of Himachal Pradesh Report No. 2 of the Year 2021]

Jasrotia, A., & Gangotia, A. (2020). A case study of Dharamshala. In *The Routledge Handbook of Community Based Tourism Management: Concepts, Issues & Implications* (pp. 39-52). Routledge. https://doi.org/DOI: 10.4324/9781003011840-3

Joshi, R., & Ahmed, S. (2016). Status and challenges of municipal solid waste management in India: A review. *Cogent Environmental Science*, 2(1), 1139434. DOI: 10.1080/23311843.2016.1139434

Juvan, E., & Dolnicar, S. (2014). The attitude–behaviour gap in sustainable tourism. *Annals of Tourism Research*, 48, 76–95. DOI: 10.1016/j.annals.2014.05.012

Kahhat, R., Kim, J., Xu, M., Allenby, B., Williams, E., & Zhang, P. (2008). Exploring e-waste management systems in the United States. *Resources, Conservation and Recycling*, 52(7), 955–964. DOI: 10.1016/j.resconrec.2008.03.002

Kandel, A. (2022, November 11). A report on e-waste management. *E-Waste Management.*

Koliotasi, A. S., Abeliotis, K., & Tsartas, P. G. (2023). Understanding the impact of waste management on a destination's image: A stakeholders' perspective. *Tourism and Hospitality*, 4(1), 38–50. DOI: 10.3390/tourhosp4010004

Kumar, S., Smith, S. R., Fowler, G., Velis, C., Kumar, S. J., Arya, S., Rena, , Kumar, R., & Cheeseman, C. (2017). Challenges and opportunities associated with waste management in India. *Royal Society Open Science*, 4(3), 160764. DOI: 10.1098/rsos.160764 PMID: 28405362

National Green Tribunal (NGT). (2022). Environmental assessment of tourism in the Indian Himalayan Region [A report submitted to MoEFCC in compliance of order of Honorable National Green Tribunal (NGT) Principal Branch, New Delhi].

Perkumienė, D., Atalay, A., Safaa, L., & Grigienė, J. (2023). Sustainable waste management for clean and safe environments in the recreation and tourism sector: A case study of Lithuania, Turkey and Morocco. *Recycling*, 8(4), 56. DOI: 10.3390/recycling8040056

Perkumienė, D., Atalay, A., Safaa, L., & Grigienė, J. (2023). Sustainable waste management for clean and safe environments in the recreation and tourism sector: A case study of Lithuania, Turkey, and Morocco. *Recycling*, 8(4), 56. DOI: 10.3390/recycling8040056

Pham Phu, S. T., Fujiwara, T., Hoang Minh, G., & Pham Van, D. (2019). Solid waste management practice in a tourism destination–The status and challenges: A case study in Hoi An City, Vietnam. *Waste Management & Research*, 37(11), 1077–1088. DOI: 10.1177/0734242X19862042 PMID: 31319773

Reno, J. (2015). Waste and waste management. *Annual Review of Anthropology*, 44(1), 557–572. DOI: 10.1146/annurev-anthro-102214-014146

Sharma, A. (2023). Physico-chemical analysis of municipal solid waste in Smart City Dharamshala, Himachal Pradesh. In *Aspects of Engineering and Technology in Human Life* (pp. 117–136). Springer.

Sharma, A., Ganguly, R., & Gupta, A. K. (2018). Matrix method for evaluation of existing solid waste management system in Himachal Pradesh, India. *Journal of Material Cycles and Waste Management*, 20(3), 1813–1831. DOI: 10.1007/s10163-018-0703-z

Sharma, R., Kaur, R., Rana, N., Poonia, A., Rana, D. C., & Attri, S. (2021). Termite's potential in solid waste management in Himachal Pradesh: A mini review. *Waste Management & Research*, 39(4), 546–554. DOI: 10.1177/0734242X20957394 PMID: 32938331

Singh, J., Laurenti, R., Sinha, R., & Frostell, B. (2014). Progress and challenges to the global waste management system. *Waste Management & Research*, 32(9), 800–812. DOI: 10.1177/0734242X14537868 PMID: 24938296

Thakur, D., & Gill, G. S. (2013). Some geo-environmental hazards in Dharamshala area of Himachal Pradesh and their mitigation. *International Journal of Advanced Earth and Environmental Science*, 1, 41–53.

Thakur, S. S. (n.d.). Growth & development of tourism sector in Himachal Pradesh: Issues & concerns.

The Tribune. (2024, May 15). Living underground. *The Tribune* (Jalandhar), 4.

Torres-Delgado, A., & Palomeque, F. L. (2014). Measuring sustainable tourism at the municipal level. *Annals of Tourism Research*, 49, 122–137. DOI: 10.1016/j.annals.2014.09.003

Tsai, F. M., Bui, T. D., Tseng, M. L., Lim, M. K., & Tan, R. R. (2021). Sustainable solid-waste management in coastal and marine tourism cities in Vietnam: A hierarchical-level approach. *Resources, Conservation and Recycling*, 168, 105266. DOI: 10.1016/j.resconrec.2020.105266

Ujala, A. (2024, April 9). Environmental change and its impact on the right to equality: Perspectives from the Supreme Court. *Amar Ujala,* 14.

Ujala, A. (2024, April 29). Top 56 multinational companies account for 50% of plastic pollution. *Amar Ujala,* 14.

Ujala, A. (2024, May 4). Carbon emissions from 57 major producers. *Amar Ujala,* 16.

United Nations World Tourism Organization (UNWTO). (n.d.). Global-tourism-plastics-initiative. *UNWTO*. https://www.unwto.org/sustainable-development/global-tourism-plastics-initiative

Van den Berg, A. E. (2005). *Health impacts of healing environments: A review of evidence for benefits of nature, daylight, fresh air, and quiet in healthcare settings*. University Medical Center Groningen., DOI: 10.1007/978-3-319-11242-6_4

Vij, D. (2012). Urbanization and solid waste management in India: Present practices and future challenges. *Procedia: Social and Behavioral Sciences*, 37, 437–447. DOI: 10.1016/j.sbspro.2012.03.309

Chapter 9
Navigating The Path to Sustainability and Overcoming Environmental Barriers in the Egyptian Hotel Industry

Eslam Ahmed Fathy
Faculty of Tourism and Hotel Management, Pharos University, Alexandria, Egypt

Amr Mohamed Fouad
https://orcid.org/0000-0002-2911-6379
Faculty of Tourism and Hotel Management, Pharos University, Alexandria, Egypt

Abdallalh Mohamed Kahly
Faculty of Tourism and Hotel Management, Pharos University, Alexandria, Egypt

ABSTRACT

This chapter presents the specific environmental practices of Egyptian hotels and identifies potential barriers to implementing sustainable initiatives. The aim was to gain a holistic perspective on the factors affecting hotel environmental sustainability practices rather than merely discussing ecological problems. The analysis was based on qualitative data from 42 semi-structured interviews with top management from a sample of four- and five-star hotels in Egypt. The chapter reveals seven internal barriers: staff capabilities, physical limitations, the attitudes of top management, cost concerns, access to resources, owner attitudes, and day-to-day activities.

DOI: 10.4018/979-8-3693-6110-8.ch009

External factors include government regulations, incentives, and guest attitudes. This chapter also examines the theoretical implications and discusses how future research might address environmental challenges in hospitality.

1. INTRODUCTION

The hospitality industry, which plays a significant role in producing and sustaining economic growth and employment worldwide (Statista, 2023), is also concerned about its ecological influence (Sachs et al., 2019). This is particularly true in Egypt, a country with a well-established history, a growing tourism industry (Sanikos, 2021), and many environmental problems (Abdelhafez et al., 2020). Hotels use large amounts of resources that negatively impact the environment (Kuo et al., 2022). Sustainable development and viability challenges affect this sector's stability (Sachs et al., 2019). These issues are relevant in the hospitality industry and require immediate attention. The research is conducted in the context of the United Nations Sustainable Development Goals (SDGs), which seek to promote sustainable economic growth while protecting the environment and its resources per the global development agenda (United Nations, 2015). Thus, this chapter explores the challenges of environmental practices in the Egyptian hotel industry, emphasising the potential problems that may hamper the adoption of sustainable projects. This research is relevant to the Egyptian context and contributes to the global understanding of environmental sustainability and its implications for the hospitality industry.

Hospitality is one of the largest industries, and hotels, one of the most important subsectors, negatively affect the environment by utilising vast amounts of resources and producing large amounts of waste (Moosa & He, 2023). Energy consumption, water, and waste are high for the industry and severely impact air, water, and soil pollution (Zhang et al., 2022). Various studies have revealed that the hospitality industry significantly contributes to environmental degradation despite various international interventions, obligations, initiatives, agreements, and policies to mitigate the issue (Abdou et al., 2020; Liu et al., 2022). Research on the effects of the Egyptian hotel industry on sustainable development has revealed that despite growing awareness of environmental concerns, many companies are reluctant to embrace sustainable development (Suliman et al., 2023; Taha & Espino-Rodríguez, 2020), which calls for more research on factors that may impede the adoption of sustainable practices.

This study aims to identify internal and external factors that limit the application of environmental sustainability in the Egyptian hotel industry, arguing that a closer look into these barriers will help understand the real causes, as such barriers remain unexplored (Fathy, 2019). An internal issue that presents a significant challenge is hotel employees' lack of knowledge, orientation, and awareness (Ashaal et al., 2024; Elkhwesky et al., 2022). Most employees are unaware of environmental and natural resource conservation (Yucedag et al., 2018). Employees do not receive training on how to put these into practice and develop skills that will facilitate the change process (Cop et al., 2020). While there is recognition of sustainable practices and the adoption of these processes into operation, resistance to change hinders the implementation of sustainable practices (Gull & Rehman, 2022). Management is central to changing behaviours to embrace sustainable practices (Damaianti et al., 2023). Lack of commitment from top management, a focus on short-term goals and revenue optimisation, and a failure to delegate adequate resources and encouragement hinder the achievement of green programs (Sachdeva et al., 2024). It is also possible to develop perceptions that result in employee disinterest and lack of commitment to improving sustainability (Alipour et al., 2019).

Some of the challenges that organisations face in the implementation of sustainable technologies include the high initial capital cost in the installation of sustainable technologies and practices, laws that act as barriers to green IT adoption, organisational culture and structure, incentives and policies that support or hinder green computing and ICT technologies, conflicting priorities between green IT and business goals, strategies and objectives, and lack of research and development initiatives to support sustainable development in organisations, especially those operating in the emerging solar panel installation, increasing energy-efficient systems, and implementing new effective ways of waste management (Ambaye et al., 2023). While some variations may be due to other factors, the existing infrastructure of many hotels presents a critical challenge to incorporating sustainable renovations (Honchar et al., 2024). For example, retrofit measures may have limitations, such as a lack of accommodation for solar panels in old structures or an inability to incorporate efficient lighting and HVAC systems due to infrastructural constraints (Ahmed, 2024).

Most tourists focus on their pleasures, which does not allow them to care for the environment; even the hotels they visit may try to be environmentally friendly (Li et al., 2023). For instance, guests are likely to avoid changes in the service delivered, such as environmentally friendly towel reuse programs, or they may not readily accept paying a premium price for an eco-friendly option (Sucheran & Moodley, 2019). Where legal frameworks for environmentalism are not robust or are poorly enforced, hotels will not face much pressure to become green (Chen et al., 2021; Giang et al., 2020). The lack of strict regulations can also demotivate hotels to implement environmentally sustainable measures, leading to disparities in policy standardisation on sustainable issues (Khonje & Leonard, 2019). Some governments have implemented policies to foster sustainable tourism; however, a lack of sufficient financial motivation or encouragement or a lack of well-defined legal provisions to govern environmental conservation can pose a challenge to green investment (Novacka et al., 2019).

Environmental sustainability has received ever more attention from Egyptian hotel businesses, yet it is still regularly very patchy, with significant challenges to green practice adoption. Despite improved awareness and pressures worldwide to be greener in operations, little has been done effectively to put concrete solutions on the ground in Egyptian hotels. This gap is of particular concern, given that Egypt faces other environmental issues—most notably, water shortages and pollution—that will directly impact this sector's long-term viability. This research covered the explicit need to understand the barriers to implementing environmental practices among four- and five-star hotels in Egypt, a segment with significant influence and resource capacities. It will, therefore, outline both the internal organisational barriers and external pressures as significant steps toward a clearer understanding of the complexities facing environmental sustainability in Egyptian hotels.

2. LITERATURE REVIEW

2.1 Barriers to Environmental Practices Adoption

Scholars and researchers have concentrated on categorising environmental adaptation obstacles according to different patterns since the 1990s. Some authors categorised the barriers as industry and organisational barriers (Chan et al., 2020; Khatter et al., 2021). However, other authors divided the barriers into internal and external categories, and the current study combined the two categories (Bacchiocchi et al., 2024).

2.1.1 Internal Environmental Barriers

Several internal factors within hotels can act as roadblocks to sustainable development.

- ***Staff Capabilities***

One of the biggest obstacles to implementing environmental standards is the high cost of personnel qualification and training (Barakagira & Paapa, 2024). Furthermore, according to Al-Sabi et al. (2024), inadequate training on environmental issues leads to mistrust among employees, deters them from the need to adopt environmentally friendly practices, and makes them resistant to doing so because it takes more time and effort to adapt additional environmental practices to fulfil their regular responsibilities. These factors ultimately impact employee performance (Sun et al., 2024). Dekoulou et al. (2023) also noted two additional challenges: staff members' environmental obligations are not specified, and the whole workforce is not participating.

- ***Physical Limitations***

According to Liu et al. (2022), obstacles include the hotel's location, age, amenities, and space available, especially for older structures. Installing solar panels on the hotel roof, for instance, is challenging, even if the panels may be utilised for other hotel operations. Hotel amenities and utilities may need to change to implement environmental policies (Abdou et al., 2020).

- ***Attitude of Top Management***

Hotel managers ' awareness greatly influences staff commitment to exploring and implementing environmental policies for long-term performance (Fayed & Fathy, 2022; Tariq et al., 2020). There are several reasons to recognise top management as a potential source of barriers (Tanveer et al., 2024). Firstly, some managers think traditionally (Oriade et al., 2021). Secondly, they do not seek to mandate additional financial obligations on owners and carry out inexpensive environmental projects that do not require their consent (Sun et al., 2022). Thirdly, according to Donovan and McElligott (2000), they do not want to execute more administrative responsibilities, and integrating environmental practices into the hotel strategy would take too much time away from their regular duties (Schilling-Vacaflor & Gustafsson, 2024). Fourthly, managers would not be motivated to implement environmental regulations (Graafland & Bovenberg, 2020). Finally, hotel managers' concerns about environmental practices conflict with guests' requirements and preferences (Chan, 2021; Fathy, 2023; Fayed & Fathy, 2022).

- ***Cost Consideration***

According to research by Legrand et al. (2014), the primary obstacle to environmental adoption is cost. The relatively high cost of creating and implementing environmental practices is caused by the high cost of goods, maintenance, and equipment, which makes management reluctant to embrace environmental practices (Anh et al., 2020; Chang et al., 2021; Tan et al., 2021).

- ***Lack of Resources***

The lack of personnel skills and environmental equipment, among other resources, is needed to adopt environmental standards (Munawar et al., 2022). Staff time is another barrier to adopting environmental practices because of the increased time required for operating processes and training (Anwar et al., 2020). According to Chan (2021), the lack of environmental advice is a significant barrier. Therefore, hotel managers struggle to locate environmental specialists who can help them (López-Gamero et al., 2023).

- ***Attitude of Owners***

Most owners of hotel businesses have a short-term and revenue-focused mentality, considering sustainable practices a costly extra (Demetriou, 2022). They may avoid investing in energy-efficient technologies like solid-wall houses or eco-friendly products because they may see high costs and uncertain benefits (Abdel et al., 2019; Filimonau et al., 2019). This "cost-reduction versus added value" approach may minimise the possibilities for long-term profitability, customer satisfaction, and positive alterations to the brand identity (Molina-Azorín et al., 2009; Nassar & Fouad, 2022). It is possible that some of them do not have a favourable understanding of sustainability and its entrails. Some do not understand the environmental impact, particularly regarding hotel operations, nor do they understand the value of embracing sustainable practices. This lack of understanding contributes to complacency and resistance to change (Aragon-Correa et al., 2015; Filimonau et al., 2019).

- ***Daily Operations***

Time management is a significant issue in the hospitality industry because time equals money (Demetriou, 2022). Hotels heavily rely on meeting guests' demands, optimising guest interactions, and ensuring quick cycle times. This pressure can often increase or disprove environmental concerns to some extent. For this reason, staff might take more straightforward options or short-term solutions, for example, using disposable items when it is easier or not turning off lights and switches when they think the process would slow the service (Chan et al., 2017; Deraman et al., 2017). Additionally, successful implementation of sustainable practices training programs may be perceived as time-consuming activities that are not required to promptly meet the needs of guests (Espino-Rodríguez & Gil-Padilla, 2015). Routine activities involve meeting the existing needs of guests and stakeholders. They are inclined to optimise the operational processes, which may lead to poor planning for sustainable development initiatives (Filimonau et al., 2020).

2.1.2 External Environmental Barriers

Many external constraints make it difficult for the hotel industry to adopt sustainable management practices. These barriers are beyond individual hotels' primary control, are often complex, and require a system solution.

- ***Government Regulations and Incentives***

One major challenge is the absence of adequate standardised guidelines and policies from the government (Alemão, 2019; Bohdanowicz, 2007). An essential function of governments includes facilitating conditions that encourage sustainability through establishing guidelines, offering funding, and raising people's awareness. However, the lack of standardisation in rules and policies can prevent hotels from investing in green technologies and solutions since such initiatives do not guarantee tangible and short-term profits. It is imperative in areas where policies may be disjointed or there is no single vision of how to progress sustainably in tourism (Demetriou, 2022).

- ***Guest Attitude***

Another significant barrier is the lack of a well-defined market call for sustainable practices. However, today's consumers are more aware of environmental problems but are unwilling to sacrifice financial value for environmental quality by choosing environmentally friendly hotels (Altinay et al., 2020; Filimonau et al., 2019). Hotels may avoid or limit their commitment to sustainable practices if they believe consumers are uninterested. The hotel anticipates that customers may switch to other providers if sustainability becomes a priority. This lack of a clear market may explain why it is challenging for hotels to invest in practices that would increase the cost of delivering their services. However, it is crucial to understand that the customer base interested in sustainable tourism is not numerous or vociferous enough to enforce radical shifts in the industry's approaches (Mandić & Petrić, 2021).

- ***Certification Costs***

Another potential challenge is the cost of obtaining certifications for sustainability initiatives and strategies (Bilbao-Terol & Bilbao-Terol, 2020). The absence of an accredited certification program may also create ambiguity and challenges in defining sustainable hotel recognition. They pointed out that certification is also a costly aspect that may prove very costly to many hotels already struggling financially, especially the small ones (Calisto et al., 2021; Bianco et al., 2023).

- ***Sustainable Supply Chain Management***

The actual purchase of green products and services from suppliers that are easily accessible and inexpensive is often a challenge (Espino-Rodríguez & Taha, 2022). The nature of supply chains makes it difficult to achieve sustainability as suppliers vary regarding the quality of their output and compliance with industry standards (Modica et al., 2020). Establishing and developing a pool of trusted suppliers who can provide environmental supplies at reasonable prices could remain a challenge for hotels, especially when seeking specialised products such as biodegradable guest amenities and earth-friendly sanitising products (Alreahi et al., 2023).

- ***Competitive Hotel Industry***

Lastly, due to the intense competition in the hotel industry and persistent pressure to cut costs and increase revenues, it may be challenging to justify the implementation of sustainable practices as they may not necessarily yield an immediate financial return (Filimonau et al., 2019; Molina-Azorín et al., 2009). Sustainable business strategies can be expensive, and hotels may struggle to manage these costs, particularly within highly competitive markets or if consumer attitudes toward travel expenses are relatively conservative (Yousaf et al., 2021).

2.2 Current Status of Sustainable Initiatives in the Egyptian Hotel Industry

The Egyptian hotel industry, which forms a significant aspect of the Egyptian tourism sector, has agreed with the need to adopt sustainable practices for sustainable benefit in the future (Shaalan, 2005). Due to the rising demand in the global market for various initiatives that are environmentally friendly, hotels in Egypt are implementing environmentally friendly strategies in their operations to reduce the adverse impacts on the environment and promote their products to guests who seek green products (Abdou et al., 2022).

2.2.1 Energy Efficiency and Renewable Energy

Currently, most of the hotels being built in Egypt have incorporated energy-efficient systems to produce renewable energy (Abdou et al., 2020). This comprises solar systems on roofs, energy-efficient bulbs, and smart energy management systems in buildings (Shehata & Elfeel, 2017). For example, several luxury resorts in the Red Sea area have invested in solar water heating and, to some extent, solar electricity, thereby minimising the use of fossil fuels (Gohar & Kondolf, 2020).

2.2.2 Water Conservation

Water shortages characterise Egypt; therefore, hotels work to conserve water at all costs (Gössling et al., 2012). These include low-flow fittings, water recycling facilities, and Indigenous plants, significantly reducing the watering rate (Han et al., 2018). Some hotels also apply graywater recycling to repurpose the water from showers or sinks for irrigation systems (Han et al., 2018).

2.2.3 Waste Management

Waste management is another aspect of how Egyptian hotels gradually build their sustainability profiles. Almost all hotels have adopted extensive recycling policies and are eliminating disposable items (Abdou et al., 2020). These consistencies include recycling waste products through composting organic materials, engaging local waste management companies, and involving guests in recycling (Abu Elnasr et al., 2021). Furthermore, some hotels are beginning to implement measures to reduce food waste, such as serving reasonable portions and offering leftovers to the needy (Soliman, 2020).

2.2.4 Sustainable Sourcing and Local Engagement

In Egypt, the emerging perspective of sustainable sourcing is slowly becoming appealing to hotel businesses (Yousaf et al., 2021). These include getting most supplies, such as food, even from local producers as a strategy to embrace the environment (Mohamed et al., 2020). This also involves ensuring that local farmers purchase the food organically and seasonally (Salem et al., 2022). This makes the food fresh and of good quality for guests. Also, they interact with the locals through various activities, including arts and culture, which include partnering with local people, purchasing locally-made souvenirs, and learning about the local history (Salem et al., 2022).

2.2.5 Green Building Certifications

Most hotels in Egypt have adopted the green building concept and are seeking LEED and Green Star certifications, among other things (Mosaad et al., 2016). These certifications have strict guidelines concerning energy and water usage, indoor environmental quality, and sustainable development. Getting such certifications also helps cut operational expenses, making the hotel more attractive (Wei et al., 2015).

3. METHODOLOGY

This research adopted semi-structured in-depth interviews to elicit informative responses on the subject under study using qualitative research methods to gain a holistic understanding of the issue. Overall, this design provides the possibility of using triangulation as a method to draw robust inferences about this study. This research was conducted on four- and five-star hotels with the recognition that they most often possess relatively more resources and are under greater pressure to consider sustainable issues. A purposive sampling technique was used to select 50 hotels across all regions of Egypt to ensure representative operational contexts. Initial contact was made with 50 hotels to explain the study's objectives and to seek participation; the response rate was 60%.

Semi-structured interviews were held with forty-two senior management staff of the hotels. Those targeted included the general managers, heads of departments, and those at other levels who could make influential decisions on matters related to sustainability. The topics covered in our interview guide, informed by relevant literature, included staff training and management perceptions about sustainability, cost factors, and guest attitudes. The average length of the in-depth and online interviews was around two hours, and all of these were recorded with the consent of the participants. In conjunction with this, transcripts of the recordings were made while the accuracy of the information was strictly maintained.

There are three sections to the interview. First, the interview started with a warming-up section that included the study's objectives and demographic information to create a more relaxed, trustworthy setting and encourage the participants to respond easily. Additionally, the researchers guaranteed the confidentiality and anonymity of the data gathered and used for scientific reasons. Second, the researchers spoke with industry participants about the challenges to environmental implementation in the hotel sector. Ultimately, the conversations concluded with expressions of gratitude intended to elicit helpful recommendations. Time management, voice tone, posture, mobility, and non-bias are some elements that increase efficacy. Other measures include not disapproving of respondents' replies. Moreover, the respondents allowed the researcher to record their discussions to prevent participant interruptions. The researcher held interviews on discussion themes and handled inappropriate topics diplomatically by ignoring them. Following 42 interviews, the researcher concluded that there was too much data and that the respondents' responses did not significantly improve the outcomes. The transcripts and talks that were recorded were analysed using the NVIVO software. Codes for the interviews ranged from INT1 to INT4.

The sample used for this study was stratified random sampling, a variant of random sampling with the added element of refinement by the researcher, as articulated by Saunders et al. (2016). This technique refines simple random sampling and breaks the target population into clear and meaningful strata from which the sample is selected - the area or destination where the hotel is located. The stratified random sampling technique was adopted to inform considerable accuracy, as it involves the dividing of the population into subgroups or classes called strata (Taherdoost, 2016; Etikan & Bala, 2017), and every individual element of the population has an equal opportunity to get selected by the sampling process. Each stratum has distinct outcomes; this sampling technique makes data collection easier (Alvi, 2016; Sharma, 2017). There are three phases in which the sampling technique was stratified for the present research study. In the first phase, the hotel was categorised into five destinations: Cairo, the North-West Coast, the Canal Zone and Sinai, the Red Sea, and Upper Egypt. These represent Egypt's five main tourist destinations, as Salem (2015) stated. The next phase was to select hotels from these five destinations. This technique was used to get a small sample of the respondents or managers, whereby the result would be generalised to the total population.

According to Kock and Hadaya (2018), the minimum acceptable sample size for descriptive studies is 10%. This study used a sample of 50 five-star hotels out of 157, representing 32% of all hotels. Before visiting the hotel, the researcher contacted the human resources director or the training manager. The letter from the Faculty of Tourism and Hotels, Alexandria University, was also displayed, seeking help and clarifying that the study will benefit the hospitality sector. Measures were taken to decrease any response bias, including anonymous and confidential responses and using previously validated interview tools. Personal identifying information was not collected for the participants, and the respondents' data was coded so that the responses could not be tracked back to any one individual.

4. RESULTS

The primary types of environmental obstacles, including the subcategories for impediments, have been developed using qualitative data. As a result, the researcher constructed the study's theoretical model based on qualitative data. Most respondents stated that employees lack environmental awareness, knowledge, training, skills, desire, and motivation (INT 5: INT 37). For example, in INT 25, "the employees do not apply environmental practices on their own but must be assigned tasks." A few respondents also mentioned the difficulty of installing solar panels in ancient hotels due to infrastructural modifications. The statement "the hotel roof may be occupied as a café or other activities" is an example of an INT 7.

Most respondents indicated that department managers and hotel general managers play a crucial role in encouraging the adoption of environmental standards (INT1: 32). As a result, hotel managers must be willing to accept the financial costs and exert the necessary effort to implement environmental policies, as well as possess the expertise and willingness to do so. A few managers pointed out that implementing environmental practices is decreasing due to high costs and overcapacity. "It takes more work to put environmental practices into effect." The owner will be agreeable if a detailed feasibility study of environmental adoption benefits is provided, according to one respondent (INT 17). For instance, INT 35, "I am afraid to adopt environmental practices that affect customer satisfaction, and therefore the hotel's profits and the current guest do not have sufficient environmental awareness," was mentioned by most respondents.

A small proportion of participants mentioned that owners may influence the adoption of environmental practices by agreeing to provide a budget specifically for this purpose. This is attributed to the consciousness of environmental stakeholders and their belief in the significance of environmental measures in long-term cost reduction. In addition, a few respondents said that adopting environmental practices in the range of 5 to 15 is limited by a shortage of human resources, physical resources, and time. For instance, the lack of a comprehensive feasibility assessment hinders the hotel's ability to implement environmental measures.

A few respondents indicated that environmental legislation creates obstacles to adopting environmental practices (INT 3, 7, 11, 13, 15, 16, 19, 20, 22, 33, and 34). When asked about the government's role in limiting the adoption of environmental practices, most respondents stated that the government has a minor influence on hotel environmental practices. For instance, the waste disposal vehicle combines the waste the hotel staff has separated, negatively affecting the staff's environmental conduct.

5. DISCUSSION

The book chapter examines the obstacles to implementing environmental practices to protect this crucial asset for the hotel industry's profitability and survival. Identifying challenges to environmental practices enables hotels to enhance their strategies for overcoming them. The research on the internal obstacles indicates that staff capabilities, the attitude of top management, and cost concerns may concurrently restrict the application of environmental efforts. These results are compatible with those of Amankwah-Amoah (2020), Haldorai et al. (2022), and Khan et al. (2023). It demonstrated that hotel employees are unwilling to implement environmental practices since they have not defined environmental duties, incentives, or willingness to apply and participate in hotel environmental efforts. Also, the results revealed that senior management is not enthusiastic about using environmental practices and is not convinced that these practices will decrease operational costs. Furthermore, they do not perceive it as a top priority and restrict environmental adoption, as demonstrated by the findings of Martínez-Peláez et al. (2023). Possible reasons for these outcomes may be that they lack adequate knowledge of all environmental activities, inadequate environmental knowledge, and the advantages of environmental practices (Fathy, 2021). Moreover, the results show that the cost has a minor impact on environmental practice implementation. This result is consistent with earlier studies that claimed the cost was a limitation to environmental practice implementation, especially for small hotels (Musavengane, 2019).

Examining external barriers to applying environmental practices demonstrates that guest attitudes are the most critical external barriers. The findings align with Chan et al. (2017) conclusion that visitors may impede environmental adoption because they search for high water pressure and increased lighting utilisation, which conflicts with environmental practices. Moreover, the environmental practices may negatively impact their comfort, delight, and experience. This can be attributed to the fact that the tourist market in Egypt has suffered since 2011 (Fathy, 2018), leading to the current guests being characterised by poor expenditure, education, and, subsequently, less environmental consciousness. Moreover, the hotel management did not adequately express its commitment to the environment and did not address visitor inquiries. Most hotels employ these sustainable activities to achieve cost savings and generate revenue rather than environmental commitment (Rahman & Reynolds, 2016). Moreover, the obstacles to environmental legalisation and governmental bodies are the second and third critical external barriers. The present research effectively highlights the challenges to environmental practice implementation.

6. THEORETICAL IMPLICATIONS

In terms of theoretical contribution, this book chapter effectively enhances the knowledge of its readers regarding sustainable development in the Egyptian hotel industry. This can be attributed to the fact that it contributes to the existing literature by categorising adoption barriers in a more detailed manner, i.e., going beyond internal and external factors. To facilitate a more elaborate description of the challenges faced, the chapter goes further to analyse specific factors such as staff capabilities, management attitude, costs, guests, environmental policies, and government support so that one can examine the spurious sector and come up with the suitable approaches to the challenges.

The study focuses on internal factors, such as staff knowledge, management responsiveness, and cost considerations, as another important determinant of environmental practice adoption. It highlights the role of HRM in leading environmental sustainability change within organisations, and those hotels must redesign internal capacities and start changing managerial perceptions to align with sustainable development objectives (Fathy & Fouad, 2022).

Nevertheless, the chapter does not disregard the impact of internal factors; simultaneously, it acknowledges the effects of outside factors such as guest attitude, environmental legislation, and government assistance. This underlines the importance of industries' and other stakeholders' efforts to undertake a multifaceted approach to fighting environmental threats because hotels alone cannot realise environmental sustainability in hospitality.

Though the chapter discusses potential barriers, it provides practical solutions to the issues that help bridge the theory-practice divide. It provides detailed action plans for staff training, senior management engagement, guest relations, regulatory compliance, and ecological advocacy, including requests for federal assistance. As a result, these practical insights provide direction on how hotel owners intending to establish sustainable development can conduct their procedures. This highlights the advantage of concentrating on the Egyptian hotel industry, as it gives a sense of the specific geographic area and context of the environments within which these hotels operate, to the extent that there are more generalisable lessons to be drawn from the surroundings. This type of behaviour determines barriers to adopting environmental practices, which is advantageous due to its specificity over a general one.

7. PRACTICAL IMPLICATIONS

The findings of this study provide valuable recommendations for hotels in Egypt to cultivate a sustainable environment. The following represent some of the vital common factors that must be addressed. There should be sophisticated staff training programs that raise employee awareness, promote general knowledge of sustainability practices, increase skills development to enable them to achieve green programs and provide rewards to encourage sustainable practices. In the same way, senior management must actively get involved by supporting sustainability and environmentalist measures, embedding sustainability into the organisational strategic objectives, and providing a desirable level of resource support to drive sustainable development. All hotel managers and authorities should focus on acquiring knowledge about strategies that can be sustainable and simultaneously affordable, which give certain advantages like cost reduction and brand enhancement. Subsequently, they can make gradual adjustments depending on the financial situation, starting with inexpensive options and looking for funding sources for capital-heavy alternatives.

Consequently, it could be suggested that cooperation is the primary strategy for managing external barriers and acknowledging and delivering lessons on sustainability to guests and involving the guests in sustainability through actions such as encouraging those to use towels wisely, providing environmentally friendly products, and informing the guests about any environmentally conscious initiatives nearby. Furthermore, they need to engage with other stakeholders in the supply chain, particularly authorities in the tourism sector, governmental organisations, and NGOs, to encourage them to enforce more robust environmental legislation, create resources for disseminating sustainability initiatives, and disseminate optimal practices. Another requirement is government incentives and grants to support the initiative in realising the approaches. Politics is complex, and it is possible to collaborate with government agencies to participate in policymaking and convince the authorities to adopt more friendly legislation.

To support these practical strategies, specific steps that hotels could consider in terms of a sustainability agenda should include developing a clear sustainability plan, identifying goals, an action plan, and measurable objectives with sustainability. The plan should address the following viable points: energy, water, waste, and the supply chain. Ensuring that the hotel invests in energy-efficient measures such as solar panels, LED lights, smart thermostats, efficient plumbing fittings, and transporting guests with efficient means while planning green transport services, among others, are essential strategies for the hospitality industry. Monitoring the impacts of tourism on local stakeholders and ensuring that partnerships and engagement with such communities are done to develop sustainable, productive projects for the hotel and the surrounding area carries considerable importance (Khalil & Fathy, 2021; Shawky et al., 2019). It can be engagement in environmental conservation, promoting cultural practices and beliefs, and supporting domestically produced goods.

Lastly, hotels should continually assess their environmental impact by measuring data at a specified interval. Further, this data can be used for decision-making and to track advancements toward the intended sustainability objectives. Education, transparency, and communication should be the critical components of organisational strategies. It is recommended that hotels make their guests, employees, and stakeholders aware of their sustainable initiatives by posting information about them on their Internet sites, booklets, and social networks. This type of transparency instils confidence in guests because the hotel conveys a sense of responsibility for the environment.

8. LIMITATIONS AND FUTURE RESEARCH

Although contributing to knowledge about environmental barriers in the hotel sector in Egypt, the present chapter is restricted by certain constraints. One appreciable limitation is that the research sample was limited to only four- and five-star hotels. This may suggest that the results are not generalisable to the remaining diverse sectors of the industry, including small or budget hotels that can face other forms of challenges and opportunities. Subsequent studies should uncover the barriers to adopting environmental practices in various types of hotels to get a better picture of the hospitality industry (Musavengane, 2019).

Overdependence on the managerial point of view is also a limitation when information becomes unavailable. Although the insights from management are informative, the inclusion of employee and guest perceptions generates a more complete picture of how these barriers affect applying sustainable practices. To capture a wider variety of experiences and views, future research could adopt both quantitative surveys and qualitative interviews using the mixed method (Chan et al., 2017; Filimonau et al., 2019).

Finally, the Egyptian context may limit the generalizability of the findings to other developing countries. In the future, distinct obstacles and facilitators to embracing environmental practices must be more deeply studied in various cultural and economically diverse settings for sustainability tourism. This would enhance a more nuanced understanding of the international sustainable development map in the hospitality industry (Altinay et al., 2020).

9. CONCLUSION

This chapter has outlined the various problems and challenges the Egyptian hotel industry encounters when seeking to embrace and accomplish environmental sustainability. Analysing the interviews with the top management of the different hotels allows us to identify some internal and external barriers to the implementation of sustainability. Factors such as staff ability and experience, available space, corporate hierarchy and belief systems, costs, resource availability, owner beliefs, and certifications contribute to internal constraints. There are also forces external to the organisation that serve to make environmental sustainability a challenge to implement; these include the government, the attitude of the guests toward sustainability, the costs of sustainability certification, sustainable procurement, and rivalry from other hospitality organisations. However, the chapter also demonstrates how society is gradually making progressive changes within the industry. New policies indicate that Egyptian hotels have started adopting energy-efficient, water-efficient waste management systems, green purchasing, and using and involving the local community. These practices are being pursued to increase environmental performance and unlock competitive advantage in the marketplace.

The information presented in this paper emphasises the need for a multi-faceted approach to addressing these challenges. According to the analysis, some of the proposed recommendations include the following: increasing personnel training and awareness, promoting the principles of sustainable development among managers and employees, receiving financial and legislative support from the government, and involving guests in the sustainability process. Furthermore, sustainable supply chain management and the search for green accreditation could begin to shift the industry in a more sustainable direction. Future studies should conduct further research on these factors, focusing on the challenges and potential solutions relevant to the Egyptian hotel industry. This will allow for the dissemination of the idea of environmental sustainability both inside and outside organisations to support the development of long-term strategies for the regional industry's competitiveness.

REFERENCES

Abdel Rassoul, S. E., & Fathy, E. A. (2019). Assessment of Food Safety Practices in Siwa's Hotels. *The Second International Conference of the Faculty of Hotel Tourism, Matrouh University, entitled" The Balanced Development of Siwa Oasis on the World Tourism Map, A Future Vision for Sustainability" from March*, (pp. 26-28).

Abdelhafez, A. A., Metwalley, S. M., & Abbas, H. H. (2020). Irrigation: Water Resources, Types and Common Problems in Egypt. In *E.-S. E. Omran, & A. M. Negm, Technological and Modern Irrigation Environment in Egypt Best Management Practices & Evaluation* (pp. 15–34). Springer. DOI: 10.1007/978-3-030-30375-4_2

Abdou, A. H., Hassan, T. H., & El Dief, M. M. (2020). A Description of Green Hotel Practices and Their Role in Achieving Sustainable Development. *Sustainability (Basel)*, 12(22), 9624. DOI: 10.3390/su12229624

Abdou, A. H., Shehata, H. S., Mahmoud, H. M., Albakhit, A. I., & Almakhayitah, M. Y. (2022). The Effect of Environmentally Sustainable Practices on Customer Citizenship Behavior in Eco-Friendly Hotels: Does the Green Perceived Value Matter? *Sustainability (Basel)*, 14(12), 7167. DOI: 10.3390/su14127167

Abu Elnasr, A. E., Aliane, N., & Agina, M. F. (2021). Tackling Food Waste in All-Inclusive Resort Hotels in Egypt. *Processes (Basel, Switzerland)*, 9(11), 2056. DOI: 10.3390/pr9112056

Ahmed, M. E. (2024). Development of a Management Framework for Energy Efficiency for Sustainability of Hotels in Jeddah, Saudi Arabia, Using Structural Equation Modelling. *Journal of Advanced Research in Applied Sciences and Engineering Technology*, 36(1), 159–175.

Ahmed Fathy, E. (2021). Potentials of Agricultural Heritage Systems Tourism (AHST): A case study of date palm uses in gastronomy, Siwa Oasis–Egypt. . , 10(10), 3-50.

Al-Sabi, S. M., Al-Ababneh, M. M., Al Qsssem, A. H., Afaneh, J. A., & Elshaer, I. A. (2024). Green human resource management practices and environmental performance: the mediating role of job satisfaction and pro-environmental behaviour. *Cogent Business & Management, 11*(1).

Alemão, C. P. (2019). *Environmental sustainability practices in the supply chain of hotels and the consumers' perception.* Master dissertation.

Alipour, H., Safaeimanesh, F., & Soosan, A. (2019). Investigating Sustainable Practices in Hotel Industry-from Employees' Perspective: Evidence from a Mediterranean Island. *Sustainability (Basel)*, 11(3), 6556. DOI: 10.3390/su11236556

Alreahi, M., Bujdosó, Z., Dávid, L. D., & Gyenge, B. (2023). Green Supply Chain Management in Hotel Industry: A Systematic Review. *Sustainability (Basel)*, 15(7), 5622. DOI: 10.3390/su15075622

Altinay, F., Altinay, Z., Altinay, M., & Dagli, G. (2020). Evaluation of the Barrier-Free Tourism and Sustainability of the Barrier-Free Society in Cyprus. *European Journal of Sustainable Development*, 9(4), 137. DOI: 10.14207/ejsd.2020.v9n4p137

Alvi, M. H. (2016). *A Manual for Selecting Sampling Techniques in Research.* Munich Personal RePEc Archive.

Amankwah-Amoah, J. (2020). Stepping up and stepping out of COVID-19: New challenges for environmental sustainability policies in the global airline industry. *Journal of Cleaner Production*, 271, 123000. DOI: 10.1016/j.jclepro.2020.123000 PMID: 32834564

Ambaye, T. G., Djellabi, R., Vaccari, M., Prasad, S., Aminabhavi, T. M., & Rtimi, S. (2023). Emerging technologies and sustainable strategies for municipal solid waste valorisation: Challenges of circular economy implementation. *Journal of Cleaner Production*, 423, 138708. DOI: 10.1016/j.jclepro.2023.138708

Anh, N. H., Soa, N. L., Hanh, H. H., & McMillan, D. (2020). Environmental accounting practices and cost of capital of enterprises in Vietnam. *Cogent Economics & Finance*, 8(1), 1790964. DOI: 10.1080/23322039.2020.1790964

Anwar, N., Mahmood, N. H., Yusliza, M. Y., Ramayah, T., Faezah, J. N., & Khalid, W. (2020). Green Human Resource Management for organisational citizenship behaviour towards the environment and environmental performance on a university campus. *Journal of Cleaner Production*, 256, 120401. DOI: 10.1016/j.jclepro.2020.120401

Aragon-Correa, J. A., Martin-Tapia, I., & Torre-Ruiz, J. (2015). Sustainability issues and hospitality and tourism firms' strategies: Analytical review and future directions. *International Journal of Contemporary Hospitality Management*, 27(3), 498–522. DOI: 10.1108/IJCHM-11-2014-0564

Ashaal, A., Bakri, A., AlKhatib, N., Chreif, M., & Zakhem, N. B. (2024). Harvesting sustainability: Navigating green practices and barriers in Lebanon's hospitality sector. *Business Strategy & Development*, 7(2), e369. DOI: 10.1002/bsd2.369

Bacchiocchi, A., Bellocchi, A., & Giombini, G. (2024). Green Investment Challenges in European Firms: Internal vs. External Resources. *Sustainability (Basel)*, 16(2), 496. DOI: 10.3390/su16020496

Barakagira, A., & Paapa, C. (2024). Green practices implementation for environmental sustainability by five-star hotels in Kampala, Uganda. *Environment, Development and Sustainability*, 26(4), 9421–9437. DOI: 10.1007/s10668-023-03101-7 PMID: 37362970

Bianco, S., Bernard, S., & Singal, M. (2023). The impact of sustainability certifications on performance and competitive action in hotels. *International Journal of Hospitality Management*, 108, 103379. DOI: 10.1016/j.ijhm.2022.103379

Bilbao-Terol, A., & Bilbao-Terol, C. (2020). Measuring the Economic Impact of a Voluntary Sustainable Tourism Certification. *Sustainability (Basel)*, 12(13), 5465. DOI: 10.3390/su12135465

Bohdanowicz, P. (2007). A Case Study of Hilton Environmental Reporting as a Tool of Corporate Social Responsibility. *Tourism Review International*, 11(2), 115–131. DOI: 10.3727/154427207783948937

Calisto, M., Umbelino, J., Gonçalves, A., & Viegas, C. (2021). Environmental Sustainability Strategies for Smaller Companies in the Hotel Industry: Doing the Right Thing or Doing Things Right? *Sustainability (Basel)*, 13(18), 10380. DOI: 10.3390/su131810380

Chan, E. S. (2021). Influencing stakeholders to reduce carbon footprints: Hotel managers' perspective. *International Journal of Hospitality Management*, 94, 102807. DOI: 10.1016/j.ijhm.2020.102807

Chan, E. S., Okumus, F., & Chan, W. (2021). The Applications of Environmental Technologies in Hotels. *Journal of Hospitality Marketing & Management*, 26(1), 23–47. DOI: 10.1080/19368623.2016.1176975

Chang, H.-Y., Liang, L.-W., & Liu, Y.-L. (2021). Using Environmental, Social, Governance (ESG) and Financial Indicators to Measure Bank Cost Efficiency in Asia. *Sustainability (Basel)*, 13(20), 11139. DOI: 10.3390/su132011139

Chen, Y.-C., Lee, C.-S., Hsu, Y.-C., & Chen, Y.-J. (2021). Why Is Green Hotel Certification Unpopular in Taiwan? An Analytic Hierarchy Process (AHP) Approach. *ISPRS International Journal of Geo-Information*, 10(4), 255. DOI: 10.3390/ijgi10040255

Cop, S., Alola, U. V., & Alola, A. A. (2020). Perceived behavioral control as a mediator of hotels' green training, environmental commitment, and organizational citizenship behavior: A sustainable environmental practice. *Business Strategy and the Environment*, 29(8), 3495–3508. DOI: 10.1002/bse.2592

Damaianti, I., Farradia, Y., Shafi, S., Arzaman, A. F., & Salleh, H. S. (2023). Sustainable Practices in West Java Hotels: Perspectives from the Hotel Association, Hotel Management, and Visitors. *International Journal of Sustainable Development and Planning*, 18(11), 3575–3584. DOI: 10.18280/ijsdp.181122

Dekoulou, P., Anastasopoulou, A., & Trivellas, P. (2023). Employee Performance Implications of CSR for Organizational Resilience in the Banking Industry: The Mediation Role of Psychological Empowerment. *Sustainability (Basel)*, 15(15), 11946. DOI: 10.3390/su151511946

Demetriou, P. (2022). Hotel food waste in Cyprus: An exploratory case study of hotels in Limassol. *Cogent Social Sciences*, 8(1), 2026556. DOI: 10.1080/23311886.2022.2026556

Deraman, F., Ismail, N., Arifin, A. I., & Mostafa, M. I. (2017). Green practices in hotel industry: Factors influencing the implementation. *Journal of Tourism* [JTH-CA]. *Hospitality & Culinary Arts*, 9(2), 305–316.

Elkhwesky, Z., Salem, I. E., Varmus, M., & Ramkissoon, H. (2022). Sustainable practices in hospitality pre and amid COVID-19 pandemic: Looking back for moving forward post-COVID-19. *Sustainable Development (Bradford)*, 30(5), 1426–1448. DOI: 10.1002/sd.2304

Espino-Rodríguez, T. F., & Gil-Padilla, A. M. (2015). The structural and infrastructural decisions of operations management in the hotel sector and their impact on organizational performance. *Tourism and Hospitality Research*, 15(1), 3–18. DOI: 10.1177/1467358414553866

Espino-Rodríguez, T. F., & Taha, M. G. (2022). Supplier innovativeness in supply chain integration and sustainable performance in the hotel industry. *International Journal of Hospitality Management*, 100, 103103. DOI: 10.1016/j.ijhm.2021.103103

Etikan, I., & Bala, K. (2017). Sampling and Sampling Methods. *Biometrics & Biostatistics International Journal*, 5(6), 1–3. DOI: 10.15406/bbij.2017.05.00149 PMID: 28111614

Fathy, E. A. (2018). Do the Environmental Management System Practices correlate to demographic factors in the First Class Hotels in Alexandria? *Minia Journal of Tourism and Hospitality Research*, 181-214.

Fathy, E. A. (2019). Modeling the barriers of Environmental barriers on environmental practices application in Egypt. *Journal of Tourism Research*, 23, 31–56.

Fathy, E. A. (2021). Exploring Barriers of Knowledge Management Implementation KMI in Egyptian Hotel Industry. *Research Journal of the Faculty of Tourism and Hotels Mansoura University*, 10(10), 3–35.

Fathy, E. A. (2023). From Mundane to Marvelous: How Transformational Leadership Boosts Front Office Employees' Creativity. *Pharos International Journal of Tourism and Hospitality*, 2(2), 29–44. DOI: 10.21608/pijth.2023.243782.1004

Fathy, E. A., & Fouad, A. (2022). Testing the Theory of Planned Behavior TPB in Determining Intention and Behavior to hire People with Disabilities in Egyptian Hotels. *Research Journal of the Faculty of Tourism and Hotels Mansoura University*, 11(2), 747–819.

Fayed, H., & Fathy, E. A. (2022). The Impact of Transformational Leadership on Front Office Employees' Satisfaction and Organisational Commitment. *Pharos International Journal of Tourism and Hospitality*, 1(1), 49–63. DOI: 10.21608/pijth.2022.264824

Filimonau, V., Fidan, H., Alexieva, I., Dragoev, S., & Marinova, D. D. (2019). Restaurant food waste and the determinants of its effective management in Bulgaria: An exploratory case study of restaurants in Plovdiv. *Tourism Management Perspectives*, 32, 100577. DOI: 10.1016/j.tmp.2019.100577

Filimonau, V., Zhang, H., & Wang, L.-e. (2020). Food waste management in Shanghai full-service restaurants: A senior managers' perspective. *Journal of Cleaner Production*, 258, 120975. DOI: 10.1016/j.jclepro.2020.120975

Giang, K. B., Diep, P. B., Minh, H. V., Huong, N. T., Lam, N. T., Nga, P. Q., & Park, K. (2020). Improvement in Compliance With Smoke-Free Environment Regulations at Hotels and Restaurants in Vietnam After an Administrative Intervention. *Environmental Health Insights*, 14, 14. DOI: 10.1177/1178630220939927 PMID: 32684748

Gohar, A., & Kondolf, G. M. (2020). How Eco is Eco-Tourism? A Systematic Assessment of Resorts on the Red Sea, Egypt. *Sustainability (Basel)*, 12(23), 10139. DOI: 10.3390/su122310139

Gössling, S., Peeters, P., Hall, C. M., Ceron, J.-P., Dubois, G., Lehmann, L. V., & Scott, D. (2012). Tourism and water use: Supply, demand, and security. An international review. *Tourism Management*, 33(1), 1–15. DOI: 10.1016/j.tourman.2011.03.015

Graafland, J., & Bovenberg, L. (2020). Government regulation, business leaders' motivations and environmental performance of SMEs. *Journal of Environmental Planning and Management*, 63(8), 1335–1355. DOI: 10.1080/09640568.2019.1663159

Gull, S., & Rehman, H. (2022). Impact of Green Training on Ecological Sustainability in the Presence of Employee Resistance as a Moderator: Evidence from Textile Sector of Pakistan. *Pakistan Journal of Engineering and Applied Sciences*, 30, 101–110.

Haldorai, K., Kim, W. G., & Garcia, R. F. (2022). Top management green commitment and green intellectual capital as enablers of hotel environmental performance: The mediating role of green human resource management. *Tourism Management*, 88, 104431. DOI: 10.1016/j.tourman.2021.104431

Han, H., Lee, J.-S., Trang, H. L., & Kim, W. (2018). Water conservation and waste reduction management for increasing guest loyalty and green hotel practices. *International Journal of Hospitality Management*, 75, 58–66. DOI: 10.1016/j.ijhm.2018.03.012

Honchar, L., Batchenko, L., Komarnitskyi, I., Verezomska, I., Zemlina, Y., & Danylenko, O. (2024). Peculiarities of the Formation of Investment Mechanisms in the Hotel and Restaurant Sphere in the Conditions of Global Transformations. *Economic Affairs*, 69(1s), 159–167. DOI: 10.46852/0424-2513.1.2024.18

Khan, S. A., Yu, Z., & Farooq, K. (2023). Green capabilities, green purchasing, and triple bottom line performance: Leading toward environmental sustainability. *Business Strategy and the Environment*, 32(4), 2022–2034. DOI: 10.1002/bse.3234

Khatter, A., White, L., Pyke, J., & McGrath, M. (2021). Barriers and drivers of environmental sustainability: Australian hotels. *International Journal of Contemporary Hospitality Management*, 33(5), 1830–1849. DOI: 10.1108/IJCHM-08-2020-0929

Khonje, L. Z., & Leonard, L. (2019). Examining governance and growth of the hotel sector in Malawi: An environmental metabolic rift perspective. *African Journal of Hospitality, Tourism and Leisure*, 8(3).

Kock, N., & Hadaya, P. (2018). Minimum sample size estimation in PLS-SEM: The inverse square root and gamma-exponential methods. *Information Systems Journal*, 28(1), 227–261. DOI: 10.1111/isj.12131

Kuo, F.-I., Fang, W.-T., & LePage, B. A. (2022). Proactive environmental strategies in the hotel industry: Eco-innovation, green competitive advantage, and green core competence. *Journal of Sustainable Tourism*, 30(6), 1240–1261. DOI: 10.1080/09669582.2021.1931254

Legrand, W., Winkelmann, D., Sloan, P., & Simons-Kaufmann, C. (2014). Empirical Investigation on Environmental Management Systems in German Hotels. *Advances in Hospitality and Leisure*, 10, 1–20. DOI: 10.1108/S1745-354220140000010001

Li, L., Li, C., Tang, Y., & Xie, S. (2023). Going sustainable: Exploring the attribution of unusual tourist behavior in hotels. *International Journal of Hospitality Management*, 115, 103607. DOI: 10.1016/j.ijhm.2023.103607

Liu, Q., Browne, A. L., & Iossifova, D. (2022). A socio-material approach to resource consumption and environmental sustainability of tourist accommodations in a Chinese hot spring town. *Sustainable Production and Consumption*, 30, 424–437. DOI: 10.1016/j.spc.2021.12.021

López-Gamero, M. D., Molina-Azorín, J. F., Pereira-Moliner, J., & Pertusa-Ortega, E. M. (2023). Agility, innovation, environmental management and competitiveness in the hotel industry. *Corporate Social Responsibility and Environmental Management*, 30(2), 548–562. DOI: 10.1002/csr.2373

Mandić, A., & Petrić, L. (2021). The impacts of location and attributes of protected natural areas on hotel prices: Implications for sustainable tourism development. *Environment, Development and Sustainability*, 23(1), 833–863. DOI: 10.1007/s10668-020-00611-6

Martínez-Peláez, R., Ochoa-Brust, A., Rivera, S., Félix, V. G., Ostos, R., Brito, H., Félix, R. A., & Mena, L. J. (2023). Role of Digital Transformation for Achieving Sustainability: Mediated Role of Stakeholders, Key Capabilities, and Technology. *Sustainability (Basel)*, 15(14), 11221. DOI: 10.3390/su151411221

Modica, P. D., Altinay, L., Farmaki, A., Gursoy, D., & Zenga, M. (2020). Consumer perceptions towards sustainable supply chain practices in the hospitality industry. *Current Issues in Tourism*, 23(3), 358–375. DOI: 10.1080/13683500.2018.1526258

Mohamed, M. E., Hewedi, M. M., Lehto, X., & Maayouf, M. (2020). Egyptian food experience of international visitors: A multidimensional approach. *International Journal of Contemporary Hospitality Management*, 32(8), 2593–2611. DOI: 10.1108/IJCHM-02-2020-0136

Molina-Azorín, J. F., Claver-Cortés, E., Pereira-Moliner, J., & Tarí, J. J. (2009). Environmental practices and firm performance: An empirical analysis in the Spanish hotel industry. *Journal of Cleaner Production*, 17(5), 516–524. DOI: 10.1016/j.jclepro.2008.09.001

Moosa, A., & He, F. (2023). Impact of environmental management practices on corporate sustainability: Evidence from the Maldives hospitality industry. *International Journal of Emerging Markets*, 18(9), 2869–2889. DOI: 10.1108/IJOEM-06-2020-0700

Mosaad, G., Darwish, N. R., & Tarabieh, K. (2016). Proposal of a rating system: For a touristic green community in Egypt. *Journal of Al-Azhar University Engineering Sector*, 11(39), 642–655. DOI: 10.21608/auej.2016.19437

Munawar, S., Yousaf, H. Q., Ahmed, M., & Rehman, S. (2022). Effects of green human resource management on green innovation through green human capital, environmental knowledge, and managerial environmental concern. *Journal of Hospitality and Tourism Management*, 52, 141–150. DOI: 10.1016/j.jhtm.2022.06.009

Musavengane, R. (2019). Small hotels and responsible tourism practice: Hoteliers' perspectives. *Journal of Cleaner Production*, 220, 786–799. DOI: 10.1016/j.jclepro.2019.02.143

Nassar, M. A., & Fouad, A. (2022). Tomorrow Taste Comes Today: Exploring Customers' Intention to Buy 3D Printed Food in Egyptian Restaurants. *Pharos International Journal of Tourism and Hospitality*, 1(1), 31–48. DOI: 10.21608/pijth.2022.264346

Novacka, L., Pícha, K., Navratil, J., Topaloglu, C., & Švec, R. (2019). Adopting environmentally friendly mechanisms in the hotel industry: A perspective of hotel managers in Central and Eastern European countries. *International Journal of Contemporary Hospitality Management*, 31(6), 2488–2508. DOI: 10.1108/IJCHM-04-2018-0284

Oriade, A., Osinaike, A., Aduhene, K., & Wang, Y. (2021). Sustainability awareness, management practices and organisational culture in hotels: Evidence from developing countries. *International Journal of Hospitality Management*, 92, 102699. DOI: 10.1016/j.ijhm.2020.102699

Rahman, I., & Reynolds, D. (2016). Predicting green hotel behavioral intentions using a theory of environmental commitment and sacrifice for the environment. *International Journal of Hospitality Management*, 52, 107–116. DOI: 10.1016/j.ijhm.2015.09.007

Sachdeva, G., Taneja, S., & Gupta, R. (2024). The mediating role of green human resource management: Analyze the impact of green commitment and intellectual capital in hotel environmental performance. *Energy Efficiency*, 17(5), 17. DOI: 10.1007/s12053-024-10229-6

Sachs, J., Schmidt-Traub, G., Kroll, C., Lafortune, G., & Fuller, G. (2019). *Sustainable Development Report 2019. New York: Bertelsmann Stiftung and Sustainable Development Solutions Network (SDSN).* Retrieved May 2024, from https://s3.amazonaws.com/sustainabledevelopment.report/2019/2019_sustainable_development_report.pdf

Salem, I. (2015). Transformational leadership: Relationship to job stress and job burnout in five star hotels. *Tourism and Hospitality Research*, 15(4), 240–253. DOI: 10.1177/1467358415581445

Salem, I. E., Elkhwesky, Z., & Ramkissoon, H. (2022). A content analysis for government's and hotels' response to COVID-19 pandemic in Egypt. *Tourism and Hospitality Research*, 22(1), 42–59. DOI: 10.1177/14673584211002614

Sanikos. (2021). *Sanikos ESG Repor.* Retrieved May 2024, from https://ikosresorts.com/wp-content/uploads/2022/11/2021-SANI-IKOS-GROUP-ESG-REPORT1.pdf

Saunders, M., Lewis, P., & Thornhill, A. (2016). *Research Methods for Business Students*. Pearson.

Schilling-Vacaflor, A., & Gustafsson, M.-T. (2024). Towards More Sustainable Global Supply Chains? Company Compliance with new Human Rights and Environmental Due Diligence Laws. *Environmental Politics*, 33(3), 422–443. DOI: 10.1080/09644016.2023.2221983

Shaalan, I. M. (2005). Sustainable tourism development in the Red Sea of Egypt threats and opportunities. *Journal of Cleaner Production*, 13(2), 83–87. DOI: 10.1016/j.jclepro.2003.12.012

Sharma, G. (2017). Pros and cons of different sampling techniques. *International Journal of Applied Research*, 3(7), 749–752.

Shawky, I., Abo El Enen, M., & Fouad, A. (2019). Examining Customers' Intention and Attitude Towards Reading Restaurants' Menu Labels by Using the Theory of Planned Behaviour. *International Tourism and Hospitality Journal*, 2(4), 1–14.

Shehata, H. S., & Elfeel, S. (2017). Going Green in Egyptian Hotels: Importance and Implementation of Water and Energy Practices. *Journal of Tourism Research*, 16, 6–21.

Soliman, S. A. (2020). Food Waste Management: Exploratory Study of Egyptians chefs. *Hospitality, Journal of Association of Arab Universities for Tourism and, 18*(3), 177-198.

Statista. (2023). *Travel, Tourism & Hospitality*. Retrieved May 2024, from Statistica: https://www.statista.com/statistics/233223/travel-and-tourism-total-economic-contribution-worldwide/

Sucheran, R., & Moodley, V. (2019). Guest dynamics and perceptions towards environmentally-friendly practices in hotels in KwaZulu- Natal, South Africa. *African Journal of Hospitality, Tourism and Leisure*, 8(3).

Suliman, M. A., Abdou, A. H., Ibrahim, M. F., Al-Khaldy, D. A., Anas, A. M., Alrefae, W. M., & Salama, W. (2023). Impact of Green Transformational Leadership on Employees' Environmental Performance in the Hotel Industry Context: Does Green Work Engagement Matter? *Sustainability (Basel)*, 15(3), 2690. DOI: 10.3390/su15032690

Sun, H., Bahizire, G. M., Pea-Assounga, J. B., & Chen, T. (2024). Enhancing Employee Green Performance through Green Training: The Mediating Influence of Organizational Green Culture and Work Ethic in the Mining Sector. *Journal of Cleaner Production*, 449, 141105. DOI: 10.1016/j.jclepro.2024.141105

Sun, H., Samad, S., Rehman, S. U., & Usman, M. (2022). Clean and green: The relevance of hotels' website quality and environmental management initiatives for green customer loyalty. *British Food Journal*, 124(12), 4266–4285. DOI: 10.1108/BFJ-09-2021-1002

Taha, M. G., & Espino-Rodríguez, T. F. (2020). The Impact of the Organizational Culture on Hotel Outsourcing and Sustainable Performance an Empirical Application in the Egyptian Hotel Sector. *Sustainability (Basel)*, 12(22), 9687. DOI: 10.3390/su12229687

Taherdoost, H. (2016). Sampling Methods in Research Methodology; How to Choose a Sampling Technique for Research. *SSRN*, 5(2), 18–27. DOI: 10.2139/ssrn.3205035

Tan, Y., Li, H. X., Wang, J., Jiang, B., Song, Y., & Wang, X. (2021). Cost and environmental impact estimation methodology and potential impact factors in offshore oil and gas platform decommissioning: A review. *Environmental Impact Assessment Review*, 87, 106536. DOI: 10.1016/j.eiar.2020.106536

Tanveer, M. I., Yusliza, M. Y., & Fawehinmi, O. (2024). Green HRM and hospitality industry: Challenges and barriers in adopting environmentally friendly practices. *Journal of Hospitality and Tourism Insights*, 7(1), 121–141. DOI: 10.1108/JHTI-08-2022-0389

Tariq, M., Yasir, M., & Majid, A. (2020). Promoting employees' environmental performance in hospitality industry through environmental attitude and ecological behavior: Moderating role of managers' environmental commitment. *Corporate Social Responsibility and Environmental Management*, 27(6), 3006–3017. DOI: 10.1002/csr.2019

United Nations. (2015). *Transforming our world: The 2030 Agenda for Sustainable Development.* Retrieved March 2024, from https://sdgs.un.org/publications/transforming-our-world-2030-agenda-sustainable-development-17981

Wei, W., Ramalho, O., & Mandin, C. (2015). Indoor air quality requirements in green building certifications. *Building and Environment*, 92, 10–19. DOI: 10.1016/j.buildenv.2015.03.035

Yousaf, Z., Radulescu, M., Sinisi, C. I., Serbanescu, L., & Paunescu, L. M. (2021). Harmonization of Green Motives and Green Business Strategies towards Sustainable Development of Hospitality and Tourism Industry: Green Environmental Policies. *Sustainability (Basel)*, 13(12), 6592. DOI: 10.3390/su13126592

Yucedag, C., Kaya, L. G., & Cetin, M. (2018). Identifying and assessing environmental awareness of hotel and restaurant employees' attitudes in the Amasra District of Bartin. *Environmental Monitoring and Assessment*, 190(60), 60. DOI: 10.1007/s10661-017-6456-7 PMID: 29305729

Zhang, H., Omhand, K., Li, H., Ahmad, A., Samad, S., Gavrilut, D., & Badulescu, D. (2022). Corporate Social Responsibility and Energy-Related Pro-Environmental Behaviour of Employees in Hospitality Industry. *International Journal of Environmental Research and Public Health*, 19(23), 16141. DOI: 10.3390/ijerph192316141 PMID: 36498214

Zorpas, A. A., Voukkali, I., & Loizia, P. (2015). The impact of tourist sector in the waste management plans. *Desalination and Water Treatment*, 56(5), 1141–1149. DOI: 10.1080/19443994.2014.934721

Chapter 10
Exploring Stakeholder Engagement in Tourism Waste Management:
Self-Help Groups' Perspectives

Shreeansh Mishra
https://orcid.org/0009-0004-8847-6587
Central University of Tamil Nadu, India

Jitendra Mohan Mishra
https://orcid.org/0000-0002-5840-9701
Central Tribal University of Andhra Pradesh, India

ABSTRACT

This chapter explores the motivations driving stakeholder engagement and the challenges faced in waste management initiatives within the tourism sector, focusing on the perspectives of self-help groups (SHGs). Through qualitative research methods, including in-depth interviews, this study examines the diverse motivations that prompt SHGs to participate in waste management projects, ranging from environmental concerns to socioeconomic empowerment. Moreover, it identifies and analyses the multifaceted challenges SHGs encounter, such as limited resources, inadequate infrastructure, and bureaucratic hurdles. This study contributes to a deeper understanding of stakeholder dynamics in tourism waste management by elucidating these motivations and challenges. The findings suggest that while SHGs play a crucial role in fostering sustainable practices, effective collaboration and support mechanisms are essential to address the challenges encountered and maximize the potential benefits of stakeholder engagement in waste management projects within the tourism sector.

DOI: 10.4018/979-8-3693-6110-8.ch010

1. INTRODUCTION

Self-help groups (SHGs) are grassroots-level institutions formed by individuals who voluntarily come together to address common issues often related to economic, social, and environmental concerns (Mishra & Mishra, 2023). These groups are characterised by mutual aid, support, and shared responsibility, and they typically operate on principles of collective action and empowerment (Levy, 2000; Pritiprada et al., 2023). The mechanism of SHGs involves regular meetings in which members contribute savings, discuss issues, plan activities, and provide loans or support to members in need (Ravindran & Vinodan, 2009; Vukovic et al., 2021). This structure fosters financial independence and encourages social cohesion and community development (Rinaldi & Salerno, 2020).

In tourism, SHGs have emerged as significant players, contributing to the sustainable development of tourism destinations (Vinodan & Meera, 2022). Their roles in tourism are multifaceted; they often promote local culture, provide authentic experiences to tourists, and participate in community-based tourism initiatives (Vukovic et al., 2021). Moreover, SHGs can act as custodians of their local environments, ensuring that tourism development does not come at the cost of ecological degradation (Bwala et al., 2016). By leveraging their local knowledge and organisational capacity, SHGs can create a sustainable tourism model that benefits the community and the environment. (Vinodan & Meera, 2022).

One critical area where SHGs have demonstrated considerable impact is waste management in the tourism sector (Djaelani, 2021; Vinodan & Meera, 2022). The concept of collective initiative is exemplified by SHGs, which operate on the principles of cooperation to achieve specific goals, shared responsibility, and active community participation in decision-making processes. Several factors drive the intervention of SHGs in waste management. Environmental concerns are a primary motivator, as unmanaged waste can lead to severe pollution and degradation of natural resources, which are often attractions that attract tourists. Additionally, socioeconomic empowerment through waste management activities, such as recycling and composting, provides SHGs with opportunities for income generation and skill development (Vankar et al., 2013). By engaging in these activities, SHGs contribute to cleaner and more appealing tourist destinations, attracting more visitors and enhancing local livelihoods.

However, SHGs' involvement in waste management in tourism is challenging. This study identifies several obstacles SHGs face, including limited resources, inadequate infrastructure, and bureaucratic hurdles. Despite these challenges, the potential benefits of SHG participation in waste management remain significant. Effective stakeholder engagement is crucial to overcoming these barriers. Collaboration with local governments, tourism operators, and other stakeholders can provide the necessary support and resources to enhance the effectiveness of SHG initiatives. Such collaboration can develop robust, sustainable, and inclusive waste-management systems.

The research conducted through qualitative methods, including in-depth interviews with SHG members, revealed the complex dynamics of stakeholder engagement in tourism waste management. This highlights the diverse motivations that drive SHG participation and the multifaceted challenges they encounter. By understanding these dynamics, this study aims to contribute to a deeper understanding of how SHGs can be supported to maximise their impact on sustainable tourism practices.

SHGs play a crucial role in promoting sustainable tourism through active involvement in waste management. Their grassroots approach and community-based initiatives are essential for fostering environmental stewardship and socioeconomic development. However, to fully realise their potential, there must be a concerted effort to address the challenges they face through effective collaboration and support mechanisms. This research highlights the importance of recognising and empowering SHGs as crucial stakeholders in pursuing sustainable tourism and effective waste management strategies.

1. LITERATURE REVIEW

1.1 Tourism Sector and Waste Management

With the global increase in tourist numbers and their expanding demand, the resultant tourism waste poses a substantial threat to the sustainability of local tourism (Xu et al., 2024). Although increasing tourist footfall can stimulate economic growth, managing tourism-related waste is a significant concern (Xu et al., 2021). Jiang et al. (2009) concluded that solid waste management in Tibet faces considerable practical challenges, including inadequate infrastructure and a lack of appropriate disposal methods. Wang et al. (2018) improved the understanding by assessing the ecological footprint of hotel, restaurant, and café food waste. They found significant environmental impacts, indicating the need for more sustainable food waste practices in the region's hospitality industry. The government manages, monitors, and operates municipal solid waste (MSW), resulting in ineffective and uncompetitive management (Xu et al., 2024). These challenges highlight the need for community intervention and collaborative efforts to enhance waste management efficiency and effectiveness.

SHGs have been widely recognised for their role in fostering community development and socioeconomic empowerment, particularly in developing countries (Ahemad & Katoch, 2022; Bwala et al., 2016). These groups typically consist of

10-20 members who meet regularly to save money, provide loans to one another, and discuss community issues (Komati, 2018; Pritiprada et al., 2023). The core mechanism of SHGs revolves around collective savings and credit activities, which enhance financial security and promote mutual support and social cohesion (Nichols, 2021). Through their activities, SHGs empower members, especially women, by providing them with financial literacy, entrepreneurial skills, and a platform for social interaction (Ban et al., 2020; Deshpande & Khanna, 2021).

2. 2 The Role of SHGs in Tourism

SHGs have been instrumental in promoting sustainable tourism practices and enhancing community participation (Vukovic et al., 2021). This can be further understood in Figure 1. According to Mishra and Mishra (2023), SHGs contribute to community-based tourism by facilitating the development of local tourism enterprises, preserving cultural heritage, and ensuring that tourism benefits are equitably distributed within the community. SHGs also play a crucial role in environmental conservation by engaging in activities such as maintaining cleanliness in tourist spots and promoting eco-friendliness (Kayongo-Male & Mbithi, 1979; Vinodan & Meera, 2022). These contributions are significant in creating a sustainable tourism model prioritising local well-being and environmental sustainability.

Figure 1. Intervention areas of self-help groups in tourism

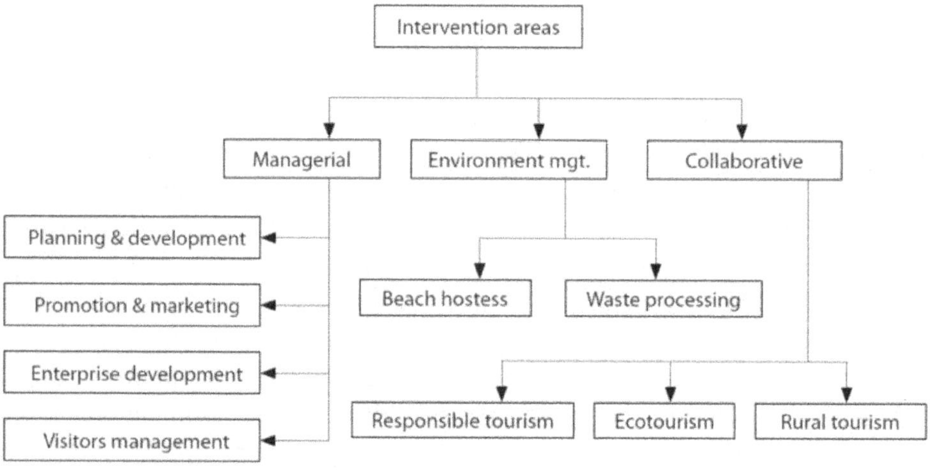

Source: Adapted from Vinodan & Meera, 2022

2. 3 SHGs and Waste Management in Tourism

The success of management is influenced by the internal community's attributes, such as collective initiative, effective leadership, and institutional transparency (Susilowati et al., 2022). Involving community members in waste management is significant because it fosters active participation, promotes shared responsibility, and enhances the effectiveness of waste management initiatives (Fajarwati et al., 2020; Lunag et al., 2021). The involvement of SHGs in waste management within the tourism sector is a relatively recent but growing area of interest (Kumar & Agrawal, 2020). Waste management is a critical issue in tourism, as poor waste handling can lead to environmental degradation and negatively impact the attractiveness of tourist destinations (Mair & Jago, 2010). SHGs, with their strong community networks and organisational skills, are well-positioned to address these challenges (Raghunathan et al., 2023; Vinodan & Meera, 2022). They engage in waste segregation, recycling, and composting, reducing tourists' environmental footprint.

2. 4 Research Gap and Significance

Despite the potential of SHGs in waste management, there is a dearth of research examining their specific roles and the challenges they face in this context. Existing literature primarily focuses on the economic and social aspects of SHG activities, with limited attention to their environmental interventions, particularly in tourism-related waste management (Mishra & Mishra, 2023). This gap highlights the need for more comprehensive studies that explore the motivations, challenges, and impacts of SHG-led waste management initiatives in the tourism sector. The current body of literature reveals a significant research gap in understanding the mechanisms and challenges of SHGs in tourism waste management. While the socioeconomic benefits of SHGs are well-documented, their environmental contributions remain underexplored, especially in managing tourism waste.

Involving SHGs in waste management through credit support and demand creation via demonstrations and pilot projects is critical for encouraging innovative startups and community support. This study aims to fill this gap by providing an in-depth analysis of SHG activities in waste management within tourism, focusing on their motivations, challenges, and the outcomes of their interventions. The significance of this study lies in its potential to inform policy and practice in sustainable tourism and waste management. By highlighting the critical role of SHGs, the research can guide the development of supportive frameworks that enhance SHG capacities and foster effective collaboration among stakeholders. This, in turn, can lead to more sustainable tourism practices that benefit both the environment and local communities.

2. RESEARCH METHODOLOGY

This study employs qualitative research methods to explore the motivations, challenges, and impacts of self-help groups involved in waste management within the tourism sector. Data collection primarily involves in-depth interviews with members of SHGs, which provide detailed insights into their experiences and perspectives. These semi-structured interviews allow for flexibility in responses while ensuring that critical topics are covered. The qualitative approach is chosen to capture the nuanced motivations driving SHG participation in waste management initiatives and understand their multifaceted challenges. The study involves compiling the outcomes that emerge from the responses. This methodology enables a comprehensive examination of the dynamics of stakeholder engagement in tourism waste management. It contributes to a deeper understanding of how SHGs can be effectively supported to maximise their impact on sustainable tourism practices.

The study area includes SHG members from various tourism-famous states of India, including Kerala, Tamil Nadu, Maharashtra, Karnataka, and Uttar Pradesh. The interviewed members were asked about the perceived benefits of SHG membership and their motivation to join the group and participate in waste management activities. In addition to their motivation, they were asked to express the challenges that limit their participation or performance during waste management activities.

3. RESEARCH FINDINGS

Community engagement has the potential to unite various sectors, including non-governmental organisations (NGOs), community-based organisations (CBOs), private entities, local government bodies, etc., in pooling resources and generating favourable outcomes (Vinodan & Meera, 2022). SHGs is a powerful approach that works at the grassroots level to manage tourism destinations and enhance destination sustainability. The research findings highlight the motivations and challenges for SHG's engagement as a tourism stakeholder in enhancing the effectiveness and sustainability of tourism waste management practices, as shown in Figure 2.

Figure 2. Motivation and challenges for SHGs' intervention in waste management practices

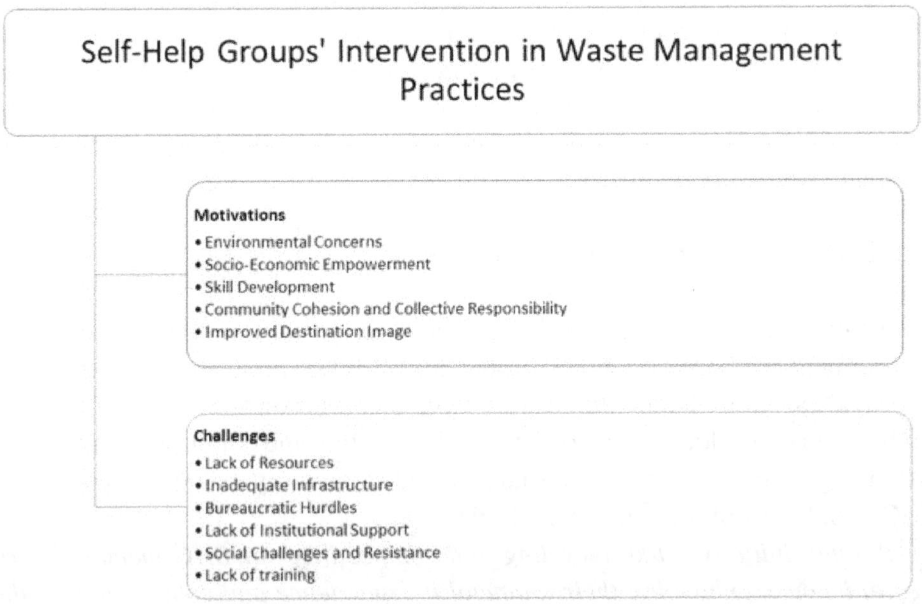

Source: Authors' Compilation

4. 1 The Motivations for Participation in Waste Management Practices

4.1.1 Environmental Concern

While the community possesses knowledge of straightforward waste processing methods, their practical implementation in everyday scenarios remains infrequent (Kandpal & Saizen, 2022). SHG members are driven by a solid commitment to preserving their local environment (Freudenberg & Steinsapir, 1991). They engage in waste management to mitigate the adverse effects of tourism on natural surroundings, ensuring ecological balance and enhancing the aesthetic appeal of their locales (Vinodan & Meera, 2022). SHGs coming forward as leaders not only educate and make community people aware of waste management problems (Kalra, 2019) but also educate the tourists to be more responsible in disposing of waste and managing the destination (Mishra & Mishra, 2023). A SHG Member from Kochi, Kerala, opined,

Our community recognises the importance of preserving our local environment, and we are committed to taking action to reduce the negative impact of tourism on our surroundings. By actively engaging in waste management initiatives, we aim to maintain the natural environment and enhance the beauty of our region. ...we believe that SHGs play a crucial leadership role in educating both the local community and tourists about waste management issues...through our efforts, we hope to raise awareness and promote responsible waste disposal practices among residents and tourists.

4.1.2 Socio-Economic Empowerment

Most community members' primary motivation to join an SHG is economic empowerment and family welfare (Alrefaei et al., 2022). Waste management activities provide SHGs with opportunities for income generation and skill development. Members can earn additional income through recycling and composting initiatives, contributing to their financial independence and community welfare. According to a SHG member from Idukki, Kerala, India:

...through initiatives like recycling and composting, our SHG members earn extra income, enhancing their financial independence and contributing to the family welfare ...this additional income promotes self-reliance among members and improves the community's overall prosperity. Additionally, participating in waste management activities helps us acquire valuable skills, empowering us to take on various other jobs and make meaningful contributions to our society.

4.1.3 Skills Development

SHGs can construct or repair houses and sanitation units, enhancing various aspects of living, including improved hygiene practices (Venkatraja, 2019). Participation in waste management projects allows SHG members to acquire valuable skills and training. These skills enhance their employability and entrepreneurial capabilities through networking and enrolling in awareness programs, providing long-term socioeconomic benefits. Through a mentoring program, waste management has successfully increased critical awareness among group members and provided them with a technical understanding of waste management, specifically in plastic waste sorting and organic waste processing (Tallei et al., 2013). A SHG member from Theni, Tamil Nadu, expressed

...I can attest to the significant skill development that occurs through our involvement in waste management projects. Engaging in these initiatives has equipped us with valuable skills, which not only improve our living habits and hygiene practices but also enhance our employability... through networking and partici-

pating in awareness programs, we gain training that boosts our entrepreneurial capabilities and awareness about modern methods of waste management... this, in turn, provides long-term socioeconomic benefits for us and our community, fostering overall development and self-reliance.

4.1.4 Community Cohesion and Collective Responsibility

Waste management projects foster a sense of unity and shared purpose among SHG members. Engagement in SHG initiatives fosters direct interaction between members and various stakeholders, such as government officials, bank personnel, and representatives from non-governmental organisations (NGOs), thereby facilitating the accumulation of social capital (Kandpal & Saizen, 2022). Working together on these initiatives strengthens social bonds, creating a community consciousness and positive behavioural changes that cannot happen overnight. Therefore, members of SHGs engage in waste management initiatives due to their collective sense of responsibility toward the community's well-being.

4.1.5 Improved Destination Image

With the rise in the number of tourists, the demand for accommodations also grows; without proper planning and effective waste management services, this can lead to substantial environmental issues (Kerber & Kramm, 2021). Proper waste management reduces health hazards, improves residents' living conditions, and attracts tourists. By enhancing community well-being, SHGs make their areas more attractive for tourists and locals, improving overall quality of life. This improvement in living conditions itself brings intrinsic motivation to the SHG members. One of the SHG members from Varanasi, India, stated,

> *...participating in proper waste management reduces health risks and enhances our residents' living standards, making our areas more appealing to tourists and locals alike. ...through our efforts in the SHG, we strive to enhance the well-being of our community, thereby contributing to overall improvements in the quality of life. This intrinsic motivation among our SHG members stems from witnessing the tangible improvements in our living conditions.*

4.2 Challenges Faced by SHGs in Waste Management

4.2.1 Lack of Resources

Many SHGs operate with limited financial means, restricting their ability to invest in necessary tools, infrastructure, and training. Their small scale, accompanied by restricted financial and personnel capacities, often necessitates their dependence on promoter organisations to access crucial resources (Pati, 2008). This lack of financial and technical resources and their dependency on other institutions hamper their effectiveness and scalability in waste management activities.

4.2.2 Inadequate Infrastructure

Despite extensive exploration of providing credit to people experiencing poverty, particularly in rural areas, significant infrastructure constraints and supply-side limitations persist in India (Mukherjee et al., 2019). The absence of proper waste management facilities, such as recycling centres and waste segregation units, poses a significant obstacle. Funding for operational management, maintenance fees, and additional infrastructure investments and facilities is needed (Tallei et al., 2013). This infrastructural deficit forces SHGs to rely on ad-hoc and often suboptimal methods.

4.2.3 Bureaucratic Hurdles

Navigating the regulatory landscape and obtaining necessary permits from local authorities can be cumbersome. SHGs frequently encounter red tape and administrative delays, which impede their progress and discourage their initiatives. The formal establishment of connections between rural local bodies and Self-Help Groups (SHGs) will significantly influence the willingness and capability of members in both entities (Kandpal & Saizen, 2019). A SHG member from Mysore Karnataka explained,

> *...dealing with the complicated administrative processes often becomes a major hurdle, slowing down our progress significantly. Getting the necessary permits, approvals, and support from local government bodies can be very time-consuming and frustrating, as it involves much paperwork and procedural delays... these bureaucratic obstacles not only delay our initiatives but also discourage many members from actively participating in waste management activities... Some of us are involved in any other activity for income generation but not waste management.*

4.2.4 Lack of Institutional Support

There is often a disconnect between SHG activities and broader municipal waste management strategies (Kandpal & Saizen, 2022). Encouraging collaboration between the formal and informal sectors is seen as beneficial in advancing environmentally friendly disposal methods while simultaneously boosting the income of waste pickers (Marello & Helwege, 2017). The lack of institutional support and recognition for SHG efforts limits their impact and integration into larger waste management frameworks and the income they can generate from such initiatives and activities. Coordination activities can be facilitated during monitoring and evaluation activities once a year through meetings and regularly exchanging opinions between self-help groups and the government if the government provides space and time for self-help groups (Djaelani, 2021). It will strengthen the destination's waste management practices and aesthetic beauty.

4.2.5 Social Challenges and Resistance

Various cultural factors may underlie resistance toward participation in waste management. These factors encompass the persistence of a semi-caste system wherein socio-economically disadvantaged individuals are relegated to low-paying, low-status occupations such as waste collection, transportation, and informal trade in recyclable materials (Mongkolnchaiarunya, 2005). Waste disposal responsibilities are frequently delegated to domestic servants and maids at the household level. Overcoming entrenched behaviours and attitudes within the community can be challenging. Persistent community engagement and education are required to change mindsets and raise awareness about the importance of proper waste management, which can be resource-intensive and slow to yield results.

4.2.6 Lack of Training

SHG members tend to gravitate towards social entrepreneurship over formal entrepreneurship due to their limited acquisition of advanced skills and entrepreneurial resources (Vukovic et al., 2021). Many SHGs are multi-purpose, which aids in their sustainability; however, with proper technical guidance, some may focus on conservation over extended periods to achieve more comprehensive conservation and destination management objectives (Kayongo-Male & Mbithi, 1979). A SHG member from Kasargod shared...

...largely due to our limited access to advanced skills and entrepreneurial resources, we are not able to take up big projects or large-scale tasks... our efforts are limited to local regions...the lack of specialised training often hinders our

ability to engage in more formal entrepreneurial ventures. However, with appropriate technical guidance, I believe SHGs could focus on long-term conservation efforts and destination management practices.

4. ANALYSIS

The analysis of SHG motivations reveals a complex interplay of environmental, socioeconomic, and community-focused drivers. SHGs' vital environmental concern aligns with broader global sustainability goals, underscoring the importance of grassroots movements in addressing ecological issues. This intrinsic motivation is pivotal in maintaining the ecological balance and enhancing the aesthetic appeal of tourist destinations, which are critical for sustaining tourism. The socioeconomic empowerment motivation highlights the dual benefits of waste management activities - economic gain and skill development. SHGs utilise waste management to generate additional income and as a platform for skill acquisition. This dual focus enhances members' financial independence and equips them with valuable entrepreneurial skills, contributing to long-term socioeconomic stability.

Community cohesion and collective responsibility emerge as significant motivators, demonstrating the social capital inherent in SHGs. The collaborative nature of waste management projects fosters unity and strengthens social bonds among members. This collective effort is crucial in building a shared responsibility toward maintaining clean and healthy environments, essential for community resilience and well-being. The drive to improve quality of life emphasises the holistic approach SHGs take towards community development. Proper waste management reduces health hazards and enhances living conditions, making the area more attractive to tourists and residents. This improvement in quality of life has a reinforcing effect, as cleaner environments attract more tourists, thereby creating a positive feedback loop that benefits the community economically and socially.

SHGs' challenges highlight systemic issues that need addressing to maximise their impact. The lack of resources is a significant barrier, limiting SHGs' ability to scale up their activities. Financial constraints restrict access to tools and infrastructure critical for effective waste management. Addressing this challenge requires targeted financial support and investment in SHG activities. Inadequate infrastructure further compounds the problem. The absence of essential facilities such as recycling centres and waste segregation units hampers the efficiency and effectiveness of SHG initiatives. This infrastructural gap necessitates coordinated efforts from local governments and stakeholders to develop and provide the necessary waste management infrastructure.

Bureaucratic hurdles present another significant obstacle. The complexity of navigating regulatory frameworks and obtaining permits can discourage SHGs and impede their progress. Streamlining administrative processes and providing clear guidelines can mitigate these challenges, enabling SHGs to focus on their core activities without undue bureaucratic burden. The lack of institutional support is a critical issue that limits the integration of SHG efforts into broader waste management strategies. Recognition and support from municipal authorities and other institutions are essential for scaling up SHG initiatives and ensuring their sustainability. Creating formal channels for collaboration and support can enhance the effectiveness of SHG activities.

Social challenges, including community resistance and lack of awareness, hinder SHG efforts. Changing entrenched behaviours and attitudes requires sustained community engagement and education. This process can be resource-intensive and slow, but it is essential for the long-term success of waste management initiatives. Developing targeted awareness campaigns and educational programs can help address these social challenges.

5. DISCUSSION

The findings confirm the pivotal role of SHGs in promoting sustainable waste management practices within the tourism sector. Their diverse motivations reflect a comprehensive approach to community development, balancing environmental stewardship, socioeconomic empowerment, and community cohesion. This multifaceted engagement is crucial for creating sustainable tourism models that benefit the environment and local communities. However, SHGs' challenges highlight the need for systemic support and collaboration. Addressing financial constraints through targeted funding and investment is essential for enhancing SHG capacities. Developing adequate infrastructure and streamlining bureaucratic processes can facilitate effective waste management practices. Institutional support and recognition are critical for integrating SHG efforts into broader waste management frameworks. Moreover, addressing social challenges requires sustained community engagement and education. Building awareness and changing attitudes towards waste management are long-term processes that necessitate persistent effort and resources. Stakeholders can help SHGs overcome these challenges and maximise their impact by fostering a supportive environment and providing the necessary resources.

6. CONCLUSION

This study highlights the significant role of SHGs in waste management within the tourism sector, driven by motivations rooted in environmental concern, socio-economic empowerment, skill development, community cohesion, and quality of life improvements. SHGs emerge as vital agents of change, leveraging their grass-roots organisational structures to address the challenges of tourism-related waste. Their efforts contribute to the sustainability of tourist destinations, ensuring that ecological integrity and community well-being are maintained. However, the study also reveals substantial challenges that hinder the effectiveness of SHG-led waste management initiatives, such as financial constraints, inadequate infrastructure, bureaucratic hurdles, lack of institutional support, and social resistance. These challenges highlight the need for a supportive framework that includes targeted financial aid, infrastructure development, streamlined administrative processes, and enhanced institutional recognition and collaboration.

The findings suggest that addressing these challenges requires a multifaceted approach. Financial support and investment can empower SHGs to scale up their activities and access necessary resources. Developing proper waste management infrastructure is essential for facilitating efficient operations. Streamlining bureaucratic processes can reduce administrative burdens, allowing SHGs to focus on their core activities. Furthermore, fostering institutional support and formalising collaboration channels can enhance SHG efforts' integration into broader waste management strategies. Community engagement and education are also crucial for overcoming social resistance and raising awareness about the importance of proper waste management. Sustained efforts in these areas can gradually change entrenched behaviours and attitudes, fostering a culture of environmental responsibility within the community. SHGs are pivotal in promoting sustainable waste management practices in tourism. To maximise their impact, they must address their systemic challenges through coordinated support, investment, and policy interventions. By empowering SHGs and enhancing stakeholder collaboration, sustainable tourism and effective waste management can benefit the environment and local communities.

REFERENCES

Ahemad, S., & Katoch, O. R. (2022). Role of self-help groups (SHGs) in enhancing incomes of rural women in J&K, India. *South Asian Journal of Social Studies and Economics*, 16(1), 24–32. DOI: 10.9734/sajsse/2022/v16i1602

Alrefaei, N., Aquinas, P., & Al-Maamari, O. A. (2022). Self-help group (SHG) in India: A path toward empowerment and poverty reduction. *Social Work with Groups*, 46(3), 249–263. DOI: 10.1080/01609513.2022.2159612

Ban, R., Gilligan, M. J., & Rieger, M. (2020). Self-help groups, savings and social capital: Evidence from a field experiment in Cambodia. *Journal of Economic Behavior & Organization*, 180, 174–200. DOI: 10.1016/j.jebo.2020.09.029

Bwala, H. B., Oladosu, R. O., & Dukku, S. J. (2016). Role of self-help groups: A needful approach to sustainable community development. *Journal of Culture and Sustainable Development, 21*(2002), 94–98. https://iiste.org/Journals/index.php/JCSD/article/view/31887

Deshpande, A., & Khanna, S. (2021). Can weak ties create social capital? Evidence from self-help groups in rural India. *World Development*, 146, 105534. DOI: 10.1016/j.worlddev.2021.105534

Djaelani, M. (2021). Social community participation in household waste management. *Journal of Social Science Studies (JOS3), 1*(1), 37–39. https://doi.org/DOI: 10.56348/jos3.v1i1.3

Fajarwati, A., Setyaningrum, A., Rachmawati, R., & Prakoso, B. S. E. (2020). Keys of sustainable community-based waste management (lesson learnt from Yogyakarta City). *E3S Web of Conferences, 200*. https://doi.org/DOI: 10.1051/e3sconf/202020002018

Freudenberg, N., & Steinsapir, C. (1991). Not in our backyards: The grassroots environmental movement. *Society & Natural Resources*, 4(3), 235–245. DOI: 10.1080/08941929109380757

Jiang, J., Lou, Z., Ng, S., Luobu, C., & Ji, D. (2009). The current municipal solid waste management situation in Tibet. *Waste Management (New York, N.Y.)*, 29(3), 1186–1191. DOI: 10.1016/j.wasman.2008.06.044 PMID: 18845428

Kalra, N. (2020). Community participation and waste management. In Ghosh, S. (Ed.), *Sustainable waste management: Policies and case studies* (pp. 183–203). Springer., DOI: 10.1007/978-981-13-7071-7_10

Kandpal, R., & Saizen, I. (2019). Enabling collaborative action between self-help groups and rural local bodies through an exploratory workshop: The case of a peri-urban village in Mumbai Metropolitan Region, India. *Sustainability (Basel)*, 11(6), 1680. Advance online publication. DOI: 10.3390/su11061680

Kandpal, R., & Saizen, I. (2022). Self-help group participation towards sustainable solid waste management in peri-urban villages: Evidence from Mumbai Metropolitan Region, India. *Environment, Development and Sustainability*, 24(3), 3791–3814. DOI: 10.1007/s10668-021-01588-6

Kayongo-Male, D., & Mbithi, P. (1979). Local environmental perceptions and conservation practices. *Agricultural Administration*, 6(4), 299–304. DOI: 10.1016/0309-586X(79)90005-0

Kerber, H., & Kramm, J. (2021). On- and offstage: Encountering entangled waste–tourism relations on the Vietnamese Island of Phu Quoc. *The Geographical Journal*, 187(2), 98–109. DOI: 10.1111/geoj.12376

Komati, L. N. (2018). Problems and prospects of women self-help groups. Serials. https://kalnet.kshec.kerala.gov.in/vufind/Record/uoc.299896

Kumar, A., & Agrawal, A. (2020). Recent trends in solid waste management status, challenges, and potential for the future Indian cities – A review. *Current Research in Environmental Sustainability*, 2, 100011. DOI: 10.1016/j.crsust.2020.100011

Levy, L. H. (2000). Self-help groups. In *Handbook of community psychology* (pp. 591–613). Springer US., DOI: 10.1007/978-1-4615-4193-6_25

Lunag, M. N.Jr, Elauria, J. C., & Burguillos, J. D. (2021). Community-based bin design approach: An initial stage toward urban composting at a hill station, Philippines. *Environment, Development and Sustainability*, 23(3), 3832–3852. DOI: 10.1007/s10668-020-00746-6

Mair, J., & Jago, L. (2009). The development of a conceptual model of greening in the business events tourism sector. *Journal of Sustainable Tourism*, 18(1), 77–94. DOI: 10.1080/09669580903291007

Marello, M., & Helwege, A. (2017). Solid waste management and social inclusion of waste pickers: Opportunities and challenges. *Latin American Perspectives*, 45(1), 108–129. DOI: 10.1177/0094582X17726083

Mishra, S., & Mishra, J. M. (2023). Self-help groups and community-based tourism: A study of interrelationship and convolutions. *Journal of Tourism*, 24, 71–81. https://www.jothnbgu.in/journal_file/FINAL_JOTH.pdf#page=79

Mongkolnchaiarunya, J. (2005). Promoting a community-based solid-waste management initiative in local government: Yala municipality, Thailand. *Habitat International*, 29(1), 27–40. DOI: 10.1016/S0197-3975(03)00060-2

Mukherjee, S., Mallik, S. S., & Thakur, D. (2019). Tracking financial inclusion in India: A study of SHG initiatives. *Indian Journal of Human Development*, 13(1), 32–46. DOI: 10.1177/0973703019839807

Nichols, C. (2021). Self-help groups as platforms for development: The role of social capital. *World Development*, 146, 105575. DOI: 10.1016/j.worlddev.2021.105575 PMID: 34602707

Pati, A. P. (2008). Subsidised micro financing financial sustainability of SHGs. *Indian Journal of Commerce*, 61, 137–150. https://papers.ssrn.com/sol3/papers.cfm?abstract_id=1630271

Pritiprada, P., Satpathy, I., Patnaik, B., Patnaik, A., & Khang, A. (2023). Role of the internet of things (IoT) in enhancing the effectiveness of the self-help groups (SHGs) in smart city. In CRC Press eBooks (pp. 183–203). https://doi.org/DOI: 10.1201/9781003376064-11

Raghunathan, K., Kumar, N., Gupta, S., Thai, G., Scott, S., Choudhury, A., Khetan, M., Menon, P., & Quisumbing, A. (2023). Scale and sustainability: The impact of a women's self-help group program on household economic well-being in India. *The Journal of Development Studies*, 59(4), 490–515. DOI: 10.1080/00220388.2022.2154151

Ravindran, A. M., & Vinodan, A. (2009). Implication of microfinance on tourism development and inclusive growth in India: Emerging models. *South Asian Journal of Tourism & Heritage*, 1(2), 56–67.

Rinaldi, A., & Salerno, I. (2020). The tourism gender gap and its potential impact on the development of the emerging countries. *Quality & Quantity*, 54(5–6), 1465–1477. DOI: 10.1007/s11135-019-00881-x

Susilowati, N., Mahmud, A., Latifah, L., Susanti, A., & Sari, P. N. (2022). Analysing the environmental sustainability of rural water management in community self-help group (CSHG) - A case study of Gunungpati district. *IOP Conference Series. Earth and Environmental Science*, 1098(1), 012039. Advance online publication. DOI: 10.1088/1755-1315/1098/1/012039

Tallei, T. E., Iskandar, J., Runtuwene, S., & Filho, W. L. (2013). Local community-based initiatives of waste management activities on Bunaken Island in North Sulawesi, Indonesia. *Research Journal of Environmental and Earth Sciences*, 5(12), 737–743. DOI: 10.19026/rjees.5.5730

Vankar, P. S., Sarswat, R., Dwivedi, A. K., & Sahu, R. S. (2013). An assessment and characterisation for biosorption efficiency of natural dye waste. *Journal of Cleaner Production*, 60, 65–70. DOI: 10.1016/j.jclepro.2011.09.021

Venkatraja, B. (2019). Are SHGs catalysts of inclusive sustainable rural development? Impact assessment of SKDRDP interventions in India. *International Journal of Sustainable Development and World Ecology*, 26(4), 302–312. DOI: 10.1080/13504509.2019.1579766

Vinodan, A., & Meera, S. (2022). Case study on self-help groups intervention and destination sustainability in Kerala: Input-output-outcome model. *Tourism (Zagreb)*, 70(2), 201–222. DOI: 10.37741/t.70.2.4

Vukovic, D. B., Petrovic, M., Maiti, M., & Vujko, A. (2021). Tourism development, entrepreneurship and women's empowerment – Focus on Serbian countryside. *Journal of Tourism Futures, 1*–21. https://doi.org/DOI: 10.1108/JTF-10-2020-0167

Wang, L., Xue, L., Li, Y., Liu, X., Cheng, S., & Liu, G. (2018). Horeca food waste and its ecological footprint in Lhasa, Tibet, China. *Resources, Conservation and Recycling*, 136, 1–8. DOI: 10.1016/j.resconrec.2018.04.001

Xu, X., Hou, Y., Zhao, C., Shi, L., & Gong, Y. (2021). Research on cooperation mechanism of marine plastic waste management based on complex network evolutionary game. *Marine Policy*, 134, 104774. DOI: 10.1016/j.marpol.2021.104774

Xu, X., Luo, L., Lai, Q., Dong, C., & Feng, C. (2024). A network perspective on tourism waste management: An evolutionary game analysis of stakeholders' participation in the Third Pole region. *Journal of Cleaner Production*, 445, 141378. DOI: 10.1016/j.jclepro.2024.141378

Chapter 11
Urban Solid Waste Management and Public Awareness in Bhubaneswar

Narayana Maharana
https://orcid.org/0000-0003-2765-0866
GVP College of Engineering, Kommadi, India

Lingaraj Prasad Patnaik
Utkal University, Bhubaneswar, India

Suman Kalyan Chaudhury
https://orcid.org/0000-0003-3206-6090
Berhampur University, India

Girish Prasad Das
Parala Maharaja Engineering College, Berhampur, India

Chandra Sekhar Patnaik
https://orcid.org/0000-0002-9487-2496
Omega Degree and PG College, Hyderabad, India

B. Udaya Bhaskar Ganesh
https://orcid.org/0000-0002-9122-4565
Lovely Professional University, Punjab, India

ABSTRACT

This chapter aims to assess the current state of solid waste management in the city and the residents' awareness and participation levels. A stratified random sampling method was employed to gather 1,447 responses from Bhubaneswar residents. The findings reveal that only 4.35% of households segregate their waste. The primary reasons for non-segregation are that many people are not accustomed to this practice and believe it is futile since municipal workers often mix the waste after collection. On a positive note, many residents sell a significant portion of their solid

DOI: 10.4018/979-8-3693-6110-8.ch011

waste to the Kawadiwalas. The study also found a strong relationship between general awareness and willingness to participate in solid waste management (SWM). Interestingly, female respondents were found to have a lower essential awareness of SWM. In contrast, senior citizens and individuals from lower socio-economic groups displayed significantly higher awareness and a greater willingness to engage in SWM practices.

1. INTRODUCTION

India is the world's fastest-growing economy and is the world's second-most populous country. The current yearly volume of solid waste created in Urban areas has risen from 30.6 million tons in 1990 and 58.41 million tons in 2020-21 (Central Pollution Control Board, 2021), and it has been projected to go beyond 165 million tons by the end of 2030 (Government of India, 2016). The gigantic rise in the volume of solid waste is due to changes in the urban population's lifestyle, food preferences, and standard of living (Talyan et al., 2008). According to MSW categorisation, the waste contains 30–45 per cent organic matter, 6 to 10 per cent recyclables, and the remainder is inert stuff (Kumar et al., 2009). The municipalities are primarily responsible for waste collection in the Indian cities. On the other hand, many municipalities have contracted with private parties to collect and transport garbage to disposal sites in different locations. In many Indian cities, the residents have been charged a collection fee by the municipality (Mohan & Yadav, 2019; Raj & Chutima, 2018), NGOs and other private organisations who have been appointed to segregate and collect waste from the urban households (Basu & Punjabi, 2020; Fredrick et al., 2018). In most Indian towns, the community bin or a standard dustbin has been placed on the roadsides with two to three bins in a kilometre radius where the residents have to put the garbage(Deswal & Laura, 2018; Rohit et al., 2018; Saha et al., 2017). This system is a very cost-effective way of collecting community or household waste, which is, in turn, transported by the municipality vehicles to the disposal site (Kumar et al., 2009).

The residents of Bhubaneswar generate about 300 tons of solid waste per day (Mohanty et al., 2014) which is expected to increase up to 1,000 tons per day by 2030. Roughly 50–70% of produced Waste is being collected, with the remaining discarded in drainage channels in city streets. About 10-20% of the collected waste has been treated using composting, the only treatment available. In contrast, the remaining garbage has been disposed of in open dumping sites on the outskirts of town. Many recycling facilities are currently not performing to their full potential. There is no doubt that these dumping sites are polluting the groundwater as well as causing air pollution in the city due to the unavailability of the waste disposal techniques like the leachate and landfill gas collection systems (Talyan et al., 2008). The higher content of inert materials and construction debris in solid waste makes it unsuitable for incineration and energy production. As a result, effective waste management at the point of generation has now been considered a challenge for the municipality (Kumar, 2014).

The government introduced the MSW Rules 2000 and the *Bhagidari* (a citizen-government partnership) approach to segregate waste at the source. This program aimed to increase awareness in the community and encourage them to segregate their waste at home (Talyan et al., 2008). Unfortunately, due to lack of community participation, it has been still a distant dream to achieve (Singh & Ramanathan, 2010). A more comprehensive creation of awareness and development of motivation is required to promote the involvement of people (Anand, 2010).

Bhubaneswar has grown into a business centre with a profusion of educational and other national and international institutions. According to the 2011 census, the population of Bhubaneswar was 8.37 lakhs (Census of India, 2011). The city comprises 60 wards that span 150 square kilometres and house roughly 1.6 lakh people (The Annual Report of OUIDF, 2017). Solid waste generation in the Bhubaneswar municipality is about 360 grams per capita per day (GPCD). The sources of these solid wastes may be divided into eight categories primarily produced in any urban city in India (Table 1).

Table 1. Types of waste generated in the urban cities in India

S. No.	Source	Type of Waste Generated
1	Households and Institutions	Plastics, metals, glass, hazardous waste and inert materials
2	Schools and Colleges	Papers and other stationary waste, mainly plastic
3	Vegetable and fruit Markets and Restaurants	Food wastes
4	Commercial Centers	Paper and plastics
5	Healthcare and Hospital, Pharmacy	Infectious includes organic waste, cotton etc. And non-infectious wastes, including glass, plastic, metal
6	Industries	Leather wastes, plastics and metals
7	Slaughterhouses	Bones, intestines, blood, carcasses
8	Animal Husbandry, Diaries, Agriculture	Straw and dung, other animal feed wastes

Source: Bhawal Mukherji et al. (2016)

Table 2. Percentage of waste generated from households in Bhubaneswar City

Type of solid waste	Composition percentage
Biodegradable organic fractions (leaves, wood, grass, etc.)	50.88
Paper, cardboard	5.96
Plastic, polyethene	5.92
Metal	1.02
Glass, Ceramics	0.68
Inert materials (Ash, Stone, Bricks, Soil, Silt etc.)	27.37
Rubber, leather	2.32
Coconut	2.42
Rags	3.43

2. REVIEW OF LITERATURE

Communities are Bhubaneswar's primary source of solid waste creation (Lalit et al., 2020), and complying with municipal solid waste management guidelines is problematic (Central Pollution Control Board, 2013). In such circumstances, it is therefore essential to determine people's awareness of solid waste management. Studies in various countries, like Brazil (Bringhenti & Günther, 2011), Italy (Castagna et al., 2013; De Feo & De Gisi, 2010), and Spain (Junquera et al., 2001), have demonstrated the significance of community awareness and sufficient infrastructure for community involvement in waste management. Furthermore, social and psychological aspects may also be necessary in encouraging community involvement in waste management. Social aspects have been observed to impact attitudes, and willingness to pay for solid waste management has been claimed in a study done on a community on a Greek island (Jones et al., 2008). Similarly, a study by Maeda and Hirose (Maeda & Hirose, 2009) Japan claimed that the anticipation of self-actualisation, a sense of self-efficacy, and solidarity influence the citizens to participate in waste management.

Several researchers across different countries have extensively used the theory of planned behaviour to explain the individual attitude, intent, subjective norms and perceived behavioural control towards solid waste management (Botetzagias et al., 2015; Davis et al., 2006; Ghani et al., 2013; Pakpour et al., 2014; Tonglet et al., 2004). Individuals have a favourable motive to participate in waste segregation at home, provided they are given the necessary knowledge, opportunity, and facilities, as claimed by a sociological survey performed in Malaysia (Ghani et al., 2013). According to research conducted in Iran, learning methods which focus on moral obligation and responsibility with action planning could be more successful in promoting the reduction of solid waste at source by households (Pakpour et al., 2014). Research conducted in the United Kingdom also indicated that having a pro-recycling mindset in people is a crucial element leading to recycling behaviour, which is not only affected by the prior experience of recycling but also by the knowledge and the availability of facilities to do so (Davis et al., 2006; Tonglet et al., 2004).

In India, especially in Bhubaneswar, little research has been done on community awareness, knowledge, attitudes, and behaviours concerning solid waste management (Phapal et al., n.d.; Samal & Sahoo, 2011; Sitakanta et al., 2008). Further, many studies overlooked the impact of socioeconomic variables on the attitude of individuals towards SWM. As per a study done in a town in south Tamilnadu, a household's income significantly impacted the disposal method of vegetable waste. Again, the income level of the households is negatively related to the maintenance of a dustbin and the reuse of waste. Similarly, the respondent's age is negatively related to the waste segregation practices (Indhira et al., 2015).

Household income positively affected the respondents' willingness to pay for waste management in Bangalore (Kumar & Nandini, 2013). Likewise, according to a study on the urban poor community in Jammu, the respondents have high concern towards the environment, and their awareness level is significantly different among the male and female respondents (Khan et al., 2015). Residents' perceptions were recorded in a survey done in specific regions of Coimbatore city in India, but the variations across socio-economic groups were not explored (Jayasubramanian et al., 2015). Several of these findings highlighted the necessity of communities better understanding waste management. Most of the studies, however, identified only a few aspects, including individuals' knowledge and attitudes towards solid waste management. Furthermore, the citizens of Bhubaneswar, the capital city of Odisha, have a completely different mindset, habits, and culture from those portrayed in the previous research.

Knowledge has been identified as a significant factor influencing environmental behaviour (Dhokhikah et al., 2015; Kallgren & Wood, 1986). Increased levels of ecologically relevant information have a significant influence on altering pro-environmental behaviour (Oskamp et al., 1991; Vining & Ebreo, 1990). A study on environmental awareness by Schahn & Holzer (1990) claimed that abstract knowledge focuses on awareness towards the environment, and concrete knowledge focuses on awareness towards recycling services. Knowledge is a crucial determinant of human behaviour since the absence of knowledge lowers self-efficacy, which leads to the creation of a belief that the individual is unable to perform due to a lack of necessary knowledge (Barr & Gilg, 2005). Thus, both concepts of abstract knowledge and concrete knowledge have been used in the present study.

3. OBJECTIVES

The following are the objectives of the study

- To explicate the situation of segregation, storage, collection, and disposal of solid waste in Bhubaneswar
- To verify the relationships between community awareness level and their willingness to participate in solid waste management.
- To explore how the awareness level of the residents of Bhubaneswar varies according to their socio-economic profile.

4. METHODOLOGY

A self-administered, semi-structured questionnaire survey was conducted in Bhubaneswar, dividing the city into three municipal regions: southeast, southwest, and north. The questionnaire was available in English and Odia to ensure that all randomly selected respondents could understand the questions. The survey occurred during the winter season (October-January) of 2020-21, following the lifting of COVID-19 lockdown restrictions. A stratified random sampling method was used to select 1,447 households for the survey, with only one male or female member from each household encouraged to respond.

The survey measured awareness levels and willingness to participate in waste management services through a series of questions. Cronbach's alpha was calculated to ensure the reliability of these derived variables. The correlation method established relationships between general awareness, specific awareness, and willingness to participate in waste management. Since the data was not normally distributed, non-parametric tests like the Mann-Whitney U and Kruskal-Wallis tests were used to verify respondents' awareness levels. In these tests, demographic variables served as the grouping variables, while awareness levels were the test variables. All statistical analyses were conducted using Jamovi, an open-access, third-generation statistical software based on R programming.

5. RESULTS AND DISCUSSION

5.1 Characteristics of the Respondents

The demographic profile of the respondents is presented in Table 3. The survey sample consists of 56% male and 44% female respondents. Regarding age distribution, most respondents fall within the 20-40 age group, comprising 49% of the sample. Approximately 7% of respondents are below 20, while around 18% are above 60. Educational level among respondents varies, with 12% having completed education up to the high school level, 31% with higher secondary education, 29% holding a graduate degree, and 28% possessing postgraduate or professional qualifications. Family size is categorised into four groups, with most respondents (42%) having 5-8 members, 33% with 3-4 members, and 11% with families larger than 8 members. Socio-economic status was assessed using monthly household income and asset ownership (house and vehicles). The respondents were classified into four socio-economic classes. Class-1, representing the highest socio-economic status, comprises 22% of the sample households. Conversely, Class-4, representing the lowest socio-economic status, includes 18% of respondents, while Class-3 and Class-2 account for 21% and 39% of respondents, respectively. Geographically, the sample distribution reflects the division of Bhubaneswar into three municipal zones: 31% of the sample was collected from the Southeast zone, 32% from the Southwest zone, and 37% from the North zone.

Table 3. Demographic profile of the respondents (N = 1447)

Category	Classification	Frequency	Percentage Census
Gender	Male	812	56.12
	Female	635	43.88
Age	Below 20	104	7.19
	20–40	709	49.00
	40–60	375	25.92
	Above 60	259	17.90
Household Size	Less than 3	204	14.10
	3–4	476	32.90
	5–8	609	42.09
	More than 8	158	10.92
Education	Up to High School (10th)	175	12.09
	Higher Secondary (+2)	447	30.89
	Graduate	424	29.30
	Post Graduate and Professional	401	27.71

continued on following page

Table 3. Continued

Category	Classification	Frequency	Percentage Census
Socio-Economic status (Based on Income/ Asset Holding)	Class-1(Highest)	317	21.91
	Class-2	563	38.91
	Class-3	308	21.29
	Class-4 (Lowest)	259	17.90
Municipality	South-East	450	31.10
	South-West	463	32.00
	North	534	36.90

5.2 Waste Storage, Collection and Disposal Practices

Table 4 illustrates Bhubaneswar's waste storage, collection, and disposal practices. The data reveals that only 4.35% of households engage in the segregation of solid and wet waste. The primary deterrent to this practice is that approximately 43.96% of respondents perceive it as futile, as the municipality often re-mixes segregated waste during collection due to most residents not participating in segregation. Additionally, 31.43% of respondents reported unintentionally mixing dry and wet waste, as this has been a long-standing habit. Furthermore, around 6% of households lack the knowledge to identify and segregate solid and wet waste correctly. Regarding waste storage practices, most households store all waste in a polybag or dustbin within their homes before handing it over to the municipality. Notably, 30% of respondents admitted directly disposing of all their waste into the drain. Nearly 22% of respondents reported placing their waste in public dustbins near their homes. Many households also participate in selling certain waste items to Kawadiwalas (waste collectors) to earn extra income. Paper is the most commonly sold waste product among the various items sold, with 39.88% of respondents selling newspapers, magazines, old notebooks, and similar items. Cardboard boxes are the least sold, with only 3.59% of respondents engaging in this practice. Additionally, 30.89% of respondents sell plastic waste items, and 9.33% sell glass items, such as bottles, to Kawadiwalas. Waste metals are sold by 7.81% of respondents. Interestingly, 8.50% of households do not sell any waste items to Kawadiwalas.

Table 4. Waste storage, collection and disposal practices

Categories	Questions	Percentage
Segregation Practices	Do you segregate Dry (Solid) and Wet (Organic) waste at home (Yes)	4.35
What prevents you from segregating the waste	I do not know how to do it	5.56
	Due to lack of time	11.78
	Do not have an extra dustbin	7.37
	Unconsciously putting all waste in one bin due to habit	31.43
	It is useless as the municipality mixes it up again as everyone is not practicing the same.	43.86
Storage Practices	Store in the standard bin outside my house	21.91
	Storing the waste in a Polybag and keep it inside the dustbin in my house	48.03
	Throwing in the Drain	30.06
Collection and Disposal practices	Does the municipality collect waste from your house regularly (Yes)	72.70
	Is the common bin near your house cleared regularly (Yes)	61.58
	Is roadside open dumping practiced in the locality (Yes)	68.76
Selling waste to Kawadiwala	Paper (Newspapers, Magazines, Books etc.)	39.88
	Corrugated boxes and Cardboard	3.59
	Glass bottles	9.33
	Plastic (Bottles and other containers)	30.89
	Metal (Containers, Tins, etc.)	7.81
	Not Selling anything to Kawadiwala	8.50

5.2 Door-to-door Waste Collection Services

Table 5 provides insights into the door-to-door waste collection practices in the city under study. The data reveals that nearly 73% of respondents confirmed that waste is collected from their doorstep. Municipal workers are responsible for waste collection in approximately 63% of households, while private contractors service 8% of households. Additionally, around 12% of respondents reported that waste is collected by individuals not affiliated with any formal organisation, and in about 7% of households, private servants handle waste collection. Regarding the frequency of service, 67% of households receiving door-to-door waste collection are served daily, while 10% are served on alternate days. Approximately 15% of households reported receiving the service on an as-needed basis. Furthermore, 7% of households receive waste collection once a week, and 1% only occasionally. When respondents were asked about their willingness to pay for door-to-door waste collection, 68% expressed a willingness to contribute. Of these, 43.80% are willing to pay a nominal fee of less than 30 rupees per month, equivalent to less than a rupee per day. Similarly, 29.17% of respondents are willing to pay between 31-60 rupees per month, followed by 20.12% who are willing to pay 61-90 rupees per month. Nearly 7% of respondents indicated they could afford to pay more than 90 rupees per month for the service. Further analysis shows that a significant portion (34.25%) of households pay waste collection fees directly to the local sweeper, usually appointed by the municipality. Additionally, 19.11% of respondents pay directly to the municipality, around 21% pay indirectly through their landlords, and 25.71% make payments through representatives of the waste collectors.

Table 5. Doorstep waste collection system in Bhubaneswar

Context	Questions	Percent
Waste Collection Service	Does anyone collect waste from your house (Yes)	72.98
Who collects the waste from your house	Municipality	63.35
	Private Contractor	8.24
	A person not associated with any organisation	11.93
	A servant appointed by the house owner	6.72
	I do not know, as the house owner looks after this	4.07
	Other	5.68
How frequently the waste is collected?	On daily basis	67.05
	On Alternate days	10.13
	As and when we need it	14.77
	Once in a week	7.01
	Occasionally	1.04
Payment for waste collection service	Do you pay for doorstep waste collection service (Yes)	68.00
Amount paid for doorstep waste collection per month	< Rs. 30	43.80
	Rs. 31-60	29.17
	Rs. 61-90	20.12
	>Rs. 90	6.91
Whom do you pay for the waste collection service	Local sweeper	34.25
	Directly to the municipality	19.11
	Paying to the house/apartment owner	20.93
	Representative of the waste collector	25.71

5.3 General Awareness, Specific Awareness and Willingness to Participate in Solid Waste Management

General awareness refers to the basic understanding of waste management, while specific awareness pertains to more in-depth and concrete knowledge about solid waste management practices. In this study, respondents were asked a series of questions to assess their general and specific awareness regarding waste management, and these were then correlated with their willingness to participate in solid waste management (SWM) activities. The willingness of Bhubaneswar residents to engage in SWM was evaluated by querying their readiness to participate in waste segregation, reduction, reuse, and handling practices.

As presented in Table 6, only 34% of respondents know the distinction between biodegradable and non-biodegradable waste, commonly referred to as 'wet' and 'dry' waste. The municipality has facilitated this segregation by providing separate dustbins for each type of waste in the city. This segregation is crucial, as biodegradable or organic waste can be composted and used as manure or fertiliser, as emphasised in the Solid Waste Management Rules of 2016 (Government of India). However, only 58% of respondents know the adverse effects of improper waste management, and 78% recognise that burning waste contributes to air pollution. Regarding specific awareness, the data indicates that 92% of respondents understand that "Kawadi-walas" and waste pickers play a significant role in recycling municipal solid waste. Conversely, only 43% are aware that SWM is a costly endeavour, necessitating financial contributions from the community. Approximately 32% of respondents are informed about the municipality's methods for handling solid waste to maintain city cleanliness, and only 31% recognise the critical importance of proper waste disposal for public health.

Given that these constructs—willingness to participate, general awareness, and specific awareness—are derived from multiple questions, a reliability check is warranted. The Cronbach's alpha for willingness to participate is 0.712; for general awareness, it is 0.733; and for specific awareness, it is 0.698. These values exceed the acceptable threshold of 0.6, indicating sufficient reliability and making these variables suitable for further statistical analysis.

Table 6. Respondent's willingness and awareness of waste management

Constructs	Statement questions	Response option	Percentage	Cronbach's alpha
Willingness to Participate	I like to segregate waste to help the municipality in SWM.	(Yes)	72.01	0.712
	I like to reuse waste material (like glass or plastic bottles, paper, clothes, etc.) as much as possible.	(Yes)	76.99	
	I am willing to hold back bio-degradable waste for composting.	(Yes)	70.97	
	I like to reduce the generation of solid waste at home.	(Yes)	57.98	
	I am willing to store waste and hand it over to the municipality instead of throwing it outside.	(Yes)	85.97	
	I like to reduce the use of plastic or polyethene bags.	(Yes)	61.02	
General Awareness	I know the difference between biodegradable (wet) and non-biodegradable (dry) waste.	(Yes)	34.00	0.733
	I understand that improper disposal of non-biodegradable waste causes severe health and environmental problems.	(True)	57.98	
	The burning of waste causes air pollution; therefore, it should be appropriately disposed of.	(True)	78.02	
	Improper dumping of solid waste management causes pollution.	(Yes)	44.99	
	Reusing more things is better than buying new things.	(Yes)	66.97	
Specific Awareness	I understand that proper disposal of solid waste is suitable for the city.	(Yes)	31.03	0.698
	Solid waste management is a costly affair. Therefore, citizens should also contribute financially.	(Yes)	42.99	
	I know how the municipality is handling the waste produced in the city.	(Yes)	32.00	
	Kawadiwalas and waste pickers recycle most municipal solid waste generated.	(True)	91.98	

Figure 1 illustrates the correlation between the awareness levels of respondents and their willingness to participate in solid waste management (SWM). The analysis reveals a significant positive correlation between general awareness and community willingness to participate, as indicated by Spearman's correlation coefficient ($r=0.346$, $n=1447$, $p < 0.001$). This suggests that higher general awareness levels are strongly associated with an increased willingness to engage in SWM activities. In contrast, while the correlation between specific awareness and willingness to participate is also positive, it is not statistically significant at the 0.05 level. This finding implies that while general awareness significantly influences community participation in SWM, a higher level of specific knowledge does not exhibit a similar association with participation.

*(r = Spearman's correlation coefficient; value in bracket is the level of significance)

Figure 1. Relationship between general awareness, specific awareness and willingness to participate

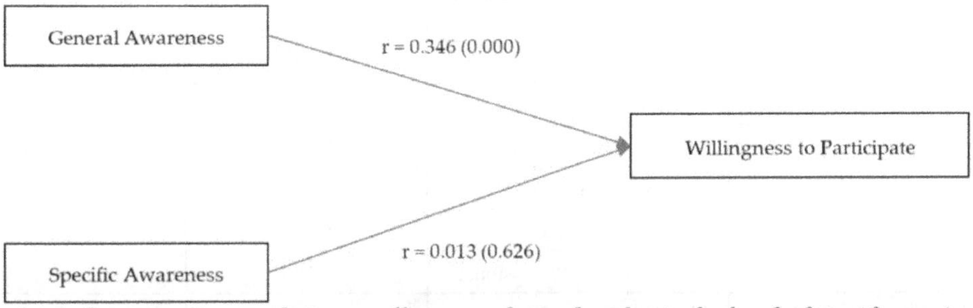

**(r = Spearman's correlation coefficient; value in bracket is the level of significance).*
Source: Interpretation of Primary Data

5.4 Difference of Awareness among Different Socio-Economic Groups

Table 7 analyses the differences in general attitudes toward solid waste management (SWM) among respondent groups categorised by socio-economic profiles. The Mann-Whitney U test was applied to groups with two categories, while the Kruskal-Wallis test was employed for groups with more than two categories to compare the mean values. The results indicate that male respondents exhibit significantly higher general awareness of SWM than female respondents, with the difference being significant at $p < 0.001$. This finding contrasts with previous studies, which have often reported that females tend to have a higher general awareness of SWM (Du, 1995; Beall, 1997; De Feo & De Gisi, 2010; Banga, 2011; Babaei et al., 2015). The discrepancy could be attributed to socio-cultural factors, as women in many societies, including the one under study, are predominantly engaged in household work. The lack of basic knowledge about SWM among women presents a significant challenge for municipal efforts to manage solid waste effectively. It underscores the need for targeted awareness campaigns to enhance women's understanding and participation in SWM initiatives, fostering greater community involvement.

Table 7. Difference of awareness among the socio-economic groups

Category	Classification	General Awareness	Test (Sig.)
Gender$	Male	0.513	185167.5 (0.000)
	Female	0.628	
Age#	Below 20	0.457	119.545 (0.000)
	20–40	0.563	
	40–60	0.513	
	Above 60	0.680	
Household Size#	Less than 3	0.566	2.221 (0.528)
	3–4	0.574	
	5–8	0.544	
	More than 8	0.569	
Education#	Schooling Education	0.570	7.307 (0.063)
	Higher Secondary	0.560	
	Graduate	0.540	
	Postgraduate/ Professional	0.583	
Socio-Economic status#	Class-1(Highest)	0.630	251.184 (0.000)
	Class-2	0.487	
	Class-3	0.505	
	Class-4 (Lowest)	0.718	
Municipality#	South-East	0.566	0.074 (0.964)
	South-West	0.560	
	North	0.564	

Note: #- Kruskal-Wallis Test, $- Mann-Whitney U test.

The analysis reveals that younger and older age groups exhibit significantly higher general awareness than the middle-aged cohort. For older individuals, this increased awareness may stem from their inclination towards traditional and environmentally sustainable practices. Meanwhile, younger individuals, particularly students, are concerned about environmental conservation. These findings are consistent with previous research demonstrating a direct relationship between age and knowledge of waste management (Swami et al., 2011; Pearson et al., 2012; Pakpour et al., 2014; Babaei et al., 2015). However, De Feo and De Gisi's (2010) study reported lower awareness levels among younger individuals, suggesting some divergence in results. Therefore, municipalities should prioritise middle-aged groups when developing targeted awareness programs on solid waste management.

Significant differences in awareness levels regarding socio-economic status are observed across the four socio-economic classes. The lower socio-economic class shows notably higher basic solid waste management awareness than other groups. This higher awareness may be due to their increased involvement in household activities, frequent reuse of resources, and lower waste generation resulting from economic constraints. Conversely, the higher socio-economic class demonstrates substantial awareness, likely influenced by their education and elevated living standards. However, no significant differences in awareness levels are observed when respondents are categorised by education level, family size, or municipality zone. These factors do not substantially affect waste management awareness across different municipal zones.

6. IMPLICATIONS OF THE STUDY

The study's findings elucidate significant gaps in community awareness regarding solid waste management, offering valuable insights for municipal planning. By identifying middle-class, middle-aged individuals, especially females, as priority groups for targeted awareness programs, the municipality can more effectively address these gaps and foster greater engagement in waste management practices. The study reveals that many households produce substantial kitchen waste, including plastics, food wrappers, metals, glass, and organic materials. Educating women on waste segregation and composting can be particularly beneficial. Training programs to teach composting techniques could be implemented, as this helps manage organic waste and empowers females with practical skills. To further enhance community participation, those with higher general awareness levels should be encouraged to monitor, educate, and assist others in waste segregation and composting. Focusing on these segments, which are more receptive to behavioural change, could significantly improve community involvement in solid waste management. Such targeted interventions are crucial for advancing public health and environmental sustainability (Barr, 2012).

Based on the study findings, the following recommendations are made:

a. The primary challenge in solid waste management in Bhubaneswar is the lack of waste segregation at the source and improper household disposal practices. Many residents are unaware of the correct methods for segregating solid (non-biodegradable) and wet (biodegradable) waste. To address this, it is essential for each household to consistently separate dry and wet waste, which would streamline the municipality's operations. Additionally, to mitigate the issue of waste mixing, the municipality should provide an additional collection bin specifically for households that inadvertently combine the two types of waste.

b. Municipal workers could play a crucial role in educating households on proper waste segregation. Observations indicate that many residents sell various items to "kawadiwalas," a practice that supports recycling. Installing standard dustbins within 100 to 200 meters of congested streets inaccessible to municipal vehicles can facilitate waste collection.
c. Awareness programs should be implemented at the street and neighbourhood levels, complemented by effective monitoring mechanisms to promote environmental cleanliness. Offering incentives, such as discounts on Solid Waste Management (SWM) payments, for households that consistently segregate their waste could encourage wider adoption of this practice. Observing the benefits of segregation—such as recycling dry waste and composting wet waste for use in gardens—may further motivate residents.
d. Involving all stakeholders is critical for enhancing waste management in Bhubaneswar. Non-governmental organisations (NGOs) can support municipalities by conducting awareness programs and coordinating regular doorstep waste collection. The municipality should consider increasing the number of bins based on public requests or as identified through surveys. Additionally, private waste management companies can assist with waste transportation and the disposal of hazardous domestic waste.
e. Enhancing community participation, particularly in source segregation, would significantly alleviate the municipality's burden, including the challenges of transporting hazardous waste to centralised processing and landfill facilities and composting organic waste. Further research is necessary to develop cost-effective and efficient waste management strategies explicitly tailored for Bhubaneswar.

7. CONCLUSION

The study conducted in Bhubaneswar on community awareness and participation in effective solid waste management (SWM) has yielded several insights. The findings reveal that only 4.35% of residents actively engage in and understand the importance of waste segregation. This low rate of participation, coupled with inadequate awareness among citizens and municipal workers, exacerbates the challenges associated with waste management. A positive aspect identified is the practice of selling selected types of solid waste to the municipality, which represents a constructive element in the current SWM practices. The study highlights that general awareness within the community plays a more significant role than specific awareness in fostering participation in SWM. Enhanced general awareness is crucial for increasing community involvement in waste management efforts. Moreover, the study highlights the need to improve awareness among females, whose participation in waste segregation could substantially alleviate SWM challenges. It was observed that senior citizens (above 60 years) and individuals from lower socio-economic backgrounds demonstrate higher levels of basic awareness and a greater willingness to engage in SWM activities. Addressing these factors could be vital in improving overall waste management practices in the city.

REFERENCES

Anand, S. (2010). *Solid waste management.* Mittal Publications.

Babaei, A. A., Alavi, N., Goudarzi, G., Teymouri, P., Ahmadi, K., & Rafiee, M. (2015). Household recycling knowledge, attitudes, and practices towards solid waste management. *Resources, Conservation and Recycling*, 102, 94–100. DOI: 10.1016/j.resconrec.2015.06.014

Banga, M. (2011). Household knowledge, attitudes, and practices in solid waste segregation and recycling: The case of urban Kampala. *Zambia Social Science Journal*, 2(1), 4–10.

Barr, S. (2012). *Environment and society: Sustainability, policy, and the citizen.* Ashgate Publishing, Ltd. https://www.google.co.in/books/edition/_/HVWjbSUC3SAC?hl=en&gbpv=0

Barr, S., & Gilg, A. W. (2005). Conceptualizing and analyzing household attitudes and actions towards a growing environmental problem: Developing and applying a framework to guide local waste policy. *Applied Geography (Sevenoaks, England)*, 25(3), 226–247. DOI: 10.1016/j.apgeog.2005.03.007

Basu, A. M., & Punjabi, S. (2020). Participation in solid waste management: Lessons from Mumbai's Advanced Locality Management (ALM) programme. *Journal of Urban Management*, 9(1), 93–103. DOI: 10.1016/j.jum.2019.11.002

Beall, J. (1997). Thoughts on poverty from a South Asian rubbish dump: Gender, inequality, and household waste. *IDS Bulletin*, 28(3), 73–90. DOI: 10.1111/j.1759-5436.1997.mp28003006.x

Botetzagias, I., Dima, A.-F., & Malesios, C. (2015). Extending the theory of planned behavior in the context of recycling: The role of moral norms and demographic predictors. *Resources, Conservation and Recycling*, 95, 58–67. DOI: 10.1016/j.resconrec.2014.12.004

Bringhenti, J. R., & Günther, W. M. R. (2011). Participação social em programas de coleta seletiva de resíduos sólidos urbanos. *Engenharia Sanitaria e Ambiental*, 16(4), 421–430. DOI: 10.1590/S1413-41522011000400014

Castagna, A., Casagranda, M., Zeni, A., Girelli, E., Rada, E. C., Ragazzi, M., & Apostol, T. (2013). 3R's from citizens' point of view and their proposal from a case-study. *UPB Scientific Bulletin*, 75(2), 253–264. DOI: 10.1080/14747731.2014.959092

Census of India. (2011). Retrieved from https://censusindia.gov.in/2011-common/censusdata2011.html

Central Pollution Control Board. I. (2013). *Status report on municipal solid waste management*. Central Pollution Control Board. Retrieved from https://cpcb.nic.in/status-of-implementation-of-solid-waste-rules/

Central Pollution Control Board. (2021). *Annual report 2020-21 on solid waste management rules implementation, 2016* (pp. 288–293). Retrieved from https://cpcb.nic.in/uploads/MSW/MSW_AnnualReport_2020-21.pdf

Davis, G., Phillips, P. S., Read, A. D., & Iida, Y. (2006). Demonstrating the need for the development of internal research capacity: Understanding recycling participation using the Theory of Planned Behaviour in West Oxfordshire, UK. *Resources, Conservation and Recycling*, 46(2), 115–127. DOI: 10.1016/j.resconrec.2005.07.001

De Feo, G., & De Gisi, S. (2010). Domestic separation and collection of municipal solid waste: Opinion and awareness of citizens and workers. *Sustainability (Basel)*, 2(5), 1297–1326. DOI: 10.3390/su2051297

Deswal, M., & Laura, J. S. (2018). Selection of collection sites and optimization of waste bins of Rohtak City, Haryana (India). *International Journal of Engineering Research and Development*, 14(4), 29–38.

Dhokhikah, Y., Trihadiningrum, Y., & Sunaryo, S. (2015). Community participation in household solid waste reduction in Surabaya, Indonesia. *Resources, Conservation and Recycling*, 102, 153–162. DOI: 10.1016/j.resconrec.2015.06.013

Du, P. T. (1995). *The determinants of solid waste generation, reuse, and recycling by households in Ho Chi Minh City, Vietnam* [PhD Thesis]. Asian Institute of Technology.

Fredrick, M., Oonyu, J. C., & Sentongo, J. (2018). Influence of education on the solid waste management practices of communities in Kampala city. *Journal of Environmental Waste Management*, 5, 261–274. DOI: 10.1080/20421338.2018.1518850

Ghani, W. A. W. A. K., Rusli, I. F., Biak, D. R. A., & Idris, A. (2013). An application of the theory of planned behaviour to study the influencing factors of participation in source separation of food waste. *Waste Management (New York, N.Y.)*, 33(5), 1276–1281. DOI: 10.1016/j.wasman.2012.09.019 PMID: 23415709

Government of India. (2016). *Solid waste management rules*. Retrieved from https://bbmp.gov.in/documents/SWM-Rules-2016.pdf

Indhira, K., Senthil, J., Vadivel, S., Appl, A., & Res, S. (2015). Awareness and attitudes of people perception towards household solid waste disposal: Kumbakonam Town, Tamilnadu, India. *Archives of Applied Science Research*, 7(3), 6–12.

Jayasubramanian, P., Saratha, M. M., & Divya, M. (2015). Perception of households towards waste management and its recycling in Coimbatore. *International Journal of Multidisciplinary Research and Development*, 2(1), 510–515.

Jones, P., Jones, A., Packham, G., & Miller, C. (2008). Student attitudes towards enterprise education in Poland: A positive impact. *Education + Training*, 50(7), 597–614. Advance online publication. DOI: 10.1108/00400910810909054

Junquera, B., del Brío, J. Á., & Muñiz, M. (2001). Citizens' attitudes to reuse of municipal solid waste: A practical application. *Resources, Conservation and Recycling*, 33(1), 51–60. DOI: 10.1016/S0921-3449(01)00057-X

Kallgren, C. A., & Wood, W. (1986). Access to attitude-relevant information in memory as a determinant of attitude-behavior consistency. *Journal of Experimental Social Psychology*, 22(4), 328–338. DOI: 10.1016/0022-1031(86)90018-1

Khan, S., Kumar, A., & Malaviya, P. (2015). Assessment of environmental behavior among the urban poor of Panjtirthi slum, Jammu, India. *Current World Environment*, 10(3), 801–810. DOI: 10.12944/CWE.10.3.10

Kumar, A. (2014). *Waste to energy opportunities and challenges* [Report of the Task Force on Waste to Energy (Volume II)]. Retrieved from https://niti.gov.in/planningcommission.gov.in/docs/reports/genrep/rep_energyvol2.pdf

Kumar, M., & Nandini, N. (2013). Community attitude, perception, and willingness towards solid waste management in Bangalore city, Karnataka, India. *International Journal of Environmental Sciences*, 4(1), 87–95.

Kumar, S., Bhattacharyya, J. K., Vaidya, A. N., Chakrabarti, T., Devotta, S., & Akolkar, A. B. (2009). Assessment of the status of municipal solid waste management in metro cities, state capitals, class I cities, and class II towns in India: An insight. *Waste Management (New York, N.Y.)*, 29(2), 883–895. DOI: 10.1016/j.wasman.2008.04.011 PMID: 18595684

Lalit, D., Adyasha, D., & Sitikantha, M. (2020). Application of multi-criteria decision making in adopting a suitable solid waste management model for an urban local body: Case study of Bhubaneswar City of Odisha, India. *Journal of Environmental Management & Tourism*, 11(3 (43)), 741–754. DOI: 10.14505//jemt.v11.3(43).28

Maeda, H., & Hirose, Y. (2009). Expectation of empowerment as a determinant of citizen participation in waste management planning. *The Japanese Psychological Research*, 51(1), 24–34. DOI: 10.1111/j.1468-5884.2009.00385.x

Mohan, G., & Yadav, A. K. (2019). Multi-stakeholder partnership for managing solid waste: A case study of ITC Limited. In *Waste Management and Resource Efficiency* (pp. 1087–1098). Springer., DOI: 10.1007/978-981-10-7290-1_90

Mohanty, C. R., Mishra, U., & Beuria, P. R. (2014). Municipal solid waste management in Bhubaneswar, India – A review. *International Journal of Latest Trends in Engineering & Technology : IJLTET*, 3(3), 10–14. DOI: 10.21172/1.33.002

Oskamp, S., Harrington, M. J., Edwards, T. C., Sherwood, D. L., Okuda, S. M., & Swanson, D. C. (1991). Factors influencing household recycling behavior. *Environment and Behavior*, 23(4), 494–519. DOI: 10.1177/0013916591234005

Pakpour, A. H., Zeidi, I. M., Emamjomeh, M. M., Asefzadeh, S., & Pearson, H. (2014). Household waste behaviors among a community sample in Iran: An application of the theory of planned behavior. *Waste Management (New York, N.Y.)*, 34(6), 980–986. DOI: 10.1016/j.wasman.2013.10.028 PMID: 24252373

Pearson, H. C., Dawson, L. N., & Radecki Breitkopf, C. (2012). Recycling attitudes and behavior among a clinic-based sample of low-income Hispanic women in Southeast Texas. *PLoS One*, 7(4), e34469. DOI: 10.1371/journal.pone.0034469 PMID: 22493693

Phapal, T., Bhosle, S., & Kulkarni, A. (n.d.). Municipal solid waste management in various cities: A literature review.

Raj, A. P., & Chutima, P. (2018). Determination of waste treatment fee pricing mechanism for municipal solid waste by mechanical biological treatment method utilizing the public-private partnership model in Thailand. [ISJET]. *International Scientific Journal of Engineering and Technology*, 2(1), 11–19.

Rohit, G. S., Chandra, M. B., Saha, S., & Das, D. (2018). Smart dual dustbin model for waste management in smart cities. In *2018 3rd International Conference for Convergence in Technology (I2CT)* (pp. 1–5). IEEE. https://doi.org/DOI: 10.1109/I2CT.2018.8529600

Saha, H. N., Auddy, S., Pal, S., Kumar, S., Pandey, S., Singh, R., Singh, A. K., Banerjee, S., Ghosh, D., & Saha, S. (2017). Waste management using Internet of Things (IoT). In *2017 8th Annual Industrial Automation and Electromechanical Engineering Conference (IEMECON)* (pp. 359–363). IEEE. https://doi.org/DOI: 10.1109/IEMECON.2017.7986797

Samal, S. D., & Sahoo, J. Y. (2011). Developing solid waste management plan: Assessment of current waste management system by BMC (Bhubaneswar Municipal Corporation) in the State of Odisha. *Kushagra International Management Review, 1*(1), 62–72.

Schahn, J., & Holzer, E. (1990). Studies of individual environmental concern: The role of knowledge, gender, and background variables. *Environment and Behavior*, 22(6), 767–786. DOI: 10.1177/0013916590226003

Singh, J., & Ramanathan, A. L. (2010). *Solid waste management: Present and future challenges*. IK International Pvt Ltd.

Sitakanta, K. M. S. A. H., Ghosh, P. M., & Patra, H. S. (2008). Threat to ambient air quality due to poor municipal solid waste management: A case study in and around Vuasuni dumping site, Bhubaneswar, Odisha, India. *Environmental Science and Pollution Research International*, 15(3), 241–250. DOI: 10.1007/s11356-008-0054-3

Swami, V., Chamorro-Premuzic, T., Snelgar, R., & Furnham, A. (2011). Personality, individual differences, and demographic antecedents of self-reported household waste management behaviours. *Journal of Environmental Psychology*, 31(1), 21–26. DOI: 10.1016/j.jenvp.2010.08.001

Talyan, V., Dahiya, R. P., & Sreekrishnan, T. R. (2008). State of municipal solid waste management in Delhi, the capital of India. *Waste Management (New York, N.Y.)*, 28(7), 1276–1287. DOI: 10.1016/j.wasman.2007.05.017 PMID: 17692510

The Annual Report of OUIDF. (2017). Retrieved from http://www.ouidf.in/pdf/OUIDF_Annual-Report_2016-17.pdf

Tonglet, M., Phillips, P. S., & Read, A. D. (2004). Using the Theory of Planned Behaviour to investigate the determinants of recycling behaviour: A case study from Brixworth, UK. *Resources, Conservation and Recycling*, 41(3), 191–214. DOI: 10.1016/j.resconrec.2003.11.001

Vining, J., & Ebreo, A. (1990). What makes a recycler? A comparison of recyclers and non-recyclers. *Environment and Behavior*, 22(1), 55–73. DOI: 10.1177/0013916590221003

Chapter 12
Waste Management Practices in Kerala's Premier Tourist Destinations: A Review

Tinto Tom
https://orcid.org/0000-0002-2320-6015
Kristu Jayanti College, India

S. J. G. Preethi
https://orcid.org/0000-0002-6980-6542
Kristu Jayanti College, India

ABSTRACT

Kerala is known as "God's Own Country" because of its varied tapestry of lush landscapes, tranquil backwaters, and lively cultural heritage. The state's commitment to sustainable tourism demonstrates how serious it is about protecting its abundant natural and cultural resources for the enjoyment and benefit of future generations. Waste management in Kerala is a significant and urgent problem because of its high population density, fast urban growth, and flourishing tourism sector. It is crucial to tackle this problem to reduce environmental deterioration and protect public health. The waste management system in Kerala combines ancient practices with modern technology to effectively tackle the intricate difficulties brought about by urbanization and population expansion. Kerala aims to accomplish effective trash collection, segregation, recycling, and disposal through community engagement, decentralized initiatives, and innovative technologies. This will contribute to devel-

DOI: 10.4018/979-8-3693-6110-8.ch012

oping a more sustainable and environmentally conscious society.

1. INTRODUCTION

Kerala, situated along the southwest coast of the Indian Peninsula, spans 360 miles along the Malabar Coast and has gained a reputation as a tourism paradise. It's considered a top-tier destination for travellers in the Indian subcontinent, earning praise from the National Geographic Channel in 2004 as one of the fifty must-visit destinations in a lifetime. Its pleasant climate, vibrant art scene, lively festivals, and a wide array of natural and cultural attractions contribute to its popularity. Kerala stands out among Indian states for its remarkable geographical diversity in a relatively small area. It offers tourists many experiences, from beaches and backwaters to wildlife sanctuaries and lush forests teeming with diverse flora and fauna. It's often touted as the "Green Gateway" to India. (Edward & George, 2008). Eco-tourism is a big deal in Kerala's tourism scene, boasting 60 forest destinations and 12 nominated sites. This aspect of tourism is crucial for driving visitor numbers in the state. Kerala has 15 wildlife sanctuaries and five national parks, covering about 5.5 percent of its area. The state's eco-tourism projects, focusing on sustainability, play a significant role in promoting eco-friendly tourism. The recent recognition of Western Ghats eco-tourism sites as World Heritage sites is expected to draw even more global visitors. Kerala stands out with 12 of the 39 decorated sites in the Ghats, making it an attractive destination for domestic and international travellers (Rajasenan et al., 2012). Tourism is a significant catalyst for economic expansion in most global regions. The vast potential of tourism is especially pertinent for emerging nations such as India. India stands apart from other areas of the world due to its plentiful and serene natural landscapes and numerous sacred places. This advantageous circumstance is particularly relevant to the state of Kerala in the Indian subcontinent, commonly known as 'God's own Country' due to its picturesque landscapes and ideal topographical, climatic, and other factors. India has had significant growth in tourism over the past two decades, and the tourism industry has strong growth potential.

Kerala has been recognised as an exemplary growth model, and various internal and external factors might influence its performance. The government and bureaucracy promote tourism as a 'catalyst for economic growth,' a valuable means of generating foreign exchange and employment opportunities to revitalize the local economy. Environmental and labour advocates frequently address the resistance against extensive tourism via non-democratic methods (J. K. Sharma, 2020). Although rural tourism undeniably delivers advantages to local communities, it is crucial not to disregard its notable disadvantages. Tourists' conduct towards local inhabitants is typically favourable, but the detrimental impacts on society and the environment raise significant apprehensions over the enduring viability of tourism initiatives. Approximately 61.67 per cent of community members have concerns about rural tourism causing harm to the environment. This emphasizes the immediate requirement for measures to address and reduce these adverse effects to protect the environment's quality. Tourists' improper waste disposal is recognized as the leading cause of environmental damage, with pollution of water bodies and destruction of native plant and animal life being close behind. In addition, whereas rural tourism brings about societal advantages, it also causes problems and nuisances for residents. Privacy infringement is the most significant concern, along with issues like waste management, noise pollution, and water pollution. Around 10% of the local population reports encountering all of these problems. To promote the long-term development of rural tourism, it is necessary to mitigate these adverse impacts (P K, 2016).

2. TOURISM AND WASTE MANAGEMENT

Effective waste management is needed as tourism becomes one of the world's most significant industries. International beach destination Kovalam in southwestern India has seen considerable visitor growth since the 1960s. Tourism has been prevalent for almost 40 years. The "zero waste" concept was created in Kovalam to address the growing trash problem in the early 21st century. Inefficient waste treatment can generate many issues for humans and the environment. Open burning, landfilling, and composting are the most frequent traditional waste management practices. Burning reduces waste and produces ash and gases but harms the environment by emitting toxic gases. "Landfill" is a standard waste management method that involves collecting rubbish from one location and dumping it in a less crowded area. Waste is dropped into a specially designed pit in the earth or built on top of it. Zero waste has evolved from recycling, the 1980s' most successful environmental effort. The zero-waste idea prioritizes efficiency, economics, and ethics over the 3Rs (recycling, reducing, and reusing) of the 1990s integrated waste management model (Dileep, 2007).

According to Thanal, Kovalam hotels and restaurants create 6.7 tonnes of biodegradable garbage and an equal amount of non-degradable waste daily, such as pet bottles, plastic bags, milk covers, and linen. Waste was dumped or burned nearby. The NGO and Kerala Hotel and Restaurant Association worked together to eliminate rubbish in Kovalam. Thanal supported the 'localization of chemical-free food through gardening' and waste management by turning biodegradable debris into compost and addressing pesticide issues in Kerala. Local Azhakulam homemakers and pensioners were targeted. The NGO advised members to grow chemical-free vegetables at home using Zero Waste Centre compost. Farmers learned to make compost, panchagavya, jeevamrita, and five-leaf extract and manage pests and diseases. Limited-space farmers were taught to cultivate vegetables in pots and sacks on terraces. Landless farmers received compost and subsidized organic inputs from Thanal. Due to increased vegetable output and home surpluses, the NGO founded the Organic Bazaar in 2003 to assist urban people who had no time for farming but wanted a nutritious, safe diet. Farmers produced for themselves; therefore, there was no third-party certification. The internal control technique includes NGO field officers inspecting farmers' fields twice weekly to monitor and advise. Organic Bazaar began on weekends in the evening and expanded to Wednesdays and Fridays. Producers got more excellent prices without haggling because consumers wanted quality. Organic Bazaar grew from 20 farmers in 2003 to 300 producers and 500 consumers in 7 years (P K, 2016).

Pookodu Lake is a 13-acre freshwater lake in Wayanad, in the Western Ghats. This location is known for its natural beauty, blue water ecology, and evergreen canopy, which provides shelter for much wildlife. Boating on the lake offers a beautiful experience for visitors. Tourist destinations attract diverse individuals with diverse product usage. Garbage entry and buildup are the primary sources of garbage in these places. Controlling garbage accumulation is a top priority for the location. Pookodu Lake encourages visitors to decrease garbage by rewarding their efforts and labelling plastic products. The Kovalam plan is being hampered by the appointment of external agencies to sanitize the area. Authorities appointed cleaning staff. Visitors will not be held accountable, but it will create employment opportunities (Das & Diwakar, 2019). A certain level of control can be implemented to minimise waste. It will reduce the trash in the tourist spot, and others will appreciate those vital steps to protect the environment.

Kerala pioneered India's *Responsible Tourism Policy*; Kochi, Kerala, held the Second International Conference on Responsible Tourism in Destinations in March 2008. Kerala implemented a state and local Responsible Tourism plan with cooperation. By linking local communities with the hotel business and government, the Kumarakom Responsible Tourism project has empowered and developed them while encouraging eco-friendly tourism. A three-pronged strategy, including economic responsibility, guaranteed all stakeholders received tourist benefits. Social responsibility: Researched local art and culture; women and children formed cultural associations and performed in hotels and resorts with RT Cell support. Environmental responsibility: With the Gramme Panchayat, the RT Cell organized and trained for the 'Zero waste Kumarakom' and vermicomposting projects to encourage trash management and treatment. Resort owners received seeds to plant in backwaters, while local clubs, students, and activists raised awareness to regenerate mangroves (Narikkodan, 2024).

Kerala houseboats (Riceboats) or kettuvallam were historically used to convey commodities from remote backwater villages to places like Alleppey, Kollam, Cochin, and Kottayam. As land transit improved, kettuvalloms disappeared.

They are now a popular tourist attraction, as is the modern moving boat house. They generate significant cash and prosperity for the region. Approximately 100 of the 600+ houseboats on Vembanad Lake and neighbouring canals and rivers have valid Pollution Control Certificates (PCC) due to onboard sewage treatment systems. Houseboats that meet essential standards and eco-friendly procedures will receive the Green Palm Certificate, the highest grade from the Department of Tourism. Eco-tourism sustainability involves four key areas: Environment, Energy, Water, and Propulsion. Reduced fuel/running costs, no oil/diesel contamination, no untreated sewage, and reduced noise pollution are crucial to achieving a sustainable future for Kerala backwaters and future generations (Mathen, 2012). The houseboats have to meet the basic standards of eco-friendly procedures. If they make any compromise, direct action must be taken against them to protect the spot.

Keeping the area clean and having efficient waste management systems is essential to influence tourists' perceptions of the region. Furthermore, the study indicates that tourists may struggle to comply with carrying capacity requirements and strict waste management techniques in natural areas. Although most tourists desire to revisit eco-tourism destinations, some may choose not to return because of the stringent restrictions and regulations on waste management and environmental conservation. Implementing sustainable waste management methods in eco-tourism locations cannot be overstated. Such measures are crucial for preserving the ecosystem and improving visitors' overall experience (Rajasenan et al., 2012). The waste management challenges in tourism arise from the increased waste generation resulting from tourism activities, including solid waste, wastewater, and littering. Insufficient garbage disposal infrastructure in well-visited tourist destinations can result in ecological damage and pose health hazards. Inadequate waste management practices can harm ecosystems, wildlife, and marine life, reducing the overall attractiveness of a location. The significance of implementing efficient waste management practices in the tourism industry lies in its ability to maintain cleanliness, hygiene, and the aesthetic appeal of tourist locations. Implementing sustainable waste management strategies helps to preserve natural resources, mitigate pollution, and protect biodiversity. Efficiently organised waste management systems not only enhance visitors' satisfaction but also promote the long-term viability of the destination and strengthen the local economy (P K, 2016).

When considering tourist destinations in underdeveloped countries like Tanzania, it becomes evident that these nations face significant challenges in managing the waste generated at popular sites such as Kilimanjaro. Studies indicate that while tourism is flourishing in these areas, waste generation is also increasing, and they struggle to manage this waste effectively (Kaseva & Moirana, 2010). In contrast, India has implemented numerous initiatives to manage waste to a certain extent, and many of these activities have been successful. Compared to other underdeveloped countries, India has demonstrated a better capacity for managing waste generated at tourist spots. On the other hand, even developed countries face difficulties in this regard (Mmereki et al., 2016). Therefore, countries can benefit from adopting waste management methods that have proven successful in other places, such as India.

3. SUSTAINABLE TOURISM DEVELOPMENT

A fair perspective on sustainability that recognises the role of human intervention in nature, rather than just idolising it, must be adopted. The study emphasizes the significance of effectively handling the expansion of tourism while considering the requirements of tourists, the conditions of the destinations, and the local communities. Furthermore, it necessitates formulating pragmatic strategies and actions to narrow the disparity between abstract concepts and practical execution, guaranteeing the pertinence and efficacy of sustainable tourism endeavours (Liu, 2003). These actions can bring more sustainability to tourism. States like Kerala have to give due consideration to sustainable tourism development plans. Tourism is an up-and-coming and rapidly evolving industry worldwide. Strategically and sustainably planning and developing tourism is essential, as it aims to balance society's environmental, economic, and social goals. The management of sustainable tourism development must prioritize the satisfaction of tourists' needs, ensure a meaningful experience for consumers, enhance their awareness of sustainability issues, and promote sustainable tourism behaviours among them. The comprehensive systematic literature assessment yielded some intriguing findings: The business participants are keen on adopting new technologies in tourism services that positively impact the environment and local communities. However, there are numerous challenges in transforming the environment, increasing consumer motivation for sustainable tourism services, and changing their behaviour to be more sustainable (Streimikiene et al., 2021).

The only approach to prevent disorderly tourism development and combat everyday changes in volatile environments is to plan sustainable tourism development. Sustainable development implies preserving natural, cultural, and other tourist resources for future generations. Sustainable development is increasingly crucial because tourism depends on natural and artificial attractions in the area. If these resources are destroyed or degraded, the destination's attractiveness decreases. Still, tourism in this region is at risk because tourists want to visit clean, attractive destinations with high-quality services. However, a poor atmosphere is undesirable to tourists and uncomfortable for locals. Sustainable tourism development planning involves environmental preservation and research and analysis before making a development choice. All these activities prevent intensive resource extraction in specific locations without regard for resource preservation (Angelevska-Najdeska & Rakicevik, 2012).

The difficulties in sustainable tourism development are primarily linked to the priorities of national economic policy, the structure of public administration, the emergence of environmental issues, over-commercialization, and the structure of the international tourism system. The study indicates that while the principles of sustainable tourism development are advantageous, their implementation is a highly challenging endeavour due to the existing socio-economic and political circumstances in underdeveloped countries. Therefore, implementing principles of sustainable tourist growth requires making difficult political and economic decisions, considering intricate socio-economic and environmental trade-offs. Additionally, it asserts that the execution of these difficult choices may be unattainable unless international organisations actively support and cooperate with governments of poor nations to follow the principles of sustainable tourist development (Tosun, 2001). Entrepreneurship stimulates growth, innovation, and change, vital to economic prosperity. The entrepreneurial sector must diversify tourism products and services to meet the rising demand for new tourism. There are sustainable tourism opportunities. The case study, Sweden's Söderslätt, is a new tourist destination with many natural and cultural attractions. One of Sweden's most agricultural districts, it has rural enterprise opportunities. This is due to rural social pressure. This article presents a SWOT analysis of Söderslätt tourism entrepreneurship development to assess their existing status. Successful, innovative farm tourism entrepreneurs include "Healthy Pig Farm," a family-owned business. Based on the findings and analyses, numerous recommendations are made to overcome rural tourism entrepreneurial development barriers (Lordkipanidze et al., 2005).

The Government of Kerala has incorporated sustainability considerations into its tourism policy framework, as stated in the "Vision 2020" statement, which emphasises the importance of implementing sustainable tourism practices. Kerala has implemented eco-tourism initiatives, like the one in Thenmala, to integrate tourism development with environmental preservation. These programmes provide opportunities for activities like jungle trekking and staying in tree-top huts. Moreover, stringent quality control procedures for prominent tourism attractions in Kerala, such as houseboats and ayurvedic rejuvenation centres, have been implemented. These establishments are awarded certifications like 'Green Leaf' and 'Olive Leaf' based on specific criteria such as genuineness, staff education, facilities, and service excellence. The aim is to encourage sustainable tourism practices. The government has acknowledged the need for rules to guide the sustainable expansion of tourism. They emphasise the comprehensive development of infrastructure to protect the environment and establish Kerala as a top-notch, high-quality destination (Edward & George, 2008).

India has envisioned leveraging tourism as a catalyst for developing village and local communities in Kumarakom, Kerala. This initiative promotes sustainable practices that benefit the local economy and environment. The Responsible Tourism Mission, in collaboration with various local and state organizations, has spearheaded efforts to ensure sustainable growth. A comprehensive survey was conducted to devise a strategic approach for tourism development. This survey identified unsustainable practices and provided insights into creating a roadmap for establishing Kumarakom as a model for responsible tourism. The mission emphasized the importance of capacity-building programs and the innovation of locally sourced products. These programs aimed to enhance their ability to engage in and benefit from tourism by equipping local communities with the necessary skills and resources. The emphasis was on creating unique, authentic experiences for visitors while preserving the cultural and natural heritage of the region. As a result of these concerted efforts, a new strategy was adopted to develop responsible and sustainable destinations. This strategy focused on capacity building and fundraising specifically for implementing sustainable practices. The initiatives raised awareness about the importance of sustainability and provided the financial support needed to bring these practices to fruition. This transformation has profoundly impacted Kumarakom, setting a precedent for other regions. This initiative's success highlights tourism's potential as a powerful tool for sustainable development. By fostering responsible tourism, Kumarakom has demonstrated how local communities can thrive economically while preserving their cultural and environmental assets. The Responsible Tourism Mission's work in Kumarakom is a blueprint for other destinations aiming to balance tourism growth with sustainability. Through strategic planning, capacity building, and innovative practices, tourism can become a driving force for the sustainable development of local communities (Sustainable Tourism India, n.d.).

4. PROBLEMS FACED CONCERNING RESPONSIBLE TOURISM

The Department of Tourism must be a forceful moderator to promote tourism in Kerala's backwater regions. Sustainable tourism requires community support, particularly in backwaters. The government faces limits as local community members can oppose some initiatives government entities offer. Local community members often raise significant concerns when planned development plans are made. Due to their attitude, the Department of Tourism has obstacles when implementing development projects. Without the assistance of local community members, the department cannot achieve meaningful change in the region. Backwater Tourism is plagued by unlicensed houseboats. Many people start houseboat businesses by borrowing large amounts from banks. Some houseboat operators sell services without proper licences to generate rapid profits. No effective waste management process exists for houseboats, with operators often throwing rubbish directly into backwaters. One of the most enormous challenges is the devastation it causes to the backwaters, which are vital to the local community (Joseph et al., 2020)

Cherai Beach's expansion is hampered by inadequate infrastructure. Even though the region's infrastructure has improved in recent years, it is not enough to attract a reasonable number of domestic and foreign visitors. Recently, the government lengthened and modified road networks under the 'Musiris Heritage Project'. Lack of parking is another issue. Lack of lodging is another obstacle. Cherai Beach's tourism potential will depend on local accommodations in the future. Accommodation is crucial to tourism infrastructure. Travellers need a cheap place to stay. There is room for upcountry hotels and restaurants because the nearby hotels are not enough. Beach sanitation is also an issue. Toilets and other sanitation facilities are inadequate for guests. E-toilets can remedy this problem by providing sanitation facilities for tourists and cleaning the environment. Garbage and plastic dumping threaten beaches. Beach trash disposal is prevalent. Lack of scientific beach preservation and trash handling worsens the problem. However, plastic and other litter litter the shore, making it unsightly (Suresh .L, 2019). Based on environmental considerations, solid waste management is an essential component of the ecological management of institutions.

On the other hand, most of the environmental management measures implemented have been carried out by the larger group of houseboat owners. Smaller properties may not often have the financial means or administrative structure to make it viable. Many houseboat owners are less concerned about managing the environment in general and garbage management in particular. This is because smaller properties are convinced that they only produce a tiny amount of waste (Community-Based Waste Management Backwater Tourism as a Case Example.Pdf, n.d.).

Kerala, situated on the Malabar Coast in the southwestern region of India, is promoted as "God's own country" and is renowned for its eco-tourism, backwaters, and distinctive culture. In 1986, the state of India declared "industry status" for tourism, making it one of the first states to acknowledge its potential. Kerala has also established a framework for the Responsible Tourism Mission to prioritise economic, social, and environmental responsibilities. The primary conclusions derived from this research indicate that corporations attribute various interpretations to sustainability. Houseboats are less familiar with sustainability, whereas resorts associate it with their operational survival and commitment to being environmentally benign. Homestays differ in familiarity, but they value the sustainability of social and cultural aspects and the durability of their business. Regarding the sustainability challenges, all participants identified infrastructure as the primary obstacle (Steinhaug, 2023).

Despite a generally good attitude towards responsible tourism efforts, firms are not allocating resources to implement them. The report also indicates that expenses related to responsible tourist practices, competitive market conditions, and insufficient government backing impede the widespread implementation of responsible tourism initiatives in developing countries. Although various studies have been conducted to investigate the role of responsible tourism and its effects on different outcomes, to the best of the authors' knowledge, no studies to date have examined the influence of responsible tourism on local communities' perception of sustainability and, consequently, their perception of quality of life. It is crucial to comprehend that the actual advantages of responsible tourism can only be realized if the local communities in these tourist destinations perceive that the accountable tourism initiatives contribute to destination sustainability and enhance their quality of life (Mathew & Sreejesh, 2017).

The tourism industry's economic impact on the local economy is negligible, whereas tourism developments in the state have significant environmental consequences. The state of Kerala has exhibited a considerable lack of sensitivity towards this issue and frequently implements policies that exacerbate the detrimental effects of tourism. The state's political ecology of tourism development reveals the strong connection between bureaucracy, legislators, and the hotel business. Furthermore, the expression of the economic storyline of the advantages of tourism is effectively employed to suppress political resistance against financially unsustainable and ecologically detrimental tourism initiatives (Sreekumar & Parayil, 2002).

Tourism has both beneficial and adverse effects on destinations and local communities. Management, staff, and the local community have identified significant sustainability problems, including water contamination and fish shortage. Despite being a popular attraction, backwater tourism indirectly contributes to water contamination by causing three layers of waste: plastic, cooking oil, and petroleum. The government should monitor houseboats to reduce water pollution. Fishing is a significant occupation in the local community. As tourism grows, fish populations have decreased due to poor waste management in destinations.

Additionally, the government should solicit community input for tourism business growth. Identifying findings and sustainability concerns enhances the research agenda for green growth. Water contamination is a severe issue of the destination. The concept of green growth and responsible tourism can be expanded to promote environmental sustainability (George, n.d.).

Kerala, India, advertises tourism as "God's own country." It has unusual "backwaters" that flow parallel to the Arabian Sea inland from Kochi and Kollam. Vembanad Lake is the core of ecologically vulnerable backwaters. Locals depend on these backwaters for their livelihood and garbage disposal. Houseboat trips draw thousands of tourists to the backwaters. Since 2002, backwaters have been Ramsar sites, but opposing interests threaten overuse, pollution, and reclamation. The instance shows how tourism, development, and livelihood activities harm backwaters. Several authorities do uncoordinated backwater management. Houseboat operators encounter challenges and non-adoption of sustainable techniques. Thus, the example illustrates public policy challenges in tourism management and CPR protection and intends to educate students about responsible tourism (Sharma & Sannabhadti, 2024).

5. IMPACTS OF WASTE GENERATION IN TOURIST SPOTS

The tourism industry depends on the environment's quality, including natural and artificial elements. On the other hand, the relationship between tourism and the environment is challenging to understand. It involves various activities, some of which may negatively affect the environment. The construction of general infrastructure, such as roads and airports, and tourism amenities, such as resorts, hotels, restaurants, shops, golf courses, and marinas, all contribute to many of these impacts. The potential adverse effects of tourism development can lead to the progressive depletion of the environmental resources essential to its success. In contrast, tourism can positively impact the environment by contributing to preserving and protecting the natural environment. This is a method for increasing awareness of the importance of environmental values. It may also be used to finance the protection of natural regions and boost the economic significance of these areas (Sunlu, n.d.).

From the point of view of sustainability, the current academic discussion on climate change science centres on achieving greater efficiency in managing environmental waste. Due to the exponential growth that has occurred in the tourist business over the past several decades, it is currently regarded as one of the most active economic activities everywhere in the globe. When it comes to tourism, solid waste is a commonly recognized and ever-increasing issue. The poor management of this trash can result in significant and permanent direct and indirect repercussions on the environment, the economy, and society. Despite this, managing solid waste in communities primarily dependent on tourism is particularly challenging. This is because of the climatic conditions, topography, financial constraints, planning issues, shifting consumption patterns, transient population, and seasonal variations in the quantity and composition of solid waste. Additionally, there is frequently a lack of momentum to implement new initiatives and programs because parties engaged in the design, construction, and administration of tourist resorts have conflicts of interest because of the nature of the resorts (Ezeah et al., 2015).

While tourism is a significant contributor to the development of solid waste in various nations and regions, there is a scarcity of empirical studies that examine the specific factors responsible for this contribution. In areas with a low gross domestic product per capita, the focus on tourism intensifies the impact of tourist visits on the production of municipal solid trash. However, in regions with a high gross domestic product per capita, the emphasis on tourism lessens this impact. This study enhances our understanding of the intricate correlation between tourism and the production of municipal solid trash at a provincial scale. The study also offers valuable insights to strengthen management methods and regulations regarding municipal solid waste in the context of tourism in China and other regions (Yuxi et al., 2023).

Economic expansion, population increase, technology advancements, and institutional capabilities can impact environmental changes. Tourism in protected areas has three environmental effects: direct, indirect, and cumulative. The three impact groups can be categorised according to their characteristics: positive and negative, random and predictable, local and extensive, temporary and permanent, and short- and long-term. These effects manifest within the confines of protected areas in Kenya. Safeguarding wildlife in Kenya is a formidable task due to the menace of poaching and the illicit trade of endangered species. The ecological ramifications of tourism pose the most challenging aspect of the sector. Ecological upheavals encompass a range of detrimental effects, such as the loss of ecosystems, an increase in visitors, shortages of water, contamination of the ground, the extinction of species, harm to flora, uncontrolled urban expansion, and the process of desertification resulting from a scarcity of water. Land erosion has significant implications for the flora and fauna (Belsoy & Korir, 2012).

In light of growing tourist pressure, the research attempts to evaluate the environmental conditions in two significant Northwestern Himalayan locations: Kullu and Manali. The text addresses the issues of excessive hotel development, deforestation, air pollution, solid waste, and water pollution. It also proposes several solutions to reduce the impact of these different types of pollution. The 1980s saw the most significant increase in hotel development in this region, at roughly 38%, as a result of the terrorism in Jammu and Kashmir. 25% of the hoteliers and tourists regard deforestation as one of the foremost issues of unregulated tourism. For four years (1996-99), research on air pollution indicates that the level of suspended particulate matter (SPM) in Manali during summers has increased to 112 $\mu g/m^3$. An extensive analysis of the overall trash produced reveals that the percentage of garbage that decomposes (about 63% in Kullu and 72% in Manali) is greater than that of waste that does not biodegrade. Indiscriminately disposing of municipal solid garbage into rivers is prevalent in many tourist destinations, leading to river water pollution. To address these issues, it is necessary to implement a comprehensive prohibition on constructing further large hotels, regulate the influx of tourists and rigorously enforce pollution limits for automobiles. Implementing trash segregation at the point of origin and converting biodegradable waste into compost can be a significant approach to managing solid waste (Kuniyal et al., 2003).

Like the rest of India, Kerala has been placing more emphasis on tourism growth. Due to the promising economic opportunities presented by tourism, particularly in a state abundant in tourism resources like Kerala, eco-tourism and other sustainable tourism models have significant growth potential. Recently, the government's initiatives have successfully transformed 'Kumbalangi' into an eco-tourism community that stands out all over India. The importance of embracing eco-tourism is increasing, not just to meet legislative requirements but also to fully embody its principles and values. This necessitated a collective effort from all parties involved to promote sustainable tourism development. Due to the government's limited resources, it is necessary to involve the private sector to achieve a comprehensive, integrated, and sustainable strategy for tourist development. This highlights the importance of the Public-Private Partnership (PPP) method (Nambath, 2012).

6. VARIOUS INITIATIVES TO MANAGE WASTE

Waste management has emerged as a worldwide concern. Proper waste disposal or recycling ethically and responsibly minimizes the adverse environmental impacts. The situation in India is such that approximately 42 million tons of solid trash are generated yearly. The 3Rs strategy is the most appropriate approach for waste management due to its incorporation of a sustainable framework. The primary factors contributing to the significant waste generation are urbanization, privatization, and a large population. This trash contains a variety of hazardous components. To effectively handle it, we require a more system-centric approach. We must implement the potential ways to address the limitations of our system gradually and systematically. Thus, these methods will be economically efficient and feasible for our system. Enforcing stringent regulations and promoting awareness about the issue can contribute to a certain degree to trash control and management (Reddy & Tom, 2023). Initiatives such as Smart Cities and the Clean Development Mission were launched by the government of India to facilitate the development of intelligent and sustainable cities. Solid waste management, often known as SWM, has emerged as one of the primary areas of concentration for efforts of this kind. The shifting dynamics of urban waste features due to unplanned urbanization and rapid expansion in urban populations (as a result of migration) have been a critical concern for local authorities in urban areas to establish an effective solid waste management plan. To design a long-term solid waste management strategy that aligns with the objectives of government efforts, it is necessary to be aware of the waste amounts, characteristics, and practices currently in place for waste management (Cheela et al., 2021).

The tremendous increase in trash and waste management in India has gotten little policy attention. Reduce, recycle, and reduce (3R) is India's environmental protection framework for trash management, which has limitations. Beyond standard methods are needed to combat increased waste caused by population, urbanization, and globalization. We offer a paradigm that includes the zero-waste notion, which has grown widely in recent decades. Few Indian and global towns and countries have taken the zero-waste approach to the next level by establishing a system prioritising garbage generation over waste management. Decentralized waste management is vital to zero waste success. Indian judiciary and policy frameworks have yet to embrace the zero-waste philosophy fully. National and international stakeholders must address this zero-waste law, policy, and practice gap. Zero waste demands transforming our linear economy into a circular economy and viewing trash as a negative externality that must be handled. Over the past two decades, environmental consciousness and judicial insensitivity to zero waste adoption have been highlighted in waste management legislation and policy (Kumar & Bhati, 2022).

Tourism's relevance to national development is well known. Numerous measures, including increased funding and infrastructure development, aim to attract tourists. However, policymakers and implementing agencies continue to overlook the need to provide sanitation facilities at tourist destinations. Tourism development in India has been hampered by sanitation. The research analyses the influence of cleanliness on tourism in Karnataka, using 600 sample travellers from 12 selected tourist destinations. Besides Dharmasthala, selected tourist destinations lack basic sanitation services, including toilets, dust bins, and drainages. The disposal of plastic waste, such as water bottles, food packages, and cups, has worsened cleanliness in tourist areas. Poor cleanliness led to tourist disease and income loss in the tourism industry (Kulkarni, 2022).

India's economic development depends on tourism, which is a crucial industry. One-tenth of India's gross domestic product and 42 million employment opportunities were created in 2018 as a result of tourism. India has initiated several steps to encourage and advance the tourist business. The Clean India Mission appeared to have a good impact on the tourism industry, which is one of the many initiatives and programs undertaken by the government of India. What is known as the Clean India Mission is an initiative that aims to address many types of solid waste, sanitation and hygiene concerns, and other types of pollution. What is known as the Clean India Mission is a program that aims to restore the natural beauty of India by making the country cleaner and free of pollution (Aquino et al., 2021).

7. CONCLUSION

In Kerala's premier tourist destinations, waste management practices play a crucial role in maintaining the sustainability and attractiveness of these areas. Unfortunately, inadequate waste management practices have been observed in these tourist destinations, leading to adverse environmental and tourism industry impacts. To address this issue, it is essential to focus on waste minimisation and recycling efforts. The amount of waste generated can be significantly reduced by implementing effective prevention strategies and promoting collaborative actions with residents and tourism actors. Additionally, conducting tourist awareness campaigns before their visit can help educate and encourage responsible waste disposal practices. These measures can improve the clean-coastal index, creating a more appealing and sustainable tourism experience. To create a sustainable waste management system in Kerala's premier tourist destinations, it is crucial to focus on waste minimisation and recycling. By implementing best practices in waste management, such as recycling and waste minimisation, Kerala's premier tourist destinations can limit the amount of unsorted waste generated per guest, significantly reducing the environmental impact. Furthermore, integrating local culture in tourism packages can help promote sustainability and reduce material and energy use. By implementing these waste management practices and creating a culture of environmental responsibility, Kerala's premier tourist destinations can ensure these areas' long-term sustainability and attractiveness to tourists and locals. A multi-faceted approach is necessary to address the waste management challenges in Kerala's premier tourist destinations.

REFERENCES

Angelevska-Najdeska, K., & Rakicevik, G. (2012). Planning of Sustainable Tourism Development. *Procedia: Social and Behavioral Sciences*, 44, 210–220. DOI: 10.1016/j.sbspro.2012.05.022

Aquino, P. G., Justin, M. S. M., & Jalagat, R. C. (2021). Clean India Mission and Its Impact on Cities of Tourist Importance in India. In Balsalobre-Lorente, D., Driha, O. M., & Shahbaz, M. (Eds.), *Strategies in Sustainable Tourism, Economic Growth and Clean Energy* (pp. 89–107). Springer International Publishing., DOI: 10.1007/978-3-030-59675-0_5

Belsoy, J., & Korir, J. (2012). *Environmental Impacts of Tourism in Protected Areas. 2.*

Cheela, V. R. S., Ranjan, V. P., Goel, S., John, M., & Dubey, B. (2021). Pathways to sustainable waste management in Indian Smart Cities. *Journal of Urban Management*, 10(4), 419–429. DOI: 10.1016/j.jum.2021.05.002

Community-Based Waste Management Backwater Tourism as a Case Example.pdf. (n.d.). Retrieved April 23, 2024, from https://thuvienso.hoasen.edu.vn/v/web/viewer.html?file=/bitstream/handle/123456789/14869/Community-Based%20Waste%20Management%20Backwater%20Tourism%20as%20a%20Case%20Example.pdf?sequence=1&isAllowed=y

Das, A. S., & Diwakar, G. D. (2019). *Solid Waste Management for Sustainable Tourism.*

Edward, M., & George, B. P. (2008). *Destination Attractiveness of Kerala as an International Tourist Destination: An Importance-Performance Analysis.* Service Quality.

Eswara Reddy, E., & Tom, T. (2023). A Review on Economic Assessment on Solid Waste Management. In Aloysius Edward, J., Jaheer Mukthar, K. P., Asis, E. R., & Sivasubramanian, K. (Eds.), *Current Trends in Economics, Business and Sustainability* (pp. 26–35). Springer Nature., DOI: 10.1007/978-981-99-3366-2_4

Ezeah, C., Fazakerley, J., & Byrne, T. (2015). Tourism Waste Management in the European Union: Lessons Learned from Four Popular EU Tourist Destinations. *American Journal of Climate Change*, 4(5), 5. Advance online publication. DOI: 10.4236/ajcc.2015.45035

George, A. (n.d.). *Impacts of Responsible Tourism in Environmental sustainability issues: A Case Study in Kerala, India.*

Joseph, E., Kallarakal, T. K., & Varghese, B. (2020). Sustainable tourism development in the backwaters of South Kerala, India: The local government perspective. *Geo Journal of Tourism and Geosites*, 33(4), 1532–1537. DOI: 10.30892/gtg.334spl13-604

Kaseva, M. E., & Moirana, J. L. (2010). Problems of solid waste management on Mount Kilimanjaro: A challenge to tourism. *Waste Management & Research*, 28(8), 695–704. DOI: 10.1177/0734242X09337655 PMID: 19723827

Kulkarni, A. (2022). *Sanitation and its impact on tourism in Karnataka.*

Kumar, S., & Bhati, H. V. (2022). Waste management to zero waste: Global perspectives and review of Indian law and policy. In Hussain, C. M., Singh, S., & Goswami, L. (Eds.), *Emerging Trends to Approaching Zero Waste* (pp. 79–101). Elsevier., DOI: 10.1016/B978-0-323-85403-0.00013-X

Kuniyal, J. C., Jain, A., & Shannigrahi, A. (2003). Environmental Impacts of Tourism in Kullu-Manali complex in North Western Himalaya, India. Part 1: The Adverse Impacts. *International Journal of Fieldwork Studies, 1.*

Liu, Z. (2003). Sustainable Tourism Development: A Critique. *Journal of Sustainable Tourism*, 11(6), 459–475. DOI: 10.1080/09669580308667216

Lordkipanidze, M., Brezet, H., & Backman, M. (2005). The entrepreneurship factor in sustainable tourism development. *Journal of Cleaner Production*, 13(8), 787–798. DOI: 10.1016/j.jclepro.2004.02.043

Manoj, P. K. (2016). Impact of rural tourism on the environment and society: Evidence from Kumbalangi in Kerala, India. *International Journal (Toronto, Ont.)*, 4(2).

Mathen, R. (2012). Houseboats in Kerala—Constructional Features and Environmental Issues. *IOSR Journal of Environmental Science, Toxicology and Food Technology*, 1(6), 31–43. DOI: 10.9790/2402-0163143

Mathew, P. V., & Sreejesh, S. (2017). Impact of responsible tourism on destination sustainability and quality of life of community in tourism destinations. *Journal of Hospitality and Tourism Management*, 31, 83–89. DOI: 10.1016/j.jhtm.2016.10.001

Mmereki, D., Baldwin, A., & Li, B. (2016). A comparative analysis of solid waste management in developed, developing and lesser developed countries. *Environmental Technology Reviews*, 5(1), 120–141. DOI: 10.1080/21622515.2016.1259357

Nambath, S. (2012). Problems and Prospects of Ecotourism in Kerala: Some Empirical Evidence from "Kumbalangi" Panchayat In Ernakulam District (India). *Journal of Global Economy (ISSN Print-0975-3931, Online -2278-1277), 8*, 293–311. DOI: 10.1956/jge.v8i4.259

Narikkodan, V. (2024). *Climate Action For Sustainable Tourism In Kerala: A Perspective.*

Rajasenan, D., Manaloor, V., & Abraham, B. G. (2012). *Tourist Profiles and Characteristics vis-à-vis Market Segmentation of Ecotourism Destinations in Kerala.*

Sharma, J. K. (2020, January 9). *Trends in Tourism Development of Kerala in India.* https://www.semanticscholar.org/paper/Trends-in-Tourism-Development-of-Kerala-in-India-Sharma/a9275ca5dc76290aea83b34618a7fea47cb44cff

Sharma, R., & Sannabhadti, R. S. (2024). Backwater Tourism and Houseboat Operations in Vembanad, Kerala: Competing Uses and Management Conflicts. In M. Sigala, M. Fang, A. Yeark, J. N. Albrecht, & O. Vorobjovas-Pinta (Eds.), *Case Based Research in Tourism, Travel, and Hospitality: Rethinking Theory and Practice* (pp. 243–263). Springer Nature. DOI: 10.1007/978-981-97-1891-7_14

Sreekumar, T. T., & Parayil, G. (2002). Contentions and contradictions of tourism as development option: The case of Kerala, India. *Third World Quarterly*, 23(3), 529–548. DOI: 10.1080/0143659022013841 1

Steinhaug, A. M. (2023). *Sustainable Tourism in Kerala*. Potentials and Pitfalls. [Master thesis, Norwegian University of Life Sciences], https://nmbu.brage.unit.no/nmbu-xmlui/handle/11250/3078690

Streimikiene, D., Svagzdiene, B., Jasinskas, E., & Simanavicius, A. (2021). Sustainable tourism development and competitiveness: The systematic literature review. *Sustainable Development (Bradford)*, 29(1), 259–271. DOI: 10.1002/sd.2133

Sunlu, U. (n.d.). *Environmental Impacts Of Tourism.*

Suresh, L. (2019, December 25). *A Study on Waste Management to Enhance Green Tourism-An Empirical Observation of Cherai Beach in Kerala | Thematics Journal of Geography.* https://journals.edupub.org/index.php/tjg/article/view/15718

Sustainable Tourism India. (n.d.). Retrieved July 5, 2024, from https://sustainable.tourism.gov.in/case-study-Responsible-Tourism-Mission.html

Tosun, C. (2001). Challenges of sustainable tourism development in the developing world: The case of Turkey. *Tourism Management*, 22(3), 289–303. DOI: 10.1016/S0261-5177(00)00060-1

Yuxi, Z., Filimonau, V., Ling-en, W., & Linsheng, Z. (2023). The impact of tourism on municipal solid waste generation in China. *Journal of Cleaner Production*, 427, 139255. DOI: 10.1016/j.jclepro.2023.139255

Chapter 13
Cruise Ship Waste Management:
A Systematic Review and Research Gap Analysis

Reshma Sucheran
Durban University of Technology, South Africa

ABSTRACT

Large cruise ships, capable of accommodating over 6,000 passengers and 1,500 crew members, generate substantial waste, necessitating effective management strategies to mitigate environmental impacts. This study employs a systematic literature review to identify key themes and research gaps in cruise waste management research. Using the PRISMA flow method, 47 relevant research articles were selected based on relevance. The data from these articles were analysed using bibliometric and content analysis techniques. The key themes identified include plastic waste, microplastic waste, food waste, wastewater, and waste management strategies. The study revealed significant research gaps that demand immediate attention and action concerning the effectiveness of waste management strategies, socio-economic considerations, interdisciplinary approaches, regulatory compliance, food waste disposal methods, regional collaborations, policy initiatives, and integrated waste management approaches.

1. INTRODUCTION

Cruises are currently one of the fastest-growing segments in the tourism sector. In 2009, the global ocean cruise industry carried approximately 17,8 million passengers. As of 2019, this figure peaked at 29,7 million (Statista, 2020.). Cruise

DOI: 10.4018/979-8-3693-6110-8.ch013

ships operate in all the earth's oceans, often in pristine coastal waters and sensitive marine ecosystems. They tend to concentrate their activities in specific coastal areas and repeatedly visit the same ports, thereby creating a significant cumulative impact at the local scale due to overcrowding, which can strain local resources and infrastructure (Toneatti et al., 2020). These ships are often described as floating cities that accommodate thousands of passengers, and there is a growing trend to increase ship sizes, which exacerbates the issue by further intensifying the environmental and infrastructural pressures on the visited areas (Cerveny et al., 2020). The scale of waste generation by a large cruise ship, which can accommodate more than 6000 passengers and 1500 crew members, is overwhelming, necessitating effective waste management strategies to mitigate its environmental impact.

The waste generated by cruise ships primarily falls into five categories: sewage, greywater, oily bilge water, solid waste, and hazardous waste (EPA, 2008). Sewage, often called "black water" on cruise ships, consists of human waste, whereas greywater includes wastewater from sinks, baths, showers, laundry facilities, and kitchen areas. Oily bilge water, which is a mixture of water, oily fluids, and lubricants from the ship's engine rooms, and solid waste, which encompasses food scraps, garbage, refuse, sludge, rubbish, and other discarded materials, contribute significantly to the waste output of cruise ships. Additionally, hazardous waste refers to materials that risk human health or the environment. Copeland (2011) calculated the average waste production for a cruise ship with 3000 people during a week-long cruise. It generates 795 m^3 of sewage, 3785 m^3 of greywater, 95 m^3 of oily water, and 8 tons of solid waste. In an earlier study, Butt (2007) reported that although cruise ships represented less than 1% of the world's merchant fleet, they would account for 25% of all waste generated by these ships. This problem caused by cruise-generated waste is expected to become even more relevant due to the current growth trend of the cruise industry (Xu, 2016).

The escalating concerns surrounding the environmental impact of leisure cruises have become a focal point in recent discussions, particularly regarding the excessive generation of waste, leading to consequences on both ports and primary sea routes (Butt, 2007; Klein, 2011). These cruise ships tend to concentrate their activities within specific coastal regions, frequently revisiting the same ports, thereby amplifying the cumulative impact at a local level. Furthermore, isolated yet substantial waste discharge incidents, accidental or intentional, can trigger considerable adverse consequences (Krenshaw, 2009). As reported by the Environmental Protection Agency in 2008, it has been noted that daily, a substantial cruise ship has the potential to produce a significant amount of waste. This includes 74,000 gallons of sewage, 249,000 gallons of grey water, 5300 gallons of bilge water, 50 tons of garbage, 12,000 bottles, 12,000 cans, and 10 tons of hazardous waste.

The initial recognition of the potential environmental impacts stemming from the uncontrolled disposal of waste at sea emerged globally during the late 1960s, ultimately leading to the establishment of the London Dumping Convention (LDC) in 1972. Before this crucial development, the world's oceans and seas were commonly perceived as the most significant and easily accessible 'dumping site' on Earth, with waste generated on land often being transported beyond a nation's territorial sea boundary (which, at that juncture, stood at merely 3 nautical miles for UK waters) and being discarded into the sea (IMO, 2011). Concurrently, during this era, the maritime industry displayed limited concern for averting pollution of marine environments beyond the stipulations outlined in the International Convention for the Prevention of Pollution of the Sea by Oil (OILPOL) of 1954, primarily targeting oil-related pollution stemming from tanker operations and the general discharge of oily waste by other vessels. Handling waste aboard ships at this time typically involved a careless approach of simply disposing of it overboard, reflecting a nonchalant attitude and a blatant disregard for the subsequent ecological harm. This lackadaisical disposition and failure to recognise the environmental repercussions are now vividly captured in images showcasing deep-sea floors strewn with items like toilet fixtures, containers, paint drums, and many other 'foreign' objects. The shift in perspectives regarding waste management, notably post the Rio Earth Summit, and the acknowledgement of the necessity for crafting a sustainable waste management strategy have prompted the enactment of regulations at various levels – local, national, and international – that encompass not only the disposal of waste generated on land but also waste originating from maritime activities (IMO, 2012).

The International Maritime Organization (IMO) established the MARPOL protocol in 1973, a crucial global agreement concerning cruise ship pollution. This protocol is particularly significant as it addresses the environmental impact of cruise ship operations and aims to regulate and reduce pollution levels in marine environments. The primary focus of these initiatives revolves around compliance with the essential documents and regulations governing waste management in the maritime sector, particularly Annex V of the MARPOL Convention (MARPOL 73/78). This regulatory framework, which came into effect on December 31st, 1988, and has since undergone several amendments and updates, serves as the cornerstone of environmental protection efforts in the maritime industry. The most recent revision of the Annex was ratified in July 2011, reflecting the ongoing commitment to enhancing environmental sustainability practices within the sector. Additionally, the latest guidelines, approved in July 2017, signify a significant milestone in the evolution of waste management regulations, showcasing the industry's dedication to continuous improvement and compliance with international standards. A critical aspect underscored in these regulations is adopting a rational and responsible approach towards general waste management practices. From the inception of these guidelines, there has been a clear emphasis on shifting away from the historical practice of indiscriminate waste disposal at sea. The regulations explicitly affirm that ‘‘Annex V overturns the historical assumption that trash could be dumped into the ocean depending on the characteristics of the trash and designated distances from land” (Copeland, 2011; Klein, 2011; MEPC 295 71, 2017), highlighting a change in direction towards more sustainable and environmentally conscious waste management strategies in the maritime sector. The regulations outlined in Annexes V and VI of the International Convention for the Prevention of Pollution from Ships (MARPOL) set by the International Maritime Organization (IMO) dictate the permissible types of wastes that vessels can release into the ocean and mandate ships to have mechanisms in place for waste reception. To adhere to the MARPOL stipulations, most cruise companies have established environmental management systems to minimise, segregate, and effectively handle onboard waste.

Implementing ‘‘green policies” for managing these wastes is of utmost importance and has become a significant topic in the literature (Guo et al., 2012; Lu et al., 2009). In response to this growing concern, the cruise sector has taken proactive measures by initiating various initiatives to enhance pollution prevention measures. These efforts involve the adoption of specific guidelines and procedures, as well as the exploration of innovative technologies that align with the existing regulatory requirements and industry standards. Moreover, the contemporary approach to sustainable waste management embodies the foundational doctrine of 'prevention is better than the cure'. It adheres to the waste management hierarchy, emphasising waste prevention, reuse, recycling, and other forms of recovery (Gharfalkar et al., 2015). The primary objective is to diminish the volume of waste disposed of in landfills, all within the framework of a circular economy. This approach is equally applicable to waste generated by ships (Argüello, 2020); however, the responsibility for upholding sustainable practices onboard primarily reflects the environmental dedication of ship operators and the conscientiousness of the crew (Butt, 2007). Prominent players in the cruise ship sector actively promote and endorse heightened levels of environmental consciousness, implementing diverse management strategies to mitigate their impact, serving as an added incentive in the promotion of their cruises (Butt, 2007; CLIA, 2020). Conversely, there have been reports indicating that, except for a few major ocean cruise companies, others fail to disclose or furnish details regarding their environmental endeavours and protocols, potentially overlooking the emerging environmentally conscious consumer demographic that prioritises value and considers various factors beyond price during their decision-making process (Ramoa et al., 2018; Jones et al., 2019).

Despite the substantial expansion of the cruise sector and its associated waste generation, managing waste generated by cruise ships has gathered limited attention in academic discourse (Sanches et al., 2020). Similarly, Ramoa et al. (2018) highlighted the scarcity of research on cruise tourism, particularly studies on environmental sustainability aboard ships. This chapter, therefore, addresses the following research questions. "What are the key themes, trends, and gaps in the existing body of research on sustainable cruise tourism and waste management and what potential directions for future research can be identified?" Despite the substantial expansion of the cruise sector and the resulting increase in waste generation, academic discourse on managing waste from cruise ships has been limited. This lack of attention is concerning, considering the magnitude of waste produced by cruise ships and its potential environmental impact. Secondly, the scarcity of research on cruise tourism, particularly studies focusing on environmental sustainability aboard ships, further highlights the need for a comprehensive review of the existing literature. The environmental aspect of cruise tourism, especially waste management, remains unexplored. This gap is alarming, given the environmental challenges of ship-generated waste pollution. Furthermore, the limited coverage of ship-generated waste pollution in scientific literature compared to land-based waste management issues is another reason why conducting a study to review the literature on cruise waste management is essential.

2. METHODS

The process of selecting relevant literature for a systematic review of waste in the cruise sector was meticulously conducted using the PRISMA (Preferred Reporting Items for Systematic Reviews and Meta-Analyses) flowchart. The PRISMA flow diagram is crucial for systematic literature reviews, providing a structured framework to document the progression from the initial search to the final inclusion of studies. It enhances thoroughness, clarity, and replicability by visually displaying the research process, helping to replicate the search strategy, and maintaining the review's integrity through predetermined selection and data extraction criteria. This methodical approach reduces bias and leads to reliable conclusions. The standardised format ensures consistent reporting, enabling comparisons across studies and identifying critical review elements. Adhering to PRISMA guidelines increases the credibility and reliability of reviews, advancing knowledge within specific fields.

Figure 1. PRISMA flow diagram

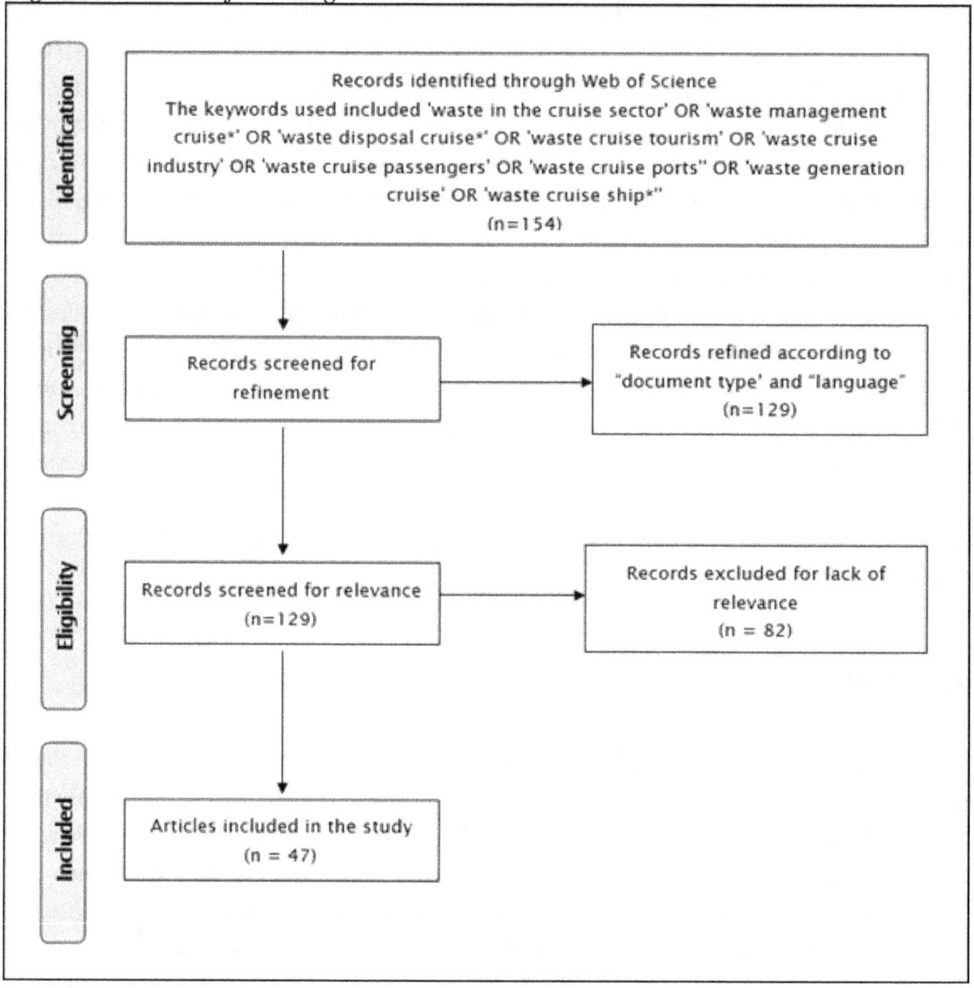

Source: Author's Compilation

The initial step involved conducting a thorough search on the Web of Science database. Specific search terms were employed to capture various aspects of waste management in the cruise sector literature. The keywords used included 'waste in the cruise sector' OR 'waste management cruise*' OR 'waste disposal cruise*' OR 'waste cruise tourism' OR 'waste cruise industry' OR 'waste cruise passengers' OR 'waste cruise ports' OR 'waste generation cruise' OR 'waste cruise ship*'. This search yielded 154 documents. The literature search was further refined by filtering for 'document type' to include only research articles published in English, resulting in 129 papers. Each of these 129 documents was meticulously examined for relevance to the study, with 47 papers ultimately deemed pertinent based on their content.

The selected literature was rigorously analysed using two primary tools: Bibliometrix-R and SciSpace. Bibliometrix-R, a widely utilised tool in bibliometric analysis, facilitated the examination of publication trends, including the annual number of publications, the top ten journals in which papers were published, and the authors' countries of origin. This tool's 'biblioshiny' function enabled the analysis of bibliometric data through a web interface, obviating the need for manual coding. In contrast, SciSpace provided advanced visualisation tools to explore and comprehend co-occurrence network data. According to Khan et al. (2019), the SciSpace platform offers a comprehensive overview of data exchanged across multiple High-Performance Computing (HPC) data centres located in various geographical regions through a unified workspace, thereby enhancing efficiency in data storage or retrieval within the appropriate data centre namespace. A content analysis was conducted to identify the main research themes within the publications. This qualitative method involved systematically organising and examining the content of the articles to uncover recurring themes, patterns, and areas necessitating further research. The study's primary objective was to provide valuable insights into current research trends, thematic patterns, and knowledge gaps in the field through a systematic approach to data collection, analysis, and synthesis. The anticipated results will advance academic discussions and inform future research efforts and policy actions in sustainable tourism development.

3. RESULTS

3.1 Overview of the Data

Table 1. Data overview

Timespan	**1994:2024**
Sources (Journals)	35
Documents	47
Annual Growth Rate %	4.73
Average citations per doc	32.04
Keywords Plus (ID)	147
Author's Keywords (DE)	179
Authors	127
International co-authorships %	17.02
Article	47

The systematic literature review on cruise waste dates from 1994 to 2024, encompassing thirty years of research and development. During this time, research on this topic was published in thirty-five different journals, reflecting its relevance to various disciplines, including environmental science, marine biology, waste management, and tourism. The dataset consists of forty-seven documents. The annual growth rate of 4.73% indicates a consistent increase in scholarly attention towards cruise waste, suggesting a growing recognition of its environmental impact and the need for effective management strategies. The average of 32.04 citations per document highlights this research's significant academic impact and influence within the field.

The content of these documents is extensive and varied, with 147 Keywords Plus (terms derived from article references) and 179 Author's Keywords (terms chosen by the authors themselves). This diverse range of keywords suggests the complexity and breadth of the subject matter, covering multiple themes, methodologies, and specific issues related to cruise waste. Additionally, this variety of keywords highlights the multidimensional nature of the topic, comprising technical, scientific, policy, and management aspects. The literature on cruise waste includes contributions from 127 authors, which shows a dedicated research community. International collaboration is a significant feature of this research field, with 17.02% of documents involving co-authors from different countries. This international co-authorship highlights the global importance of managing cruise waste and the shared interest among researchers worldwide in exploring various perspectives and innovative approaches. All forty-seven documents in the dataset are articles, signifying peer-reviewed research for advancing understanding, providing rigorous analysis, and offering credible and reliable findings that inform policy and practice.

Figure 2. Publication trend

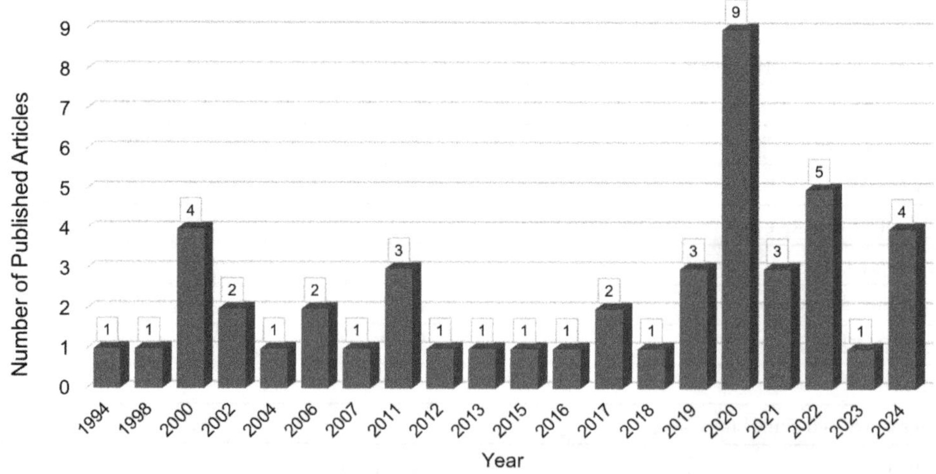

The data shows a clear upward trend in research on cruise waste, particularly over the last decade. Specific spikes in certain years indicate periods of intensified focus, likely driven by broader environmental movements or major regulatory and policy changes affecting the industry. However, several factors have contributed to the relatively low volume of publications on cruise waste over the years. The niche focus of cruise waste management narrows the pool of researchers interested in the topic, leading to fewer publications than more broadly scoped environmental issues. Additionally, obtaining accurate and comprehensive data on waste generation and management within the cruise industry is often challenging. Cruise companies may be reluctant to share detailed operational data due to competitive concerns or fear of regulatory repercussions, limiting researchers' access to crucial information. Moreover, the cruise industry's strong influence on regulatory practices and environmental standards can result in less stringent regulations, reducing the urgency for independent research. Resistance from the industry to external scrutiny may further restrict research opportunities and the publication of findings, which could negatively affect the sector. Despite these challenges, the recent surge in publications since 2020 suggests that growing environmental awareness and concern are beginning to influence this research area. As sustainability becomes increasingly central to global discussions, research on specific sectors like cruising is receiving more attention.

3.2 Key Themes in the Literature

3.2.1 Plastic Waste

The literature on plastic waste management in the cruise sector offers a comprehensive understanding of the sources, regulatory frameworks, and mitigation strategies for addressing plastic pollution, with a focus on both ocean- and river-based cruise activities. Čulin and Bielić (2016) provide a foundational analysis of the environmental impact of plastic pollution from ships, emphasising ongoing concerns despite regulations such as MARPOL Annex V, which prohibits the discharge of plastics into the sea. They highlight that inconsistent compliance with these regulations and frequent discharge violations remain significant barriers to effective plastic waste management. This view is echoed by van Klink et al. (2022), who identify similar regulatory gaps and inconsistencies in the context of river cruises along the Rhine River. The authors point to substantial policy omissions and misalignment between environmental protection and tourism policies, which hinder the effective management of plastic waste.

Both studies assert that legislation alone cannot address plastic pollution and advocate for stronger policy frameworks and educational initiatives to raise awareness of its harmful effects. Adding a quantitative perspective to this discussion, Deville et al. (2023) use Material Flow Analysis (MFA) to trace plastic waste flows from ocean-based sources within Peru's Economic Exclusive Zone (EEZ). Their study identifies critical sources of plastic pollution, such as packaging from onboard consumption and emissions from antifouling polymers. The data provided by Deville et al. (2023) offer a solid foundation for more targeted mitigation strategies, complementing the policy-focused findings of the other studies.

While Čulin and Bielić (2016) and Klink et al. (2022) emphasise the need for better policy integration and standardised waste collection systems in distinct geographical contexts—ocean and river cruises—Deville et al. (2023) highlight the importance of understanding specific plastic waste sources to develop effective regulations. Moreover, all three studies stress the critical role of stakeholder engagement and education in combating plastic pollution. Čulin and Bielić (2016) advocate for educational initiatives to raise awareness, aligning with Klink et al.'s (2022) recommendation to expand policy discussions on plastic waste. Though Deville et al. (2023) adopt a more data-driven approach, their findings implicitly support the need for education by providing factual insights to inform stakeholders about the sources and impacts of plastic pollution.

3.2.2 Food Waste

The literature on food waste management in the cruise sector offers diverse perspectives on the regulatory, operational, and cultural factors shaping waste generation and mitigation strategies. This review synthesises findings from key studies, highlighting the issue's complexity and the range of approaches suggested to address it. Polglaze (2003) provides a foundational analysis of the regulatory framework governing food waste management on ships, particularly in areas where MARPOL regulations restrict disposal at sea. The study underscores the importance of compliance with these regulations to prevent environmental harm, emphasising the need for alternative waste management methods such as incineration or diversion to sewage treatment plants. Polglaze also highlights the significant challenge of accurately estimating food waste generation, which complicates the implementation of effective management practices.

In contrast, Wilewska-Bien et al. (2020) focus on operational practices and communication within the cruise industry, exposing shortcomings in current food waste management, particularly regarding waste separation and measurement. They identify buffet-style restaurants as key sources of food waste and suggest that targeted reduction programs could yield substantial improvements. The authors propose a sustainable solution based on separation, measurement, and reduction and emphasise the crucial role of crew engagement and training in promoting more sustainable waste management practices.

Li and Wang (2020) introduce a cultural perspective, examining how behavioural factors influence food waste among Chinese cruise passengers. Their research reveals the impact of cultural norms, such as an emphasis on abundance and fear of missing out, in exacerbating food waste. By identifying behaviours like over-ordering and plate waste, the study suggests that cultural interventions—such as offering authentic cuisine and organising educational events—could encourage more sustainable consumption practices among passengers. This cultural lens adds complexity to the issue, underscoring the need for tailored strategies for diverse passenger demographics. Filho et al. (2024) address the challenge of offshore food waste by highlighting the lack of reliable data and the importance of implementing best practices in ship waste management. They advocate for an integrated approach that includes proper incineration, waste segregation, and public awareness campaigns to mitigate environmental impacts. The study emphasises the need for coordinated efforts, regulatory enforcement, and public engagement to implement food waste management strategies successfully.

3.2.3 Microplastics

Microplastic (MP) pollution from cruise ships is a significant environmental concern, with greywater discharge identified as a primary source. Jang et al. (2024) conducted an in-depth analysis of MPs in greywater from the RV Onnuri, finding that laundry greywater had the highest mean load of MPs, predominantly consisting of fibre-type polyester. This underscores the substantial contribution of ship-generated greywater to marine MP pollution and highlights the urgent need for effective treatment solutions. Similarly, Peng et al. (2022) examined the environmental impact of greywater discharge from ships, identifying it as a major MP source due to existing regulations that permit direct discharge into marine environments. Their findings suggest that cruise ships may release significant amounts of MPs annually, necessitating accurate estimation and better management to reduce pollution.

In addition to greywater, Folbert et al. (2022) investigated other sources and pathways of MPs in cruise ship wastewater, identifying personal care products, cleaning agents, and synthetic microfibers from textiles as significant contributors. They pointed out that untreated greywater and the overboard discharge of biosludge are key pathways through which MPs enter marine ecosystems, emphasising the need for comprehensive wastewater management practices. Although advanced wastewater treatment systems (AWTS) showed some effectiveness in reducing MPs, the authors recommended broader measures, such as phasing out single-use plastics and adopting non-synthetic textile alternatives, to address MP pollution more effectively.

Despite these findings, gaps remain in current waste management practices on cruise ships. While studies like Jang et al. (2024) and Peng et al. (2022) focus on understanding the occurrence and environmental impacts of MPs, Folbert et al. (2022) focus deeper on identifying specific sources and pathways, offering practical recommendations for mitigation. However, the partial effectiveness of AWTS in reducing MPs reveals the complexity of addressing MP pollution through technology alone. Collaborative efforts, including policy reforms, changes in industry practices, and shifts in consumer behaviour, are essential for effectively tackling the issue. Moreover, the discrepancy between regulations allowing greywater discharge and the environmental consequences highlighted by Peng et al. (2022) underscores the need for updated regulations that align with ecological sustainability goals.

3.2.4 Wastewater

Loehr et al. (2006) report that Alaska's Department of Environmental Conservation (ADEC) sampled various effluents from large vessels in response to growing public concerns about cruise ship discharges. Their analysis revealed that marine sanitation devices (MSDs) designed to treat black water (sewage) failed to meet expected standards. In contrast, untreated grey water contained high levels of typical pollutants and unexpectedly high concentrations of bacteria. On several occasions, black and grey water discharges exceeded state water quality thresholds for toxic substances. Considering these findings, the state convened a Science Advisory Panel to assess the potential impacts of wastewater discharges from cruise ships. The release of this effluent data, amplified by extensive media coverage, heightened public concerns, leading to swift legislative actions at both the state and federal levels—even before the Panel completed its evaluation. This sequence of events underscores the powerful role public perception and environmental monitoring data play in shaping policy decisions on maritime pollution.

Pérez et al. (2017) explored the key drivers behind oily waste from cruise ships, identifying ship size as the main contributor. This finding suggests that revising the tariff system to reflect objective indicators directly related to waste generation, with differential tariffs for each type of waste, could improve waste management. Meanwhile, Perić et al. (2016) demonstrated how cruise ships' predetermined itineraries, typically planned a year, provide realistic data inputs for their model predicting cruise traffic in the Adriatic Sea. Their model, designed with MARPOL Annex IV regulations in mind, enables the prediction of cruise ship movements and retention times in areas with limited wastewater discharge allowances. Using only two input parameters—cruise capacity and itinerary—the model can estimate the generated grey and black water volume. This data allows for traffic distribution or limitation while aiding future environmental pollution trend analysis.

Özkaynak et al. (2023) offer an in-depth analysis of bilge water waste, emphasising the importance of precise calculations to prevent illegal discharges and ensure the availability of adequate waste reception facilities. Their study highlights the critical role of advanced modelling techniques in estimating waste generation and facilitating effective waste management across regions and ship types. Accurate predictions and management of waste outputs are essential for reducing the environmental impact of cruise ships, particularly concerning bilge water, which can be a significant source of microplastic pollution if not properly handled. Similarly, Şahin and Vardar (2020) studied wastewater discharge from large passenger ships, identifying their substantial contribution to marine pollution. Their research involves developing an Artificial Neural Network (ANN) model to estimate dilution factors, which improves the accuracy of waste dilution predictions and supports implementing more effective waste management practices.

Gollasch et al. (2000) broaden the conversation by examining the survival of tropical plankton organisms in ballast water during transoceanic voyages. Their study sheds light on the ecological risks of ballast water discharges, particularly introducing non-native species into new environments. These studies emphasise the importance of accurate data and predictive models for managing waste in the cruise sector. Özkaynak et al. (2023) and Şahin and Vardar (2020) stress the need for advanced estimation techniques to improve waste management, while Gollasch et al. (2000) draw attention to the broader ecological implications of cruise ship operations. Integrating accurate waste estimation methods and stringent ecological management practices is crucial to addressing the complex challenges of microplastic pollution and other environmental impacts from cruise ships.

3.2.5 Waste Management Practices

Toneatti et al. (2020) and Popa (2021) emphasise the necessity of comprehensive waste management plans in cruise operations, highlighting their importance in mitigating environmental impacts and reducing operational costs. Toneatti et al. (2020) specifically focus on various types of waste, such as glass, paper, and cellulosic materials, exploring innovative strategies like syngas production from cellulosic waste. This aligns with Popa's (2021) focus on the detailed procedures required for waste reduction, collection, storage, processing, and disposal, as mandated by MARPOL regulations. While both studies stress the importance of waste segregation and categorisation, Toneatti et al. (2020) delve into technological solutions for waste treatment. Popa (2021) underscores the role of regulatory compliance, regular inspections, and monitoring in ensuring effective waste management. Şahin and Vardar (2015) provide a more focused analysis of sewage treatment systems, contrasting with other studies' broader waste management strategies. They examine the environmental challenges posed by sewage, comparing the effectiveness of Marine Sanitation Devices (MSD) with Advanced Wastewater Treatment (AWT) systems. Their conclusion—that AWT systems are more efficient but require additional measures, such as rapid effluent dilution—complements the comprehensive waste management approaches by addressing a specific aspect of environmental impact.

Sanches et al. (2024) take a broader view of solid waste management on cruise ships, stressing the issue's interconnected social, economic, and political dimensions. Their call for multidisciplinary research and collaboration aligns with the recommendations of Kotrikla et al. (2021), who focus on the challenges posed by inadequate land-based waste management infrastructure in the Caribbean, advocating for modern technological solutions and regional cooperation. Both studies underscore the complexity of managing waste across different regions and emphasise the importance of stakeholder collaboration. Tsai and Bui (2020) introduced a hierarchical framework for sustainable consumption on cruise ships, identifying key attributes influencing sustainable performance, such as waste reduction and recycling. This methodological approach complements the practical implementations discussed by Toneatti et al. (2020) and Mănoiu and Crăciun (2018), highlighting Disney Cruise Line's comprehensive waste management practices. Both studies acknowledge the significance of emission controls and alternative energy sources in enhancing sustainability.

Svaetichin and Inkinen (2017) explore waste management systems in the Baltic Sea and advocate for unified environmental legislation across EU ports. Vaneeckhaute and Fazli (2020) focus on handling nutrient-rich organic waste in the Baltic Sea and highlight the role of the No Special Fee system in encouraging ships to deliver waste ashore. Both studies stress the importance of regional policies and collaboration in improving waste management practices. Francis (2021) and Vilotijević et al. (2020) address policy and regulatory challenges. Francis advocates for stricter regulations and better enforcement to combat marine pollution, while Vilotijević et al. examine the ongoing evolution of the MARPOL Convention in response to the increasing waste generated by larger vessels. Both studies call for stronger regulatory frameworks to ensure sustainable maritime operations and highlight persistent issues with marine pollution. Strazza et al. (2015) and Di Vaio et al. (2024) explore technological and digital solutions to enhance environmental performance in waste management. Strazza et al. (2015) focus on reducing paper use through green practices and Life Cycle Assessment, while Di Vaio et al. (2024) investigate the role of digital technologies and sustainability disclosures. Both studies reveal the potential for significant environmental benefits and cost savings through technological innovations and improved stakeholder engagement.

4. BEST PRACTICES

Kortrikla et al. (2021) investigated waste management practices on a cruise ship operating in the Caribbean Sea, revealing that educating and training passengers and crew is a key step in reducing pollution. The ship implements several eco-friendly practices, including banning plastic cups, straws, plates, and cutlery for sustainable materials like wood. Collaborations with suppliers facilitate the return of glass bottles and other materials, while usable items no longer needed are donated to charitable organisations. Passengers are also encouraged to conserve water and energy by reusing towels, and signs around the deck remind passengers and crew to avoid discarding waste overboard, adhering to MARPOL Annex V guidelines (MARPOL Practical Guide, 2015).

The ship surpasses International Maritime Organization (IMO) requirements by treating oily bilge water with an Oily-Water Separator, reducing oil content to 5 ppm or less, and discharging it at least 12 nautical miles from land at a minimum speed of 6 knots. Food waste is processed by a pulper, reducing it to particles of 20 mm or less, which are more easily degraded by marine microorganisms. The soft food components are separated from hard parts, such as bones and seafood shells, to protect the pulper, with galley personnel managing maintenance. The processed food waste is discharged into the sea under strict conditions, while hard food waste is bagged and stored for transport to port reception facilities. The garbage storage area is above the engine space, and food waste is kept in a cold room at 4°C or lower to prevent odours and spoilage. Special wastes like cooking oil, chemicals, and medical waste are also stored in cold rooms (Kortrikla et al., 2021).

Through its 4GOODFOOD program, Costa Crociere has devised novel methods to redistribute surplus food, aiming to halve food waste by the cruise line and its parent company, Carnival Corporation. Since its launch in 2017, the initiative has expanded to cities in Italy, France, Spain, Guadeloupe, and Martinique, resulting in over 200,000 meals donated to those in need. The Costa Crociere Foundation supports this effort, with over 920,000 meals distributed to local communities in 15 destinations since 2015, bringing the total to over 1.1 million meals donated (UN Tourism, 2023b).

In 2023, Royal Caribbean became the first company to introduce next-generation waste management systems aboard their newest ships, Icon of the Seas and Silver Nova, which use land-based waste-to-energy technology to convert solid waste into gas for energy demands in hotel facilities. The company also launched phase three of its plastics reduction program, aiming to eliminate single-use plastics by 2025. Royal Caribbean's longstanding Save the Waves program, now in its 30th year, focuses on reusing and recycling materials, and its WIN on Waste initiative seeks to halve food waste within five years. By 2022, 26 ships will employ AI systems to predict food production needs. Food waste was processed, incinerated, or handed to authorised waste providers, with all waste logged and audited by regulatory bodies. Celebrity Apex, part of the Royal Caribbean Group, installed a Hydrothermal Oxidation Technology prototype in 2021 to treat food waste without discharging it at sea, further enhancing energy efficiency (UN Tourism, 2023a).

Disney Cruise Line also champions sustainable waste management, diverting more than 600 tonnes of metal, glass, plastic, and paper from waste streams annually through shipboard recycling. The ships recycle condensation from air conditioning systems, saving over 30 million gallons of fresh water annually. Used cooking oil is recycled at ports like Vancouver, Miami, and Port Canaveral, with partners like Bahamas Waste Management converting it into biodiesel. These efforts have removed over 6,400 tonnes of waste from traditional streams and cleared 31,000 pounds of trash from beaches and waterways (Disney Cruise Lines, 2016).

5. GAPS IN THE RESEARCH

The current literature on plastic waste management in the cruise industry identifies several areas that require further research. One key area is the effectiveness of mitigation strategies for addressing specific sources of plastic pollution across different cruise routes and regions, as Deville et al. (2023) noted. While their research provides valuable data on plastic waste flows, a more in-depth exploration of how well these strategies work is essential. Another research gap lies in the socio-economic aspects of plastic waste management. Future studies could investigate the economic impacts of alternative waste management methods on cruise ships, such as recycling and waste-to-energy conversion. By examining the cost-effectiveness and feasibility of these approaches, researchers can offer guidance to cruise operators and policymakers.

A more interdisciplinary approach that integrates environmental science, policy studies, and social sciences is also necessary to develop comprehensive solutions for plastic waste management in the cruise sector. For example, considering the socio-cultural factors that influence plastic consumption and disposal behaviours among passengers and crew could lead to more effective educational campaigns tailored to the unique dynamics of the cruise environment. Despite regulations like MARPOL Annex V, which prohibits plastic discharge at sea, challenges persist in ensuring compliance. Research by Čulin & Bielić (2016) and Klink et al. (2022) highlights ongoing issues with discharge violations, and future studies could explore the reasons behind these violations and propose practical strategies to improve regulatory adherence.

In food waste management, studies by Wilewska-Bien et al. (2020) and Na and Jiayu (2020) provide insights into the operational and cultural factors contributing to food waste on cruise ships. However, there is limited research on the environmental impacts of current food waste disposal methods, such as incineration, composting, or sewage treatment. Future research could assess the ecological consequences of these methods, focusing on aspects like greenhouse gas emissions, nutrient recycling, and marine pollution. Additionally, studies must explore synergies between plastic and food waste management strategies. Given the interconnected nature of these waste streams, integrated approaches that tackle both could lead to more sustainable solutions. For example, researchers could investigate the potential for using food waste as a resource for plastic recycling or biofuel production on cruise ships.

Although existing literature emphasises regulatory compliance, operational changes, and cultural interventions in food waste management, comprehensive data on offshore food waste is still lacking, as Filho et al. (2024) highlighted. Future studies could address this by thoroughly evaluating food waste generation rates and disposal practices during cruise voyages. Research on greywater discharge and its contribution to marine microplastic (MP) pollution acknowledges the limited effectiveness of current wastewater treatment systems. However, no consensus exists on the most effective methods to reduce MP contamination. Future research could explore innovative technologies and practices designed specifically for greywater treatment in the cruise industry. While studies such as Folbert et al. (2022) identify MPs' main sources and pathways, further research is needed to develop targeted interventions, such as phasing out single-use plastics and promoting non-synthetic textile alternatives, to combat MP pollution from cruise ships effectively.

Wastewater management research has primarily focused on estimating waste volumes, creating predictive models, and understanding ecological impacts. However, more detailed studies are needed to explore wastewater treatment technologies' practical application and efficiency on cruise ships. Despite some attention to the ecological risks associated with ballast water discharge, there is still a gap in research on developing and implementing strict ballast water management protocols to mitigate these risks. While the literature provides valuable insights into various waste management strategies and technologies, there is a lack of holistic studies evaluating these practices' environmental and economic impacts. Future research could use a life cycle assessment framework to thoroughly assess the sustainability of different waste management systems on cruise ships. Additionally, there is a need for further exploration of regional collaborations and policy initiatives aimed at improving waste management, particularly in regions with limited land-based infrastructure, such as the Caribbean, as noted by Sanches et al. (2020) and Kotrikla et al. (2021). Research into emerging technologies and digital solutions in waste management could also provide valuable information to enhance environmental performance and sustainability in the cruise industry.

6. CONCLUSION

The literature review highlights the challenges of waste management in the cruise industry despite the regulatory framework provided by MARPOL Annex V. Ongoing problems like plastic pollution, microplastic discharge, solid waste, and wastewater indicate stronger policies. Consistent enforcement of regulations across different regions is critical for reducing these issues. Involving stakeholders such as cruise operators, regulatory bodies, environmental groups, and the public is also key to improving waste management practices. Education and awareness programs are important for promoting sustainable waste practices among all involved parties. Collaboration among stakeholders is essential for progressing, complying with stricter regulations, and using advanced technologies. These efforts are necessary to protect marine environments and ensure the cruise industry's long-term sustainability. Technological innovations, including better waste treatment systems and monitoring tools, are crucial for handling the complex waste cruise ships produce. Ongoing research and monitoring of cruise waste are needed to develop effective solutions and improve current waste management practices. The research also identifies gaps in understanding, pointing to the need for further study and innovation. Addressing these gaps will require collaboration between researchers, policymakers, and industry stakeholders. The growing interest in this field among researchers and journals is promising, as it can lead to better waste management strategies that benefit both the environment and the cruise industry's sustainability.

ACKNOWLEDGEMENT

This work is based on the research supported by the National Institute for the Humanities and Social Sciences REF NO: BRI22/1215.

REFERENCES

Argüello, G. (2020). Environmentally sound management of ship wastes: Challenges and opportunities for European ports. *Journal of Shipping and Trade*, 5(1), 1–21. DOI: 10.1186/s41072-020-00068-w

Butt, N. (2007). The impact of cruise ship generated waste on home ports and ports of call: A study of Southampton. *Marine Policy, 31*(5), 591–598. Elsevier Sci Ltd. https://doi.org/DOI: 10.1016/j.marpol.2007.03.002

Caniëls, M. C. J., Cleophas, E., & Semeijn, J. (2016). Implementing green supply chain practices: An empirical investigation in the shipbuilding industry. *Maritime Policy & Management*, 43(8), 1005–1020. DOI: 10.1080/03088839.2016.1182654

Cerveny, L. K., Miller, A., & Gende, S. (2020). Sustainable Cruise Tourism in Marine World Heritage Sites. *Sustainability (Basel)*, 12(2), 611. DOI: 10.3390/su12020611

CLIA. (2020). State of the cruise industry outlook 2020, economic impact studies, cruise lines international association. Available at https://cruising.org/en-gb/news-and-research/research/2019/december/state-of-the-cruise-industry-outlook-2020. Accessed on March 2021.

Copeland, C. (2011). Cruise ship pollution: Background, laws and regulations, and key issues. In *Maritime Law: Issues, Challenges and Implications* (pp. 155–183). Nova Science Publishers Inc.

Čulin, J., & Bielić, T. (2016). Plastic pollution from ships. *Pomorski Zbornik*, 51, 57–66.

Deville, A., Vazquez-Rowe, I., Ita-Nagy, D., & Kahhat, R. (2023). Ocean-based sources of plastic pollution: An overview of the main marine activities in the Peruvian EEZ. *Marine Pollution Bulletin, 189*. Pergamon-Elsevier Science Ltd. https://doi.org/DOI: 10.1016/j.marpolbul.2023.114785

Di Vaio, A., Dell'Amura, G., & Varriale, L. (2024). *Biodiversity disclosure in the cruise industry: Empirical research challenges. Current Issues in Tourism. Routledge Journals*. Taylor & Francis Ltd., DOI: 10.1080/13683500.2024.2350586

Disney Cruise Line. (2016. Disney Cruise Line Environmental Overview. https://disneycruiselineblog.com/wp-content/uploads/2016/06/DCL-Environmental-Fact-Sheet-2015.pdf

Folbert, M. E. F., Corbin, C., & Lohr, A. J. (2022). Sources and leakages of microplastics in cruise ship wastewater. *Frontiers in Marine Science, 9*. Frontiers Media SA. https://doi.org/DOI: 10.3389/fmars.2022.900047

Gharfalkar, M., Court, R., Campbell, C., Ali, Z., & Hillier, G. (2015). Analysis of waste hierarchy in the European waste directive 2008/98/EC. *Waste Management (New York, N.Y.)*, 39, 305–313. DOI: 10.1016/j.wasman.2015.02.007 PMID: 25725949

Gollasch, S., Lenz, J., Dammer, M., & Andres, H. (2000). *Survival of tropical ballast water organisms during a cruise from the Indian Ocean to the North Sea. Journal of Plankton Research, 22(5), 923–937*. Oxford Univesity Press., DOI: 10.1093/plankt/22.5.923

Guo, T. L., Zhang, H. X., & Dai, H. J. (2012). Analysis of green ships design and manufacturing technology. *International Conference on Mechanics and Manufacturing Systems*, 489–493. https://doi.org/DOI: 10.4028/www.scientific.net/amm.109.489

IMO. (2011). IMO and environment. IMO's response to current environmental challenges.

IMO. (2012). Guidelines for implementation of MARPOL annex V. Resolution MEPC.219 (63).

Jang, Y. L., Jeong, J., E, S., Hong, S. H., & Shim, W. J. (2024). Occurrence and characteristics of microplastics in greywater from a research vessel. *Environmental Pollution, 341*. Elsevier Sci Ltd. https://doi.org/DOI: 10.1016/j.envpol.2023.12294

Jones, P., Comfort, D., & Hillier, D. (2019). Sustainability and the world's leading ocean cruising companies. *Journal of Public Affairs*, 19(1), e1609. DOI: 10.1002/pa.1609

Khan, A., Kim, T., Byun, H., & Kim, Y. (2019). SciSpace: A scientific collaboration workspace for geo-distributed HPC data centers. *Future Generation Computer Systems*, 101, 398–409. DOI: 10.1016/j.future.2019.06.006

Klein, R. A. (2011). Responsible cruise tourism: Issues of cruise tourism and sustainability. *Journal of Hospitality and Tourism Management*, 18(1), 107–116. DOI: 10.1375/jhtm.18.1.107

Kotrikla, A. M., Zavantias, A., & Kaloupi, M. (2021). Waste generation and management onboard a cruise ship: A case study. *Ocean & Coastal Management, 212*. Elsevier Sci Ltd. https://doi.org/DOI: 10.1016/j.ocecoaman.2021.105850

Krenshaw, O. G. (2009). *Cruise ship pollution*. Nova Science Publishers Inc.

Li, N., & Wang, J. (2020). Food waste of Chinese cruise passengers. *Journal of Sustainable Tourism*, 28(11), 1825–1840. DOI: 10.1080/09669582.2020.1762621

Loehr, L. C., Beegle-Krause, C.J., George, K., McGee, C. D., Mearns, A. J., & Atkinson, M. J. (2006). The significance of dilution in evaluating possible impacts of wastewater discharges from large cruise ships. *Marine Pollution Bulletin, 52*(6), 681–688. Pergamon-Elsevier Science Ltd. https://doi.org/DOI: 10.1016/j.marpolbul.2005.10.02

Lu, D., Wang, J., Liu, Y., & Wang, X. (2009). The important factors for the implementation of green shipbuilding. *International Conference on Computer Applications in Shipbuilding*, 999–1022.

Mănoiu, V.-M., & Crăciun, A.-I. (2018). Waste management on board Disney Cruise Line Company ships. *Papers of the "Dimitrie Cantemir" Geographic Seminar, 46*(2), 77–99. http://dx.doi.org/DOI: 10.15551/lsgdc.v46i2.06

MEPC. (2017). 2017 guidelines for implementing MARPOL Annex V, Resolution MEPC, vol. 295. Available at https://www.liscr.com/sites/default/files/liscr_imo_resolutions/Res%20MEPC.295(71)%20-%20Guidelines%20MARPOL%20Annex%20V_0.pdf

Peng, G., Xu, B., & Li, D. (2022). Grey water from ships: A significant sea-based source of microplastics? *Environmental Science & Technology*, 56(1), 4–7. DOI: 10.1021/acs.est.1c05446 PMID: 34898179

Peric, T., Komadina, P., & Racic, N. (2016). Wastewater pollution from cruise ships in the Adriatic Sea. *Promet - Traffic & Transportation, 28*(4), 425–433. Sveučilište u Zagrebu, Fakultet prometnih znanosti. https://doi.org/DOI: 10.7307/ptt.v28i4.2087

Polglaze, J. (2003). Can we always ignore ship-generated food waste? *Marine Pollution Bulletin*, 46(1), 33–38. DOI: 10.1016/S0025-326X(02)00324-7 PMID: 12535966

Ramoa, C. E. A., Flores, L. C. S., & Stecker, B. (2018). The convergence of environmental sustainability and ocean cruises in two moments: In the academic research and corporate communication. *Revista Brasileira de Pesquisa em Turismo*, 12(2), 152–178. DOI: 10.7784/rbtur.v12i2.1432

Royal Caribbean Group. 2023. Seastainability 2023 Environmental, Social and Governance Report. https://www.royalcaribbeangroup.com/wp-content/uploads/2024/06/Sustainability-Report-2023.pdf

Sahin, V., & Vardar, N. (2020). Determination of wastewater behaviour of large passenger ships based on their main parameters in the pre-design stage. *Journal of Marine Science and Engineering, 8*(8). MDPI. https://doi.org/DOI: 10.3390/jmse8080546

Sanches, V. L., Aguiar, M. R. D. C. M., de Freitas, M. A. V., & Pacheco, E. B. A. V. (2020). Management of cruise ship-generated solid waste: A review. *Marine Pollution Bulletin*, 151, 110785. DOI: 10.1016/j.marpolbul.2019.110785 PMID: 32056584

Sanches, V. M. L., Gobbi, C. N., Calderari, M. R. da C. M., & Pacheco, E. B. A. V. (2024). Assessment of cruise ship waste management performance: A case study of the Rio de Janeiro port, Brazil. *Marine Pollution Bulletin, 201*. Pergamon-Elsevier Science Ltd. https://doi.org/DOI: 10.1016/j.marpolbul.2024.116216

Statista. (2020). Leisure travel. Available at https://www.statista.com/markets/420/topic/493/leisure-travel/#overview. Accessed on March 2021.

Strazza, C., Del Borghi, A., Gallo, M., Manariti, R., & Missanelli, E. (2015). Investigation of green practices for paper use reduction onboard a cruise ship-a life cycle approach. *International Journal of Life Cycle Assessment, 20*(7), 982–993. Springer Heidelberg. https://doi.org/DOI: 10.1007/s11367-015-0900-0

Svaetichin, I., & Inkinen, T. (2017). Port waste management in the Baltic Sea area: A four port study on the legal requirements, processes and collaboration. *Sustainability, 9*(5). MDPI. https://doi.org/DOI: 10.3390/su9050699

Toneatti, L., Deluca, C., Morgera, A. F., Piller, M., & Pozzetto, D. (2022). Waste to energy onboard cruise ships: A new paradigm for sustainable cruising. *Journal of Marine Science and Engineering, 10*(4). MDPI. https://doi.org/DOI: 10.3390/jmse10040480

Tourism, U. N. (2023a). Case Study from Royal Caribbean Group: Win On Waste. https://www.oneplanetnetwork.org/knowledge-centre/resources/case-study-royal-caribbean-group-win-waste

Tourism, U. N. (2023b). Case Study From Costa Cruises: Responsible Food Consumption, Waste Tracking And Community Support. https://www.oneplanetnetwork.org/knowledge-centre/resources/case-study-costa-cruises-responsible-food-consumption-waste-tracking-and

Tsai, F.-M., & Bui, T.-D. (2020). Assessing sustainable consumption practices on cruise ships. *Maritime Business Review, 5*(2), 229–247. Emerald Group Publishing Ltd. https://doi.org/DOI: 10.1108/MABR-12-2019-0059

van Klink, D., Wiering, M., van Eerd, M., & Schoor, M. (2022). Travelling plastics: Exploring river cruise companies' practices and policies for the environmental protection of the Rhine. *Water, 14*(12). MDPI. https://doi.org/DOI: 10.3390/w14121978

Vaneeckhaute, C., & Fazli, A. (2020). Management of ship-generated food waste and sewage on the Baltic Sea: A review. *Waste Management (New York, N.Y.)*, 102, 12–20. DOI: 10.1016/j.wasman.2019.10.030 PMID: 31654875

Wilewska-Bien, M., Granhag, L., & Andersson, K. (2020). Pathways to reducing and efficiently handling food waste on passenger ships: From Baltic Sea perspective. *Environment Development and Sustainability, 22*(1), 217–230. Springer. https://doi.org/DOI: 10.1007/s10668-018-0192-1

Xu, K. (2016). China's cruise industry: Progress, challenges and outlook. *Maritime Affairs*, 12(1), 38–45. DOI: 10.1080/09733159.2016.1175129

Chapter 14
Reducing Single-Use Plastic and Waste:
Fantasy or Reality at Disney Theme Parks

Michelle Whitmore
https://orcid.org/0000-0001-7721-059X
Western Sydney University, Australia

ABSTRACT

The tourism industry contributes to and is impacted by plastic waste. The Walt Disney Company (Disney) is one of the largest media and entertainment companies contributing to global travel and waste. Disney theme parks represent a microcosm of society and the conflict between consumerism and environmentalism in the effort to reduce single-use plastic and waste to landfills. Disney has targets to reduce single-use plastic and waste at its wholly owned and operated parks and resorts in France and the United States (U.S.). As a multinational company that operates without a legally binding international agreement on plastic waste, Disney is subject to the plastic policies of each jurisdiction. The policy context in which Disney established its waste targets and the actions taken by Disney to reduce plastic waste in theme parks are discussed in this chapter. Moreover, opportunities to improve waste management practices and shared learning were identified.

1. INTRODUCTION

Disney theme parks represent a utopian holiday destination for many, primarily middle-class families. For some, it is a once-in-a-lifetime experience; for others, it becomes an integral part of the family tradition. Whether you love or hate it, deny-

DOI: 10.4018/979-8-3693-6110-8.ch014

ing Disney's impact on popular culture is challenging. The author visited a Disney theme park for the first time in her early twenties and has since made numerous visits, alone and with family. The author finds herself conflicted between her love for Disney and their commitment to environmental sustainability. Apart from the carbon emissions from her long-haul flights, the author found it challenging to limit the waste she generated during her trips to Disney resorts in 2019 and 2020, before COVID-19. The author observed conservation messaging and solar arrays at the parks yet noticed a disconnect between these efforts and the amount of waste produced. This observation led the author to investigate the factors contributing to Disney's stance on plastic and waste. Disney's messaging is environmentally positive; however, Disney makes decisions based on financial incentives as a corporation. The Disney Executive is legally obligated to act in the interests of shareholders, not broader society. However, to continue to make a profit, Disney must maintain favour with its customers and staff, or in Disney theme park parlance, its guests and cast. Lipton (2024) provides an example of Disney taking a stand against state policies in response to cast pressure but notes that in doing so, it was subject to repercussions from the state. This finding reinforces the fact that Disney operates in a political environment. Social, political, and economic factors influence Disney's waste targets and its actions to achieve them. This chapter explores plastic waste, Disney theme parks, and the assemblage of factors that influence decisions regarding plastic use and disposal within the theme parks.

2. THE PROBLEM WITH PLASTIC

Plastic is a term used to describe thousands of materials made from polymers that can be moulded when hot and then set to rigid or elastic form (Meikle, 1995; Freinkel, 2011). Plastics can be used to imitate any material. Some of its early applications include replacing natural products threatened by overconsumption, such as tortoise shells, elephant ivory, and seal fur (Bensaude Vincent, 2013). As a cheaper alternative, plastic products are affordable and widely available. Plastics are daily, and many applications are often unnoticed (Gabrys et al., 2013; Hawkins, 2018). From barely measurable quantities in 1940, it is estimated that approximately 400 million tons of plastic waste is produced each year, half of which is designed to be single-use (United Nations Environment Program [UNEP], 2021).

Every stage of the plastic life cycle, from production to transport, use, and disposal, has negative environmental impacts. Plastics contribute to over 3.4% of global greenhouse gas emissions (Organisation for Economic Co-operation and Development [OECD], 2022) primarily from their production, which relies on the use of residues from oil refineries (Freinkel, 2011). The plasticiser chemicals used in the manufacturing process that leach from manufacturing plants and the products we consume have been shown to contribute to endocrine disruption and other health concerns (Liboiron, 2013). The excess plastic products and their combination of disposability and durability have made plastic one of today's most significant environmental issues.

When plastic is disposed of at the end of its useful life, approximately 9% is recycled, but approximately 22% is mismanaged (OECD, 2022). Mismanagement means the waste is not captured in an appropriate waste stream and contributes to plastic pollution. The OECD (2022) estimated that 140 million tons of plastic waste has accumulated in our waterways. Plastic waste has disastrous effects on marine wildlife through its ingestion and entanglement. The United Nations (UN) (2017) estimated that by 2050, there would be more plastic in the ocean than fish (by weight). In addition to the visible macro-plastic pollution on land and waterways when plastic waste is mismanaged, microplastics also break down and wash out of synthetic clothes and other products (4Ocean, 2020). Beaumont et al. (2019) estimated that plastic pollution contributes to an annual loss of $2.5 trillion annually from marine ecosystem services. However, they noted that this may be underestimated, as the broader social and economic costs, such as the impact on tourism, transport, fisheries, and human health, could not be quantified.

Affluent communities produce the greatest amount of plastic waste, but lower socioeconomic communities often bear the brunt of this problem. As Gabrys et al. (2013, p.3) state, "...plastic has become emblematic of economies of abundance and ecological destruction". In Fiji, which relies on tourism for its economic contribution, tourists generate seven times more plastic waste than locals, and the local waste management system cannot cope (World Wildlife Fund [WWF], 2023). Kerber and Kramm (2021) described the inequity of plastic waste disposal on a Vietnamese island, where the waste generated by tourism accumulates in local areas outside the tourist gaze.

Tourism is both a contributor to and directly affected by plastic pollution (Global Tourism Plastics Initiative [GTPI], 2021). Many single-use plastic items used throughout the tourism value chain result in marine pollution, such as water bottles, bags, straws, disposable plates, and cutlery. Plastic waste in the Mediterranean increases by 30% during the peak tourist season, resulting in an annual loss of up to Euro 268 million in the tourism sector (GTPI, 2021). Qiang et al. (2020) studied beach tourism along the East China Sea. They found a correlation between the amount of marine litter and the length of visitors' stays, with tourism revenue increasing by up to 32% when plastic was removed from the beach. At all scales and locations, tourism organisations play a role in reducing the use of single-use plastics and improving the collection and recycling of these items in circulation.

3. DISNEY THEME PARKS AND PLASTIC

The entity, now known as the Walt Disney Company (Disney), was started in 1923 by the brothers Roy and Walt Disney. Over the last 100 years, it has grown from a family business to one of the world's largest media and entertainment companies. Disney's mission is to "… entertain, inform, and inspire people around the globe through the power of unparalleled storytelling…" (Disney, n.d.). It is one of the world's most recognised and successful brands (Cliffe, 2020; Jones, 2021). Disney's values are well recognised (Wasko et al., 2001) and positively associated with customers. It is a multinational company with approximately 220,000 employees and earned over $ 80 billion in revenue in the 2022 fiscal year (Disney, 2023a).

Disney has three divisions: parks, experiences and products; media and entertainment distribution; and content groups. These three divisions control hundreds of subsidiaries. Disney is engaged in tourism through many operations, such as promoting destinations on its National Geographic channel, travelling aboard the Disney Cruise Line, staying overnight at its resorts, and making souvenir purchases at Disney theme parks. To explore Disney's engagement with plastic waste, the focus was on Disney theme parks, which function as a microcosm of society. It also provides a case study to observe the struggle between consumerism and environmental values.

In 2024, there were 12 theme parks across six Disney resort destinations and three continents (Table 1). Disney owns and operates theme parks and resorts in California, Florida, the U.S., and Paris, France. Disney theme parks and resorts also exist in Tokyo, Japan, Hong Kong, and Shanghai, China, but Disney does not have complete ownership or operational rights for those parks. Most Disney theme park guests are domestic visitors, many holding annual passes enabling regular trips to their preferred Disney resort. In 2019, Disney theme parks received approximately 150 million visits globally. The world's top three visited theme parks are the Magic Kingdom in Florida, Disneyland Park in California, and Tokyo Disneyland (TEA/AECOM, 2020). The 2019 figures were used, as attendance at all theme parks was significantly impacted by COVID-19 restrictions, and visitation is only now starting to rebound (TEA/AECOM, 2022).

Table 1. Chronology of Disney theme park development and location

Date Opened	Theme Park	Resort and Location
July 1955	Disneyland	Disneyland Resort Anaheim, California, U.S.
October 1971	Magic Kingdom	Walt Disney World Resort Orlando, Florida, U.S.
October 1982	EPCOT	Walt Disney World Resort Orlando, Florida, U.S.
April 1983	Tokyo Disneyland	Tokyo Disney Resort Tokyo, Japan
May 1989	Disney's Hollywood Studios (previously MGM Studios)	Walt Disney World Resort Orlando, Florida, U.S.
April 1992	Disneyland Park (originally Euro Disney)	Disneyland Paris Resort Chessy, Ile-de-France, France
April 1998	Disney's Animal Kingdom	Walt Disney World Resort Orlando, Florida, U.S.
February 2001	California Adventure Park	Disneyland Resort Anaheim, California, U.S.
September 2001	Tokyo DisneySea	Tokyo Disney Resort Tokyo, Japan
March 2002	Walt Disney Studios Park (soon to be renamed - Disney Adventure World)	Disneyland Paris Resort Chessy, Ile-de-France, France
September 2005	Hong Kong Disneyland	Hong Kong Disneyland Resort Hong Kong, SAR China
June 2016	Shanghai Disneyland	Shanghai Disney Resort Pudong, Shanghai, China

Disney theme parks are regularly described as happy places that invite people to relive memories and imagine a brighter future (King, 1981; Gottdiener, 1982). Research in environmental psychology suggests that positive images and emotions inspire people to create solutions for positive change (Patrick, 2019). At the same time, Disney theme parks are recognised as exemplars of fundamental commercialisation principles adopted across the tourism and hospitality sectors (Bryman, 2004). The term 'Disneyization' was devised by Bryman (1999) to refer to the commercial principles of theming, hybrid consumption, merchandising, and performative labour that Disney theme parks exemplify (Bryman, 2004), and which have been adopted in shopping malls, restaurants, and sports stadia (Budd & Kirsch, 2005). Disney's commercial influence through Disneyization, combined with its strong storytelling power, positive messaging, and global reach with an audience of millions, suggests that Disney could be a theoretically powerful agent of change. If Disney's actions to reduce single-use plastics and waste in theme parks were successful, it could create a shift to more environmentally responsible practices across a range of commercial settings. Nevertheless, Disney is also entangled with plastic.

Disney would not be the company that it is today without plastics. Modern plastics emerged during the post-World War II economic boom, as did the first Disney theme park. Disney theme parks are structurally and metaphorically plastic landscapes. The parks are organised into themed areas that use scaled simulations to depict a utopian past or future as an escape from the present. Meikle (1995) describes how plastic has been used in Disney theme park structures, including the all-plastic 'Monsanto House of the Future', an attraction at Disneyland California from 1957 to 1968. Plastics are also instrumental in set designs and costumes, electrical equipment and wiring, merchandise, food and drink packaging, and toiletries in its resorts.

3.1 Stakeholder Influence

Clément (2018) discusses the significant role of sponsors, like Monsanto, in establishing and operating Disney theme parks. He states that approximately 80-90% of the funding to build EPCOT came from sponsors like ExxonMobil and General Motors/Chevrolet. These sponsors, many of whom are interested in the petrochemical industry, were provided with pavilions designed to showcase their products, and the Disney brand provided them with a trustworthy mouthpiece. The sponsors also influence messaging at Disney rides and shows. While many of the Disney attractions, particularly at Disney's Animal Kingdom and EPCOT in Florida, feature discussions of sustainability, conservation, and waste management, they are often angled towards the value of corporate innovations (Clément, 2018).

One of Disney's longest-standing sponsorship arrangements is with The Coca-Cola Company, which has been a sponsor since the opening of Disneyland in 1955 (Clément, 2018). As a significant sponsor, The Coca-Cola Company influences the available beverages and promotes its products through park signage and branding. The Coca-Cola Company has been ranked as the top plastic polluter for the sixth year in a row (Break Free from Plastic, 2024a). As a purchaser of Coca-Cola products, Disney is implicated in plastic pollution and is affected by the decisions made by The Coca-Cola Company regarding the proportion of recycled materials or the recyclability of its products. Disney's sponsors are only one factor that influences its relationship with plastics.

Disney does not operate in a vacuum; it exists within multiple assemblages at different scales, and the actors within these assemblages influence how Disney operates. Bennett (2010, pp.23-24) defines assemblages as "…ad hoc groupings of diverse elements … [for which] power is not distributed equally …". Following DeLanda's (2006) concept of assemblage, Disney theme parks can be described as assemblages of people, networks, organisations, and physical infrastructure shaped by local practices and political processes. Disney theme parks are defined by their physical borders and routine human practices. While many of the practices within Disney theme parks are globally homogenous, the plastic practices within wholly owned and operated theme parks are impacted by the political discourse on plastics, internationally and in Europe and the U.S.

4. THE GOVERNANCE OF PLASTIC WASTE

Despite the recognition that plastic pollution is a significant environmental problem, there is considerable disparity in the actions taken to reduce single-use plastics at all levels of governance. The lack of efficacy in the implementation of plastics policies is evident in the fact that, despite the creation of "28 international policies … since 2000 …[and] a 60% increase in national and subnational policies between 2017 and 2022, the total volume of plastic in the ocean has increased by more than 50%" (WWF, 2023, p.39). Consumers are traditionally responsible for addressing plastic waste through consumption and disposal practices. Consumers, whose behaviours largely reflect social norms (Bamberg et al., 2018; Borg et al., 2020), are expected to carry this burden without systemic change and against consumer marketing that positions plastics as convenient (Hawkins, 2009, 2018). Some countries have started to redress this situation through enhanced producer responsibility and regulation of single-use products.

4.1 International Plastic Policies

The UN Sustainable Development Goals (SDGs) are 17 goals, and 169 targets are designed to provide a roadmap to achieve a more sustainable future by 2030 (UN Department of Economic and Social Affairs, 2023). All 193 Member States of the UN adopted the SDGs in 2015. They are an international non-binding agreement with a few specific references to plastic pollution, but the management of plastic waste factors in achieving many of the desired outcomes (Walker, 2021).

SDG 12 is the most apparent referent for addressing plastic waste in the tourism industry, including the following targets.

- the reduction of waste by reducing food waste (Target 12.3)
- increasing recycling and reuse (Target 12.5).
- reporting of sustainable practices by large and transnational companies (Target 12.6).
- sustainable tourism (Target 12b).

Halfway through the 2030 target, we are not on track to meet the SDGs. The UN Secretary-General (2023) reported that only 12% of all SDG targets were on track, while 30% showed no progress or regression below the 2015 baseline. As a voluntary agreement, implementing SDGs relies on Member States integrating targets into legislation and public policy. Many Member States have not developed the necessary governance structures to support the successful implementation of the SDGs.

In 2022, recognising the lack of progress from voluntary measures, 175 UN Member States voted to develop an international legally binding agreement to end plastic pollution by the end of 2024 (UNEP, 2022). Unsurprisingly, there is some contention about the degree to which the agreement should impose controls across the plastic life cycle, with some dissent coming from countries attempting to protect their fossil fuel interests (Hernandez, 2024). The value of the final agreement will depend on the veracity of the language used to impose a global mandate with strict requirements and mechanisms to ensure compliance rather than enable Member States to determine the application of those limits.

4.2 National and Subnational Positions on Plastic Waste

Disney owns and operates theme parks in California, Florida (U.S.), and Paris (France). The plastic legislation in these jurisdictions is vastly different and reflects their success in implementing SDGs and recycling. In Sachs et al.'s (2022) assessment of individual countries' performance in implementing the SDGs, the top 10 countries were all in Europe, with France listed seventh. This is primarily due to the European Union establishing laws to implement the SDGs, which the Member States enforce. According to the OECD (2018a, 2018b), the provision of statutory targets in the European Union resulted in an average of 30% of recycled plastics, while in the U.S., the average plastic recycling rates were less than 10%. The recycling rates in the U.S. dropped to only 5% by 2021 after China stopped taking millions of tons of plastic waste (California Legislative Information, 2022). For the implementation of the SDGs, the U.S. was ranked 41 out of 195 countries, as it does not have any federal legislation to address the SDGs and has never contributed a voluntary national report to track progress against them (Sachs et al., 2022). Disney is a U.S. company ahead of the federal government in implementing, measuring, and reporting against the SDGs and waste targets.

There is no federal legislation in the U.S. to address single-use plastics, and the actions taken in each state vary significantly. Eleven states, including California, have enacted statewide plastic bag bans; however, 19 states, including Florida, have enacted pre-emption legislation to prevent local authorities from restricting plastic products (Wang et al., 2022). California was the first state to implement *Single-Use*

Carryout Bags legislation in 2014, banning the distribution of lightweight single-use carryout bags from large grocery stores and pharmacies. This legislation omits outlets such as Disney stores and permits the provision of heavier, 'reusable' plastic bags. Tabuchi (2024) suggested that these heavier plastic bags contributed to the disposal of more plastic bags by weight in 2023 than when the law was first passed.

For other single-use plastic products in California, the *Plastic Pollution Prevention and Packaging Producer Responsibility Act 2022 (SB54)* makes producers responsible for reducing the amount of single-use plastic used in packaging and foodware by 25%, increasing the proportion of these products that are recycled to 65%, and ensuring that 100% of these products are recyclable or compostable by 2032. This legislation will be ineffective without supporting the infrastructure and a viable market for recycled products. The 2023 amendment to the *California Beverage Container Recycling and Litter Reduction Act* works to build a market for recycled materials by dictating that major suppliers' non-refillable plastic beverage containers must contain 50% post-consumer recycled plastic by 2030. However, suppliers like The Coca-Cola Company already use 100% recycled plastic bottles for some of their products in some markets. What needs to be addressed are the barriers restricting the use of 100% of post-consumer recycled plastic products in all markets.

Currently, there are no targets for reducing single-use plastics or increasing plastic recycling in Florida. In 2008, the *Florida Statutes Title XXIX - Public Health - Chapter 403 - Environmental Control - Part IV - Resource Recovery and Management* set a general recycling target of 75% for municipal waste by 2020, including incineration for waste to energy as a contributor to this figure. However, the target was not met by the deadline. If comparable calculations were used before setting the 75% goal, the actual amount of recycling in the state would have been reduced over the period (Florida Department of Environmental Protection, 2021).

In 2021, an amendment to Section 403.7033 of this statute required the Department of Environmental Protection to submit updated recommendations on regulating single-use plastic packaging. *Section 403.7033* also prevented local and state government agencies from enacting laws regarding single-use packaging or plastic bags until the legislature adopted the department's recommendations. Plastic bag bans or fees that had been imposed by agencies prior to this amendment were nullified (Wang et al., 2022). The updated report prepared by Townsend et al. (2021) recommended restrictions on single-use carryout packaging, but the legislature has not adopted these recommendations. As such, there are no limits to using plastic bags or packaging in Florida. A review of The Florida Senate website reveals that numerous bills have been introduced since 2021 to repeal pre-emptive legislation, impose limitations on single-use packaging, and set recycling targets. However, none of these bills have passed the Florida Legislature.

In comparison, the European Parliament and the Council of the European Union (2019) established a directive that bans many single-use plastic products from 2021, set a target for the collection of 90% of plastic bottles (by weight) for recycling by 2029, and extend producer responsibilities to cover the costs of awareness-raising and clean-up for a range of single-use plastic products. Member States of the European Union were encouraged to set ambitious national reduction and recycling targets and to bring in provisions to implement the directive by 2021.

In 2020, the French government passed *Law no. 2020-105 on 10 February 2020 on the fight against waste and the circular economy.* This law reflects the provisions of the European Union. It sets targets to reduce the number of single-use plastic bottles in the market by 50% by 2030 and to end the marketing of single-use plastic packaging by 2040. These targets are significantly higher than any imposed in the U.S. to date, but the recycling target sets the level of achievement at collection rather than the recycling outcomes.

The introduction of legislation in France may have contributed to Disney's 2030 waste targets to reduce single-use plastics and divert waste from landfills in wholly owned and operated parks. Disney is captured within the definition of a 'producer' under French legislation and is required to reduce the use of single-use plastics and increase its recycling in that country. As the targets were set across theme parks in the U.S. and France, this may be an example of the 'Brussels Effect' (Bradford, 2012), whereby European Union regulations can influence practices in the U.S. and other global markets. The legislation to reduce plastic packaging and increase recycling in California followed shortly thereafter. Disney's corporate social responsibility reports provide insights into the corporate governance of single-use plastics and waste in theme parks.

5. CORPORATE SOCIAL RESPONSIBILITY

Corporate social responsibility (CSR) is a "… voluntary corporate commitment to exceeding the explicit and implicit obligations imposed on a company by society's expectations of conventional corporate behaviour" (Falck & Heblich, 2007, p.247). In thinking beyond the financial bottom line, CSR reports typically address a company's impact and actions in terms of environmental, social, and human rights issues (Luiten, 2020). McMahan (2023) reports that there has been a surge in CSR activities in response to consumer demand to act on social and environmental issues. The UN Secretary-General (2023) has reported a triple the number of companies publishing sustainability reports since 2016. However, Smith and Brisman (2021) advise that CSR be used as an indicator of systemic change, suggesting that corporations are selecting environmental actions that do the minimum necessary to satisfy consumers and enhance their brand and profitability rather than addressing the root cause of the issue.

The primary imperative of a corporation is to return a profit to its shareholders, to whom they have a legal obligation. Disney's commitment to environmental stewardship leads to continuous corporate growth. Disney states that its "...commitment to environmental stewardship focuses on using resources wisely and protecting the planet as we operate and grow our businesses" (Disney, 2021, p.3). In the absence of stringent national or subnational regulations, Whitfield et al. (2022) propose that customer expectations provide the most significant incentive for companies to take environmental action. Disney recognises that responding to consumer demand supports the financial bottom line, stating that "... doing well increasingly involves doing good..." (Disney, 2018, p.3).

Making changes to systems and processes to eliminate single-use plastics can be costly (Whitfield et al., 2022); however, placating customers and being environmentally responsible can increase a company's profits (Falck & Heblich, 2007), reputation, customer goodwill, and the evaluation of a company's products (Chernev & Blair, 2015). To result in greater brand loyalty and trust, McMahan (2023) suggested that these actions must be perceived as authentic and not just for self-interest. Maintaining a positive corporate image is especially important for brands such as Disney.

Engaging in environmentally responsible behaviours supports the vision "... that Disney be as admired as a good corporate citizen as we are for the quality of our entertainment products and experiences" (Disney, 2017, p.2). Disney (2019, p.2) states that being "... seen as one of the world's most respected and admired companies" is as vital to the company as its reputation for creativity and innovation in entertainment. Being seen as doing good is good for business, but achieving positive social and environmental changes must be about more than just appearances.

CSR reporting is voluntary, and the topics covered are not consistent. However, international standards are also available for information disclosure. Disney uses the Sustainability Accounting Standards Board (SASB) standards to disclose information on greenhouse gas emissions, waste, water, diversity, and labour conditions. Since 2008, Disney has reported social and environmental targets across its global operations.

In 2009, a consumer survey by the Boston College Center for Corporate Citizenship ranked Disney as the most reputable company in terms of CSR (Holcomb et al., 2010). Holcomb et al. (2010) proposed that other hospitality companies could learn from Disney's CSR reporting practices, as it provided the most detailed information on Florida's top three theme parks. While the content of Disney's CSR reports has fluctuated over the years, it still has more comprehensive CSR reports than other major theme parks and many other multinational companies.

In 2024, The RepTrak Company ranked Disney 34 in its report on the world's top 100 most reputable companies. The top-ranked company in this report was The LEGO Group (LEGO) (The RepTrack Company, 2024). The LEGO sustainability reports cover the themes of children, environment, people, and governance. LEGO reports that less than 1% of the waste from its factories goes to landfills, and recyclable packaging is being introduced across its products (LEGO, 2024). LEGO and Disney have strong associations with plastics, and LEGO is a Disney partner. Disney-branded LEGO sets and LEGO stores are at most Disney resort destinations.

5.1 Disney's CSR Reporting

Disney's CSR reports include commentary on its global operations' environmental, social, diversity, equity, and inclusion activities. Many environmental targets focus on specific business components. The first of Disney's reported waste targets was to achieve 60% of the waste diverted from landfills and incineration without energy recovery by 2020 (Disney, 2017). The Disney CSR reports clarified that this included operational waste from all theme parks and resorts, ESPN, enterprise-owned facilities, and Pixar. It excluded all leased properties, Disney Stores and Distribution Centers, and TV and radio stations (Disney, 2020).

In 2019, Disney committed to eliminating single-use plastic straws and stirrers at all owned and operated locations worldwide (Disney, 2020). This amounts "... to a reduction of more than 175 million straws and 13 million stirrers annually" (Disney, 2019, para. 5). Disney is also transitioning to refillable in-room amenities in its hotels and cruise ships, intending to reduce plastics in guest rooms by 80% (Disney, 2019, 2020). These actions contributed to Disney meeting its target of 60% of operational waste being diverted from landfills and incineration in 2020 (Disney, 2021). Waste production in 2020 was low (Figure 1) owing to the impact of COVID-19 restrictions on Disney operations and visitation. However, Disney could still divert 60% of the waste from landfills when recycling services in parts of the U.S. were suspended (Keough, 2021).

In its 2020 CSR report, Disney (2021) announced the first iteration of its 2030 environmental goals and started reporting against SDGs, including SDG 12. The Disney 2030 waste goals apply to wholly owned and operated parks and resorts (those in the U.S. and France) and cruise lines. The targets for theme parks and resorts are to achieve zero waste to landfills and reduce single-use plastics (Disney, 2023b).

The theme parks' zero waste to landfill target is defined as "...90% operational waste diversion, including thermal waste to energy for Walt Disney World, Disneyland Resort, Disneyland Paris..." (Disney, 2023b, p.17). Operational waste includes that generated by guests and through the day-to-day operation of parks but excludes construction waste, which is addressed under the sustainable design goal. Waste may be diverted from landfills through waste prevention, donations, recycling, composting, anaerobic digestion, or incineration to convert waste to energy (Disney, 2023c).

Figure 1. Disney's waste reporting

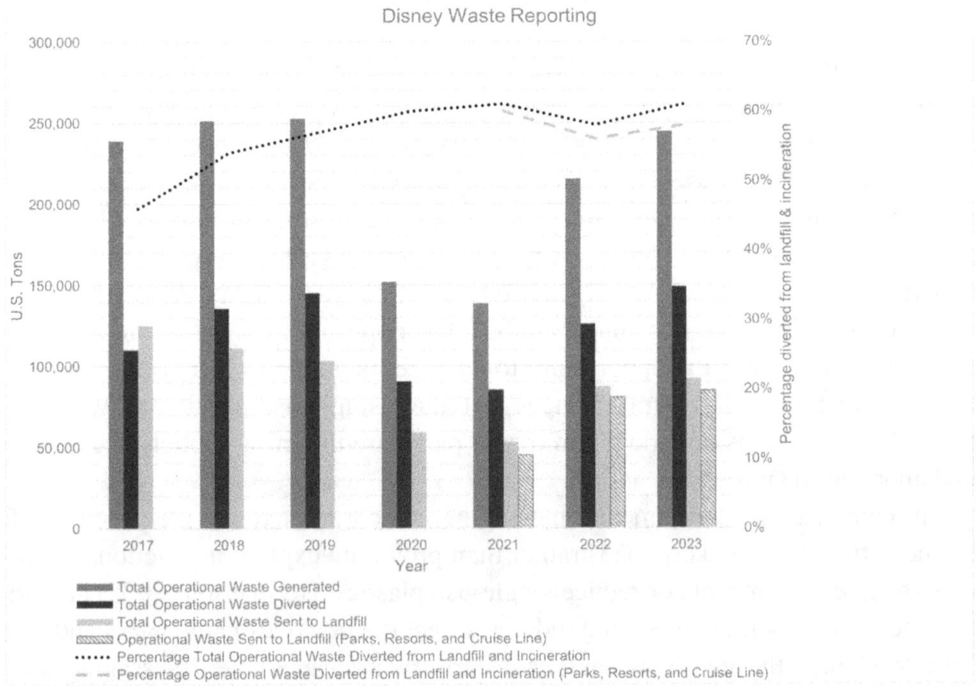

Source: Disney (2020)

The Disney CSR reports provide data on the total operational waste generated and diverted from landfills and the amount of operational waste sent to landfills from parks, resorts, and cruise lines. These reports do not provide separate figures for plastic waste or the proportion of recycled waste, which is just one of the waste diversion practices. To date, the initiatives implemented by Disney appear to have contributed to a decrease in waste generation and an increase in waste diverted from landfills (Figure 1).

The percentage of operational waste diverted from landfills has fluctuated over the last three reporting periods. The COVID-19 restrictions and subsequent recovery efforts may explain this variance. Between 2022 and 2023, the percentage of waste from parks, resorts, and cruise lines diverted from landfills increased, as did the quantity of waste sent to landfills (Figure 1). There is a risk that, despite progress, Disney's actions may not be sufficient to deliver zero waste-to-landfill goals by

2030. The following sections examine the waste and single-use plastic practices observed within wholly owned and operated Disney theme parks.

6. ADDRESSING PLASTICS IN DISNEY'S THEME PARKS

While Disney's CSR messaging is prominent on its corporate website, it is absent from its guest website, where reservations for theme parks and resorts are made. This might be because most behavioural changes are focused behind the scenes, away from the guests' gaze. Disney's CSR reports describe a wide range of activities conducted across its operations to reduce waste at the source or improve the reuse and recycling of products–actions that would not be registered as part of the guest experience. However, not engaging millions of guests in a company's efforts towards more sustainable operations appears to be a lost opportunity. Even before entering the parks, Disney has the opportunity to educate its guests to increase awareness and contribute to behavioural changes to reduce single-use plastic and waste to landfills. The actions undertaken in theme parks are the most publicly accessible and apparent to guests.

In theme parks, environmental messaging is incorporated into the narrative of themed attractions (Baker, 2023) rather than providing explicit instructions to encourage guests to recycle or reduce single-use plastics. Hermanson (2005) argued that the environmental messaging did not reflect the practices in the parks and did little to change the guests' behaviour. Nearly two decades later, messaging still reflects a more ideal situation than reality, but there have been observable changes in the plastic practices of parks.

Between October 2022 and April 2024, the author visited theme parks and associated shopping districts at the Disneyland Paris Resort, France (Disneyland Park, Walt Disney Studios Park, Disney Village), Walt Disney World Resort in Orlando, Florida (Magic Kingdom, EPCOT, Disney's Hollywood Studios, Disney's Animal Kingdom, and Disney Springs), and Disneyland Resort, Anaheim, California (Disneyland, California Adventure Park, Downtown Disney). The authors observed that although the theme parks worked towards the same waste goals, the operationalisation varied considerably. While there are many areas for improvement, positive solutions were offered in each resort location that could be replicated across theme parks and elsewhere.

6.1 Food with a Side of Plastic

Reducing single-use plastics and other disposable foodware (plates, bowls, cups, straws, cutlery, and packaging) represents one of the most significant opportunities for Disney to meet its waste targets in theme parks. The food outlets in the parks typically provide either takeaway or table service, with dramatic differences in the foodware provided. Restaurants and cafés with table services are more expensive dining options and require booking. These outlets provide meals with reusable crockery, metal cutlery, and ceramic or glass cups. Takeaway food outlets, which are more prevalent, include dining areas and mobile vendors. The food prepared at these outlets was primarily served with disposable paper-based foodware and offered the potential to reduce waste significantly.

There were differences in the materials used for takeaway foodware across the theme parks and between the U.S. and French parks. The Paris parks appeared to have replaced most traditional plastic foodware items with alternative materials. The U.S. parks still used plastic cups for frozen and alcoholic drinks and some desserts. The cup lids and cutlery in U.S. parks were also plastic, except for the Animal Kingdom in Florida, where these items were marked as biodegradable. The biodegradable materials appeared like their plastic counterparts but required specific degradation conditions. Biodegradable materials provide minimal environmental benefits over plastics if not sorted and treated appropriately. The transition from single-use plastic to an alternative material addresses Disney's goal of reducing single-use plastics in theme parks but does not address the issue of disposability.

The quantity of non-recyclable paper composite material used for foodware suggests that Disney aims to achieve its zero-waste-to-landfill target through incineration with energy generation. UNEP (2021) advocates thermal waste-to-energy as a viable option when waste management and incineration standards are high, but policy support for recycling is low. Incineration with energy generation is accepted as a solution in the European Union, where it is energy efficient and accepted as a form of recycling in Florida but not California. Although incineration reduces landfills and provides an alternative energy source, it can also result in high levels of chemical residues. A complete life cycle analysis would reveal the resulting environmental benefits of this option and its alternatives. One alternative approach is to provide reusable and/or refillable foodware to reduce the quantity of non-recyclable disposable materials contributing to landfills.

Takeaway outlets with dining areas are the most suitable for collecting reusable foodware. Reusable foodware was available in only one takeaway food outlet in the Disneyland Paris and Walt Disney World resorts (Florida) but at several locations within the Disneyland Resort (California). At the one outlet in Walt Disney World, in EPCOT, the reusable plates and bowls appeared to have been made using virgin plastic, and the cutlery was made of metal. The range of reusable foodware items offered in Disneyland Resort varied depending on the themed area. In Disneyland Paris, at the Stark Factory food outlet, meals were served on what appeared to be recycled plastic foodware. Designing foodware from recycled plastic increases the demand for recycled plastic and reduces the potential for macroplastic pollution from plastic waste.

Of the outlets offering reusable foodware options, Disneyland Paris was the only resort providing reusable hot drink cups. Hot drinks at the Stark Factory were obtained from a self-service machine using a token at the register. This approach could also be used for cold drinks, but at the time of the author's visit, the cold drink machine in this outlet was covered, precluding use, and the cold drinks were provided in plastic bottles or disposable cups. In U.S. parks, self-service drink fountains and hot drink machines were more common than in Paris, but they were always accompanied by disposable cups, even in the outlets with reusable foodware. If reusable cups for hot and cold drinks were applied across all parks, landfill waste would be significantly reduced.

All theme parks sell reusable cups and bottles in merchandise stores and Starbucks Coffee Company (Starbucks) outlets, but they cannot typically be filled at food outlets in the park. The author regularly asked the cast at Starbucks and Joffrey's Coffee and Tea Company outlets if they could fill her reusable cup. On every occasion, the author was advised that they had to use the designated disposable cups. Despite reports, Starbucks has expanded how customers can have their reusable cups filled in stores across the United States and Canada (Reiner, 2024; Valinsky, 2024). Many people have brought refillable water bottles into parks. These could be refilled at the bubblers around the park but not at soft drink fountains. Drinking from the water bubbler did not appear to be a social norm in Florida, where the author observed more people buying a large single-use disposable plastic cup filled with ice-cold water from Starbucks or Joffrey's Coffee & Tea Company than using free water bubblers.

Starbucks is one of the major sponsors and suppliers of drinks and snacks in the Disney theme parks. Starbucks outlets also appeared to be one of the most significant sources of plastic waste because of their large plastic cups and lids. Valinsky (2024) reported that aiming to halve its carbon emissions and waste by 2030, Starbucks has redesigned its plastic cups to reduce the amount of plastic used by 20%, with the rollout of these cups to commence in select locations in 2024. This falls short of the California legislated 25% reduction target by 2032 and does nothing to reduce the quantity of plastic items that end up as waste. This is a concern because, according to Starbucks (2020), more packaging waste was mismanaged than recycled waste in 2018. Polypropylene (PP) plastic, the largest source of packaging waste from Starbucks and used in cold drink cups and lids, is widely recyclable. However, less than 10% of the plastic packaging was recycled, and no recycled content was used in the plastic packaging of Starbucks (Starbucks, 2020). Reducing plastic waste from Starbucks' outlets would assist Disney to reach its waste targets.

The concept of refillable containers is not absent in parks. Numerous food outlets offered event-specific or character-based foodware items that could be purchased as part of a food or drink package and refilled later at a reduced price. This included a diversity of character-based large plastic popcorn buckets and drink bottles and a flute for alcoholic beverages as part of the EPCOT food and wine festival in Florida. For these reusable items to be a successful alternative, there needs to be incentives for their use, given the higher price points and burden of carrying them around the park. Reusable popcorn buckets appeared popular because guests were rewarded with cheaper refills, and the containers generally had straps to make them easier to carry around the parks. The production cost and lifespan still need to be considered, but this addresses the immediate issue of disposability.

Pre-packaged food items such as bottled drinks, snacks, and sweet and savoury sides provide another opportunity to reduce single-use plastics. Single-serve sauce sachets were abundant in most meals at the Disneyland Paris outlet. Sachets, such as those used for sauce, are provided for convenience but contribute significantly to plastic pollution, given their small size and non-recyclability (Break Free from Plastic, 2024b). In U.S. parks, especially in Florida, sauces were offered for self-selection at condiment stations where the most popular sauces were provided in refillable containers, reducing this waste source. Self-service options reduce the inadvertent provision of unnecessary items and can significantly reduce the amount of single-use plastics and waste generated. Another transferable learning that can be applied across theme parks to reduce plastic waste.

6.2 Disposing of Plastic and Other Waste Materials

One does not need to look far to find a garbage bin in any Disney theme park. The bins are regularly placed throughout pedestrian areas, usually in pairs, with a general waste and recycling bin available. There were occasions, such as in lines for rides, where there was only a single bin, usually for general waste. In other instances, particularly at food outlets, there are often larger waste separation stations. Finding a bin was easy, but determining which packaging items were recyclable or recycled was more difficult. When the bins were emptied, the waste was thrown into the same large wheelie bin for removal from the back-of-house area. This gave the impression that everything ended up in the same place or that sorting would occur later. Neither scenario encourages guests to try to separate their waste. As the bins were lined with different coloured plastic bin liners, depending on the waste category, the author assumed this would help with sorting. However, in the Magic

Kingdom, where there is an automated vacuum system (AVAC) to whisk the rubbish away, there is only one compactor to use.

What happened to the waste in those bags after collection was not apparent to Disney guests and reflected a missed opportunity to encourage more environmentally responsible waste practices. In 2023, the percentage of waste diverted from landfills in wholly-owned parks, resorts, and cruise lines had not reached 60% (Figure 1). More will be needed to achieve the zero-waste-to-landfill target by 2030. As Luiten (2020) suggests, Disney's ability to meet its waste goals depends on how it engages and educates its cast and guests. Given the scale of its operations and the strength of its media platform and brand, Disney is well-positioned to educate and connect with consumers about sustainability issues (Luiten, 2020). Consistent and clear signage that caters to the diversity of Disney guests may support improved waste management practices. Currently, the signage of general waste bins in U.S. parks varies between no wording, waste please, and trash, often only in English. In Paris, most general waste bins had no signage to identify their purpose, except at larger sorting stations with images and words in French and English.

The recycling bins in all theme parks regularly used the international standard universal recycling symbol (a triangle of three chasing arrows) and monochromatic images of items that could be recycled. This included images of bottles, cans, and cups in U.S. parks, but whether this referred to plastic cups or paper composite ones was unclear. In addition, there were other instances when the recycling bins only contained English words for plastic, glass, paper, and cans printed near the opening. In Paris parks, the recycling bins had images and/or words in English and French for just cans and bottles. Across all theme parks, the recycling bin usually had a different opening than the general waste bin and did not require a flap to be pushed aside to discard the rubbish. The different openings are standard for recycling bins. However, without needing to touch the bin to deposit waste, the ease of access made these bins the preferred option for some guests, regardless of the item being disposed of.

Changing the appearance of bins could make determining which bin to use easier, as bin design has been shown to affect waste management practices (Geller et al., 1979). Ecovidrio, a non-profit organisation in Spain, recognises the value of novelty bins, having joined forces with Disneyland Paris to create Disney-themed glass recycling bins (Disney, 2022). Disney (2022) noted that these bins were placed in 17 Spanish cities during the tourist season to raise awareness of glass recycling. The same degree of novelty and awareness raising was not applied to the bins within the Disney theme parks. Within the parks, the appearance of the bins varied depending on the themed land in which they were located; however, in most cases, the general waste and recycling bins were the same size, shape, and colour. There is evidence that this may be about to change.

The bins at one outlet in the Animal Kingdom, Florida, part of an organic waste collection trial (Disney, 2023c), were of different sizes, shapes, and colours. They also had differently shaped openings, full-colour images, and wording in English and Spanish. Different shaped openings and colours were also present in some of the bins in the parks at Disneyland Resort in California. These changes could significantly change guests' waste management practices while at the parks and create opportunities to influence the practices of guests and businesses beyond park boundaries. Collecting organic waste presents an opportunity to reduce waste in landfills and greenhouse gas emissions. Many of the food outlets at the California parks had separate bins for organic and liquid wastes. The Animal Kingdom was the only Florida park offering organic waste separation in the guest areas. Food waste was collected from larger bin stations in Disneyland, Paris.

Larger sorting stations reflect a commitment to improved waste management practices, but design and operational applications must also align. At the Stark Factory in Disneyland Paris, there appeared to be a lack of communication between those designing the foodware and bins. The bin stations at this outlet had a sixth waste category for plates and cups, but those bins were all marked as 'Do not use', and reusable plastic foodware was stacked on top of the bins. Some food outlets in U.S. parks had trays for guests to return reusable items. In other locations, it was up to the cast to sort items. Larger sorting stations regularly required a cast member to be present to assist guests in identifying the bin to use. This could be alleviated with improved signage; however, providing reusable foodware is still more resource-intensive, with additional labour required for sorting and cleaning. The full life cycle and economic costs of reusable foodware are unknown. Nevertheless, these instances demonstrate that reusable foodware is a workable option for reducing waste in theme parks and could be applied more broadly with appropriate infrastructure.

6.3 Taking Plastic Home

Removing single-use plastic foodware and separating waste helps Disney address its goal of reducing the amount of single-use plastic and waste in landfills from theme parks. However, the plastic disposed of within the parks is only a portion of the plastic guests engage with at the theme parks. There is much plastic that leaves the Disney parks with guests as souvenirs, while other items leave of their own volition, such as balloons that escape their tethers. Most Disney merchandise contains plastic, from plastic bubble blowers to Mickey ears, toys, and t-shirts. These items are designed for greater durability than single-use plastic packaging. However, their production still contributes to the emission of greenhouse gases and plastic waste at the end of their useful life.

Through its material goals, Disney has committed to using plastic containing at least 30% recycled content or lower-impact materials and designing packaging for reuse, recycling, or composting (Disney, 2024). In 2020, Disney used recycled bottles collected from the Walt Disney World Resort to create recycled fibres, and one Disney apparel line was made using up to 50% recycled fibres (Disney, 2021). In 2021, Disney launched 100% recyclable, plastic-free packaging for its classic dolls (Armenia, 2021; Disney, 2022), but this has not been expanded to any other toy line. In 2022, Winnie-the-Pooh plush toys, partially filled with recycled polyester, were introduced, which was expanded to over 50 plush toys in 2023 (Disney, 2024). While Disney has taken steps to reduce the amount of virgin plastic used in its merchandise, there has not been a change in the quantity of plastic merchandise produced by Disney.

In U.S. parks, visitors are given a free Disney plastic bag after selecting souvenirs and taking them to the counter or self-check-out. Some outlets had more robust reusable plastic bags, but there was little interest because free bags were still available. The free plastic bags leave the park and end up in the waste stream at another location, simply shifting the problem elsewhere. The current California plastic bag legislation does not apply to plastic bags given out in Disney stores, and in Florida, there is no restriction on the use of plastic bags. In comparison, in line with the stricter requirements in France, the Disney theme parks in Paris only offered the option to purchase a reusable plastic bag or go without it; no free plastic bags were available. Disneyland Paris had gone even further in reusing plastic, selling high-priced 'upcycled' bags made from the tarp that covered the scaffolding when Sleeping Beauty castle was being refurbished.

The difference in the approach in the French and U.S. theme parks, all of which have the same waste targets, reflects the need for regulatory change to create systemic change; however, corporate leadership can also make a difference. Disney (2024, p.47) has stated that it is '… working to eliminate single-use merchandise bags across Disney-owned parks and resorts by 2025'. An action that is not a regulatory requirement. If Disney removes all plastic bags from its parks and resorts, it will incorporate shopping districts and may have flow-on effects on the plastic practices of the brands with stores in these locations. Corporations making fundamental changes to their practices can create the momentum needed for political and systemic changes.

7. CONCLUSION

The need to reduce single-use plastics and plastic waste is internationally recognised, but the governance needed to support implementation is often lacking. SDGs are an example of a voluntary international agreement to which many countries have agreed, yet progress has been minimal. Disney provides a case study for exploring the implementation of SDGs and waste targets by a multinational company, examples of which are largely missing from the literature. Disney sources and sells products worldwide, operating across multiple jurisdictions with very different regulatory provisions for plastics. Without uniform global requirements, there is no consistency in how plastic waste is addressed in the manufacture or disposal of products. In the absence of a binding global agreement, it is up to national and subnational jurisdictions to adopt regulations to reduce plastic waste. As demonstrated by the U.S. and France, the strength of national and subnational regulations varies significantly. Without strict regulatory provisions, individual companies decide whether to take action to reduce plastic waste.

Disney has committed to reducing single-use plastics and waste to landfills across its wholly owned and operated theme parks and resorts in the U.S. and France, often without a regulatory requirement to do so. However, the level of action taken to reduce plastic waste at each location varies. This suggests the lack of regulatory requirements and supporting infrastructure has impeded progress towards waste goals. In Paris, single-use plastic foodware for meals prepared in parks and plastic bags have been removed from theme parks. This reflects the strictness of French legislation compared to the U.S., where these items are still present in theme parks. Conversely, U.S. theme parks have reduced the waste from single-serve sauce sachets by providing self-service condiment stations. Reusable foodware is more available in California, but recycled materials and cups are used in Paris. The provision of bins for appropriate waste separation has been inconsistent across all theme parks, but there appears to be a move towards an improved waste collection system.

Being seen as environmentally responsible is good for business, but outcomes require action rather than just perception. Over the last ten years, Disney has presented a strong corporate commitment to reducing single-use plastic and waste to landfills. Further improvements to its waste practices will be needed if Disney is to achieve its goal of zero waste to landfill from wholly owned and operated parks, resorts, and cruise lines by 2030.

External corporate powers may also limit efforts to address single-use plastics and waste in Disney theme parks. This includes Disney's sponsors and suppliers, who have varying levels of influence on and from Disney. For example, if Disney considered removing all single-use plastics from the parks, it would need to collaborate with The Coca-Cola Company and Starbucks, as they supply most of the beverages. Each company has sustainability commitments that may or may not support Disney's goals.

Disney guests travel across jurisdictional boundaries, bringing their plastic norms into parks and taking their Disney experiences and products. Not all the plastic that enters and leaves the park boundaries is within Disney's control, but it can control the level of plastic waste it generates and what happens to the items disposed of within the park. Quantitative data on the plastics used and disposed of in Disney theme parks would enable a more thorough analysis of Disney's actions. Where plastic cannot be stopped at the source, reduce, reuse, and recycle options are required. The success of alternative materials, such as biodegradable options, depends on appropriate collection and disposal systems. Providing guests with reusable foodware and clear messaging on waste separation could help reduce waste sent to landfills. Technological innovations should focus on eliminating unnecessary waste. A complete life cycle analysis would help to identify the most sustainable options to reduce the prevalence of single-use plastics and waste within the Disney theme parks.

Positive lessons from each Disney theme park could be applied more broadly across the Disney theme parks and in other high-visitation areas. By reducing single-use plastic and waste in its theme parks, Disney can influence the behaviour of millions of guests, corporate partners, and the broader tourism and hospitality sector. In doing so, Disney can help *inspire people around the globe* to adopt environmentally responsible waste practices.

REFERENCES

Armenia, A. (2021). First look: Disney announces new plastic-free packaging for classic dolls. Disney Parks Blog. https://disneyparks.disney.go.com/blog/2021/04/first-look-disney-announces-new-plastic-free-packaging-for-classic-dolls/

Baker, C. (2023). The prevalence of storyworlds and thematic landscapes in global theme parks. *Annals of Tourism Research Empirical Insights*, 4(1). Advance online publication. DOI: 10.1016/j.annale.2022.100080

Bamberg, S., Rees, J. H., & Schulte, M. (2018). Environmental protection through societal change: What psychology knows about collective climate action – And what it needs to find out. In S. Clayton & C. Manning (Eds.), Psychology and climate change: Human perceptions, impacts, and responses (pp. 185-213). Elsevier Academic Press.

Beaumont, N. J., Aanesen, M., Austen, M. C., Börger, T., Clark, J. R., Cole, M., Hooper, T., Lindeque, P. K., Pascoe, C., & Wyles, K. J. (2019). Global ecological, social and economic impacts of marine plastic. *Marine Pollution Bulletin*, 142, 189–195.

Bennett, J. (2010). *Vibrant matter: A political ecology of things*. Duke University Press.

Bensaude Vincent, B. (2013). Plastics, materials and dreams of dematerialization. In Gabrys, J., Hawkins, G., & Michael, M. (Eds.), *Accumulations: The material politics of plastic* (pp. 17–29). Routledge.

Borg, K., Curtis, J., & Lindsay, J. (2020). Social norms and plastic avoidance: Testing the theory of normative social behaviour on environmental behaviour. *Journal of Consumer Behaviour*, 19(6), 594–607.

Bradford, A. (2012). The Brussels effect. *Northwestern University Law Review*, 107(1), 1–67.

Break Free from Plastic. (2024a). Branded 6: Holding the world's worst plastic polluters accountable since 2018.

Break Free from Plastic. (2024b). Branded: The sachet scourge in Asia: Exposing the top sachet polluting companies with brand audits.

Bryman, A. (1999). The Disneyization of society. *The Sociological Review*, 25–47.

Bryman, A. (2004). *The Disneyization of society*. SAGE Publications Ltd.

Budd, M., & Kirsch, M. H. (Eds.). (2005). *Rethinking Disney: Private control, public dimensions*. Wesleyan University Press.

California Legislative Information. (2022). Plastic pollution prevention and packaging producer responsibility act 2022. https://leginfo.legislature.ca.gov/faces/codes_displayexpandedbranch.xhtml?tocCode=PRC&division=30.&title=&part=3.&chapter=3.&article=

Chernev, A., & Blair, S. (2015). Doing well by doing good: The benevolent halo of corporate social responsibility. *The Journal of Consumer Research*, 41, 1412–1425. DOI: 10.1086/680089

Clément, T. (2018). 'They all trust Mickey Mouse': Showcasing American capitalism in Disney theme parks. InMedia [Online], 7(1). https://journals.openedition.org/inmedia/1021

Cliffe, C. (2020). Top 10 most recognisable brands in the world. https://www.ig.com/au/trading-strategies/what-are-the-top-10-most-recognisable-brands-in-the-world--200217

DeLanda, M. (2006). *A new philosophy of society: Assemblage theory and social complexity*. Continuum International Publishing Group Ltd.

Disney. (2017). 2016 Disney citizenship data table. https://impact.disney.com/resources/2016-csr-report/

Disney. (2018). Corporate social responsibility update 2017. https://impact.disney.com/resources/2017-csr-report/

Disney. (2019). Corporate social responsibility update 2018. https://impact.disney.com/resources/2018-csr-report/

Disney. (2020). 2019 corporate social responsibility update. https://impact.disney.com/app/uploads/2022/01/CSR2019Report.pdf

Disney. (2021). 2020 corporate social responsibility report. https://impact.disney.com/resources/2020-csr-report/

Disney. (2022). 2021 corporate social responsibility report. https://impact.disney.com/app/uploads/Current/2021-Disney-CSR-Report.pdf

Disney. (2023a). Fiscal year 2022 annual financial report. Author.

Disney. (2023b). 2030 environmental goals: Updated February 2024. https://impact.disney.com/resources/2030-environmental-goals-white-paper/

Disney. (2023c). 2022 corporate social responsibility report. https://impact.disney.com/resources/2022-csr-report/

Disney. (2024). 2023 sustainability and social impact report. https://impact.disney.com/resources/2023-ssi-report/

Disney. (n.d.). About the Walt Disney Company. https://thewaltdisneycompany.com/about/

European Parliament and the Council of the European Union. (2019). Directive (EU) 2019/904 of the European Parliament and of the Council of 5 June 2019 on reducing the impact of certain plastic products on the environment. Official Journal of the European Union. https://eur-lex.europa.eu/legal-content/EN/TXT/PDF/?uri=CELEX:32019L0904

Falck, O., & Heblich, S. (2007). Corporate social responsibility: Doing well by doing good. *Business Horizons*, 50, 247–254.

Florida Department of Environmental Protection. (2021). Florida and the 2020 75% recycling goal: Final report.

Freinkel, S. (2011). *Plastic: A toxic love story*. The Text Publishing Company.

Gabrys, J., Hawkins, G., Michael, M., & Taylor, F. (Eds.). (2013). *Accumulation: The material politics of plastic*. Routledge.

Geller, E. S., Brasted, W. S., & Mann, M. F. (1979). Waste receptacle designs as interventions for litter control. *Journal of Environmental Systems*, 9(2), 45–160. DOI: 10.2190/5P46-8H2N-41JR-C2EJ

Gottdiener, M. (1982). Disneyland: A utopian urban space. *Journal of Contemporary Ethnography*, 11(2), 139–162.

GTPI. (2021). *Addressing pollution from single-use plastic products: A life cycle approach: Key messages for tourism businesses*. Author.

Hawkins, G. (2009). The politics of bottled water: Assembling bottled water as brand, waste and oil. *Journal of Cultural Economics*, 2(1-2), 183–195.

Hawkins, G. (2018). The skin of commerce: Governing through plastic food packaging. *Journal of Cultural Economics*, 11(5), 386–403.

Hermanson, S. (2005). Truer than life: Disney's Animal Kingdom. In Budd, M., & Kirsch, M. H. (Eds.), *Rethinking Disney: Private control, public dimensions* (pp. 199–227). Wesleyan University Press.

Hernandez, V. (2024, April 24). Here's who is holding the UN plastics treaty hostage. Canada's National Observer. https://www.nationalobserver.com/2024/04/24/opinion/holding-un-plastics-treaty-hostage

Holcomb, J., Okumus, F., & Bilgihan, A. (2010). Corporate social responsibility: What are the top three Orlando theme parks reporting? *Worldwide Hospitality and Tourism Themes*, 2(3), 316–337. DOI: 10.1108/17554211011052230

Jones, K. (2021). The top 50 most valuable global brands. https://www.visualcapitalist.com/top-50-most-valuable-global-brands/

Keough, S. B. (2021). Masks and materiality in the era of COVID-19. *Geographical Review*, 11(4), 558–570. DOI: 10.1080/00167428.2021.1897813

Kerber, H., & Kramm, J. (2021). On- and offstage: Encountering entangled waste-tourism relations on the Vietnamese island of Phu Quoc. *The Geographical Journal*, 187, 98–109. DOI: 10.1111/geoj.12376

King, M. J. (1981). Disneyland and Walt Disney World: Traditional values in futuristic form. *Journal of Popular Culture*, 15(1), 116–140.

LEGO. (2024). 2023 sustainability progress report. https://www.lego.com/cdn/cs/sustainability/assets/blt032d7f4d14a4682d/The_LEGO_Group_Sustainability_Progress_Report_2023_Final.pdf

Liboiron, M. (2013). Plasticizers: A twenty-first century miasma. In Gabrys, J., Hawkins, G., & Michael, M. (Eds.), *Accumulations: The material politics of plastic* (pp. 134–149). Routledge.

Lipton, A. M. (2024). Boden lecture: Of chameleons and ESG. *Marquette Law Review*, 107(3), 597–631.

Luiten, S.-C. (2020). Corporate social responsibility and consumerism: Case examples from Hyatt, Scandic and Walt Disney. *Worldwide Hospitality and Tourism Themes*, 12(5), 547–562. DOI: 10.1108/WHATT-06-2020-0045

McMahan, C. (2023). Long term implications of a pandemic on strategic brand marketing and corporate social responsibility (CSR). *Quarterly Review of Business Disciplines*, 10(2), 98–117.

Meikle, J. L. (1995). *American plastic: A cultural history*. Rutgers University Press.

OECD. (2018a). *Improving plastics management: Trends, policy responses, and the role of international co-operation and trade, OECD Environment Policy Paper No. 12*. OECD Publishing., DOI: 10.1787/c5f7c448-

OECD. (2018b). *Improving markets for recycled plastics: Trends, prospects and policy responses*. OECD Publishing., DOI: 10.1787/9789264301016-

OECD. (2022). Plastic pollution is growing relentlessly as waste management and recycling fall short, says OECD. https://www.oecd.org/newsroom/plastic-pollution-is-growing-relentlessly-as-waste-management-and-recycling-fall-short.htm

Patrick, K. M. (2019). *How to Save the World*. Hello World Labs.

Qiang, M., Shen, M., & Xie, H. (2020). Loss of tourism revenue induced by coastal environmental pollution: A length-of-stay perspective. *Journal of Sustainable Tourism*, 28(4), 550–567. DOI: 10.1080/09669582.2019.1684931

Reiner, T. (2024). Starbucks' reusable cup initiative: Moving towards 50% waste reduction. LinkedIn Newsletter: Game Changer Trends in Packaging, 22 February 2024. https://www.linkedin.com/pulse/starbucks-reusable-cup-initiative-moving-towards-50-waste-reiner-w4r3f/

The RepTrak Company. (2024). 2024 global RepTrack 100. https://ri.reptrak.com/hubfs/2024%20Global%20RepTrak%20100/The%202024%20Global%20RepTrak%20100%20Report.pdf

Sachs, J., Lafortune, G., Kroll, C., Fuller, G., & Woelm, F. (2022). *Sustainable development report 2022: From crisis to sustainable development: the SDGs as roadmap to 2030 and beyond*. Cambridge University Press.

Smith, O., & Brisman, A. (2021). Plastic waste and the environmental crisis industry. *Critical Criminology*, 29, 289–309. DOI: 10.1007/s10612-021-09562-4

Starbucks. (2020). 2018 environmental baseline report. https://stories.starbucks.com/uploads/2020/01/Starbucks-Environmental-Baseline-Report-2.pdf

Tabuchi, H. (2024). California tried to ban plastic grocery bags: It didn't work. The New York Times, 15 February 2024. https://www.nytimes.com/2024/02/15/climate/california-plastic-bagban.html

TEA/AECOM. (2020). *2020 theme index and museum index: The global attractions attendance report*. Themed Entertainment Association.

TEA/AECOM. (2022). *2022 theme index and museum index: The global attractions attendance report*. Themed Entertainment Association.

Townsend, T. G., Anshassi, M., Robey, N. M., Steffen, N., Gonzales, F., Ricketts, A., & Bielecki, J. (2021). *Update of the 2010 retail bags report*. Florida Department of Environmental Protection.

UN. (2017). Feature: UN's mission to keep plastics out of oceans and marine life. https://www.un.org/sustainabledevelopment/blog/2017/04/feature-uns-mission-to-keep-plastics-out-of-oceans-and-marine-life/

UN Department of Economic and Social Affairs. (2023). The 17 goals. https://sdgs.un.org/goals

UNEP. (2021). Our planet is drowning in plastic pollution – it's time for change. https://www.unep.org/interactive/beat-plastic-pollution/

UNEP. (2022). Historic day in the campaign to beat plastic pollution: Nations commit to develop a legally binding agreement. https://www.unep.org/news-and-stories/press-release/historic-day-campaign-beat-plastic-pollution-nations-commit-develop

Secretary-General, U. N. (2023). *Progress towards the sustainable development goals: Towards a rescue plan for people and planet, Advanced unedited version, Economic and Social Council*. Author.

Valinsky, J. (2024). Starbucks is giving its plastic cups a makeover. CNN. https://edition.cnn.com/2024/04/18/food/starbucks-new-plastic-cups/index.html

Walker, T. R. (2021). (Micro)plastics and the UN sustainable development goals. *Current Opinion in Green and Sustainable Chemistry*, 30. Advance online publication. DOI: 10.1016/j.cogsc.2021.100497

Wang, Q., Tweedy, A., & Wang, H. G. (2022). Reducing plastic waste through legislative interventions in the United States: Development, obstacles, potentials, and challenges. *Sustainable Horizons*, 2. Advance online publication. DOI: 10.1016/j.horiz.2022.100013

Wasko, J., Phillips, M., & Meehan, E. R. (Eds.). (2001). *Dazzled by Disney? The global Disney audiences project*. Leicester University Press.

Whitfield, J., Svobodova, P., & Webber, D. J. (2022). How can we reduce conference venues' use of single-use plastics? *Event Management*, 26(7), 1577–1590.

WWF. (2023). Who pays for plastic pollution? https://www.worldwildlife.org/publications/who-pays-for-plastic-pollution

4. Ocean. (2020). What are microplastics? And why you should care. https://www.4ocean.com/blogs/notebook/what-are-microplastics-and-why-you-should-care

Chapter 15
Constructed Wetlands as a Greywater Management Alternative in the Tourism Industry

Reena Esther Rani. J
https://orcid.org/0000-0002-3871-0054
Pondicherry University, India

Anu Chandran
https://orcid.org/0009-0000-7468-5274
Pondicherry University, India

ABSTRACT

The exponential increase in tourism escalates greywater production from laundry, gardening, bathing, and sewage treatment, leading to water scarcity. Simple physical and biological processes can purify greywater for reuse, leading to water conservation. Constructed wetlands offer new approaches to dealing with greywater, especially in polluted tourist areas, through plant, soil and microbial activity filtration processes. These systems are feasible for tropical regions with limited water resources. This chapter examines the suitability of constructed wetlands for greywater treatment, highlighting techniques, performance evaluations and environmental benefits. By integrating these systems, tourism facilities can reduce their ecological footprints with significant savings on water consumption. Constructed wetlands aid in improving greywater management and promoting sustainable water use because they have numerous advantages, such as low maintenance costs and adaptability, which are crucial in ecological preservation amidst rising tourism numbers.

DOI: 10.4018/979-8-3693-6110-8.ch015

1. INTRODUCTION

Freshwater is vital for sustaining life, ecosystems, and human activities. It is essential for drinking water, agriculture, commerce, and recreation. Among these, tourism is an important sector that relies heavily on freshwater resources (Goncharuk & Goncharuk. 2019). Recreational tourism, especially around beaches, rivers, and lakes, requires much clean water for many purposes, including shelters, land for recreational activities, and maintenance of grounds and gardens. The rapid expansion of international tourism is placing increasing pressure on freshwater resources, often leading to significant ecological impacts. Sustainable water management is essential because the industry uses significant quantities of fresh water (Jones et al., 2005) due to many activities such as tourism, skiing and golf, spas, swimming pools, fishing, landscaping, and other infrastructure (Gössling et al., 2012). This trend is seen as a significant ecological problem. Moreover, many types of tourism also involve water-based, counting landscapes with snow winter, farming, or nature travel.

Backwater tourism forms the foundation of the tourism industry in Kerala (Safoora & Devadas, 2014). These interconnected channels serve as a vital rural tourism resource in many regions, attracting numerous tourists and representing one of the state's distinctive products. Alleppey and Kumarakom are the primary centres of this industry (Rajan et al., 2011). Houseboats on still waters use water for various purposes, including drinking, cooking, cleaning, hygiene, and maintenance. In the hotel sector, water consumption is categorised into aquatic use, hydrology, water usage, and best practices (Antonova et al., 2021). Several factors influence water use in restaurants. Some research identifies the tangible attributes of hotels as determinants of drinking water usage (Gabarda et al., 2017; Gössling et al., 2012). Other factors include environmental certification and the use of water-saving products. Studies have also examined the function and composition of water prices (Antonova et al., 2021). Pereira et al. (2021) focus on a luxury five-star hotel in Portugal, detailing water-saving measures and strategies for differentiating itself from competitors.

The Mediterranean region is the most visited tourism destination globally, attracting 300 million tourists annually (Lanquar, 2011). It is described as a location where tourism induces a crisis in water supply, competing with local users for resources (Kent et al., 2002; Tortella & Tirado, 2011; Rico et al., 2009). Consequently, the Mediterranean has become the most extensively studied area regarding water consumption within the tourism sector. However, similar stresses on local water supplies have been observed in Indonesia, particularly in Bali, where the tourism industry reportedly consumes 65% of the area's water, leading to noticeable tensions between the hotel industry and neighbouring communities.

Previous research has indicated that the accumulation of tourists over time and space and certain types of water use exert significant pressure on local water supplies (Toth et al., 2018). Businesses in the tourism sector require substantial amounts of fresh water not only for essential needs such as maintaining restrooms, refilling swimming pools, and preserving green spaces but also for food preparation and oil production (Gössling et al., 2012). It has been observed that tourists consume approximately three times more water than residents. This is particularly problematic in regions with water scarcity and high tourist numbers, where water demand severely threatens sustainable development (Garcia & Servera, 2003; Tortella & Tirado, 2011).

Additionally, significant greenhouse gas emissions from fossil fuel sources hinder sustainable tourism progress. The tourism industry relies heavily on energy sources such as coal, oil, and gas for transportation, services, and various tourism activities (Lin, 2010; Gössling et al., 2002; Gössling et al., 2005). Studies indicate that transportation modes such as buses, private cars, and aircraft are major energy consumers within the tourism sector (Kuo & Chen, 2009). Furthermore, research by Yoon et al. (2022) established a positive relationship between water use and the uninterrupted electricity supply for recreational activities in restaurants and tourist attractions. However, excessive energy consumption also contributes to increased greenhouse gas emissions.

Thus, examining tourism development from the water and energy constraints perspective is essential for implementing sustainable and low-carbon tourism practices. Although studies have revealed water footprints for key tourism elements—lodging, food, travel, transport, entertainment, and retail—there has been limited focus on the comprehensive measurement of energy and water use. Researchers have developed several methods to measure water consumption in tourism, including assessments of resources, transportation, and energy use through direct and indirect sources, highlighting the complexity and importance of water use in recreational activities. For example, Huang et al. (2015) attempted to provide a virtual water force to quantify the total water footprint of tourism. He and Wang (2021) developed an energy-water footprint calculation based on the tourism lifecycle theory, estimating the water footprint of Xinjiang based on daily energy consumption. Other researchers have measured the energy-water footprint from primary and secondary water sources. Research by Li and Wang (2015) and Li (2018) developed an energy-water footprint that includes energy used through resources and transportation.

This chapter explores the potential of constructed wetlands as an ecologically sustainable solution for greywater processing in tourism areas. Tourists often increase water consumption, which can exert significant pressure on local water supplies, particularly in regions already experiencing water shortages. The heightened demand can lead to competition between tourists and local communities for limited water resources, resulting in stress and conflict. Additionally, the mismanagement of wastewater from tourism, including greywater from showers, sinks, and toilets, poses further environmental challenges. If not properly treated, this wastewater can contaminate local water supplies, pollute the water table, and diminish the availability of clean water for other uses.

Thus, freshwater management and tourism intersection necessitate sustainable practices to ensure the long-term use of water resources. Addressing these issues requires the tourism industry to explore innovative water conservation and wastewater treatment methods. By implementing sustainable water management practices, such as using constructed wetlands for greywater purification (Chapagain & Hoekstra, 2008), tourism destinations can mitigate their environmental impact and help preserve freshwater resources. The integration of these practices not only maintains the health of local ecosystems but also enhances the travel experience by preserving the natural beauty and integrity of the area.

This chapter examines the effectiveness and sustainability of constructed wetlands in greywater treatment, proposing practical solutions that promote water conservation while improving environmental management in the tourism industry. It also provides a detailed analysis of wetland use to offer insights into sustainable greywater management strategies for tourism destinations.

2. METHODOLOGY

A thorough methodology was adopted to systematically select papers for this study on water scarcity and the roles of greywater management and wetland preservation. The research scope and objectives were initially defined precisely, focusing on the effectiveness and sustainability of constructed wetlands in greywater management, especially within the tourism sector. Key terms such as "water scarcity," "greywater management," "constructed wetlands," and "sustainable tourism" were identified to guide the search. A range of academic databases, including PubMed, Google Scholar, JSTOR, Scopus, and Web of Science, as well as grey literature sources like government reports and NGO publications, were utilised to gather relevant papers. A structured search strategy using Boolean operators combined the keywords to ensure comprehensive topic coverage. Inclusion criteria focused on papers published within the last 20 years, peer-reviewed articles, and studies relevant to the research objectives, while exclusion criteria filtered out non-English papers and those not directly related to greywater management or constructed wetlands. The screening process began with a review of titles and abstracts to determine relevance, followed by a full-text review of selected papers to ensure depth and applicability. Quality assessment criteria included the clarity of objectives, methodological robustness, relevance of findings, and source credibility. Data extraction focused on critical points such as study objectives, methodologies, results, conclusions, and case studies. The extracted data were thematically organised to align with the research objectives. This rigorous methodology ensured that the selected papers provided a credible and comprehensive foundation for the study's analysis and conclusions.

3. GREYWATER

Greywater is defined as wastewater that excludes sewage. It includes effluent from domestic activities such as laundry, bathing, and dishwashing but excludes toilet water. The composition of greywater varies depending on lifestyle, household fixtures, and weather conditions. Household wastewater is categorised into blackwater and greywater, with blackwater referring to toilet water. Greywater sources include showers, bathtubs, kitchen and bathroom sinks, washing machines, dishwashers, and utility sinks. Bathroom greywater typically contains toothpaste, shampoos, soaps, other personal care products, and traces of bodily excretions. Laundry greywater contains chemicals from detergents, while kitchen greywater may contain food particles, fats, oils, and washing detergents(Casanova et al., 2001).

Despite containing microbes from human contact, the microbial concentrations in greywater are significantly lower than those in blackwater, making greywater a safe option for recycling and reuse. Although generally less contaminated than other types of wastewater, greywater can still contain suspended solids, pathogenic microbes, and other household particles (Antoine & Stefan, 2006). Despite being less polluted, the substantial volume of greywater contributes significantly to the overall effluence of household wastewater in drainage systems. Greywater accounts for approximately 50% to 80% of household water use, 30% to 50% of the organic load, and 9% to 20% of the essential minerals in household sewage (Friedler & Hadari, 2006). In Indonesia, domestic wastewater contributes approximately 60% of the pollution in natural water systems (Qomariyah et al., 2018).

Increasing concerns about water shortage, utilisation of resources, and obsolete or unusable physical frameworks in municipal water schemes have led to increased investigation into alternative water reuse strategies. The abundance of greywater that carries organic matter and nutrients, low levels of faecal matter, and fewer bacteria have received much attention, and greywater is a candidate for water reuse. It is important to remember that greywater is present in human activities and that equipment can meet the needs with an appropriate buffering capacity (Li et al., 2009). Numerous researchers have extensively examined the characteristics of greywater concerning various factors such as fixture types, lifestyle patterns, and settlement configurations (Vigneswaran & Sundaravadivel. 2004; Alsulaili & Hamoda, 2015; do Couto et al., 2013).

Water management is intrinsically linked to public well-being, food security, ecosystem services, education, human rights, and the improvement of aquatic supply systems, thereby making it a critical developmental objective pursued by many nations (Spencer, 2021). Reusing greywater is essential as it reduces the need for long-distance water transportation, leading to significant savings and enhancing the protection and recycling of local water resources (Rabaey et al., 2020). Simple physical and biological processes can purify greywater and other wastewater for subsequent use in irrigation or ground cleaning, thereby promoting water conservation through recycling practices (Su, 2023). Consequently, high-volume water can be considered for recycling and potential use with low-pressure systems.

The practice of reusing greywater, which has been used for quite some time, continues today, especially in regions facing water scarcity. By prioritising greywater reuse, we can significantly reduce our dependence on freshwater reservoirs and decrease pollution from the unprocessed discharge of greywater into natural water sources. This pollution reduction is a crucial environmental benefit of greywater reuse. Additionally, greywater reuse provides a valuable supplementary water source, particularly in regions experiencing severe water shortages or in arid climates. Repurposed greywater can be applied to various water-intensive tasks, including potable and non-potable uses such as flushing toilets and agricultural irrigation. Despite its potential, challenges persist, notably regarding public health perceptions and the appropriateness of technologies for implementing greywater reuse systems.

Greywater reuse, when combined with natural solutions, can serve as a transformative approach, turning a significant portion of wastewater from waste into a valuable water resource. Natural solutions, inspired by nature and sustained by it, are a source of inspiration and optimism. They efficiently utilise resources to address various societal challenges in a resource-efficient and adaptable manner, delivering economic, social, and environmental benefits. Their diverse applications, including constructed wetlands, green walls, green roofs, and urban green spaces, offer a promising path towards sustainable water management.

The potential of greywater reuse to conserve drinking water is significant. International projections indicate that by 2025, 2.7 billion people will face water scarcity issues (Godfrey et al., 2009). Greywater reuse emerges as a pivotal solution, potentially conserving between 29% and 47% of drinking water consumption (Humeau et al., 2011). Moreover, greywater management and recycling reduce municipal water consumption and affect the volume and quality of household wastewater released into sewage treatment facilities.

Greywater management is crucial in tourism, where water conservation and sustainability are paramount. Effective greywater management practices, such as treatment and reuse systems, hotels, resorts, and other tourism facilities, can significantly reduce freshwater consumption. This alleviates pressure on local water resources and demonstrates a commitment to environmental stewardship, enhancing the reputation and appeal of tourism destinations. Constructed wetlands have been proposed as a cost-effective and efficient method to treat greywater for reuse. These systems utilise biological, physical, and chemical processes to remove contaminants from greywater. Water treatment through constructed wetlands improves water sustainability, reduces the risk of pollution, and helps protect the ecosystem.

4. TYPES OF WETLANDS FOR GREYWATER TREATMENT IN TOURISM

Wetlands are crucial ecosystems characterised by their unique hydrology, soil, and plant life. They play a vital role in environmental health and are often known as the "kidneys of the Earth" because of their capability to filter and purify water (Zhang et al., 2010; Barducci et al., 2009). They are vital ecosystems that contribute significantly to enhancing water quality, protecting shores, refilling groundwater, mitigating the severity of floods and droughts, and offering distinct environments for a broad array of species of flora and fauna. Climate, topography, and hydrology are reasons for evolving natural marshes.

There has been an increasing focus among researchers and government officials on conserving and managing wetlands to uphold the biodiversity of aquatic systems. Recognising the multifaceted benefits of wetlands, efforts have been intensified to protect and restore these vital habitats for the well-being of the environment and society. Two types of wetlands exist - natural wetlands and man-made wetlands. The concept of wetlands encompasses a wide range of ecosystems, with coastal marshes, peatlands, estuarine wetlands, mangrove forests, marshlands, and various additional wetland varieties. Seaside wetlands serve as vital hubs for nutrient reusing, which is crucial in maintaining high water quality levels (Schmidt et al., 2003). Peatlands serve a critical function in the carbon flux by sequestering CO_2, a significant greenhouse gas, out of the atmosphere as peat. However, they also emit significant amounts of carbon dioxide and methane, the second most significant contributor to the greenhouse effect (Gorham, 1991). Mangrove ecosystems serve as essential environments for numerous fish and shellfish species while also playing pivotal functions in safeguarding sea areas and maintaining water quality (Green et al., 1998). Remote detection technology has been extensively employed in the study of wetlands over the past five decades: mapping land cover changes in the areas of wetlands, the exchange of carbon and its impact on the warming of the climate in wetland ecosystems, the emission of carbon due to peatland fires, as well as Hydrological procedures in wetlands have been a focal point of study and analysis for previous years (Schmidt & Skidmore. 2003; Gorham, 1991; Van et al., 2006; Li et al., 2009; Vörösmarty & Moorey, 1991; Holden, 2005).

Constructed wetlands are human-made or swamps and remain purposefully engineered to leverage natural phenomena like filtration, sedimentation, and biological uptake. This enables them to eliminate pollutants and nutrients from wastewater efficiently. One of the primary factors in advocating for the utilisation of constructed wetlands is their comparatively lower expenses for construction, operation, and maintenance compared to conventional treatment systems. Various types of aquatic plants, such as duckweed, are involved in this process. Duckweeds exhibit very low biomass but boast a remarkably high growth rate, necessitating frequent harvesting, sometimes as often as every week or even twice a week. Their primary function is creating anaerobic conditions, as they can effectively remove nutrients from a shallow layer of water (1-2 cm). Water hyacinth requires frequent harvesting but offers more significant potential for nutrient removal and reducing suspended particulates. In contrast, cattails, reeds, and sedges only need to be harvested once a year. These plants can be utilised in various ways. However, challenges associated with harvesting and using water hyacinths have limited their application in constructed wetlands to pilot plants only. Consequently, most constructed wetlands now rely on reeds or cattails (Brix et al., 1998).

Constructed wetlands can be classified into surface flow-constructed wetlands (SCWs) and subsurface flow-constructed wetlands based on their hydraulic characteristics. SCWs are like wet areas but located deeper due to wastewater on the substrate. They act as a unit for wastewater to pass through the substrate surface, creating an anhydrous layer that increases oxygen consumption and promotes photodegradation. Many examples of SCW include free-floating macrophytes and submerged macrophytes, floating-leaved macrophytes, emergent macrophytes, emergent macrophyte floating mats, and woody uses (Vymazal, 2022). According to the flow of direction, subsurface flow-constructed wetlands can be divided into Horizontal Flow Constructed Wetland (HFCW) and Vertical Flow Constructed Wetland (VFCW).

In these subsurface flow methods, wastewater flows below the filter surface and through the filter media. VFCW can present a water-saturated or unsaturated matrix, a sinking or rising matrix, saturated water, or reciprocating or tidal flow. In HFCW and permanently saturated VFCW, anaerobic or anoxic conditions prevail, while unsaturated, saturated, and depleted VFCWS allow good ambient oxygenation and provide the aerobic majority. In one of the most common arrangements of vertical unsaturated flow systems, the application of wastewater occurs intermittently and evenly at the surface and enters the filter media. After the wastewater passes, the filter is constantly exposed to air. This is necessary for aerobic environmental management to support the oxidation process in nitrification and organic matter oxidation problems.

Research by Li et al. (2013) investigated hybrid systems combining SCW, HFCW, and unsaturated VFCW in different ways, taking advantage of the characteristics of each type of constructed wetlands to create a better combination, with the primary purpose being nitrogen and phosphorus removal, and disease outbreaks. As a result, there has been an increase in research into the combination of chemical weapons and drugs, either alone or in conjunction with other effective medical procedures. This problem arises because chemical weapons have the advantage of low construction and maintenance costs and are widely used by developed and developing countries.

Both natural and constructed wetlands foster varied ecosystems, offering habitats for various plant and animal species. They function as efficient filters, purging pollutants and nutrients from water and enhancing the water quality of their surroundings. They play a role in sequestering atmospheric carbon and contributing to climate change mitigation. Recreational opportunities, such as birdwatching, hiking, fishing, and boating, enhance the quality of life for nearby communities.

5. MECHANISMS OF GREYWATER FILTRATION IN WETLANDS

Constructed wetlands replicate diverse chemical, physical, and biological processes observed in typical environments to manage anthropogenic discharges (Bakhshoodeh et al., 2017). This system has been demonstrated to be an efficient, economical, and maintainable method for wastewater treatment across several fields (Panrare et al., 2015). Furthermore, owing to its benefits, such as low energy demands, straightforward operation, and minimal maintenance, the constructed wetland skill has been extensively embraced as an environmentally friendly solution for addressing ecological contamination and promoting production with reduced environmental impact Top of Form

Bottom of Form

practices (Corbella et al., 2016). There are physical, chemical, and biological ways to filtrate greywater. Mechanical processes involve adsorption and volatilisation, whereas oxidation, photodegradation, chemical degradation, and hydrolysis are chemical processes. The biological process involves biological degradation, plant absorption, nitrification, denitrification, microbial activity, etc. All these activities act together to eliminate the pollutants in constructed wetlands (Hijosa et al., 2016).

Different substrates have been shown to influence the performance of constructed wetlands significantly, and researchers (Hernandez-Crespo et al., 2017; Gikas et al., 2018a) have investigated various substrates for their effectiveness in removing pollutants. The efficiency of constructed wetlands in eliminating different contaminants is significantly affected by the types of substrates employed. As the target pollutants vary, designing and operating each component of constructed wetlands with more specificity and individual consideration becomes imperative. Therefore, understanding the features and benefits of different particles is crucial for constructing a wetlands model. Substrates are divided into groups based on their appearances and material sources, including traditional chemical products, mineral materials, biomass materials, industrial and municipal waste materials, adapted functional resources, and original materials.

5.1 Natural Inorganic Resources

Inorganic materials, such as limestone, sand, natural zeolite, soil, and gravel, are among the most typically used particles in constructed wetlands. As essential and traditional materials, sand and soil mainly care for plant growth within constructed wetlands (Gill et al., 2017). Likewise, gravel is extensively utilised here and is often employed in small-scale wetland reproduction strategies (Hernandez-Crespo et al., 2017; Gikas et al., 2018a). These are sourced from the natural surroundings, rendering them plentiful, readily accessible, and cost-effective. However, because of their relatively minor precise surface area and lower concentration of active ingredients, natural materials may have limited treatment capacity for highly concentrated contaminants.

5.2 Purpose Mechanisms of Particles in Constructed Wetlands

According to Kadlec and Wallace (2009), in constructed wetlands, the particles serve several vital functions such as:

- Filtration and interception: They effectively filter out larger particles and contaminants.
- Adsorption: Substrates can adsorb various contaminants.
- Electron donor: provides donor electron functionality for metabolic processes and denitrification.
- Carrier: Substrates act as carriers used for microorganisms involved in biodegradation.
- Physical support: They provide structural support for wetland plants to thrive.

5.3 Filtration

Particles with varying sizes establish a three-dimensional network within constructed wetlands, enhancing their filtration capabilities. Initially, larger particles and pollutants are captured and removed through mechanical filtration and interception mechanisms. Subsequently, smaller substances and colloidal matter pass by the substrates and undergo removal by additional processes. Removing particulate matter in suspension primarily occurs through substrate filtration and sedimentation (Batool & Saleh, 2020). Filtration entails removing particulate matter that remains after preceding processes. Infiltration systems, both physical and biological processes, eliminate solids. However, this review focuses solely on the physical removal of solids, as it is the predominant method in most greywater treatment schemes. As for filtration, media can take various forms, such as gravel, sand, fine mesh, and numerous additional substances.

A study examined the efficacy of a system of filtration for greywater management, employing a configuration comprising 2 cm thick layers of pebbles Positioned above drain apertures, succeeded by a 12 cm intermediate layer comprising plastic filter media, and capped off with a 4 cm thick layer of peat (Gross et al., 2007). The effectiveness of a filtration system utilising pine bark, activated charcoal, polyurethane foam, and sand as filter media in treating greywater was evaluated (Dalahmeh et al., 2012). An effective coarse filtration system was investigated, which constitutes slow sand filtration with hydraulic retention times of 8 and 24 hours, respectively (Finley et al., 2009). In an experiment with coconut shells, charcoal, coarse sawdust, sand, and bricks were utilised as filter materials to evaluate the efficiency of greywater treatment (Parjane & Sane, 2011). Al-Hamaiedeh and Bino employed a four-barrel filtration unit to examine greywater treatment (Al & Bino, 2010). Arranged in series, the first three barrels were filled with gravel of 2–3 cm diameter, while the last barrel was designated for collecting the treated effluent intended for irrigation. In Israel, a hybrid filtration system was reported, which involves a 130-μm mesh filtration, a tuff filter, and a sand filter, followed by electrolysis (Gross et al., 2008). Another study undertaken by Zuma et al. (2009) employed a mulch tower system for greywater treatment in South Africa. The system consisted of layers of mulch, coarse sand, and fine and coarse gravel within a 650-mm high plastic column with a diameter of 150 mm. A stainless-steel sieve mesh was installed at the top of the column to eliminate larger particles. Among these filtration systems, only the bark filters could meet the pH criteria for reuse. Additionally, the bark and charcoal filters could meet the biochemical oxygen demand regulatory standard for reuse. Moreover, bark, charcoal, and sand filters exhibited high removal rates of total phosphorus

5.4 Adsorption Function

Pollutant treatment in constructed wetlands involves processes such as sorption, complexation, precipitation, and ion exchange, with substrate adsorption being the most significant. These processes primarily remove heavy metals, ammonium, nitrogen, and phosphorus (Mohammed & Babatunde, 2017). The metal removal efficiency increases with the substrate's adsorbent capacity and contact area. Pollutant adsorption is a three-phase process: external diffusion, internal diffusion, and surface adsorption/desorption (Wang & Guo, 2020; Karimi et al., 2019). Phosphorus removal involves complex biotic and abiotic processes (Di Luca et al., 2017), including uptake by plants and microorganisms, which rapidly address inorganic and organic forms of phosphorus. Abiotic processes, such as co-precipitation, adsorption, and burial, retain 50-70% of phosphorus (Reddy et al., 1999). Phosphorus removal efficiency depends on the substrate, with factors such as saturation time, local availability, heavy metal content, and recyclability defining suitability (Vohla et al., 2011). The simultaneous adsorption, precipitation, and complexation of phosphorus in wetlands are primarily influenced by the presence of iron, aluminium, and calcium in substrates (Vohla et al., 2011; Park et al., 2017). The oxides of calcium, aluminium, and iron in zeolite and bauxite contribute significantly to their superior phosphorus removal effect, while magnesium-containing materials further increase efficiency (Di Luca et al., 2017; Lan et al., 2018). Additionally, adsorption is the primary mode of boron removal, influenced by the organic matter content in substrates (Türker et al., 2014). The adsorption capacity of different substrates varies for various contaminants.

5.5 Electron Donor Function

Organic matter is the conversion process's primary electron donor for heterotrophic denitrification (Tan et al., 2021). Conversely, autotrophic denitrification relies on inorganic matter such as iron, hydrogen, and sulphur as electron donors. Denitrifying bacteria are mainly heterotrophic bacteria, which need additional organic carbon sources to supply nutrients and electrons for metabolism and denitrification. Certain organic materials, such as rice straw, reed, and biochar, can be incorporated into substrates as supplementary materials. These additions offer additional carbon sources and electron donors for heterotrophic microbes (Zhao et al., 2022; Bhattacharyya et al., 2020), thereby enhancing nitrogen removal through two mechanisms: (i) providing an external carbon source to increase denitrification by microorganisms and (ii) facilitating specific adsorption effects for ammoniacal nitrogen.

5.6 Microorganisms' Carrier Function

Substrates offer physical support and carry plants and microorganisms within constructed wetlands. The presence of microorganisms and biofilm within substrates enhances the hydrophobic sorption properties of the substrates. The microbial community plays a crucial role in pollutant removal, with its composition and distribution significantly impacting wetland functions (Arroyo et al., 2013). The extensive root system and substrate carry the microbial community within constructed wetlands. The diversity of microbial species and the structure of the microbial community are heavily influenced by the types of substrates utilised. Microorganisms adhere to the substrate and rhizome surfaces under the influence of chemotaxis, subsequently forming biofilms through proliferation.

5.6 Plant Support Function

Plants, including submerged, emergent, and floating species, are vital in wetlands. They influence the decontamination effectiveness of constructed wetlands (CWs) by creating conducive conditions for pollutant removal processes. The supportive function of substrates is essential for the germination and survival of most submerged and emergent plants. The rhizosphere refers to the soil region directly influenced by root secretions, providing a more conducive habitat for microorganisms and biogeochemical transformations. Plant roots release oxygen into substrates, creating an aerobic environment that enhances microbial activities (Tian et al., 2017). Certain macrophytes, such as Phragmites australis, Cyperus papyrus, and Typha latifolia, are commonly planted in constructed wetlands to improve pollutant removal and enhance landscape aesthetics (Gikas et al., 2018a; Riggio et al., 2018). Their roots spread throughout the substrates, leading to the occurrence of the rhizosphere effect in this zone, which influences the local microenvironment. Roots transfer oxygen and exudates from the upper biomass to the microbial consortium on the rhizome surface.

5.7 Biomass

Biomass materials are a heterogeneous mixture of organic and inorganic matter sourced mainly from agricultural, wood, woody, animal, human, aquatic, household, industrial, commercial and construction waste These materials serve as electron donors and organic/inorganic carbon sources for biological processes. Also, biomass materials possess an abundant pore structure and favourable adsorption performance, making them suitable substrates for constructed wetlands (Belviso, 2018).

6. PERFORMANCE EVALUATION AND CASE STUDIES

Several studies have investigated various proposals designed to treat greywater from different sources (Quispe et al., 2022; Niwagaba et al., 2014; Chithra & Dandapani, 2017; Yaseen et al., 2019). The number of developed constructed wetlands in the Middle East is limited, but exact numbers are difficult to determine. While in some countries, this number may fall below 10, in others, there may be one to two dozen pilots or different types of chemical weapons. These wastewater treatment plants differ in terms of wastewater treatment and design, succeeding in the unique conditions of the Middle East.

Figure 1. a) France has constructed a two-level wetland in the Oman desert with a Vertical Flow (VF) bed and (b) a water field for wastewater treatment

Source: Stefanakis (2019)

A constructed wetland plant was designed and built to treat wastewater from a workers' camp in a desert oil field in Oman (Bauer Nimr LLC, Muscat). This wet well is designed to treat 120 cubic meters/day of wastewater produced by 600 PE. It should not be forgotten that dust storms occur in the region every year and cause dust/sand to accumulate on the surface of the rocks. Installing textile fencing is a simple and easy way to control this phenomenon (Figure 1a). The wastewater passes through the first stage of the wetland and is then transported to the pulse second stage using a siphon meter. Treated wastewater is reused from unirrigated groundwater to irrigate adjacent areas where native trees are planted (Figure 1b). In this way, purified water remains at the bottom of the soil, eliminating the need for disinfectant and the risk of people coming into contact with purified water.

Figure 2. Aerial view of French-designed wetlands with split-level VF beds in Al Hamra, Ras Al Khaimah, UAE

Source: Stefanakis (2019)

Another example, located at Al Hamra in Ras Al Khaimah, UAE (Ingenieurbüro Blumberg in Bovenden, Germany) (Figure 2), serves an area of 100 villas with 800 PE, equivalent to the access of 216 cubic meters per day (Blumberg Engineers, 2019). The recycled water is directly flown to irrigate green areas on site. The German University of Science and Technology (GUtech) in Oman and the Research Council of the Sultanate of Oman conducted a study and demonstration in 2011. The main aim is to identify suitable local ways to build solid and sustainable buildings. EcoHaus is a two-story house with an area of approximately 250 m2, the school's guest house. As part of the ecological approach, a constructed wetland system was installed in the venture (Figure 3). This is designed to capture all sewage water (black and grey) from the building and provide appropriate and sustainable treatment so that the sewage can be reused for garden or rooftop green irrigation (Al Sulaimi et al., 2017; Knebel, 2019).

Figure 3. At the EcoHaus project demonstration site at the German University of Technology in Oman, wastewater management included horizontal flow forming a septic tank and wetland

Source: Stefanakis (2019)

The research evaluated the effectiveness of using a multilayer test system (such as gravel, charcoal, and mulch) designed to treat domestic greywater used for garden irrigation. The system showed significant removal results, reaching approximately 90.8% chemical oxygen demand (COD) and approximately 96.1% biochemical oxygen demand (BOD5) after 36 hours. However, the effect of greywater underground irrigation on plant growth is affected by seasonal factors, especially the rainy season, and canal treatment affects the soil (Niwagaba et al., 2014). Another study evaluated the effectiveness of water filters containing natural materials such as sand, pea gravel, coconut shell waste, charcoal, Lipianodiflora, and canna. Various arrangements and sequences of these materials are used. Treated greywater reduces pollution by a factor of 1-3 compared to untreated greywater, demonstrating the success of biofilters in reducing the adverse effects of domestic wastewater on the environment and groundwater pollution (Chithra & Dandapani, 2017). Additional research evaluated different sources such as eucalyptus biochar, teff straw, and sand for treating household greywater at different rates.

This study demonstrates the feasibility of using low-cost materials to treat wastewater and proposes a cost-effective and adaptable model for various situations (Yaseen et al., 2019). International Islamic University Masjid in Malaysia explored the potential of repeating ablution water after sand filtration. Laboratory water tests showed that ablution water was not very toxic even when discharged directly, with Chemical Oxygen Demand (COD) showing only a slight increase at 31 mg/L. Research shows that once sand is found, this water can be used efficiently and reused for cleaning and beautification (Al et al., 2014). A sustainable filtration system found by another group of researchers (Hassan et al., 2024) enhances the water quality in mosques by reducing the turbidity and total dissolved solids in greywater. Furthermore, the system uses durable and environmentally friendly filter material based on waste management and environmental protection. These findings significantly impact sustainable water management in religious buildings such as masjids and have broader probable applications.

Numerous studies have reported the efficiency of the constructed wetlands. It is noteworthy to recall the treatment of wetlands with low water content during the dry season (Zhang et al.,2020). The nitrogen removal efficiency of water in dry and cold soils was examined (Zhao et al., 2020). An investigation report regarding the evolution of constructed wetlands in dry and cold seasons for removing nitrogen from greywater (Zhang et al., 2021). Evaluation of tropical wetlands based on seasonal variations, pollutant load, and the types of plants present were studied effectively (Bojcevska & Tonderski, 2007). Zhao and his colleagues studied the effect of nitrogen removal on wetlands during the cold season (Zhao et al., 2018). The effect of temperature on eliminating cadmium from wastewater using wetlands was published by (Ma et al., 2017). It can be concluded from the data that the influence of weather conditions on the performance of wetlands in removing chemical compounds has not yet been investigated.

Alsubih et al. (2021) established an aerobic fluidised bed bioreactor to evaluate its behaviour. On the other hand, Khan et al. (2020) compared the treatment efficiency of membrane bioreactors and sequencing batch bioreactors for hospital wastewater. The researchers utilised horizontal flow-constructed wetlands to treat hospital wastewater. Hospital wastewater treatment using horizontal flow to create wetlands. However, these studies mainly focus on removing pollutants from hospital wastewater, especially nutrients like nitrates, phosphates, and organic matter. However, these studies mainly focus on removing pollutants from hospital wastewater, especially removing nutrients and organic matter.

The greywater treatment plants' cost-effectiveness and maintenance have already been studied. The cost of small-scale wetland development on the Great Barrier Reef is regularly reported to be 650,000 to 820,000 hectares1 (2020/21 financial year), including site planning, construction, and management costs (Australia 2016). The second study shows the preliminary cost of building farm-scale lagoons (0.02–2.5 ha) in the Wet Tropics, with an average structure cost of 34,117 ha-1 (2020/21 fiscal year) (Rust & Star,2018).

In terms of cost-effectiveness, a reference point of 110 kg-1 Dissolved Inorganic Nitrogen (DIN) (AUD 150 kg-1 DIN) at the end of the basin is suggested for the Great Barrier Reef (Rolfe & Windle, 2016). For example, we assume the water transport factor from wetland treatment to the Great Barrier Reef lagoon is 0.8. In this case, the economic efficiency of four constructed wetlands would be less than DIN 46.1, while the cost-effectiveness of six wetlands at the end of the basin would be less than DIN 92.1 kg. Both are well below the recommended end-of-watershed criteria. These results also include lower costs for bioreactors.

The optimisation method used here is consistent with other measurements of coral reef basins (Star et al., 2018). Another advantage of our accounting system is that it allows sponsors to adjust costs and working hours based on the situation elsewhere. This study concludes that productivity depends on nitrogen removal capacity rather than wetland area or cost per hectare. For example, the price per hectare of 10 hectares of one of the divided constructed wetlands considered for the study is higher. However, a third price ($29 - excluding 1 DIN), VD2, with only 0.5 hectares, has the lowest price per hectare but is the fourth most profitable. (except 36 kg - 1 DIN).

7. INTEGRATION WITH SUSTAINABLE TOURISM PRACTICES

In the 1970s, many European international organisations emerged, creating the next direction of sustainable development. The first United Nations conference on the human environment was held in Stockholm, Sweden 1972. Sustainable business development seeks to satisfy present needs while at the same time not compromising future generations' ability to meet these needs, while also considering economic, social and aesthetic considerations, preserving culture, ecology and biodiversity (UNWTO, 1949). In 2002, the UN World Summit prioritised poverty reduction and recognised the role of tourism in societal development. The UNEP-UNWTO guide introduces 12 actions that foster economic, social, and environmental sustainability (wealth creation plus human rights plus resource management) as outlined in the UNEP-UNWTO publication (UNEP-UNWTO, 2016). The Sustainable Tourism Initiative values wetlands' cultural importance (natural environments) and works hand in hand with local communities to value cultural heritage and observe traditional practices. Educational programs offered through guided tours and sign interpretation make visitors aware of the ecological importance of wetlands, their functioning in terms of wastewater treatment, purification, and flood control, and why these areas should be preserved.

8. SOCIO-CULTURAL BARRIERS

Builders and real estate groups would have initiated concrete plans regarding commercial usage of open spaces in tourism destinations, which were construed to be ideal havens for the inception of constructed wetlands (Angotti, 2011). The pronouncing ecological value of the conservation projects will be initially at loggerheads with the returns that could be generated from the commercial tourism projects. Side by side, the scales may favour highly capital-intensive tourism ventures even though they indicate adverse ecological impacts and violate land use policies. In the given situation, strong political will and good governance need to take centre stage in leveraging greywater harvesting and management as a social good irrespective of the status or level of tourism development; an overwhelming number of tourism centres worldwide have acute water shortages (Mahendru et al., 2023). Overtourism, which is becoming part and parcel of the new order of tourism post-COVID-19, results in overconsumption. Studies on the water consumption pattern by tourism properties point out discriminatory release and usage of the resource. Hence, the blight of water scarcity tentacles is causing pinching efforts in destinations (Gössling et al., 2012).

An example can be drawn from the landscaping of golf courses, lawns, and gardens of hospitality and tourism ventures. A large quantity of water must be released to meet such requirements. Considering constructed wetlands as social capital with a Scope for innovation, the wastewater is conserved efficiently and can be harvested for prospective uses (Humeau et al., 2011). Neotourism initiatives, such as agro-tourism and aqua-tourism, will hugely benefit from such conservation efforts projects. Further tourism projects that initiate greywater management will give exceptional impetus to awareness campaigns for water conservation.

The scepticism of the local community on reusing or recycling untreated grey water emanating from hospitality ventures, apart from the households owning potential health risks, acts as a deterrent in setting up constructed wetlands in many destinations (Urban Wastewater Scenario in India, n.d.). The perception of hygiene parameters varies according to the cultural dispositions of tourists and hosts. It is an unalloyed fact that tourists from some European countries, as well as North America, are highly cautious with exposure to water; the incidents reported on tourists suffering from skin diseases after they to wrists indulged in water sports activities and swimming- on the beaches lead to the formation of generalised viewpoints on the quality of water in tourism destinations (Hindley & Marmion, 2020.) The tendency of aversion towards secondary water usage, even if purified, also exists among host communities. Apart from that, an overriding ambiguity exists on issues about the sustainable management of greywater conservation in constructed wetlands. This is because several projects launched by the tourism departments confirming water conservation, like lake restoration, rainwater harvesting, etc., were nixed, citing workforce shortage and lack of funds, among Others. Eventually, the area turned out to be a dump yard. (mostly non-biodegradable) escalating the pollution levels alarmingly. The waterbodies, too, later became breeding areas for mosquitos and parasites. The resultant apprehension creates apathy among the host residents. Host communities, particularly of destinations promoting mass tourism, have started campaigns asking tourists to leave, mainly because they were concerned about sharing water with the visitors.

9. ECONOMIC FEASIBILITY OF PROJECTS AND INITIATIVES

In the long term, particularly in developing countries, identifying adequate fiscal options for constructing constructed wetlands has been recognised as a significant concern. However, integrating constructed wetlands projects with the Sustainable Development Goals (SDGs) allows them to be considered under the framework of environmental economics, thereby gaining access to funding sources within this domain. Concurrently, constructed wetlands should be developed as commercial, revenue-generating projects. The support from government agencies and multinational corporations may not be substantial. However, this can be mitigated by emphasising the quantity of water that can be recycled and reused, highlighting both conservation and its economic benefits. Consistent and continuous monitoring and effective maintenance systems are crucial for constructing wetlands' efficient operation and functioning. Mega plants must ensure that the ventures remain pollutant-free. Additionally, technically competent and skilled labour is essential, making job creation in this sector both necessary and value-adding. From a revenue-generating perspective, constructed wetlands can be promoted as tourist attractions, offering recycling mechanics and state-of-the-art conservation practices with significant educational value. Swimming pools in hotels, resorts, and clubs incur high maintenance costs, and water changes contribute to wastage as water is often channelled to drainage systems. However, if the outlets of swimming pools are connected to constructed wetlands, the economic viability of operating these facilities would be enhanced. Thus, a smaller water footprint would result in greater economic rewards.

10. CHALLENGES AND OPPORTUNITIES

The region's groundwater is being used faster than its return rate, leading to seawater intrusion and poor water quality. Subsurface water contamination from agriculture and industry can cause pollution in groundwater. The problem is exacerbated by sewage restrictions in many countries in the region and the release of untreated or partially treated sewage into agricultural waterways and elsewhere. While sewer systems are generally found in large cities, septic tanks and cesspools are used in rural or remote areas (Aleisa & Al-Zubari, 2017). For this reason, there is no wastewater treatment opportunity in rural and remote areas, or wastewater is not collected by trucks and transported to central treatment facilities. Due to various financial and economic issues, centralised healthcare is not an option in rural areas of small and medium-sized cities. In the case of Iran's urban areas, only 18% are covered with sewage; this rate rises to 45% in Saudi Arabia, which has an excellent geographical location. In some countries in the region, such as Oman, Sudan, and Palestine, it is often reported that wastewater disposal is very hazardous, possibly because the amount of wastewater collected and treated is less than the total produced. In many nations, the percentage of wastewater collected and treated does not exceed 1/3 of total production (Moghaddam et al., 2017). Overall, the region's wastewater treatment rate is a mere 40%, reflecting a common challenge. Merely 1.6 billion cubic meters are reused despite wastewater undergoing collection and treatment at secondary and tertiary levels, with less than 15% being recycled (Aleisa & Al-Zubari, 2017).

11. CONCLUSION

Poor and improper water treatment causes water pollution and damage to the ecosystem and seriously threatens public health. Of course, improperly treated wastewater is also unfit or unsuitable for reuse. It is acknowledged that water and sewage systems at tourist destinations still lack the necessary infrastructure. Although there are many initiatives to open up this industry in the private sector, most water and wastewater management services continue to be the responsibility of governments and organisations. Many factors affect the reuse of purified water, such as social barriers, energy and energy conservation, and domestic and political constraints. At the political level, there is often confusion about who is responsible for new measures, activities, and plans. Political structures frequently do not allow for the participation of other stakeholders, such as local communities, and multiple decision-making centres make the procedure even more difficult.

Lack of control over the combined water supply often results in inefficient and ineffective facilities. Moreover, although reuse guidelines and legal standards are generally more stringent than the relevant guidelines established by the World Health Organization (WHO), particularly concerning microbial infections, social and religious factors still create some opposition to wastewater reuse. In general, stringent emission standards have been adapted to the guidelines of developing countries. These standards are challenging and expensive to achieve and do not consider the importance of climate, cultural rights, and diversity.

Cost recovery in wastewater management lags behind that of water supply and water for drinking usage. The need to meet the high emission standards mentioned above has an economic and financial impact. It has become a part of many countries in the Middle East. This leads to poor operation and maintenance of wastewater treatment plants. Therefore, planning and financing wastewater treatment plants do not cover the entire life span of these plants in most cases. In general, authorities focus only on the investments that need to be made without considering a period of 15 to 20 years of maintenance and operating costs, especially in technology-using equipment. This leads to poor working conditions, overload, and frequent breakdowns, compounded by skilled workers and knowledgeable people not operating most services. Then, sometimes non-recyclable use occurs, reinforcing the perceived quality of treated water is no good and, therefore, must be given free to consumers.

However, despite many restrictions, the need for water and hygiene services continues within the required regions. Gradually but progressively, people are becoming aware of the restrictive policies aimed at protecting universities and freshwater resources. Although much remains to be completed, significant changes and reserves are strategic in the coming centuries to increase the efficiency of the entire wastewater system, such as the sewage system, behaviour technology, and increasing wastewater use. Local governments often seek out experienced companies and outside experts; Public and private partnerships are encouraged to solve these problems and share necessary information and resources. In this context, this also means the opportunity to introduce effective treatment methods. A permanent solution is needed, especially considering the nature of the problem in the region.

REFERENCES

Al-Hamaiedeh, H., & Bino, M. (2010). Effect of treated grey water reuse in irrigation on soil and plants. *Desalination*, 256(1-3), 115–119. DOI: 10.1016/j.desal.2010.02.004

Al Sulaimi, A., Barghash, H., & Stefanakis, A. I. (2017). A demo constructed wetland for domestic wastewater treatment at the EcoHaus of the German University of Technology in Oman. *Proceedings of the World Water Day Event, German University of Technology in Oman*, Halban, Muscat, Oman, 21.

Aleisa, E., & Al-Zubari, W. (2017). Wastewater reuse in the Gulf Cooperation Council (GCC) countries: The lost opportunity. *Environmental Monitoring and Assessment*, 189(1), 1–15. DOI: 10.1007/s10661-017-6269-8 PMID: 29022117

Allende, K. L., Fletcher, T. D., & Sun, G. (2012). The effect of substrate media on the removal of arsenic, boron and iron from acidic wastewater in planted column reactors. *Chemical Engineering Journal*, 179, 119–130. DOI: 10.1016/j.cej.2011.10.069

Alsubih, M., El Morabet, R., Khan, R. A., Khan, N. A., Khan, A. R., & Sharma, G. (2021). Performance evaluation of aerobic fluidized bed bioreactor coupled with tube-settler for hospital wastewater treatment. *Journal of Environmental Chemical Engineering*, 9(5), 105896. DOI: 10.1016/j.jece.2021.105896

Alsulaili, A. D., & Hamoda, M. F. (2015). Quantification and characterisation of greywater from schools. *Water Science and Technology*, 72(11), 1973–1980. DOI: 10.2166/wst.2015.408 PMID: 26606091

Angotti, T. (2011). *New York for sale: Community planning confronts global real estate*. MIT Press.

Antoine, M., & Stefan, D. (2006). Greywater management in low and middle-income countries. *Sandec Report No.14/6* (Switzerland: Eawag), 1-7.

Arroyo, P., Ansola, G., & de Miera, L. E. S. (2013). Effects of substrate, vegetation and flow on arsenic and zinc removal efficiency and microbial diversity in constructed wetlands. *Ecological Engineering*, 51, 95–103. DOI: 10.1016/j.ecoleng.2012.12.013

Bakhshoodeh, R., Alavi, N., Majlesi, M., & Paydary, P. (2017). Compost leachate treatment by a pilot-scale subsurface horizontal flow constructed wetland. *Ecological Engineering*, 105, 7–14. DOI: 10.1016/j.ecoleng.2017.04.058

Barducci, A., Guzzi, D., Marcoionni, P., & Pippi, I. (2009). Aerospace wetland monitoring by hyperspectral imaging sensors: A case study in the coastal zone of San Rossore Natural Park. *Journal of Environmental Management*, 90(7), 2278–2286. DOI: 10.1016/j.jenvman.2007.06.033 PMID: 18423842

Batool, A., & Saleh, T. A. (2020). Removal of toxic metals from wastewater in constructed wetlands as a green technology; Catalyst role of substrates and chelators. *Ecotoxicology and Environmental Safety*, 189, 109924. DOI: 10.1016/j.ecoenv.2019.109924 PMID: 31759740

Belviso, C. (2018). State-of-the-art applications of fly ash from coal and biomass: A focus on zeolite synthesis processes and issues. *Progress in Energy and Combustion Science*, 65, 109–135. DOI: 10.1016/j.pecs.2017.10.004

Bhattacharya, S., Roy, C., Mandal, S., Sarkar, J., Rameez, M. J., Mondal, N., Mapder, T., Chatterjee, S., Pyne, P., Alam, M., Haldar, P. K., Roy, R., Fernandes, S., Peketi, A., Chakraborty, R., Mazumdar, A., & Ghosh, W. (2020). Aerobic microbial communities in the sediments of a marine oxygen minimum zone. *FEMS Microbiology Letters*, 367(19), fnaa157. DOI: 10.1093/femsle/fnaa157 PMID: 32975580

Blumberg Engineers. (n.d.). Reed bed treatment systems (constructed wetlands) in the Middle East. References of Blumberg Engineers and its associates. Available online: www.blumberg-engineers.de (accessed on 16 December 2019).

Brix, H., Cooper, P. F., Green, M. B., & Haberl, R. (1998). Constructed wetlands for wastewater treatment in Europe.

Byrne, J., & Sipe, N. (n.d.). Green and open space planning for urban consolidation—A review of the literature and best practice.

Casanova, L. M., Gerba, C. P., & Karpiscak, M. (2001). Chemical and microbial characterization of household greywater. *Journal of Environmental Science and Health. Part A, Toxic/Hazardous Substances & Environmental Engineering*, 36(4), 395–401. DOI: 10.1081/ESE-100103471 PMID: 11413826

Chapagain, A. K., & Hoekstra, A. Y. (2008). The global component of freshwater demand and supply: An assessment of virtual water flows between nations as a result of trade in agricultural and industrial products. *Water International*, 33(1), 19–32. DOI: 10.1080/02508060801927812

Corbella, C., García, J., & Puigagut, J. (2016). Microbial fuel cells for clogging assessment in constructed wetlands. *The Science of the Total Environment*, 569, 1060–1063. DOI: 10.1016/j.scitotenv.2016.06.163 PMID: 27392579

Dalahmeh, S. S., Pell, M., Vinnerås, B., Hylander, L. D., Öborn, I., & Jönsson, H. (2012). Efficiency of bark, activated charcoal, foam, and sand filters in reducing pollutants from greywater. *Water, Air, and Soil Pollution*, 223(7), 3657–3671. DOI: 10.1007/s11270-012-1139-z

Di Luca, G. A., Maine, M. A., Mufarrege, M. M., Hadad, H. R., Pedro, M. D. C., Sánchez, G. C., & Caffaratti, S. E. (2017). Phosphorus distribution pattern in sediments of natural and constructed wetlands. *Ecological Engineering*, 108, 227–233. DOI: 10.1016/j.ecoleng.2017.08.038

do Couto, E. D. A., Calijuri, M. L., Assemany, P. P., da Fonseca Santiago, A., & de Castro Carvalho, I. (2013). Greywater production in airports: Qualitative and quantitative assessment. *Resources, Conservation and Recycling*, 77, 44–51. DOI: 10.1016/j.resconrec.2013.05.004

Finley, S., Barrington, S., & Lyew, D. (2009). Reuse of domestic greywater for the irrigation of food crops. *Water, Air, and Soil Pollution*, 199(1-4), 235–245. DOI: 10.1007/s11270-008-9874-x

Friedler, E., & Hadari, M. (2006). Economic feasibility of on-site greywater reuse in multi-storey buildings. *Desalination*, 190(1-3), 221–234. DOI: 10.1016/j.desal.2005.10.007

Garcia, C., & Servera, J. (2003). Impacts of tourism development on water demand and beach degradation on the island of Mallorca (Spain). *Geografiska Annaler. Series A. Physical Geography*, 85(3-4), 287–300. DOI: 10.1111/j.0435-3676.2003.00206.x

Gikas, G. D., Pérez-Villanueva, M., Tsioras, M., Alexoudis, C., Pérez-Rojas, G., Masís-Mora, M., Lizano-Fallas, V., Rodríguez-Rodríguez, C. E., Vryzas, Z., & Tsihrintzis, V. A. (2018). Low-cost approaches for the removal of terbuthylazine from agricultural wastewater: Constructed wetlands and biopurification system. *Chemical Engineering Journal*, 335, 647–656. DOI: 10.1016/j.cej.2017.11.031

Gill, L. W., Ring, P., Casey, B., Higgins, N. M., & Johnston, P. M. (2017). Long-term heavy metal removal by a constructed wetland treating rainfall runoff from a motorway. *The Science of the Total Environment*, 601, 32–44. DOI: 10.1016/j.scitotenv.2017.05.182 PMID: 28549286

Godfrey, S., Labhasetwar, P., & Wate, S. (2009). Greywater reuse in residential schools in Madhya Pradesh, India—A case study of cost-benefit analysis. *Resources, Conservation and Recycling*, 53(5), 287–293. DOI: 10.1016/j.resconrec.2009.01.001

Goncharuk, V. V., & Goncharuk, D. K. (2019). Water is everywhere. It holds everything, even a key to understanding the Universe. DI Mendeleev's law is the prototype of the Universe constitution. *Journal of Water Chemistry and Technology*, 41(6), 341–346. DOI: 10.3103/S1063455X19060018

Gorham, E. (1991). Northern peatlands: Role in the carbon cycle and probable responses to climatic warming. *Ecological Applications*, 1(2), 182–195. DOI: 10.2307/1941811 PMID: 27755660

Gössling, S. (2002). Global environmental consequences of tourism. *Global Environmental Change*, 12(4), 283–302. DOI: 10.1016/S0959-3780(02)00044-4

Gössling, S., Aall, C., Ceron, J. P., Dubois, G., Hall, C. M., Lehmann, L. V., Peeters, P., & Scott, D. (2012). Tourism and water use: Supply, demand and security—An international review. *Tourism Management*, 33(1), 1–15. DOI: 10.1016/j.tourman.2011.03.015

Gössling, S., Peeters, P., Ceron, J. P., Dubois, G., Patterson, T., & Richardson, R. B. (2005). The eco-efficiency of tourism. *Ecological Economics*, 54(4), 417–434. DOI: 10.1016/j.ecolecon.2004.10.006

Gössling, S., Peeters, P., Hall, C. M., Ceron, J. P., Dubois, G., & Scott, D. (2012). Tourism and water use: Supply, demand, and security—An international review. *Tourism Management*, 33(1), 1–15. DOI: 10.1016/j.tourman.2011.03.015

Green, E. P., Clark, C. D., Mumby, P. J., Edwards, A. J., & Ellis, A. C. (1998). Remote sensing techniques for mangrove mapping. *International Journal of Remote Sensing*, 19(5), 935–956. DOI: 10.1080/014311698215801

Gross, A., Shmueli, O., Ronen, Z., & Raveh, E. (2007). Recycled vertical flow constructed wetland (RVFCW)—A novel method of recycling greywater for irrigation in small communities and households. *Chemosphere*, 66(5), 916–923. DOI: 10.1016/j.chemosphere.2006.06.006 PMID: 16844197

Gross, A., Wiel-Shafran, A., Bondarenko, N., & Ronen, Z. (2008). Reliability of small-scale greywater treatment systems and the impact of its effluent on soil properties. *The International Journal of Environmental Studies*, 65(1), 41–50. DOI: 10.1080/00207230701832762

Hanjra, M. A., & Qureshi, M. E. (2010). Global water crisis and future food security in an era of climate change. *Food Policy*, 35(5), 365–377. DOI: 10.1016/j.foodpol.2010.05.006

He, Z. L., & Wang, R. F. (2021). Evaluation model and application of water supply and demand suitability for tourism in arid areas: A case study of Xinjiang. *Ziran Ziyuan Xuebao*, 36(12), 3215–3231. DOI: 10.31497/zrzyxb.20211215

Hernandez-Crespo, C., Gargallo, S., Benedito-Dura, V., Nacher-Rodriguez, B., Rodrigo-Alacreu, M. A., & Martin, M. (2017). Performance of surface and subsurface flow constructed wetlands treating eutrophic waters. *The Science of the Total Environment*, 595, 584–593. DOI: 10.1016/j.scitotenv.2017.03.278 PMID: 28399497

Hijosa-Valsero, M., Reyes-Contreras, C., Domínguez, C., Bécares, E., & Bayona, J. M. (2016). Behaviour of pharmaceuticals and personal care products in constructed wetland compartments: Influent, effluent, pore water, substrate and plant roots. *Chemosphere*, 145, 508–517. DOI: 10.1016/j.chemosphere.2015.11.090 PMID: 26702554

Hindley, A., & Marmion, M. (2020). Tourism and health, risks, and challenges. In Leal Filho, W., Wall, T., Azul, A. M., Brandli, L., & Özuyar, P. G. (Eds.), *Good Health and Well-Being* (pp. 728–737). Springer International Publishing., DOI: 10.1007/978-3-319-95681-7_10

Holden, J. (2005). Peatland hydrology and carbon release: Why small-scale process matters. *Philosophical Transactions of the Royal Society A: Mathematical, Physical and Engineering Sciences, 363*(1837), 2891-2913.

Huang, D., Fu, G. J., Shi, R. Z., Chen, J., & Wang, M. Q. (2015). Study on tourism water footprint assessment. *Ecologic Science*, 34, 211–218.

Humeau, P., Hourlier, F., Bulteau, G., Masse, A., Jaouen, P., Gerente, C., & Le Cloirec, P. (2011). Estimated costs of implementation of membrane processes for on-site greywater recycling. *Water Science and Technology*, 63(12), 2949–2956. DOI: 10.2166/wst.2011.617 PMID: 22049724

Kadlec, R. H., & Wallace, S. D. (2009). *Treatment wetlands*. CRC Press.

Karimi, S., Yaraki, M. T., & Karri, R. R. (2019). A comprehensive review of the adsorption mechanisms and factors influencing the adsorption process from the perspective of bioethanol dehydration. *Renewable & Sustainable Energy Reviews*, 107, 535–553. DOI: 10.1016/j.rser.2019.03.025

Katukiza, A. Y., Ronteltap, M., Niwagaba, C. B., Kansiime, F., & Lens, P. N. L. (2015). Gray water characterization and pollutant loads in an urban slum. *International Journal of Environmental Science and Technology*, 12(2), 423–436. DOI: 10.1007/s13762-013-0451-5

Knebel, N. (2019). Crossing the borders between teaching, research and practice: A university project for a zero-energy building in Oman. *KnE Social Sciences*, 389-402.

Kuo, N. W., & Chen, P. H. (2009). Quantifying energy use, carbon dioxide emission, and other environmental loads from island tourism based on a life cycle assessment approach. *Journal of Cleaner Production*, 17(15), 1324–1330. DOI: 10.1016/j.jclepro.2009.04.012

Lan, W., Zhang, J., Hu, Z., Ji, M., Zhang, X., Zhang, J., Li, F., & Yao, G. (2018). Phosphorus removal enhancement of magnesium-modified constructed wetland microcosm and its mechanism study. *Chemical Engineering Journal*, 335, 209–214. DOI: 10.1016/j.cej.2017.10.150

Li, H., Li, Y., Gong, Z., & Li, X. (2013). Performance study of vertical flow constructed wetlands for phosphorus removal with water quenched slag as a substrate. *Ecological Engineering*, 53, 39–45. DOI: 10.1016/j.ecoleng.2013.01.011

Li, J. (2018). Scenario analysis of tourism's water footprint for China's Beijing-Tianjin-Hebei region in 2020: Implications for water policy. *Journal of Sustainable Tourism*, 26(1), 127–145. DOI: 10.1080/09669582.2017.1326926

Li, S. N., Wang, G. X., Deng, W., Hu, Y. M., & Hu, W. W. (2009). Influence of hydrology process on wetland landscape pattern: A case study in the Yellow River Delta. *Ecological Engineering*, 35(12), 1719–1726. DOI: 10.1016/j.ecoleng.2009.07.009

Li, Y., & Wang, Y. T. (2015). Life-cycle water footprints assessment and prediction of the tourism sector in Beijing-Tianjin-Hebei region. [Social Science Edition]. *Journal of Hebei University of Technology*, 7, 16–22.

Liang, Y., Zhu, H., Bañuelos, G., Yan, B., Zhou, Q., Yu, X., & Cheng, X. (2017). Constructed wetlands for saline wastewater treatment: A review. *Ecological Engineering*, 98, 275–285. DOI: 10.1016/j.ecoleng.2016.11.005

Lin, T. P. (2010). Carbon dioxide emissions from transport in Taiwan's national parks. *Tourism Management*, 31(2), 285–290. DOI: 10.1016/j.tourman.2009.03.009

Mahendru, M., Arora, V., Chatterjee, R., Sharma, G. D., & Shahzadi, I. (2023). From over-tourism to under-tourism via COVID-19: Lessons for sustainable tourism management. *Evaluation Review*. Advance online publication. DOI: 10.1177/0193841X231189805 PMID: 37470672

Moghaddam, V. K., Changani, F., Mohammadi, A., Hadei, M., Ashabi, R., Majd, L. E., & Mahvi, A. H. (2017). Sustainable development of water resources based on wastewater reuse and upgrading of treatment plants: A review in the Middle East. *Desalination and Water Treatment*, 65, 463–473. DOI: 10.5004/dwt.2017.20383

Mohammed, A., & Babatunde, A. O. (2017). Modelling heavy metals transformation in vertical flow constructed wetlands. *Ecological Modelling*, 354, 62–71. DOI: 10.1016/j.ecolmodel.2017.03.012

Ottoson, J., & Stenström, T. A. (2003). Faecal contamination of greywater and associated microbial risks. *Water Research*, 37(3), 645–655. DOI: 10.1016/S0043-1354(02)00352-4 PMID: 12688699

Parjane, S. B., & Sane, M. G. (2011). Performance of gray water treatment plant by economical way for Indian rural development.

Park, J. H., Wang, J. J., Kim, S. H., Cho, J. S., Kang, S. W., Delaune, R. D., & Seo, D. C. (2017). Phosphate removal in constructed wetlands with rapid cooled basic oxygen furnace slag. *Chemical Engineering Journal*, 327, 713–724. DOI: 10.1016/j.cej.2017.06.155

Qomariyah, S., Ramelan, A. H., Setyono, P., & Sobriyah, . (2018, March). Linking climate change to water provision: Greywater treatment by constructed wetlands. [). IOP Publishing.]. *IOP Conference Series. Earth and Environmental Science*, 129(1), 012002. DOI: 10.1088/1755-1315/129/1/012002

Rajan, B., Varghese, V. M., & Pradeepkumar, A. P. (2011). Recreational boat carrying capacity of Vembanad Lake ecosystem, Kerala, South India. *Environmental Research, Engineering and Management*, 56(2), 11–19. DOI: 10.5755/j01.erem.56.2.270

Reddy, K. R., Kadlec, R. H., Flaig, E., & Gale, P. M. (1999). Phosphorus retention in streams and wetlands: A review. *Critical Reviews in Environmental Science and Technology*, 29(1), 83–146. DOI: 10.1080/10643389991259182

Riggio, V. A., Ruffino, B., Campo, G., Comino, E., Comoglio, C., & Zanetti, M. (2018). Constructed wetlands for the reuse of industrial wastewater: A case study. *Journal of Cleaner Production*, 171, 723–732. DOI: 10.1016/j.jclepro.2017.10.081

Safoora Beevi, K. H., & Devadas, V. (2014). Impact of tourism on Vembanad Lake system in Alappuzha district. *International Journal of Research*, 1(5), 542–551.

Schmidt, K. S., & Skidmore, A. K. (2003). Spectral discrimination of vegetation types in a coastal wetland. *Remote Sensing of Environment*, 85(1), 92–108. DOI: 10.1016/S0034-4257(02)00196-7

Stefanakis, A. I. (2019, June). Constructed wetlands for municipal and industrial wastewater treatment in Middle East: An overview. In *Proceedings of the 8th International Symposium on Wetland Pollutant Dynamics and Control—WETPOL*, Aarhus, Denmark (pp. 17-21).

Tan, X., Yang, Y. L., Li, X., Gao, Y. X., & Fan, X. Y. (2021). Multi-metabolism regulation insights into nutrients removal performance with adding heterotrophic nitrification-aerobic denitrification bacteria in tidal flow constructed wetlands. *The Science of the Total Environment*, 796, 149023. DOI: 10.1016/j.scitotenv.2021.149023 PMID: 34280639

Tian, T., Tam, N. F., Zan, Q., Cheung, S. G., Shin, P. K., Wong, Y. S., Zhang, L., & Chen, Z. (2017). Performance and bacterial community structure of a 10-years old constructed mangrove wetland. *Marine Pollution Bulletin*, 124(2), 1096–1105. DOI: 10.1016/j.marpolbul.2017.07.005 PMID: 28711288

Tortella, B. D., & Tirado, D. (2011). Hotel water consumption at a seasonal mass tourist destination: The case of the island of Mallorca. *Journal of Environmental Management*, 92(10), 2568–2579. DOI: 10.1016/j.jenvman.2011.05.024 PMID: 21696882

Toth, E., Bragalli, C., & Neri, M. (2018). Assessing the significance of tourism and climate on residential water demand: Panel-data analysis and non-linear modelling of monthly water consumptions. *Environmental Modelling & Software*, 103, 52–61. DOI: 10.1016/j.envsoft.2018.01.011

Türker, O. C., Vymazal, J., & Türe, C. (2014). Constructed wetlands for boron removal: A review. *Ecological Engineering*, 64, 350–359. DOI: 10.1016/j.ecoleng.2014.01.007

United Nations Educational, Scientific and Cultural Organization (UNESCO). About world heritage. Available online: https://whc.unesco.org/en/conventiontext/ (accessed on 28 January 2024).

United Nations Environment Program & United Nations World Tourism Organization (UNEP-UNWTO). (2005). *Making tourism more sustainable – A guide for policy makers*. UNWTO. Available online: http://www.unep.fr/shared/publications/pdf/DTIx0592xPA-TourismPolicyEN.pdf (accessed on 28 January 2024).

United Nations Environment Program (UNEP). Declaration of the United Nations Conference on the Human Environment. Available online: https://www.unep.org/Documents.Multilingual/Default.asp?documentid=97&articleid=1503 (accessed on 28 January 2024).

UNWTO-United Nations World Tourism Organization. (1949). *Agenda 21 for travel and tourism: Towards environmentally sustainable tourism.*

Urban wastewater scenario in India. (n.d.). https://www.niti.gov.in/sites/default/files/2022-09/Waste-Water-A4_20092022.pdf (accessed on 23 May 2024).

van der Werf, G. R., Randerson, J. T., Giglio, L., Collatz, G. J., Kasibhatla, P. S., & Arellano, A. F.Jr. (2006). Interannual variability in global biomass burning emissions from 1997 to 2004. *Atmospheric Chemistry and Physics*, 6(11), 3423–3441. DOI: 10.5194/acp-6-3423-2006

Vigneswaran, S., & Sundaravadivel, M. (2004). Recycle and reuse of domestic wastewater. In *Wastewater recycle, reuse, and reclamation* (Vol. 1).

Vohla, C., Kõiv, M., Bavor, H. J., Chazarenc, F., & Mander, Ü. (2011). Filter materials for phosphorus removal from wastewater in treatment wetlands – A review. *Ecological Engineering*, 37(1), 70–89. DOI: 10.1016/j.ecoleng.2009.08.003

Vörösmarty, C. J., & Moore, B.III. (1991). Modeling basin-scale hydrology in support of physical climate and global biogeochemical studies: An example using the Zambezi River. *Surveys in Geophysics*, 12(1), 271–311. DOI: 10.1007/BF01903422

Wang, J., & Guo, X. (2020). Adsorption kinetic models: Physical meanings, applications, and solving methods. *Journal of Hazardous Materials*, 390, 122156. DOI: 10.1016/j.jhazmat.2020.122156 PMID: 32006847

Wilhelm, K., Pam, B., Nicolas, B., Cecchi, C., Thomas, E., Marta, F., ... & Jurgen, T. (2015). Towards an EU research and innovation policy agenda for nature-based solutions & re-naturing cities. *Final report of the Horizon 2020 expert group on nature-based solutions and re-naturing cities.*

Yoon, H., Sauri, D., & Rico, A. (2022). The water-energy nexus in hotels and recreational activities of a mass tourism resort: The case of Benidorm. *Current Issues in Tourism*, 25(4), 592–610. DOI: 10.1080/13683500.2021.1893283

Zhang, L., Wang, M. H., Hu, J., & Ho, Y. S. (2010). A review of published wetland research, 1991–2008: Ecological engineering and ecosystem restoration. *Ecological Engineering*, 36(8), 973–980. DOI: 10.1016/j.ecoleng.2010.04.029

Zhao, L., Xue, L., Wang, L., Liu, C., & Li, Y. (2022). Simultaneous heterotrophic and FeS2-based ferrous autotrophic denitrification process for low-C/N ratio wastewater treatment: Nitrate removal performance and microbial community analysis. *The Science of the Total Environment*, 829, 154682. DOI: 10.1016/j.scitotenv.2022.154682 PMID: 35307420

Zuma, B. M., Tandlich, R., Whittington-Jones, K. J., & Burgess, J. E. (2009). Mulch tower treatment system Part I: Overall performance in greywater treatment. *Desalination*, 242(1-3), 38–56. DOI: 10.1016/j.desal.2008.03.030

Chapter 16
A Comprehensive Review of Food Waste Management Strategies in the Food Service Industry

Ronelle Crocker
https://orcid.org/0009-0007-9215-2333
Durban University of Technology, South Africa

ABSTRACT

Food waste presents a severe sustainability challenge. Of the food produced globally, 30% is wasted, and if no significant improvement is made, this figure is expected to double by 2050. With the continuous growth of the hospitality sector, an increasing amount of waste, particularly food waste, is generated. Globally, the hospitality industry is the third-largest contributor to food waste. Food waste mitigation is fundamental to the sustainable improvement of the food service industry, owing to its adverse consequences and ethical concerns. For this reason, there is an increasing demand for practical solutions in the food service industry. Establishments in the food service industry can make a significant contribution to global sustainability efforts by creating partnerships with stakeholders and providing educational programs aimed at waste reduction to employees and consumers. Several food service establishments actively implement measures to minimize waste and recognize and accept their responsibilities.

DOI: 10.4018/979-8-3693-6110-8.ch016

1. INTRODUCTION

Globally, the tourism sector, specifically the Food Service Industry (FSI), significantly contributes to food waste generation and has recently become a topic of interest in the FSI (Martin-Rois et al., 2022). With the popularity of eating out, food waste in the sector is escalating (Wang et al., 2017), and it produces twice as much organic waste compared to other sectors, impacting current food systems (Martin-Rois et al., 2022). While food waste is recognised as a global issue, there is no universally accepted singular definition of the term across various literature studies (Martin-Rois et al., 2018); however, for the aim of this chapter, the United Nations Environment Programme (UNEP) Food Waste Index Report (2021) definition will be utilised. UNEP (2021) defines food waste as *"food and the associated inedible parts removed from the human food supply chain."* The food wasted annually is estimated to increase by 30% by 2030 (Martin-Rois, 2020). Food wastage is 750 billion USD annually in terms of economic and environmental costs (Wu & Teng, 2023) and could nourish a global population of seven billion people (McAdams et al., 2019).

The extent and concerns of the food waste issue have globally exacerbated its related sustainability concerns; however, it is mainly indicative of the FSI as the sector employs more staff than other retail sectors (Martin-Rois et al., 2018) and generates an excessive amount of food waste (Sucheran & Olanrewaju, 2021; Martin-Rois et al., 2022). As a subsector within the food and beverage industry, the FSI includes establishments that provide meals for consumption out of the home (Martin-Rois et al., 2018; Martin-Rois et al., 2022), including restaurants, canteens, cafeterias, and other establishments that prepare and serve meals, and excludes the retail sector (Eriksson et al., 2017). Several factors contribute to the generation of food waste in the FSI, such as the demand for food production, access and availability of food resources, consumer motivations, and a lack of awareness about food purchasing and consumption habits, as well as market-related constraints (Martin-Rois et al., 2022; Knorr & Augustin, 2022). Food waste takes place mainly at the end of the food supply chain during the preparation and distribution of meals, but food waste management should be a priority because there are opportunities for waste reduction at every stage (Lins et al., 2021; Samsuddin et al., 2022; Gonçalves et al., 2023). For food waste management to be effective, it should include quantification solutions, the continuous development of new techniques, and practical approaches (Assamoi & Lawryshyn, 2012; Zhang et al., 2019; Martin-Rois et al., 2022).

There are several approaches to food waste management (Martin-Rois et al., 2018; Bilska et al., 2020; Dhir et al., 2020). Waste management practices adopted by the FSI encompass conventional best practices, whereas others implement cutting-edge innovative solutions, with most approaches taking place at the operational level of establishments (Martin-Rois et al., 2021). The food waste management strategies discussed in this chapter can be implemented at different stages of the food production process in the FSI and entail enhanced menu planning and demand forecasting, monitoring and tracking of food waste, effective education and training, revised portion sizes (Sucheran & Olanrewaju, 2021; Rady et al., 2021; Okumus et al., 2020McAdams et al., 2019; Sakaguchi et al., 2018) and encouraging customers to take home doggy bags (Sucheran & Olanrewaju, 2021). Food service establishments can promote sustainable operations by implementing efficient inventory-management systems (Clowes et al., 2019; Lins et al., 2021).

2. METHODOLOGY

This literature review focuses on a systematic search and analysis of relevant peer-reviewed journal articles and industry reports. The following databases were used for the search: Google Scholar, Semantic Scholar, Scopus, and the Web of Science. Search enquiries were based on a combination of relevant keywords and Boolean operators. The leading search terms were food waste, restaurants, hospitality industry, food service industry, sustainable practices, and food waste management. The leading search terms were supplemented with additional keywords, such as menu planning, monitoring and tracking, portion size, food waste awareness, government intervention, and surplus food donation to refine the search. For a study to be included in this paper, it had to be published in a peer-reviewed journal between 2000 and 2024, discussing strategies, challenges, or actors associated with food waste in the food service industry or serving as a reputable industry report. Studies that focused solely on household/domestic food waste were excluded. 121 sources were analysed, comprising 115 relevant peer-reviewed journals and 6 industry reports to make up this literature review. For each study included in the literature review, relevant data regarding the research aims, findings, and limitations were extracted. A content analysis approach was used to synthesise findings across relevant publications and derive insights into food waste management strategies and opportunities in the FSI. This structured approach provided a comprehensive overview of current knowledge from published research. The findings in this chapter provide reliable supporting data to inform the development of strategies and future studies aimed at reducing food waste within the FSI.

3. LITERATURE REVIEW

3.1 Food Waste in the Food Services Industry

Food waste in the global food services industry (FSI) is an extensive and significant concern (Sucheran & Olanrewaju, 2021; Tahir, 2023). Food waste is substantially higher in developed nations than less developed nations (Ishangulyyev et al., 2019). People dine out more frequently as lifestyles advance, placing additional pressure on food systems, including the FSI (McAdams et al., 2019; Patel et al., 2021). Nearly 20% of the food produced in the FSI is wasted, and 75% of this waste is avoidable and suitable for consumption (Malefors et al., 2019; Filimonau et al., 2019; Bhajan et al., 2022; Omari et al., 2023). Food service establishments significantly affect the sustainability of their environments by generating substantial waste (Martin-Rois et al., 2022; Bilska et al., 2022). Food waste generation occurs from procurement and storage to preparation, service, and post-consumer disposal (Gonçalves et al., 2023). The FSI is complex and consists of many significant operations in the global economy (Martin-Rois et al., 2022). The characteristics of food waste differ across various types of establishments within the sector (Dhir et al., 2020).

Food waste in the FSI can be classified into two primary types: kitchen and plate waste. Kitchen waste includes waste resulting from the overproduction of food, incorrect trimming of food items, and inadequate storage practices (Sucheran & Olanrewaju, 2021; Filimonau et al., 2021; Principato et al., 2021; Afzal et al., 2022). Plate waste, leftover food from consumer plates (Tekin & Ilyasov, 2017), results from personal preferences, extensive menu options, inferior food quality, and ineffective portion control (Karakas, 2021; Afzal et al., 2022; Cerrah & Yigitoglu, 2022; Saurab & Muhammad, 2022). Therefore, strategies to reduce food waste in the FSI should target kitchen and plate waste (Papargyropoulou et al., 2019). By addressing these two significant sources of food waste, the FSI can improve its sustainability and reduce its environmental impact.

3.2 The Importance of Food Waste Management in the Food Service Industry

Owing to the sustainability concerns associated with food waste (Vizzoto et al., 2021a; Rodrigues et al., 2024), it is a matter which must be urgently addressed (Read et al., 2020). Effective food waste management is possible and does not need to impact consumers' dining experience (Afzal et al., 2022; Diwate & Aditya, 2023; Vizzoto et al., 2021b). Often, managers are reluctant to employ food waste management strategies, such as adjusting portion sizes and introducing doggy bags (Vizzoto et al., 2020), because they are concerned that it may affect the dining experience of consumers (Vizzoto et al., 2021b); however, if food waste is managed effectively, production-related expenses can be lowered, the productivity of production systems can be increased, and environmental sustainability can be enhanced (Food and Agriculture Organisation [FAO], 2019). Furthermore, effective food waste management in food services contributes to social sustainability by addressing issues of food insecurity and inequality (Wang et al., 2020). While a substantial amount of food is wasted daily, many people globally experience severe food insecurity, highlighting a paradoxical situation (Diwate & Aditya, 2023). Food waste impacts food costs (Afzal et al., 2022) and worsens the already existing issue of food insecurity, presenting a missed opportunity to provide nutritious meals to those facing food insecurity (FAO, 2022), while almost 870 billion people lack access to food (Vaidyanathan & Shankar, 2020). Reducing food waste will enhance global food security and increase the availability of safe and nutritious foods (Rady et al., 2021; Harduth, 2017).

When food is wasted, resources such as finances, logistics, and labour used to grow, distribute, and prepare food are wasted (Lins et al., 2021; Nicastro & Carillo, 2021; Bhajan et al., 2022), and the FSI consumes a significant number of resources (Bulhoes et al., 2020; Martin-Rois et al., 2022). Therefore, reducing food waste can reduce the resources in food production (Principato et al., 2021). Moreover, food production accounts for almost 70% of the earth's water and 28% of the land (Bhajan et al., 2022). Food waste contributes to the emission of greenhouse gasses, water pollution and the depletion of natural resources (Stirnimann & Laura, 2022; Dhir et al., 2020; Lins et al., 2021; Bhajan et al., 2022), significantly impacting the environment with a carbon footprint of approximately 3.3 billion tons of carbon dioxide (Diwate & Aditeya, 2023). It also has environmental costs in terms of the resources that go into producing uneaten food as well as the disposal of waste (Vizzoto et al., 2021a).

Addressing the issue of food waste in the FSI presents numerous benefits. While reducing the environmental impact remains crucial (Diwate & Aditya, 2023), effective waste management is economically beneficial (McAdams et al., 2019). Since food service establishments are expected to bear the expense of ingredients to prepare meals and dispose of waste food, reducing food waste should make economic sense for the FSI (McAdams et al., 2019). Reducing food waste can lower expenses on ingredient purchases, enhance operational efficiency, and potentially generate new income by repurposing safe surplus food (Clowes et al., 2019). Donating surplus food to food distribution and vulnerable populations can alleviate hunger in local communities (Sundin et al., 2023). Moreover, managing food waste in the FSI can support innovation and partnerships within the industry (Martin-Rios et al., 2020; Martin-Rios et al., 2018) through the sharing of ideas and best practices, partnerships with suppliers, and the implementation of technological solutions (Martin-Rios et al., 2020). The sustainability benefits of food waste management alone should motivate the FSI to implement strategies to prevent and reduce food waste (Heikkilä et al., 2016). Food service establishments can provide an example for consumers by influencing societal attitudes towards more environmentally conscious consumption practices (Martin-Rios et al., 2020).

3.3 Sustainable Strategies for Waste Management in the Food Service Industry

Challenges within the FSI are increasing, and implementing effective food waste management strategies along the entire food supply chain is becoming increasingly crucial (McAdams et al., 2019). Effective food waste management requires a collective effort from managers, employees, and consumers (Wu & Teng, 2023). To promote and implement sustainable measures, all parties involved must thoroughly understand the nature and extent of food waste, its trends, and associated consequences (Parfitt et al., 2010). Additionally, effective food waste management requires improvement from managers, behavioural changes, and proper planning based on the historical patterns of an establishment (Silvennoinen et al., 2019). Substantial investments are required to address food waste (Goosens et al., 2022); however, the potential benefits are immeasurable (WRAP, 2015).

Although each stage in the food supply chain offers the potential for waste reduction and resource maximisation (Lins et al., 2021; Samsuddin et al., 2022), the management of food waste in the FSI is a complex issue that encompasses various elements and processes (Martin-Rois et al., 2018). Given the relatively high diversity within the FSI, implementing individualised reduction strategies and measures tailored to specific contexts is necessary rather than adopting a one-size-fits-all approach (Eriksson et al., 2017). Bharucha (2018) explains that reliance solely on prevention efforts is inadequate for maximising resource utilisation. Such efforts must entail recovery strategies to identify and correct any failures in the catering system (Bharucha, 2018). By implementing adequate purchasing plans, efficient stock control systems, portion control measures, and meticulous menu planning techniques (Sucheran & Olanrewaju, 2021), food services can mitigate overproduction and curb food wastage, thereby promoting sustainable operations (Clowes et al., 2019; Lins et al., 2021). Different measures have been suggested to reduce food waste during food production. These include accurately forecasting consumer demand, improving internal processes, effective leadership, the use of technology, employee training, food donation, and the introduction of 'doggy bags' (Sakaguchi et al., 2018; McAdams et al., 2019; Okumus et al., 2020; Rady et al., 2021; Sucheran & Olanrewaju, 2021). The FSI should consider these strategies as part of their continuous commitment to sustainability in the industry and not simply 'business imperatives' as suggested by Martin-Rois et al. (2018).

3.3.1 Monitoring and Tracking

Several studies have highlighted the importance of monitoring and tracking food waste in food service establishments as the first step toward reducing food waste (Rady et al., 2021; Sucheran & Olanrewaju, 2021; Okumus et al., 2020; Goosens et al., 2022). Dhir et al. (2020) recommend that researchers focus on collective waste data gathering and all-inclusive approaches to quantify food waste. Food service establishments can monitor and track food waste using traditional manual methods or technological applications such as websites and food waste applications. However, the primary advantage of using technological solutions over manual tools is that they enable data to be compiled and analysed automatically in real-time (Goosens et al., 2022). Monitoring and tracking will allow for the identification of the volume of food waste (Goosens et al., 2022) and categories of waste generated (Rady et al., 2021; Sucheran & Olanrewaju, 2021; Okumus et al., 2020; Goosens et al., 2022). Unpopular dishes can also be identified and removed from the menu to prevent waste (Martin-Rios et al., 2018; Papargyropoulou et al., 2019; Gonçalves et al., 2023). Monitoring and tracking of food waste will support establishments in identifying opportunities to meet consumer demand more accurately, reduce excessive food production, and reveal the financial implications of wasted food (Silvennoinen et al., 2019; Gonçalves et al., 2023). If policymakers direct their efforts to establish standardised food waste measures, it will ensure consistency across various establishments within the sector (Dhir et al., 2020). This may be beneficial because there is a need for standardised approaches to food waste quantification (Eriksson et al., 2018; Gonçalves et al., 2023).

Goosens et al. (2022) found that restaurants could significantly reduce leftovers by implementing tracking processes in their kitchens. Their study noted that each restaurant kitchen could cut leftovers by almost 1,800 kilograms annually by implementing tracking measures. Moreover, kitchens in the study realised substantial annual reductions in their environmental impact. Eriksson et al. (2019) revealed that 61% of food service operations that implemented tracking and monitoring of food waste were able to decrease the quantity of food waste over time. Monitoring and tracking are essential strategies for reducing food waste as this can promote the innovation of new food reduction techniques (Silvennoinen et al., 2019). In addition, information can be shared between local and national governments, as suggested by Dhir et al. (2020).

3.3.2 Menu Planning and Design

Menu planning and design is a crucial approach for the effective management of food waste (Okumus et al., 2020; Principato et al., 2021; Sucheran & Olanrewaju, 2021) because poorly planned menus and extensive menu options contribute considerably to food waste generation (Wu & Teng, 2023). Okumus (2020) explained that most food waste is attributed to establishment type and menu offerings. The menu must be clear and provide descriptions of the ingredients used in the meals (Krieger et al., 2013; Block et al., 2013). Establishments must down-size large menu items (Okumus, 2020). The menu must be designed to reduce the portion control of items that are usually leftover (Kreiger et al., 2013; Block et al., Rady et al., 2021; Goncalves et al., 2023) and must consider all other avenues for food waste prevention. For example, per capita servings must be considered when planning menus, as this ensures meal balance, facilitates purchasing, and controls food costs, production, and waste reduction (Aires et al., 2021). Including seasonal items will be more cost-effective and enable the establishment of financial sustainability (Principato et al., 2021). Careful planning of menus will help kitchens ensure food quality and avoid spoilage, ultimately reducing the amount of food wasted (Wu & Teng, 2023).

Food service establishments, such as university canteens (Aires et al., 2021) and restaurants (Sucheran & Olanrewaju, 2021; Principato et al., 2021), can also consider including healthier food options in their menus to support sustainability (Sucheran & Olanrewaju, 2021). The menu can also be amended to suit consumer preferences in buffet settings (Chang, 2022; Wu & Teng, 2023). Papargyropoulou et al. (2019) suggest that consumers pay for what they eat in buffet service establishments instead of an eat-as-much-as-you-like approach. In addition, establishments should instead consider the use of à la carte menus to reduce food waste generation (Pirani & Arafat, 2016; Papargyropoulou et al., 2019), taking into account the possibility of reusing surplus items to make dishes like soups and stews (Okumus, 2020).

3.3.3 Purchasing and Stock Control

Food service establishments must revise their current stock control and purchasing procedures to avoid over-purchasing (Sucheran & Olanrewaju, 2021; Vizzoto et al., 2021a) and reduce food waste (Clowes et al., 2019). Bharucha (2018) and Filimonau et al. (2019) highlighted the importance of regular inventory checks and frequent purchases of perishables as a preventative measure for generating kitchen waste. Buyers should prioritise items with an extended shelf life when making ingredient purchases while ensuring that product quality is not compromised (Sucheran & Olanrewaju, 2021). In addition, they must consider negotiating with their suppliers to arrange delivery schedules that better suit establishments' specific needs. Moreover, implementing stock rotation systems and maintaining proper food storage temperatures are crucial for food waste reduction efforts (Sucheran & Olanrewaju, 2021; Tomaszewska et al., 2021). Many restaurants in Pakistan employ this method, enabling them to forecast demand more accurately, resulting in less food waste (Afzal et al., 2022). Moreover, the inventory should be handled and stored appropriately (Vizzoto et al., 2021a). In a study conducted in Mauritius by Bhajan et al. (2022), restaurants implemented a colour-coding storage system to ensure the timely use of food products and to avoid wastage. The system ensured that all products were used quickly, before expiration, to reduce wastage. The "First-In, First-Out" method was implemented, ensuring that ingredients received earlier were used first, reducing any chance of food spoilage and subsequently minimising food waste generation. Furthermore, chefs washed perishable items using ozonated water, prolonged their shelf life, and decreased food spoilage and waste. These are undemanding, efficient, and economical strategies, yet they have influenced food waste reduction.

3.3.4 Demand Forecasting

Accurately predicting consumer food demand is a challenge and leads to overproduction of food, which is ultimately wasted (Garrone et al., 2014; Clowes et al., 2019). Accurate and successful forecasting of food demand is fundamental for avoiding food waste in food service establishments (Amicarelli et al., 2021; Wu & Teng, 2023). If food demand is accurately forecasted, it will reduce unforeseen food waste (Wu & Teng, 2023). Wu & Teng (2023) suggest that restaurants ensure guests make reservations to reduce waste generation, particularly in buffet settings. The style of service leads to overproduction, as it is better to oversupply than to undersupply, as the latter would lead to frustrated customers (Rady et al., 2021). Food is usually discarded if left for over three hours (Bhajan et al., 2022). Managers should implement either manual or digital systems that allow for more accurate forecasting of food orders and adopt predictive ordering technologies (Filimonau et al., 2019; Sucheran & Olanrewaju, 2021; Amicarelli et al., 2021; Blum, 2020). Historical data and real-time reservation records can enhance forecasting accuracy and the expected number of guest arrivals (Betz et al., 2015; Pinto et al., 2018; Wu & Teng, 2023). Food demand can then be estimated and prepared accordingly (Wu & Teng, 2023).

Numerous studies recommend utilising forecasting models, technological applications, and analyses of past trends to accurately predict consumer demand (Filimonau et al., 2020; Sucheran & Olanrewaju, 2021; Wu & Teng, 2023; Rodrigues et al., 2024). The effectiveness of various forecasting models can vary depending on the nature of the food service facility (Rodrigues et al., 2024). Demand forecasting methods are not always practical (Filimonau et al., 2020). However, when demand forecasting techniques are ineffective, managers resort to alternative strategies for food waste management.

3.3.5 Portion Control

Studies have suggested meticulously revising food portions to meet specific customer requirements to prevent plate waste (Principato et al., 2018; Filimonau et al., 2019; Sucheran & Olanrewaju, 2021). More significant portions are generally selling points to consumers, indicating they receive value from their money. However, a reduction in food portion sizes can reduce waste without impacting consumer satisfaction (Kallbekken & Sælen, 2013) and directly increase the profitability of restaurants (Blum, 2020). The wide range of consumer preferences, tastes, and demographic factors make it challenging for food service establishments to provide ideal portion sizes that satisfy all consumers (Hennchen, 2019; Filimonau et al., 2020). In buffet settings, it is particularly challenging to control portion sizes because employees have no control over them (Kallbekken & Sælen, 2013; Vizzoto et al., 2021a; Wu & Teng, 2023). Sucheran and Olanrewaju (2021) suggest that restaurants consider offering consumers the option of half portions. Berkowitz et al. (2016) employed this method, offering both full and half portion sizes (at a reduced rate), and observed a reduction in plate waste, even though consumers were not receptive to the reduction in portion sizes. Whilst portion control is encouraged as an effective food waste management strategy, consumers should still receive value for their money (Principato et al., 2018; Rady et al., 2021) as consumers do not welcome a reduction in portion size without a comparable price reduction (Vizzoto et al., 2021a) or unsolicited advice regarding portion size adjustment (Vizzoto et al., 2021b).

While it may be challenging to control consumer portion sizes in buffet-type settings, Wu & Teng (2023), Rady et al. (2021), and Sucheran & Olanrewaju (2021) suggest the use of smaller plates as an effective approach to plate waste in buffet settings. A study conducted by Richardson et al. (2021) in food halls of institutions of higher learning demonstrated that changing to smaller plates is an effective strategy for reducing food waste, as participants served themselves less food when using smaller plates. Implementing portion control measures such as standardised servings, flexible portions (Sucheran & Olanrewaju, 2021), and strategically presenting dishes can substantially reduce food waste in the FSI (Rady et al., 2021).

3.3.6 Education and Training

Given that employees are a business's most valuable resource, efforts should be concentrated on raising awareness and providing comprehensive training on food waste management as a fundamental step toward environmental sustainability

(Principato et al., 2021; Rady et al., 2021; Montesdeoca-Calder´on et al., 2024). The provision of such training in food service establishments is crucial for reducing food waste (Sucheran & Olanrewaju, 2021; Rady et al., 2021; Wu & Teng, 2023) and has proven to be an effective way to mitigate the amount of food waste generated during food production (Chalak et al., 2018; Filimonau et al., 2020; Rady et al., 2021; Montesdeoca-Calder´on et al., 2024; Principato et al., 2021). Sucheran and Olanrewaju (2021) suggest that managers receive this education before being appointed to their roles. Dhir et al. (2020) add that all kitchen employees should undergo food waste training and be formally certified by the government to assume their roles in the kitchen, failing which a penalty must be imposed on management.

Food waste management training should entail policies and legal responsibilities to properly manage and dispose of food waste (Rady et al., 2021; Wu & Teng, 2023). Consistent education and training programs will ensure that establishments remain current in their approaches to effectively managing food waste (Sucheran & Olanrewaju, 2021) and excel in establishing more environmentally conscious practices (Silvennoinen et al., 2019; Montesdeoca-Calder´on et al., 2024). Consistent, thorough training can improve portioning and food production methods and enhance recipe and menu management (Principato et al., 2021). Papargyropoulou et al. (2019) suggest that kitchen staff are trained in cutting skills, and perhaps staff can be incentivised to have the best skills or be food waste champions in the kitchen.

In addition, research highlights the importance of providing staff with food handling and safety training to effectively alleviate contamination and food spoilage and reduce food waste in restaurants (Principato et al., 2018; Wu & Teng, 2023). Moreover, a competent, well-trained team can provide thorough explanations for each menu item, preventing any confusion regarding the menu (Bharucha, 2018) and can additionally provide value by educating consumers to order sensibly to reduce plate waste (Principato et al., 2021; Wu & Teng, 2023). Employees who carefully fulfil orders and precisely follow consumer requirements are less likely to experience rejected food orders (Papargyropoulou et al., 2019; Okumus, 2020), reducing food waste generation. Staff must be trained to guide consumers through menus and reduce the possibility of rejected food items (Okumus, 2020).

Furthermore, providing awareness to consumers can be beneficial in motivating them to reduce plate waste (Vizzoto et al., 2021a; Vizzoto et al., 2021b; Chang, 2022; Wu & Teng, 2023) and increasing their receptiveness to any measures aimed at assisting them in reducing waste (Vizzoto et al., 2021a; Wu & Teng, 2023). Consumers who eat out often have more plate waste if they are ignorant of sustainability concerns associated with food waste (Vizzoto et al., 2021b). Therefore, media campaigns and educational awareness programs can shape and influence consumer behaviour related to food waste (Okumus et al., 2020; Wu & Teng, 2023).

In addition, awareness campaigns can motivate consumers to accept strategies to help reduce plate waste (Vizzoto et al., 2021b).

3.3.7 Technological Applications

Several studies have explored and encouraged the use of technology to analyse and measure food waste in food service establishments (Martin-Rois et al., 2018; Clowes et al., 2019; Blum, 2020; Mattila et al., 2020; Sucheran & Olanrewaju, 2021; Singh et al., 2024). Technology supports food preparation management, and many establishments have already implemented technological solutions to manage food waste effectively (Sucheran & Olanrewaju, 2021). Technology-driven solutions for managing food waste include using smart kitchen equipment, real-time inventory tracking systems, and solutions that support repurposing food waste (Singh et al., 2024). The use of technology in kitchen operations has supported chefs in achieving accuracy and reducing oversight in food control and measurement (Blum, 2020). Technological approaches can lower operational expenses and enhance sustainability initiatives (Singh et al., 2024; Jones et al., 2022). Technology can forecast future demand more precisely, allowing for a more accurate ordering of supplies (Wu & Teng, 2023). By improving the prediction of customer demand, the amount of food waste can be significantly reduced (Blum, 2020; Wu & Teng, 2023).

According to Hong et al. (2024), some of the latest technological approaches being utilised include the following:

- ***Artificial Intelligence (AI) and Machine Learning:*** Recent innovations in AI have enabled predictive analytics capabilities to forecast food demand. The analysis of historical data patterns, seasonal factors, and consumer preferences (Gupta et al., 2022). These advanced tools can assist businesses in optimising their purchasing and stock management to avoid overstocking, subsequently reducing food waste (Singh et al., 2024).
- ***Internet of Things (IoT) Sensors:*** These are essential tools for monitoring conditions, such as temperature and humidity, in food storage areas and tracking expiration dates (Santos et al., 2021). The data from IoT sensors is in real-time, allowing establishments to maintain proper quality and safety protocols for their food products. These systems trigger alerts if conditions deviate from the ideal range, preventing spoilage and reducing wastage (Santos et al., 2021).
- ***Food Waste Applications (apps):*** Mobile apps have emerged as helpful tools for businesses and consumers in the fight against food waste (Filimonau et al., 2020; Smith et al., 2023). These apps are organised

and classified according to their food-saving approach (Hong et al., 2024). They reduce the risk of overcooking food items, improve cooking methods and plating consistency, and ultimately reduce plate waste (Smith et al., 2023). Food waste apps can also facilitate sustainable collaboration between consumers (Apostolidis et al., 2021). Wu and Teng (2023) found that restaurants use smartphone apps to train their staff. The training covered food safety, waste management, and customer service relationships. This highlights the versatility of mobile apps in reducing waste and empowering the staff.

- ***Automated Food Preparation Systems:*** Commercial kitchens are progressively using automated systems and robotic equipment to boost the efficiency of food preparation tasks and achieve optimal portion sizes (Smith et al., 2023). These advanced solutions help prevent food items from overcooking, support efficient consistency, and reduce the amount of food waste generated (Hong et al., 2024).

Food waste contributes significantly to the costs of food service establishments (Singh et al., 2024). These technological applications provide solutions to the critical issue of food waste, with the most significant advantage being cost reduction (Jones et al., 2022). Technology can lower greenhouse gas emissions (Santos et al., 2021). Over and above the financial and environmental benefits of utilising technology to reduce food waste, these applications can also provide a competitive edge by supporting consumer engagement and loyalty with the establishment's sustainability efforts (Singh et al., 2024).

3.3.8 Surplus Food Donation

An environmentally friendly way to reduce food waste is to donate leftover food to charitable organisations, or to donate food that is not humanly edible as animal feed (Sundin et al., 2023; Gonçalves et al., 2023). This has become a globally embraced and practical approach to redirecting safe surplus food that would otherwise end up in landfills (Akwii et al., 2021; Corina-Mihaela et al., 2023). It is more prevalent in the food and beverage retail sector than in the food service industry (Martin-Rois et al., 2018). However, it presents an opportunity to repurpose the food waste generated by the FSI (Sundin et al., 2023; Corina-Mihaela et al., 2023) and offers nutritional benefits to those experiencing food insecurity (Sundin et al., 2023).

Food donations are typically arranged through food banks or charitable organisations, which collect excess, high-quality food and redistribute it to soup kitchens, shelters, and community pantries to support those who do not have access to nutritious food (Akwii et al., 2021). Food donations can be a complex process (Montoli et al.,

2023), and Dhir et al. (2020) explained that donating surplus food may not always be possible across all types of establishments in the FSI. For example, restaurants may be able to explore this option, but it may not be a probable avenue for healthcare facilities because of the heightened risk of spreading infections (Dhir et al., 2020). Comparably, educational institutions can repurpose or sell leftover food (Laakso, 2017). However, this approach of repurposing leftovers may not be feasible for restaurants operating with a constant flow of consumers throughout the business day (Dhir et al., 2020).

There needs to be a stronger focus on collaboration between the FSI and charitable organisations to maximise food waste reduction (Bilska et al., 2022). The FSI should implement standardised procedures to guide the identification and selection of food items suitable for donation (Gonçalves et al., 2023). Studies have highlighted that food donations offer environmental benefits (Sundin et al., 2023) and contribute to social welfare (Sundin et al., 2023; Albizatti et al., 2019; Damiani et al., 2021). Food donation provides more significant environmental benefits than anaerobic digestion (Sundin et al., 2023). Redistributing and donating food waste is acknowledged as a viable approach for recovering surplus food (Sundin et al., 2023; Afzal et al., 2022), the logistics involved in food donation (Bilska et al., 2022), and the high financial costs associated with food donation (Mantoli et al., 2023). Furthermore, many establishments choose not to donate food primarily because of the legal risks and high costs involved (Sakaguchi et al., 2018; Bharucha, 2018; Martin-Rois et al., 2018; Bilska et al., 2022).

Concerns often arise regarding the safety of surplus food (Mantoli et al.,2023). Barucha's study (2018) uncovered cases where individuals fell ill after consuming donated food, leading to substantial compensation claims. Unfortunately, such cases and the challenges associated with food donation highlight the primary reasons for the reluctance of many establishments to donate excess food. Throwing away surplus food is more accessible and cheaper (Mantoli et al., 2023). Martin-Rois et al. (2018) and Papargyropoulou et al. (2019) suggest that food service establishments have a no-liability agreement, removing any liability for food safety once food leaves the establishment. For example, once food leaves the premises in Switzerland, the establishment is no longer responsible (Martin-Rois et al., 2018). Perhaps this is an option to be explored instead of food disposal.

3.3.9 Doggy Bags

Plate waste in restaurants is extensive (Herpen et al., 2022), and some establishments allow consumers to take home food left on their plates (Bilska et al., 2022). This leftover food, referred to as "doggy bags" (Mirosa et al., 2018; Afzal et al., 2022), while relevant (Dhir et al., 2020), remains underexplored (Talwar et al., 2021) and not globally accepted (Rady et al., 2021). Doggy bags aim to reduce food waste by promoting the reuse of leftovers at restaurants and promoting more sustainable consumption (Billore et al., 2021). Providing consumers with doggy bags to take home effectively reduces waste (Vizzoto et al., 2021b). However, in many countries, social norms and feelings of embarrassment are associated with doggy bags (Stirnimann & Zizka, 2022; Herpen et al., 2022; Mirosa et al., 2018; Sirieix et al., 2017). The French, for example, hate doggy bags (Beizer, 2017). In France, 15% of food waste is generated in the FSI; however, taking home a doggy bag from a restaurant is considered embarrassing (Sirieix et al., 2017). To address this issue, the French government implemented a 2016 law requiring restaurants with more than 150 covers per day to provide doggy bags upon customer request (Mirosa et al., 2018). In Italy (Principato et al., 2018) and Japan (Billore et al., 2021), due to cultural factors, consumers feel that taking home a doggy bag is impolite and indicates a low financial status. However, research maintains doggy bags' effectiveness in reducing plate waste (Sucheran & Olanrewaju, 2021; Principato et al., 2021; Vizzoto et al., 2021b).

3.3.10 Repurposing and Reuse of Food

Food service establishments should always have a contingency in place to repurpose the excess food since most food service establishments face difficulties in accurately predicting demand, which results in excess ingredients and leftover food (Clowes et al., 2019). This food waste in the FSI is suitable for reuse (Bilska et al., 2022; Gonçalves et al., 2023), and its reuse presents a practical approach to food waste management (Filimonau et al., 2020). Surplus food is usually leftover and safe to eat after preparation but should only be held for up to 48 hours after preparation (Gonçalves et al., 2023). Chefs' expertise and creative thinking enable them to devise innovative solutions for repurposing and effectively using leftover food items rather than simply throwing them away (Filimonau & De Coteau, 2019; Pizetti et al., 2024). Clowes et al. (2019) recommend that restaurants offering a set menu should incorporate repurposing excess ingredients into their regular operations. For example, research conducted by Barucha (2018) found that restaurants creatively utilise surplus food by experimenting with new dishes that employees could sample. An upscale restaurant in the study repurposed leftover food ingredients into new menu items that were offered complimentary to regular patrons. A deli in the study served all the remaining food from the lunch service as discounted meal options on the dinner menu.

In buffet settings, leftover meals can be donated, or leftover ingredients and trimmings can be used to prepare staff meals (Papargyropoulou et al., 2019), or repurposed into garnishes (Papargyropoulou et al., 2019; Wu & Teng, 2023). If all food safety measures are in place, leftover buffet meals can be enjoyed by staff if they are removed from the buffet and immediately consumed (Papargyropoulou et al., 2019). Additionally, restaurants can make stock, soups, and sauces using peels, trimmings, skins, and bones (Clowes et al., 2019), reducing costs and food waste generation. Bhajan et al. (2022) found that chefs repurposed leftover fruits and vegetables into jams and used animal carcasses to make stew. Studies show that managers are often reluctant to introduce new menu items that incorporate repurposed food scraps and typically discard ingredients because consumers question the practicality of certain dishes showcased by chefs, mainly when they feature what are perceived as "obscure ingredients" (Pizetti et al., 2024). While innovative methods of repurposing leftover food demonstrate proactive strategies to reduce food waste and can potentially increase consumer loyalty by adding value, relatively few chefs actively explore and implement culinary techniques that transform by-products into full meals (Pizetti et al., 2024).

3.3.11 Government Intervention

Several studies have highlighted the vital role of governmental policies and regulations in the effective reduction of food waste (Chalak et al., 2018; Filimonau et al., 2019; Amicarelli et al., 2021; Dhir et al., 2020; Sucheran & Olanrewaju, 2021; Bilska et al., 2022). The significant benefits associated with waste reduction should serve as an influential motivating factor for policymakers worldwide to take concrete steps and implement measures to address the issue of food waste (WRAP, 2015). Governments must exhibit a higher degree of commitment to confronting the detrimental effects of food waste (Silvennoinen et al., 2019). Filimonau et al. (2023) suggested that governments employ wide-ranging dedicated strategies encompassing waste collection services, training programs for the industry and its consumers, and resources on waste management best practices for the FSI.

Several governments of developed nations have introduced voluntary, non-mandatory programs to control food waste generation (Chalak et al., 2018). Developed countries are usually proactive against food waste and have sophisticated infrastructure and comprehensive regulations to manage this issue (Tamasinga et al., 2022). For example, Germany's approach encompasses strategies, regulatory frameworks, and infrastructure at national and local levels (Leverenz et al., 2021). In 2019, the German Ministry of Food and Agriculture introduced a national strategy to prevent food waste, marking a significant advancement in establishing monitoring systems and working towards reducing food waste targets (Leverenz et al., 2021). Such precedence set by developed nations can guide and assist others in their food waste policy development efforts (Shen et al., 2022). One of the challenges experienced by developing countries is the lack of governmental policies (Sahoo et al.,

2024). China, for example, experiences significant challenges in managing food waste, food donation programs, waste recycling initiatives, and regulations related to food services (Feng et al., 2022; Shen et al., 2023) and can make meaningful progress in addressing food waste by tightening laws and regulations, facilitating the development of organisations such as food banks, and promoting collaborative governance involving the government (Shen et al., 2022).

Another significant concern in some countries, such as South Africa, is the possible legal implications for health and safety risks donated with food (Sakaguchi et al., 2018; Sucheran & Olanrewaju, 2021). Appropriate legislation can mitigate these risks (Bilska et al., 2022). Some countries, such as France and Italy, have introduced appropriate legal acts to encourage donations of unsold food in the FSI (Bilska et al., 2022). Dhir et al. (2020) recommend that governments formalise food donation and create a database to manually log all food waste management data of food service establishments. This enables them to save costs and overcome their resistance to food donation.

Filimonau et al. (2022) recommend extensive strategies by public authorities to reduce food waste substantially, entailing specialised waste collection for the FSI, educational training, and resources on ideal waste management practices for the sector. This approach should entail infrastructure, business education, and consumer awareness to alleviate food waste effectively. However, numerous governments exhibit deficiencies in enforcing precise regulations concerning food waste management (Salehudin et al., 2020). For example, no specific law in South Africa is dedicated to food donation (Akwii et al., 2021; Sucheran & Olanrewaju, 2021). Instead, government regulations are restrictive and include limitations on donating excess edible food (Sucheran & Olanrewaju, 2021). Such policies should be addressed so the FSI can utilise all possible avenues to reduce food wastage (Akwii et al., 2021).

In 2015, the United Nations General Assembly adopted an ambitious agenda for promoting sustainable development globally, including the Sustainable Development Goal (SDG), which targets the reduction of global food waste to halve it at the retail and consumer levels by 2030 (FAO, 2023). In alignment with SDGs, national governments have developed tailored policies to guide food waste management practices at the local level (Jere et al., 2021; Shen et al., 2022). Current policies include enacting laws and regulations, facilitating food donations, promoting waste recycling, raising awareness, providing education, and gathering food waste data (Shen et al., 2022). A South African study by Sucheran and Olanrewaju (2021) recommended that governments should prioritise food waste in the country and develop regulations in favour of the FSI. This could potentially be an approach that governments across the globe may want to consider.

4. CONCLUSION AND FUTURE PERSPECTIVES

Food waste is a complex challenge that requires a holistic approach involving several diverse strategies, and adequate food waste management strategies are crucial for sustainable development. Food services can significantly reduce their environmental footprints and operate sustainably by implementing effective inventory management, portion control, staff training, donation programs, and menu planning. Embracing the strategies discussed in this chapter benefits the environment and leads to cost savings, improved customer satisfaction, and a more substantial commitment to corporate social responsibility. The FSI can effectively reduce food waste by actively improving waste management practices with support from the government and other stakeholders. The ongoing expansion of the tourism industry will continue to exacerbate the escalation of food waste. Several governments globally still demonstrate deficiencies in the fight against food waste; however, their involvement is crucial. Unless measures are taken to alleviate the issue of food waste in the FSI, food service operations will persistently hinder sustainability. Implementing effective food waste management strategies can improve sustainable operations in the food service industry and encourage consumers to behave in an environmentally conscious manner.

Further research should be conducted on contextualisation of the strategies discussed in this literature review, such as accepting doggy bags. Research on doggy bags as a practical approach to food waste management in the FSI is scarce. Since doggy bags are usually met with a stigma of embarrassment in many countries, perhaps research should investigate the motives behind this and educate the industry and its consumers about the sustainability benefits of taking food home instead of throwing it away. Additionally, future studies should consider encouraging government intervention in formulating policies that contextualise the FSI and are specific to the needs of the industry in alignment with the internationally embraced waste management hierarchy.

REFERENCES

Afzal, N., Basit, A., Daniel, A., Ilyas, N., Imran, A., Awan, Z. A., Papargyropoulou, E., Stringer, L. C., Hashem, M., Alamri, S., Bashir, M. A., Li, Y., & Roy, N. (2022). Quantifying food waste in the hospitality sector and exploring its underlying reasons—A case study of Lahore, Pakistan. *Sustainability (Basel)*, 14(6914), 1–17. DOI: 10.3390/su14116914

Aires, C., Saraiva, C., Conceição Fontes, M., Moreira, D., Moura-Alves, M., & Gonçalves, C. (2021). Food waste and qualitative evaluation of menus in public university canteens—Challenges and opportunities. *Foods*, 10(2325), 2325. Advance online publication. DOI: 10.3390/foods10102325 PMID: 34681374

Akwii, E., Leib, E., Shapiro, M., Bechtel, K., & Friedman, R. (2021). South African legal guide to food donation law and policy. *The Global Food Donation Policy Atlas*. Harvard Law School Food Law and Policy Clinic (FLPC). Retrieved May 6, 2024, from https://www.foodbanking.org/wp-content/uploads/2023/01/South-Africa-Legal-Guide-v2.pdf

Amicarelli, V., Poladian, S., Bux, C., & Aluculesei, A. (2021). The role of education towards food waste minimisation. *BASIQ International Conference New Trends in Sustainable Business and Consumption*, June 2021, Foggia, Italy. DOI: 10.24818/BASIQ/2021/07/032

Apostolidis, C., Brown, D., Wijetunga, D., & Kathriarachchi, E. (2021). Sustainable value co-creation at the Bottom of the Pyramid: Using mobile applications to reduce food waste and improve food security. *Journal of Marketing Management*, 37(9-10), 1–31. DOI: 10.1080/0267257X.2020.1863448

Assamoi, B., & Lawryshyn, Y. (2012). The environmental comparison of landfilling vs. incineration of MSW accounting for waste diversion. *Waste Management (New York, N.Y.)*, 32(5), 1019–1030. DOI: 10.1016/j.wasman.2011.10.023 PMID: 22099926

Beizer, J. (2017). Why the French hate doggie bags. *Contemporary French Civilization*, 42(3-4), 373–389. DOI: 10.3828/cfc.2017.22

Berkowitz, S., Marquart, L., Mykerezi, E., Degeneffe, D., & Reicks, M. (2016). Reduced-portion entrées in a worksite and restaurant setting: Impact on food consumption and waste. *Public Health Nutrition*, 19(16), 3048–3054. DOI: 10.1017/S1368980016001348 PMID: 27256403

Betz, A., Buchli, J. J., Göbel, C., & Müller, C. (2015). Food waste in the Swiss food service industry—Magnitude and potential for reduction. *Waste Management (New York, N.Y.)*, 35, 218–226. DOI: 10.1016/j.wasman.2014.09.015 PMID: 25305683

Bhajan, C., Neetoo, H., Hardowar, S., Boodia, N., Driver, M., Chooneea, M., Ramasawmy, B., Goburdhun, D., & Ruggoo, A. (2022). Food waste generated by the Mauritian hotel industry. *Tourism Critiques: Practice and Theory*, 3(2), 120–137. DOI: 10.1108/TRC-04-2022-0010

Bharucha, J. (2018). Tackling the challenges of reducing and managing food waste in Mumbai restaurants. *British Food Journal*, 120(3), 639–649. DOI: 10.1108/BFJ-06-2017-0324

Billore, S., Kobayashi, O., & Wang, O. (2021). Consumer attitudes towards leftover food takeout interventions: A case study of the doggy bag in Japan. *British Food Journal*, 123(9), 1–18. DOI: 10.1108/BFJ-03-2021-0231

Bilska, B., Tomaszewska, M., & Kołożyn-Krajewska, D. (2020). Managing the risk of food waste in food service establishments. *Sustainability (Basel)*, 12(205), 1–18. DOI: 10.3390/su12052050

Bilska, B., Tomaszewska, M., & Kołozyn-Krajewska, M. (2022). The management of meals in food service establishments in the context of food waste—Results of focus group interviews with employees and owners. *Sustainability (Basel)*, 14(9258), 1–17. DOI: 10.3390/su14159258

Bilska, B., Wrzosek, M., Kołozyn-Krajewska, D., & Krajewski, K. (2016). Risk of food losses and potential of food recovery for social purposes. *Waste Management (New York, N.Y.)*, 52, 269–277. DOI: 10.1016/j.wasman.2016.03.035 PMID: 27026493

Block, J. P., Condon, S. K., Kleinman, K., Mullen, J., Linakis, S., Rifas-Shiman, S., & Gillman, M. W. (2013). Consumers' estimation of calorie content at fast food restaurants: Cross-sectional observational study. *British Medical Journal*, 346(may23 3), f2907. DOI: 10.1136/bmj.f2907 PMID: 23704170

Blum, D. (2020). Ways to reduce restaurant industry food waste costs. *International Journal of Applied Management and Technology*, 19(1), 1–12. DOI: 10.5590/IJMAT.2020.19.1.01

Bulhoes, M., Fonseca, M., Pereira, D., & Martins, M. (2022). Evaluation of factors responsible for waste and environmental impacts arising from the production of meals in food services. *Food Science and Technology (Campinas)*, 42, 1–13. DOI: 10.1590/fst.57920

Cerrah, S., & Yigitoglu, V. (2022). Determining the influential factors on restaurant customers' plate waste. *International Journal of Gastronomy and Food Science*, 27, 1–8. DOI: 10.1016/j.ijgfs.2022.100469

Chalak, A., Abou-Daher, C., & Abiad, M. G. (2018). Generation of food waste in the hospitality, retail, and wholesale sectors: Lessons from developed economies. *Food Security*, 10(5), 1279–1290. DOI: 10.1007/s12571-018-0841-0

Chang, Y. Y. C. (2022). All you can eat or all you can waste? Effects of alternate serving styles and inducements on food waste in buffet restaurants. *Current Issues in Tourism*, 25(5), 727–744. DOI: 10.1080/13683500.2020.1870939

Clowes, A., Hanson, C., & Swannell, R. (2019). The business case for reducing food loss and waste: Restaurants. Retrieved May 7, 2024, from https://champions123.org/publication/business-case-reducing-food-loss-and-waste-restaurants

Corina-Mihaela, O., Rezmuve , R., Camelia, M., Cornelia, P., & Nicoleta, R. (2023). Possibilities of recovery of waste from food processing in hospitality industry. *Facultatea de Management Agricol*, 3, 296–303.

Damiani, M., Pastorello, T., Carlesso, A., Tesser, S., & Semenzin, E. (2021). Quantifying environmental implications of surplus food redistribution to reduce food waste. *Journal of Cleaner Production*, 289, 125813. DOI: 10.1016/j.jclepro.2021.125813

Dhir, A., Talwar, S., Kaur, P., & Malibari, A. (2020). Food waste in hospitality and food services: A systematic literature review and framework development approach. *Journal of Cleaner Production*, 270(122861), 1–16. DOI: 10.1016/j.jclepro.2020.122861

Diwate, A., & Aditeya, N. (2023). Food waste management in restaurants. *IOSR Journal of Humanities and Social Sciences*, 28(112), 19–23.

Eriksson, M., Malefors, C., Callewaert, P., Hartikainen, H., Pietilainen, O., & Strid, I. (2019). What gets measured gets managed, or does it? The connection between food waste quantification and food waste reduction in the hospitality sector. *Resources, Conservation and Recycling*, 4, 1–8.

Eriksson, M., Osowski, C., Björkman, J., Hansson, E., Malefors, C., Eriksson, E., & Ghosh, R. (2018). The tree structure: A general framework for food waste quantification in food services. *Resources, Conservation and Recycling*, 130, 140–151. DOI: 10.1016/j.resconrec.2017.11.030

Eriksson, M., Osowski, C., Malefors, C., Björkman, J., & Eriksson, E. (2017). Quantifying food waste in public catering services—A case study from a Swedish municipality. *Waste Management (New York, N.Y.)*, 61, 415–422. DOI: 10.1016/j.wasman.2017.01.035 PMID: 28161338

Feng, Y., Marek, C., & Tosun, J. (2022). Fighting food waste by law: Making sense of the Chinese approach. *Journal of Consumer Policy*, 45(3), 457–479. DOI: 10.1007/s10603-022-09519-2 PMID: 35729934

Filimonau, V. (2019). Bin it and forget it! The challenges of food waste management in restaurants of a mid-sized Hungarian city. *Tourism Management Perspectives*, 37(10059), 1–12.

Filimonau, V., & De Coteau, D. A. (2019). Food waste management in hospitality operations: A critical review. *Tourism Management*, 71, 234–245. DOI: 10.1016/j.tourman.2018.10.009

Filimonau, V., & Ermolaev, V. A. (2022). Exploring the potential of industrial symbiosis to recover food waste from the food service sector in Russia. *Sustainable Production and Consumption*, 29, 467–478. DOI: 10.1016/j.spc.2021.10.028

Filimonau, V., Fidan, H., Alexieva, L., Dragoev, S., & Marinova, D. (2019). Restaurant food waste and the determinants of its effective management in Bulgaria: An exploratory case study of restaurants in Plovdiv. *Tourism Management Perspectives*, 22(11), 1–11. DOI: 10.1016/j.tmp.2019.100577

Filimonau, V., Krivcova, M., & Pettit, F. (2019). An exploratory study of managerial approaches to food waste mitigation in coffee shops. *International Journal of Hospitality Management*, 76, 48–57. DOI: 10.1016/j.ijhm.2018.04.010

Filimonau, V., Matyakubov, U., Allonazarov, O., & Ermolaev, V. (2022). Food waste and its management in restaurants of a transition economy: An exploratory study of Uzbekistan. *Sustainable Production and Consumption*, 29, 25–35. DOI: 10.1016/j.spc.2021.09.018

Food and Agriculture Organization. (2019). *The state of food and agriculture: Moving forward on food loss and waste reduction.* FAO. Retrieved May 12, 2024, from https://openknowledge.fao.org/server/api/core/bitstreams/11f9288f-dc78-4171-8d02-92235b8d7dc7/content

Food and Agriculture Organization. (2022). *Innovative approaches to reduce, recycle and reuse food waste*. FAO. Retrieved May 6, 2024, from https://openknowledge.fao.org/server/api/core/bitstreams/9daa60fc-dbcf-416e-911f-443aedb7ee80/content

Food and Agriculture Organization. (2023). *Innovative approaches to reduce, recycle and reuse food waste*. Retrieved May 6, 2024, from https://www.fao.org/3/cb8841en/cb8841en.pdf

Garrone, P., Melacini, M., & Perego, A. (2014). Opening the black box of food waste reduction. *Food Policy*, 46, 129–139. DOI: 10.1016/j.foodpol.2014.03.014

Gonçalves, C., Saraiva, S., Nunes, F., & Saraiva, C. (2023). Food waste in public food service sector—Surplus and leftovers. *Resources*, 12(10), 120. DOI: 10.3390/resources12100120

Goosens, Y., Leverenz, D., & Kuntscher, M. (2022). Waste-tracking tools: A business case for more sustainable and resource-efficient food services. *Resources, Conservation & Recycling Advances*, 15(200112), 1–14. DOI: 10.1016/j.rcradv.2022.200112

Gupta, A., Sharma, R., & Smith, E. (2022). Predictive analytics for food demand: A machine learning approach in the hospitality industry. *Journal of Foodservice Business Research*, •••, 1–19.

Harduth, N. (2017). Food donations as a mechanism to limit food waste, food insecurity, and malnutrition, economic and environmental sustainability. *Sustainability*, 12, 2892.

Heikkilä, L., Reinikainen, A., Katajajuuri, J.-M., Silvennoinen, K., & Hartikainen, H. (2016). Elements affecting food waste in the food service sector. *Waste Management (New York, N.Y.)*, 56, 446–453. DOI: 10.1016/j.wasman.2016.06.019 PMID: 27373724

Hennchen, B. (2019). Knowing the kitchen: Applying practice theory to issues of food waste in the food service sector. *Journal of Cleaner Production*, 225, 675–683. DOI: 10.1016/j.jclepro.2019.03.293

Herpen, E., De Hooge, I., De Visser-Amundson, A., & Kleijnen, M. (2021). Take or leave it: How an opt-out strategy for doggy bags affects consumer food waste behavior and restaurant evaluations. *Journal of Cleaner Production*, 325(129199), 1–11. DOI: 10.1016/j.jclepro.2021.129199

Hong, J., Jaegler, A., & Gergaud, O. (2024). Mobile applications to reduce food waste in supply chains: A systematic literature review. *British Food Journal*, 126(2), 1–23. DOI: 10.1108/BFJ-09-2022-0742

Ishangulyyev, R., Kim, S., & Lee, S. (2019). Understanding food loss and waste: Why are we losing and wasting food? *Foods*, 8(297), 1–23. DOI: 10.3390/foods8080297 PMID: 31362396

Jones, K., Martin, S., & Patel, N. (2022). Food waste reduction in the hospitality industry: A review of recent research and technological advancements. *Journal of Sustainable Tourism*, •••, 1–22.

Kallbekken, S., & Sælen, H. (2013). Nudging hotel guests to reduce food waste is a win-win environmental measure. *Economics Letters*, 119(3), 325–327. DOI: 10.1016/j.econlet.2013.03.019

Karakas, K. M. (2021). Food waste management in luxury hotels: Best practices and limitations. *Journal of Environmental Science & Engineering*, 10, 103–111.

Knorr, D., & Augustin, D. (2022). From food to gods to waste. *Critical Reviews in Food Science and Nutrition*. Advance online publication. DOI: 10.1080/10408398.2022.2153795 PMID: 36503306

Laakso, S. (2017). Creating new food practices: A case study on leftover lunch service. *Food, Culture, & Society*, 20(4), 631–650. DOI: 10.1080/15528014.2017.1324655

Leverenz, D., Schneider, F., Schmidt, T., Hafner, G., Nevarez, Z., & Kranert, M. (2021). Food waste generation in Germany in the scope of European legal requirements for monitoring and reporting. *Sustainability (Basel)*, 13(6616), 1–23. DOI: 10.3390/su13126616

Lins, M., Zandonadi, R., Raposo, A., & Ginani, V. (2021). Food waste on foodservice: An overview through the perspective of sustainable dimensions. *Foods*, 10(1175), 1–14. DOI: 10.3390/foods10061175 PMID: 34073708

Malefors, C., Callewaert, P., Hansson, P., Hartikainen, H., Pietiläinen, O., Strid, I., Strotmann, C., & Eriksson, M. (2019). Towards a baseline for food-waste quantification in the hospitality sector: Quantities and data processing criteria. *Sustainability (Basel)*, 11(13), 3541. DOI: 10.3390/su11133541

Martin-Rios, C., Demen-Meier, C., Gossling, S., & Cornuz, C. (2018). Food waste management innovations in the foodservice industry. *Waste Management (New York, N.Y.)*, 79, 196–206. DOI: 10.1016/j.wasman.2018.07.033 PMID: 30343746

Martin-Rios, C., Hofmann, A., & Mackenzie, N. (2020). Sustainability-oriented innovations in food waste management technology. *Sustainability (Basel)*, 13(1), 210. DOI: 10.3390/su13010210

Martin-Rios, C., Meier, C., & Pasamar, P. (2022). Sustainable waste management solutions for the foodservice industry: A Delphi study. *Waste Management & Research*, 40(9), 1335–1457. DOI: 10.1177/0734242X221079306 PMID: 35176923

Mattila, M., Mesiranta, N., & Heikkinen, A. (2020). Platform-based sustainable business models: Reducing food waste in food services. *International Journal of Entrepreneurship and Innovation Management*, 24(5), 249–265. DOI: 10.1504/IJEIM.2020.108258

McAdams, B., von Massow, M., Gallant, M., & Hayhoe, M. A. (2019). A cross-industry evaluation of food waste in restaurants. *Journal of Foodservice Business Research*, 22(5), 449–466. DOI: 10.1080/15378020.2019.1637220

Mirosa, M., Liu, Y., & Mirosa, R. (2018). Consumers' behaviors and attitudes toward doggy bags: Identifying barriers and benefits to promoting behavior change. *Journal of Food Products Marketing*, 24(5), 563–590. DOI: 10.1080/10454446.2018.1472699

Montesdeoca-Calderón, M., Gil-Saura, I., Ruiz-Molina, M., & Martin-Rios, C. (2024). Tackling food waste management: Professional training in the public interest. *International Journal of Gastronomy and Food Science*, 35(100863), 1–8. DOI: 10.1016/j.ijgfs.2023.100863

Montoli, P., Ares, G., Aschemann-Witzel, J., Curutchet, M. R., & Giménez, A. (2023). Food donation as a strategy to reduce food waste in an emerging Latin American country: A case study in Uruguay. *Nutrire*, 48(1), 22. DOI: 10.1186/s41110-023-00208-9 PMID: 38625195

Nicastro, R., & Carillo, P. (2021). Food loss and waste prevention strategies from farm to fork. *Sustainability (Basel)*, 13(10), 5443. DOI: 10.3390/su13105443

Okumus, B. (2020). How do hotels manage food waste? Evidence from hotels in Orlando, Florida. *Journal of Hospitality Marketing & Management*, 29(3), 291–309. DOI: 10.1080/19368623.2019.1618775

Okumus, B., Taheri, B., Giritlioglu, I., & Gannon, M. (2020). Tackling food waste in all-inclusive resort hotels in Bendegul. *International Journal of Hospitality Management*, 88(102543), 1–10.

Omari, A., Chafi, A., Mokhtari, O., Zraibi, L., Azzakhnin, I., Ben, B., Mchiouer, K., & Asri, O. (2023). Development of a food waste prevention strategy: A case study in a hotel in Eastern Morocco. *Web of Conferences, 364*, 1–5. DOI: 10.1051/e3sconf/202336402009

Papargyropoulou, E., Wright, N., Lozano, R., Steinberger, J., Padfield, R., & Ujang, Z. (2016). Conceptual framework for the study of food waste generation and prevention in the hospitality sector. *Waste Management (New York, N.Y.)*, 49, 326–336. DOI: 10.1016/j.wasman.2016.01.017 PMID: 26803473

Papargyropoulou, E., Wright, N., Lozano, R., Steinberger, J., Padfield, R., & Ujang, Z. (2019). Patterns and causes of food waste in the hospitality and food service sector: Food waste prevention insights from Malaysia. *Sustainability (Basel)*, 11(21), 6016. DOI: 10.3390/su11216016

Patel, S., Dora, M., Hahladakis, J. N., & Iacovidou, E. (2021). Opportunities, challenges, and trade-offs with decreasing avoidable food waste in the UK. *Waste Management & Research*, 39(3), 473–488. DOI: 10.1177/0734242X20983427 PMID: 33517869

Pinto, R. S., Dos Santos Pinto, R. M., Melo, F. F. S., Campos, S. S., & Cordovil, C. (2018). A simple awareness campaign to promote food waste reduction in a university canteen. *Waste Management (New York, N.Y.)*, 76, 28–38. DOI: 10.1016/j.wasman.2018.02.044 PMID: 29503053

Pirani, S. I., & Arafat, H. A. (2016). Reduction of food waste generation in the hospitality industry. *Journal of Cleaner Production*, 132, 129–145. DOI: 10.1016/j.jclepro.2015.07.146

Pizetti, M., Longo, C., & Ture, M. (2024). Embracing food well-being: Lessons from chefs' caring actions in the fight. *Journal of Public Policy & Marketing*, 43(2), 171–189. DOI: 10.1177/07439156231206791

Principato, L., Leo, A., Mattia, G., & Pratesi, C. (2021). The next step in sustainable dining is the restaurant food waste map for managing food waste. *Italian Journal of Marketing*, 189–204.

Principato, L., Pratesi, C. A., & Secondi, L. (2018). Towards zero waste: An exploratory study on restaurant managers. *International Journal of Hospitality Management*, 74, 130–137. DOI: 10.1016/j.ijhm.2018.02.022

Rady, A., Dawood, A., Taleb, M., & Adel, M. (2021). Food waste in hotel restaurants: Causes, reducing practices, and barriers. *Minia Journal of Tourism and Hospitality Research*, 12(1), 135–155. DOI: 10.21608/mjthr.2021.106361.1018

Read, Q. D., & Muth, M. K. (2021). Cost-effectiveness of four food waste interventions: Is food waste reduction a "win–win"? *Resources, Conservation and Recycling*, 168, 105448. DOI: 10.1016/j.resconrec.2021.105448

Richardson, R., Prescott, M., & Ellison, B. (2021). Impact of plate shape and size on individual food waste in a university dining hall. *Resources, Conservation and Recycling*, 168, 105293. DOI: 10.1016/j.resconrec.2020.105293

Rodrigues, R., Miguéis, V., Freitas, S., & Machado, T. (2024). Machine learning models for short-term demand forecasting in food catering services: A solution to reduce food waste. *Journal of Cleaner Production*, 435, 140265. DOI: 10.1016/j.jclepro.2023.140265

Sahoo, A., Dwivedi, A., Madheshiya, P., Kumar, U., Sharma, R., & Tiwari, S. (2024). Insights into the management of food waste in developing countries: With special reference to India. *Environmental Science and Pollution Research International*, 31(12), 17887–17913. DOI: 10.1007/s11356-023-27901-6 PMID: 37271790

Sakaguchi, L., Pak, N., & Potts, M. (2018). Tackling the issue of food waste in restaurants: Options for measurement method, reduction, and behavioral change. *Journal of Cleaner Production*, 180, 430–436. DOI: 10.1016/j.jclepro.2017.12.136

Samsuddin, Z., Zainal, N., Sulong, S., & Bakar, N. (2022). Handling food waste in the hotel industry. *Malaysian Journal of Social Sciences and Humanities*, 7(11), 1–11. DOI: 10.47405/mjssh.v7i11.1926

Santos, J., Chen, Y., & Kim, H. (2021). IoT-based food safety and inventory management system in the hospitality industry. *International Journal of Hospitality Management*, 93, 102796.

Shen, G., Hong, T., Wang, K., Gu, Y., Han, J., & Guo, Y. (2023). The status of the global food waste mitigation policies: Experience and inspiration for China. *Environment, Development and Sustainability*, 26(4), 1–29. DOI: 10.1007/s10668-023-03132-0 PMID: 37362998

Silvennoinen, K., Nisonen, S., & Pietiläinen, O. (2019). Food waste case study and monitoring development in Finnish food services. *Waste Management (New York, N.Y.)*, 97, 97–104. DOI: 10.1016/j.wasman.2019.07.028 PMID: 31447032

Singh, V., Archana, T., Singh, A., & Tyagi, P. (2024). Utilising technology for food waste management in the hospitality industry: Hotels and restaurants. *Sustainable Disposal Methods of Food Wastes in Hospitality Operations*, 287–295.

Sirieix, L., Lála, J., & Kocmanová, K. (2017). Understanding the antecedents of consumers' attitudes towards doggy bags in restaurants: Concern about food waste, culture, norms, and emotions. *Journal of Retailing and Consumer Services*, 34, 15–158. DOI: 10.1016/j.jretconser.2016.10.004

Sirieix, L., Lála, J., & Kocmanová, K. (2017). Understanding the antecedents of consumers' attitudes towards doggy bags in restaurants: Concern about food waste, culture, norms, and emotions. *Journal of Retailing and Consumer Services*, 34, 153–158. DOI: 10.1016/j.jretconser.2016.10.004

Smith, A., Johnson, M., & Davis, R. (2023). Robotic kitchen systems for food waste reduction in restaurants. *International Journal of Robotics and Automation*, 38(2), 87–103.

Stirnimann, A., & Zizka, L. (2022). Waste not, want not: Managerial attitudes towards mitigating food waste in the Swiss-German restaurant industry. *Journal of Foodservice Business Research*, 25(3), 302–328. DOI: 10.1080/15378020.2021.1942749

Sucheran, S., & Olanrewaju, O. (2021, March 7-11). Food waste management of restaurants in KwaZulu-Natal South Africa. In *Proceedings of the 11th Annual International Conference on Industrial Engineering and Operations Management*. Singapore. DOI: 10.46254/AN11.20210011

Sundin, N., Bartek, L., Osowski, C., Strid, I., & Eriksson, M. (2023). Sustainability assessment of surplus food donation: A transfer system generating environmental, economic, and social values. *Sustainable Production and Consumption*, 38, 41–54. DOI: 10.1016/j.spc.2023.03.022

Tahir, M. (2023). Determinants of food waste: TPB and moderating impact of demographics & guilt. *Journal of Global Hospitality and Tourism*, 2(2), 157–182.

Talwar, S., Kaur, P., Okumus, B., Ahmed, U., & Dhir, A. (2021). Food waste reduction and taking away leftovers: Interplay of food-ordering routine, planning routine, and motives. *International Journal of Hospitality Management*, 98, 103033. DOI: 10.1016/j.ijhm.2021.103033

Talwar, S., Kaur, P., Yadav, R., Sharma, R., & Dhir, A. (2021). Food waste and out-of-home dining: Antecedents and consequents of the decision to take away leftovers after dining at restaurants. *Journal of Sustainable Tourism*, •••, 1–26.

Tamasinga, P., Miri, T., Onyeaka, H., & Hart, A. (2022). Food waste and circular economy: Challenges and opportunities. *Sustainability*, 14(9868), 1–32.

Tomaszewska, M., Bilska, B., Tul-Krzyszczuk, A., & Kołozyn-Krajewska, D. (2021). Estimating the scale of food waste in hotel food services: A case study. *Sustainability (Basel)*, 13(1), 421. DOI: 10.3390/su13010421

United Nations Environmental Program (UNEP). (2021). *Food waste index report 2021*. Retrieved May 20, 2024, from https://wedocs.unep.org/20.500.11822/35280

Vaidyanathan, J., & Shankar, M. (2020). A study on the ecological consequences of food wastage. *International Journal of Pharma Research*, 11(1), 11–17.

Vizzoto, F., Testa, F., & Iraldo, F. (2021). Strategies to reduce food waste in the foodservices sector: A systematic review. *International Journal of Hospitality Management*, 95, 102933. DOI: 10.1016/j.ijhm.2021.102933

Wang, L., Liu, G., Liu, X., Liu, Y., Gao, J., Zhou, B., Gao, S., & Cheng, S. (2017). The weight of unfinished plate: A survey-based characterisation of restaurant food waste in Chinese cities. *Waste Management (New York, N.Y.)*, 66, 3–12. DOI: 10.1016/j.wasman.2017.04.007 PMID: 28438432

Wang, L., Xue, L., Li, Y., Liu, X., Cheng, S., & Liu, G. (2018). Horeca food waste and its ecological footprint in Lhasa, Tibet, China. *Resources, Conservation and Recycling*, 136, 1–8. DOI: 10.1016/j.resconrec.2018.04.001

Wang, Y., Yuan, Z., & Tang, Y. (2020). Enhancing food security and environmental sustainability: A critical review of food loss and waste management. *Resources. Environmental Sustainability*, 4, 100023.

Waste and Resources Action Programme (WRAP). (2015). *Strategies to achieve economic and environmental gains by reducing food waste*. Retrieved May 12, 2024, from https://newclimateeconomy.report/workingpapers/wpcontent/uploads/sites/5/2016/04/WRAP-NCE_Economic-environmental-gains-food-waste.pdf

Wen, Z., Hu, S., De Clercq, D., Beck, M. B., Zhang, H., Zhang, H., Fei, F., & Liu, J. (2018). Design, implement, and evaluate an Internet of Things (IoT) network system for restaurant food waste management. *Waste Management (New York, N.Y.)*, 73, 26–38. DOI: 10.1016/j.wasman.2017.11.054 PMID: 29242117

Wu, C. E., & Teng, C. (2023). Reducing food waste in buffet restaurants: A corporate management approach. *Multidisciplinary Digital Publishing Institute*, 12, 1–12. PMID: 36613378

Wu, Y., Tian, X., Li, X., Yuan, H., & Liu, G. (2019). Characteristics, influencing factors, and environmental effects of plate waste at university canteens in Beijing, China. *Resources, Conservation and Recycling*, 149, 151–159. DOI: 10.1016/j.resconrec.2019.05.022

Zhang, B., Lai, K., Wang, B., & Wang, Z. (2019). From intention to action: How do personal attitudes, facilities accessibility, and government stimulus matter for household waste sorting? *Journal of Environmental Management*, 233, 447–458. DOI: 10.1016/j.jenvman.2018.12.059 PMID: 30593004

Chapter 17
Revolutionizing Responsible Consumption and Production of Food In Hospitality Industry: Machine Learning and Deep Learning Approach Towards Zero Food Waste

Bhupinder Singh
https://orcid.org/0009-0006-4779-2553
Sharda University, India

Christian Kaunert
https://orcid.org/0000-0002-4493-2235
Dublin City University, Ireland & University of South Wales, UK

ABSTRACT

The hospitality industry is pivotal in addressing responsible consumption and food production challenges. The hotel sector is leading the way in tackling important issues related to responsible food production and consumption to lead revolutionary solutions using deep learning and machine learning technology. Integrating modern technical solutions becomes crucial in the current setting, where global poverty, malnutrition, and the complex dynamics of food adulteration considerably affect the industry's demand-supply mechanism. Keeping a close eye on the larger global scene,

DOI: 10.4018/979-8-3693-6110-8.ch017

this study promotes practical actions beyond simple technology implementations. In the face of global issues such as hunger, malnutrition and food adulteration, the assimilation of advanced technological strategies becomes imperative. This chapter explores the role of ML and DL in achieving zero waste within the hospitality sector, addressing challenges and presenting innovative strategies for adequate food loss-waste management.

1. INTRODUCTION

Beyond technological applications, the study highlights critical aspects such as inventory management, employee training, customer education, and collaboration with food bank institutions (Koshariya et al., 2024; Gatto & Chepeliev, 2024; Axmann et al., 2024). The primary goal is to reduce the environmental impact of food waste, address global challenges, and promote sustainable tourism practices within the hospitality sector (Lawal et al., 2024; Font et al., 2023). This chapter marks a significant step towards aligning the industry with responsible and sustainable practices, positively shaping the future of food consumption and production (Delgado et al., 2023). In pursuit of zero waste in hospitality, the study carefully examines the roles of Machine Learning (ML) and Deep Learning (DL), offering a thorough analysis of the challenges involved and proposing innovative solutions for managing food loss and waste (Kalenjuk Pivarski et al., 2022). Food waste within the hotel industry is an escalating concern, accounting for a significant portion of all food waste. The issue of hospitality waste is particularly prominent in both developed and developing nations due to the rise in dining out, fueled by increasing incomes and tourism (Lund-Durlacher et al., 2019).

The amount of food-related waste generated in the hospitality industry is often discussed in the media, yet it has received insufficient scholarly attention (Ozturkcan & Kitapci, 2023; Han et al., 2020). A consumption pattern that aligns with sustainable development principles is considered responsible (Riva et al., 2022). Essentially, it involves obtaining products and services to support the three pillars of sustainability—economic, social, and environmental—without prioritising one over the others (Baltescu, 2022; Hwang & Kim, 2021). This approach benefits the parties involved in transactions and stimulates the economy, particularly the local economy, by facilitating the exchange of goods and services (Hall, 2010). It fosters positive social impact by ensuring that the goods or services purchased are tied to a fairly compensated workforce and work in favourable conditions, enhancing customer welfare and considering their health (Han, 2021). The responsible consumer considers all aspects of a product's life cycle—from production and shipping to disposal—and strives to choose products that minimise environmental impact (Voukkali et al., 2023).

1. BACKGROUND OF STUDY

Food waste may be defined as using food meant for human consumption for non-consumptive uses, repurposing food as animal feed, or throwing away edible food (Bux & Amicarelli, 2022). This includes edible and inedible food components taken from the food chain and can be disposed of or retrieved (Chawla et al., 2022) (Alvarez-Risco et al., 2020). Three categories exist for food waste: (a) avoidable waste, which is defined as food that was edible at first but turns inedible by the time it is disposed of; (b) unavoidable waste, which includes non-edible items like eggshells; and (c) potentially avoidable food waste, which is used to describe particular wastes that are occasionally consumed, like potato skins (Amicarelli et al., 2022). The primary issue at hand is the widespread problem of food waste, which continues to occur on a fundamental level (Legrand et al., 2022) (ElShafei, 2022) (Lee et al., 2021). It is estimated that one-third of worldwide food is wasted (Chung, 2020) (Malheiro et al., 2020). Despite being a longstanding concern (Yoon et al., 2020), a permanent solution to food waste has yet to be established (Mirela et al., 2014) (Sharma et al., 2023) (Barrett et al., 2023). To address this issue, researchers are exploring the potential of new machine learning and deep learning techniques that are specifically designed to combat food waste (Labadi et al., 2021) (Araujo, 2023) (Shams et al., 2023) (NaThalang, 2019). The aim is to successfully manage and reduce food waste by implementing a model in the residence halls at our university (Fitchuk, 2021) (Sartor, 2021) (Voorhoeve, 2021). Specifically, the idea is to utilise artificial intelligence to develop innovative approaches to the problem of food waste in the hospitality sector (Singh, 2023) (Hunter et al., 2020).

"Sustainable" and "responsible" consumption are interchangeable terms, with the former having a narrower focus on environmental impact (Bisaria, 2022) (Singh, 2023). Sustainable consumption refers explicitly to the environmental consequences of purchasing products and services (Atser et al., 2019) (Sharma & Singh, 2022). According to Figure 1, it encompasses three key elements: choosing environmentally friendly products, adopting a more efficient attitude towards consumption, and disposing of items properly, with recycling in mind (Tice, 2024; Toivonen, 2020; Singh, 2022; Miconi, 2022).

Figure 1. Three aspects involved in sustainable consumption

Source: Author's Compilation

The primary cause of food waste is ineffective food supply chain management (Vind, 2023; Singh, 2022). Food is often packaged and shipped to stores without proper quality inspections, leading to the display of substandard products (Spiess-Knafl, 2022). This increases the likelihood of food waste, as consumers may unknowingly purchase inferior goods (Kilibarda et al., 2019). To address this issue, food manufacturers are exploring AI-powered inspection systems that assess food quality before consumption (Munir, 2022; Goh & Jie, 2019). These systems utilise machine learning (ML) and deep learning (DL) technologies to automatically monitor, analyse, and evaluate food quality, detecting hidden patterns that are imperceptible to the human eye (Singh, 2019; Goh et al., 2022). Additionally, manufacturers could prioritise distributing certain food items at risk of rapid spoilage to reduce waste (Aditya & Kurniawati, 2023; Kasavan et al., 2022). This chapter has two broad objectives:

- Highlight the need for responsible consumption and production in the hospitality sector.
- Examine the role of machine learning and deep learning in achieving zero food waste.

This chapter explores the various aspects of revolutionising responsible consumption and production in the hospitality industry through machine learning (ML) and deep learning (DL) to achieve zero food waste. Section 2 covers the current landscape of food waste in hospitality, while Section 3 explores the role of ML and DL in reducing food waste. Section 4 outlines specific applications of these technologies for waste reduction. Section 5 presents key findings and viable solutions, and Section 6 concludes the chapter by discussing future directions and potential scope for further research.

2. CURRENT LANDSCAPE OF FOOD WASTE IN HOSPITALITY

Many stakeholders in the tourism and hospitality industries agree that food waste is a global issue (Katuwal & Afzal, 2022; Filimonau et al., 2022). It places unnecessary strain on energy, arable land, and freshwater resources (Ozdemir & Guicer, 2018; Dhir et al., 2020). Nutritionists emphasise the significant nutrient losses caused by food waste (Kasavan et al., 2019; Luu, 2020), which exacerbates social inequality, threatens global food security, and leads to financial losses (Chalak et al., 2018; Ruiz Molina et al., 2022). Globally, over 30% of food intended for human consumption is lost, contributing to resource depletion and a substantial carbon footprint (Lee et al., 2023; Lagioia et al., 2023). As much as 40% of food is wasted or discarded, creating costly policy challenges in managing food loss (de Visser-Amundson, 2022). With over 820 million people facing hunger worldwide, this food waste is especially troubling, raising serious ethical and social equity concerns and wasting valuable resources (Islam et al., 2023).

Food waste occurs throughout the hospitality supply chain, from production and preparation to consumption and disposal, involving several key factors:

- Ineffective demand forecasting, leading to overproduction and spoilage due to inaccurate predictions (Sharma, 2020).
- Poor inventory management, resulting in overstocking and product expiration due to improper storage (Filimonau et al., 2019).
- Issues with portion control, where excessive serving sizes and inefficient disposal methods contribute to plate waste (Meier et al., 2021).
- Customer behaviour, where careless and uninformed consumption habits increase food waste (Scholz et al., 2020).

The hospitality industry is exceptionally complex regarding food waste, as it produces significantly more food than other sectors due to the need to serve hundreds, if not thousands or millions, of consumers daily. Estimates suggest that 10% of hospitality consumables are discarded, leading to substantial environmental, economic, and social consequences. Unsurprisingly, food waste in this sector results from overproduction, storage issues, fluctuations in demand during service times, and oversized portion sizes. For instance, commercial food services like buffets, catering events, and restaurants tend to overproduce to meet customer needs until closing, keeping buffets fully stocked until the last minute (Tomaszewska et al., 2020). This practice generates large amounts of edible waste, leading to financial losses and contributing to greenhouse gas emissions when food is discarded. The monetary value of food waste is significant, directly affecting the profitability of hospitality businesses. As a result, the industry has become increasingly conscious of the need to reduce food waste through improved inventory management and donation programs. Some businesses are even turning to technology-based solutions to enhance demand forecasting. Additionally, zero-waste kitchens and sustainability certifications are gaining popularity as the industry recognises the critical role of reducing food waste in building a sustainable and economically viable future (Okumus et al., 2020).

Since dining is an integral part of vacationing, it is no surprise that tourism contributes to global food waste. However, there is a lack of research explicitly examining food waste generated by tourists. Studies on food waste in popular tourist destinations' hotel, restaurant, and café sectors often fail to differentiate between waste produced by tourists and that generated by locals. This is primarily because it is operationally challenging to separate food waste from local sources and that produced by visitors (Camilleri, 2021). To better understand the issue, more research is needed on 1) the amount of food waste generated by tourists compared to residents and their consumption habits at home (Omar et al., 2023); 2) the specific types of food most wasted by tourists (Filimonau et al., 2020); and 3) the underlying causes of food wastage among tourists (Amicarelli & Bux, 2021). A deeper understanding of the scale of food waste in tourism, the main types of food waste, and the key factors driving this waste will help develop more effective strategies for reducing food waste (Principato et al., 2018).

3. MACHINE LEARNING AND DEEP LEARNING IN FOOD WASTE REDUCTION: HOW ML OR DL LEADS TO ZERO/REDUCED FOOD WASTE IN HOSPITALITY SECTOR

Food loss and waste are significant issues as they negatively impact local and global food availability, diminish farmer income, elevate food prices for consumers, undermine human health and nutrition and contribute to environmental harm through the excessive use of natural resources (Filimonau & Uddin, 2021). Food waste is initially intended for human consumption that exits the human food chain (Singh et al., 2023). It is still considered wasted food even if it is diverted to non-food uses such as animal feed or bioenergy (Singh et al., 2024). Food initially intended for human consumption may exit the food value chain at various stages (Park et al., 2024). For instance, due to low market prices, crops might remain unharvested during production (Kumar et al., 2024). During transportation, food may spoil due to elevated humidity or temperatures (Singh et al., 2022). At the industrial level, errors in packaging or labelling can lead to food wastage (Singh et al., 2022). Upon reaching distribution or retail centres, food might be discarded due to expiration dates or damage incurred during handling (Singh et al., 2023). At the household level, food can be wasted due to expiration or when consumers do not consume all the food they purchase or prepare (Singh et al., 2022).

Figure 2. Specifies the key genesis of machine learning and deep learning in food waste reduction

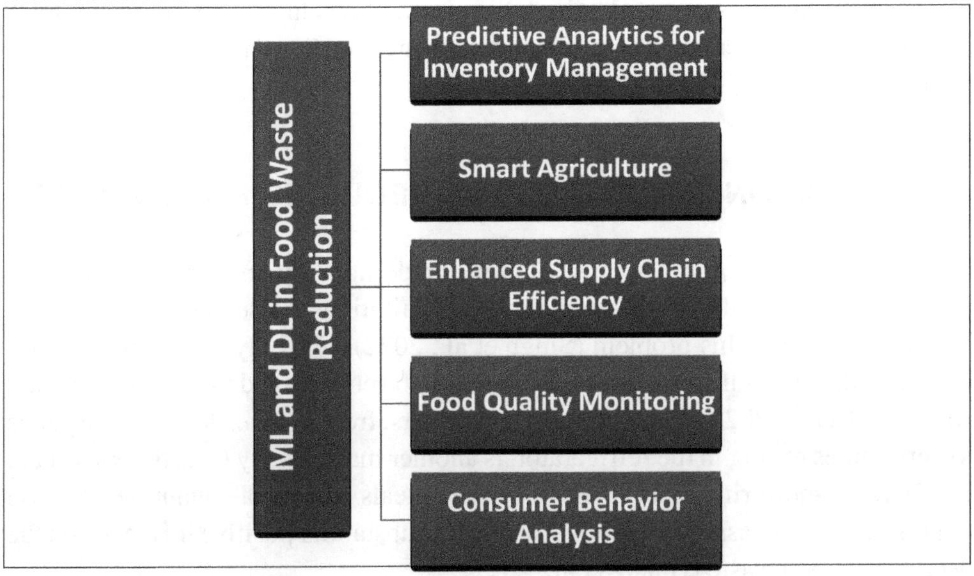

Source: Author's Compilation

ML and DL are revolutionising the fight against food waste by improving accuracy and offering predictive models to address surplus food with data-driven solutions for production and consumption. These technologies analyse large datasets, such as historical sales, customer preferences, and seasonal trends, to provide more accurate demand forecasting (Singh et al., 2021). By predicting how much food is needed, ML and DL help minimise overproduction, even preventing small amounts of excess that would otherwise be wasted. ML-powered inventory management systems also monitor use-by dates and track food usage to ensure optimal consumption, further reducing waste. Deep learning models can also analyse food waste patterns, allowing hospitality businesses to adjust practices like portion sizes and service types to align with consumption habits. Continuously evolving with each new data point, ML and DL are guiding the hospitality industry toward zero waste by offering flexible solutions that enhance sustainability and profitability (Singh et al., 2022).

Machine learning provides several approaches to reduce food waste (Figure 2). Using reinforcement learning algorithms is one way to improve retail establishments' decision-making about pricing and product ordering. This double effect can reduce waste and increase profitability simultaneously (Kumar et al., 2023). Also, by optimising many facets of food processing, ingredient use and manufacturing efficiency, machine learning approaches may lower waste and increase cost-effectiveness in the food business (Singh et al., 2021). Also, by repurposing food waste for beneficial uses, machine learning may help manage and monitor it in local food supply chains (Singh et al., 2021). The food processing business may reduce waste and improve management by utilising artificial intelligence and machine learning to optimise supply chain, logistics, crop selection, maintenance prediction and perishable food tracking (Choi et al., 2022).

4. APPLICATIONS ML AND DL FOR REDUCING FOOD WASTE

Food waste is the primary cause of global undernourishment, emphasising how urgent it is for agricultural and governmental institutions to use machines and deep learning to combat this problem (Singh et al., 2022). Not only may hunger issues be resolved, but significant cost savings can also be encouraged by minimising food waste (Park et al., 2022). Food waste by consumers, from uneaten leftovers on plates to perishables rotting in the refrigerator, is another major worry (Martin-Rios et al., 2022). By empowering individuals to prepare meals in sensible quantities, the use of personalised AI-assisted cooking can have a substantial positive influence on the decrease of food waste (Huang et al., 2021).

Figure 3. Monitoring hunger and food insecurity in the world

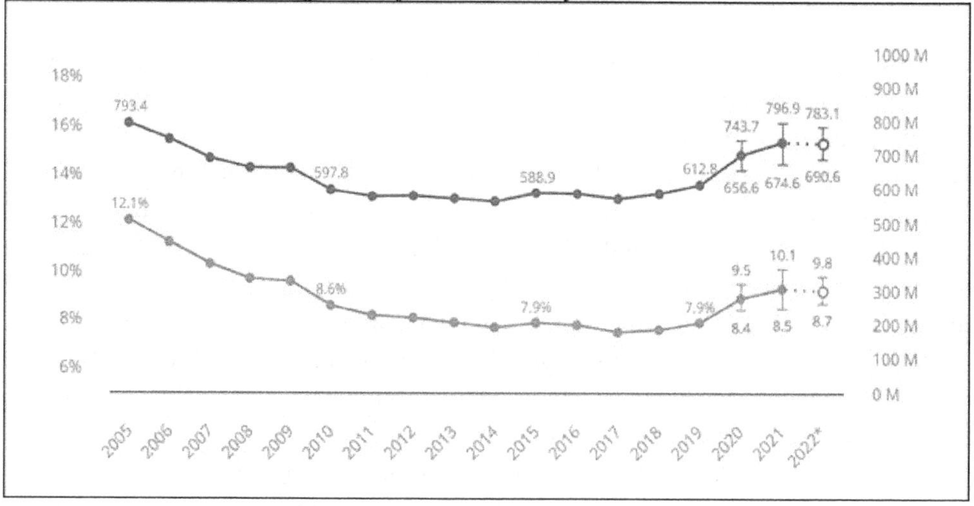

Source: Food and Agriculture Organization (2023)

Around the world, countries struggle with food shortages made worse by population growth. According to the World Hunger Statistics report in Figure 3, roughly 1 in 7 people, or 12.9% of the global population, do not have enough food. Unbelievably, a third of the food produced is said to be wasted, which results in the mismanagement of the resources used in food production (Habib et al., 2020). While public participation is essential in addressing this problem, governments must take the initiative to work with the commercial sector to establish plans that will end food waste worldwide (Despoudi et al., 2021). The important component of such solutions is the creation of strict laws to combat food waste, albeit depending only on enforcement, which could not provide benefits immediately (Synani et al., 2021).

4.1 Leveraging Machine Learning and Deep Learning to Tackle Food Waste in Hospitality Industry

Figure 4 illustrates specific applications of ml and dl in different areas of hospitality. In the hospitality industry, food waste is primarily due to customer demand that can be erratic for some establishments, often leading them into overproduction and inefficient inventory management, which then could potentially see their margins plummet. This is from our previous articles on how Machine Learning (ML)and Deep learning (DL) innovations would offer a solution. The technology is used in different parts of the food supply chain to standardise operations and eliminate waste. The most popular used case is demand forecasting. The ML algorithms predict how much food will be needed daily by analysing sales patterns, customer behaviour and external factors like weather or events from historical data. Therefore, organisations can forecast more efficiently and determine what needs replenishment (how much), helping produce the correct quantity without running into overproduction. At the same time, DL models can strictly help in more sophisticated analyses (picking up very subtle changes or customer preferences that might drive to waste) (Legrand et al., 2022).

ML and DL are also extensively used for handling inventory management. It is capable of real-time stock management for significantly reducing stocking levels, tracking expiration dates, and recommending the best possible ordering schedules. This will result in less spoilage and the ability to serve food before it turns into waste simply by automating these business processes. Also, image recognition systems powered by Deep Learning can recognise and classify food waste to learn which kinds of products are being disposed of most often. Another potential use case is refining menu offerings, modifying portion sizes, or adopting zero-waste initiatives. ML and DL solutions benefit the hospitality industry by driving better decision-making, automating processes and helping operators extract actionable insights. Thus sustaining the most cost-efficient operation (Sharma et al., 2023). Given the alarming statistics on hunger and malnutrition, quick, decisive and efficient action is required. Technologies that are advantageous to various businesses might also be crucial in tackling the problem of food waste. Machine learning and Deep learning stand out among them as potential remedies. To effectively minimise food waste, stakeholders should know how these technologies might provide value (Said et al., 2023). ML and DL significantly contribute to solving this pressing global issue by examining sectors within the food business (Hassoun et al., 2023).

Figure 4. Specific applications of ML and DL in different areas of hospitality

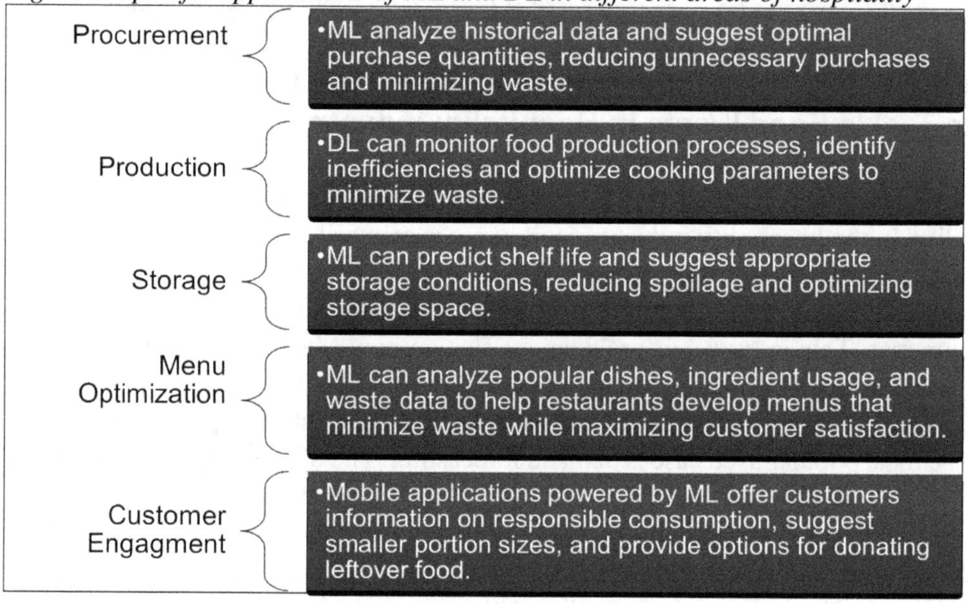

Source: Authors Compilation

Machine Learning (ML) and Deep Learning (DL) are transforming various aspects of the hospitality industry, significantly improving efficiency and sustainability. One of the most impactful applications is in demand forecasting, where these technologies analyse historical data such as sales, reservations, weather conditions, and local events to predict customer needs with remarkable accuracy. This capability allows hotels, restaurants, and catering services to forecast the amount of food and resources required for specific periods, minimising overproduction, reducing food waste, and optimising inventory management. These predictive models enhance operational efficiency and profitability by aligning procurement, staffing, and menu planning with expected demand (Jauhar et al., 2023).

In addition to demand forecasting, ML and DL are revolutionising inventory management in the hospitality sector. These algorithms automate inventory tracking, monitor stock levels in real time, and predict when supplies need replenishing based on usage trends. They can identify items nearing expiration and suggest ways to use them, such as through recipes or promotions, to prevent spoilage. This level of precision helps prevent overstocking and underutilisation, leading to more sustainable operations. As a result, hospitality businesses can cut costs, reduce waste, and maintain high-quality standards (Rane et al., 2023). Moreover, ML and DL contribute to a personalised customer experience in the hospitality industry, which indirectly helps reduce waste. By analysing customer preferences, dietary restrictions, and past behaviours, these technologies enable hotels and restaurants to tailor their offerings to individual guests. For instance, ML algorithms can recommend menu items that align with a customer's tastes, reducing the likelihood of food being ordered and left uneaten. DL models can also create personalised promotions or loyalty rewards, encouraging repeat business and fostering more predictable consumption patterns. This level of personalisation boosts customer satisfaction and ensures that food production closely matches actual demand, reducing unnecessary waste (Oriekhoe et al., 2024).

Lastly, DL-powered image recognition systems and ML algorithms are increasingly used to track and analyse food waste in the hospitality sector. These technologies categorise waste by type, identify trends in frequently discarded items, and provide actionable insights for waste reduction. For example, consistent waste of a particular dish could lead to menu adjustments, portion size modifications, or staff training to improve food handling. Automated waste tracking provides businesses with a clearer understanding of the root causes of food waste, enabling them to implement targeted interventions. This data-driven approach supports a more efficient and sustainable method of waste management (Verma et al., 2023).

In the food business, a large volume of food is frequently lost during processing, and the application of machines and deep learning might improve production efficiency. This waste results from some incidents, including spills, mismanagement, incorrect storage and damage to apparatus. Finding answers to food waste requires addressing its underlying causes. For this reason, tracking food waste in real-time becomes essential. Farmers can get up-to-date information on their farms by putting various cameras and sensors in the field. With these sensors, farmers can identify precise locations where food is being lost and take focused action to reduce those losses (De Bernardi et al., 2023). Also, government authorities and farmers have access to this real-time data, enabling it to effectively monitor and reduce food waste on a larger scale (Vargas-Sanchez, 2023).

Retail stores that sell a variety of foods must replenish their inventory as needed, but doing so before products are sold raises the possibility of insect infestations (Abatan et al., 2024). Inadequate food supply forecasting and monitoring is one of the factors leading to this problem (Delina et al., 2023). AI can improve and expedite this procedure rather than rely on manual approaches (Stewart & Gorman, 2024). Artificial intelligence algorithms can produce accurate forecasts by importing historical data on previous occurrences of stockouts, food waste and current shelf state (Mishra & Singh, 2023). This makes it possible for shops to know precisely when certain food items will run out of supply. After receiving this precise information, Retailers may place orders for just the necessary amounts of food goods (Nwokolo et al., 2023).

Maintaining sustainable practices is essential to fulfilling the rising consumer demand for sustainability in the food business and solving present problems and future difficulties (Mishra & Singh, 2023). The idea that "sustainable food and agriculture systems require significant additional efforts" is further supported by the paper's emphasis on the future of food and agriculture. This viewpoint is consistent with the need for sustainability (Rahimi-Midani et al., 2023). The consumer perspective might be seen as a barrier for food sector firms aiming for sustainable development, given the critical role that consumers play in bringing about change and that market acceptance is necessary for the success of sustainable solutions (Workie et al., 2023). The potential uses of ML and DL applications in the food industry to improve food quality and solve problems for the sector (Buhalis et al., 2023).

5. FINDINGS AND VIABLE SOLUTIONS

The responsible consumption and production goals in hospitality, driven by Machine Learning (ML) and Deep Learning (DL) to reduce food waste, are increasingly becoming attainable. These technologies optimise demand forecasting, inventory management, and waste tracking, ensuring the exact amount of food is produced with minimal surplus. Additionally, ML and DL enable more personalised customer experiences tailored to individual preferences, further minimising waste. This powerful software is automating processes, transforming daily kitchen operations, helping hospitality businesses move closer to zero waste, improving sustainability, reducing costs, and eliminating inefficient food consumption practices.

Regarding identifying defects in food production, ML and DL systems that follow a methodical approach consistently outperform human-operated systems (Ali et al., 2024). Challenges from inadequate automation in food processing chains have been exacerbated by globalisation, particularly in areas of health and safety (Hong et al., 2024). Microcontrollers, which produce output spikes when connected to sensors, offer an affordable and accessible platform for a variety of artificial neural robotics applications (Ramadass et al., 2023). This advancement could significantly benefit the food industry.

AI-driven agricultural sensors, integrating ML and big data, have been successfully applied to boost productivity, predict yields, detect diseases, and analyse other key metrics, enhancing decision-making and operational efficiency (Glebova et al., 2023). Although the UN estimates roughly one-third of all food produced for human consumption is lost or wasted—wasting valuable resources and contributing to greenhouse gas emissions (Kannan et al., 2023)—ML and DL present promising solutions. These technologies offer innovative methods for promoting sustainable consumption (Machado & Hatakeyama, 2023).

The key findings and viable solutions include:

6.1 Predictive Analysis for Demand Forecasting

Demand forecasting is an important area where artificial intelligence is making a big difference (Ordonez de Pablos et al., 2023). ML and DL algorithms can accurately estimate future demand by analysing data about market dynamics, historical trends, and consumer behaviour (Legrand et al., 2023). This feature reduces the possibility of wasting too much inventory by helping suppliers and retailers modify their purchase tactics (Gupta et al., 2023).

6.2 Increasing Food Supply Chain Efficiency

Machine-Deep Learning is crucial in optimising food supply chains from the farm to the end user (Taneja et al., 2023). With AI-powered sensors and analytics, food producers can keep a close eye on the state of their product as it is being transported and stored, preventing spoiling (Oriekhoe et al., 2024). ML may also be used to improve transportation routes, guarantee prompt delivery and keep food fresh until it reaches its destination (Abatan et al., 2024).

6.3 Smart Transportation

The transportation industry is transforming and is used in many different modes, including cars, trains, ships and aeroplanes, emphasising streamlining traffic (Stewart & Gorman, 2024). Transportation and the food sector are affected by ML promise to make modes of transportation safer, greener, smarter and more effective (Hong et al., 2024). ML-powered autonomous mobility might do away with human mistakes, a major cause of traffic accidents (Ali et al., 2024). But these potentials also come with real hazards, such as abuse and unforeseen effects, which include things like biased transportation choices and cyber-attacks (Tapo et al., 2024). Concerns about ethical issues and employment consequences arise regarding AI's accountability for judgments made without human input (Gao et al., 2024).

6.4 Smart Fridges

Enter the world of intelligent refrigerators, where ML is used for more than just observation (Koskela-Huotari et al., 2024). These cutting-edge gadgets can precisely track expiration dates, keep an eye on their contents and alert consumers before food degrades (El Zein et al., 2024). But they are considerably more intelligent than that, as imagine a refrigerator aware of your food preferences and eating habits (Menzies et al., 2024). It uses DL to suggest recipes based on the components you have on hand, which ensures that you waste as little as possible. This inventive bundle combines creative cooking with waste minimisation (Zhang et al., 2024).

6. CONCLUSION AND FUTURE SCOPE

Machine learning and Deep learning have become essential in combating food waste. Its capacity to analyse data, foresee trends and enhance processes is revolutionising food production, distribution and consumption. Embracing the capabilities of ML and DL leads us towards a future where sustainable consumption becomes more than an ambition but an actuality. In pursuing a more sustainable world, ML and DL drive transformation, steering us towards intelligent, efficient and waste-free consumption practices. AI and machine learning provide valuable ways to interact with cutting-edge technologies. Machine learning and deep learning govern many multidisciplinary systems that evaluate several criteria, including quality, appearance, texture, and overall customer acceptance. This novel method entails analysing data patterns and optimising workflows to produce precise, dependable results, minimising human intervention, guaranteeing effectiveness, and empowering operators to predict potential future situations.

These techniques are useful in dealing with the escalating problems in the food industry. Drone technology has the potential to completely transform the management of the food supply chain in the future. As AI and big data combine, sensors play a more significant and prominent role in food preparation, helping provide better, more effective results. Machine Learning and Deep Learning is at the forefront of guaranteeing food safety because ML and DL are integrated into food manufacturing and restaurant operations. The food safety business benefits from this technical breakthrough since it minimises wasted commodities and lowers production mistakes caused by human error. The advantages encompass decreased packaging and delivery expenses, more customer content, expedited services, voice search functionality and customised purchase choices. Similar benefits are not just for small businesses; big food corporations can benefit from similar developments in the long run.

REFERENCES

Abatan, A., Adeyinka, M. A., Sodiya, E. O., Jacks, B. S., Ugwuanyi, E. D., Daraojimba, O. H., & Lottu, O. A. (2024). The role of IT in sustainable environmental management: A global perspective review.

Abatan, A., Adeyinka, M. A., Sodiya, E. O., Jacks, B. S., Ugwuanyi, E. D., Daraojimba, O. H., & Lottu, O. A. (2024). The role of IT in sustainable environmental management: A global perspective review.

Aditya, A., & Kurniawati, K. (2023). Food Waste Management Challenges and Strategies in The Hotel Industry in Jakarta. *Journal of Social Research*, 2(9), 2983–2989. DOI: 10.55324/josr.v2i9.1295

Ali, R., Irfan, S., & Hamid, S. (2024). Emerging themes in food tourism: A systematic literature review and research agenda. *British Food Journal*, 126(1), 372–408. DOI: 10.1108/BFJ-11-2022-0939

Ali, R., Irfan, S., & Hamid, S. (2024). Emerging themes in food tourism: A systematic literature review and research agenda. *British Food Journal*, 126(1), 372–408. DOI: 10.1108/BFJ-11-2022-0939

Alvarez-Risco, A., Estrada-Merino, A., & Perez-Luyo, R. (2020). Sustainable development goals in hospitality management. In *Sustainable Hospitality Management: Designing Meaningful Encounters With Talent and Technology* (pp. 159-178). Emerald Publishing Limited. DOI: 10.1108/S1877-636120200000024012

Amicarelli, V., Aluculesei, A. C., Lagioia, G., Pamfilie, R., & Bux, C. (2022). How to manage and minimize food waste in the hotel industry: An exploratory research. *International Journal of Culture, Tourism and Hospitality Research*, 16(1), 152–167. DOI: 10.1108/IJCTHR-01-2021-0019

Amicarelli, V., & Bux, C. (2021). Food waste measurement toward a fair, healthy and environmental-friendly food system: A critical review. *British Food Journal*, 123(8), 2907–2935. DOI: 10.1108/BFJ-07-2020-0658

Araujo, C. M. L. D. (2023). *The effects of retail sustainable strategies on the sustainability of the food value chain* (Doctoral dissertation).

Atser, G., Tahirou, A., Dixon, A., Ekeleme, F., Alloh, J., & Akande, A. (2019). Transforming Oyo State to an Agribusiness Hub:(Our vision, Our Plan) 2019-2023.

Axmann, H. B., Guo, X., Broeze, J., Harbers, C., Viquez-Zamora, M., & Soethoudt, J. M. (2024). Mitigate+: Food Loss and Waste country profile Colombia: Estimates of Food Loss and Waste, associated GHG emissions, nutritional losses, land use and water footprints.

Băltescu, C. A. (2022). In Pamfilie, R., Dinu, V., Vasiliu, C., Ple ea, D., & Tăchiciu, L. (Eds.), *Exploring younger generation sustainable consumption in the restaurant industry* (pp. 560–566). DOI: 10.24818/BASIQ/2022/08/074

Barrett, M., Marino, M., Brkic, F., & Pratesi, C. A. (2023). *How to Create a Sustainable Food Industry: A Practical Guide to Perfect Food.* Taylor & Francis. DOI: 10.4324/9781003449744

Bisaria, P. (2022). Environmental issues and corporate social responsibility.

Buhalis, D., O'Connor, P., & Leung, R. (2023). Smart hospitality: From smart cities and smart tourism towards agile business ecosystems in networked destinations. *International Journal of Contemporary Hospitality Management*, 35(1), 369–393. DOI: 10.1108/IJCHM-04-2022-0497

Bux, C., & Amicarelli, V. (2022). Circular economy and sustainable strategies in the hospitality industry: Current trends and empirical implications. *Tourism and Hospitality Research*, •••, 14673584221119581.

Camilleri, M. A. (2021). Sustainable production and consumption of food. Mise-en-place circular economy policies and waste management practices in tourism cities. *Sustainability (Basel)*, 13(17), 9986. DOI: 10.3390/su13179986

Chalak, A., Abou-Daher, C., & Abiad, M. G. (2018). Generation of food waste in the hospitality and food retail and wholesale sectors: Lessons from developed economies. *Food Security*, 10(5), 1279–1290. DOI: 10.1007/s12571-018-0841-0

Chawla, G., Lugosi, P., & Hawkins, R. (2022). Factors Influencing Hospitality Employees' Pro-Environmental Behaviours toward Food Waste. *Sustainability (Basel)*, 14(15), 9015. DOI: 10.3390/su14159015

Choi, M., Azzaoui, A. E., Singh, S. K., Salim, M. M., Jeremiah, S. R., & Park, J. H. (2022). The future of metaverse: Security issues, requirements, and solutions. *Human-Centric Computing and Information Sciences*, 12(60), 1–14.

Chung, K. C. (2020). Green marketing orientation: Achieving sustainable development in green hotel management. *Journal of Hospitality Marketing & Management*, 29(6), 722–738. DOI: 10.1080/19368623.2020.1693471

De Bernardi, P., Bertello, A., & Forliano, C. (2023). Circularity of food systems: A review and research agenda. *British Food Journal*, 125(3), 1094–1129. DOI: 10.1108/BFJ-05-2021-0576

de Visser-Amundson, A. (2022). A multi-stakeholder partnership to fight food waste in the hospitality industry: A contribution to the United Nations Sustainable Development Goals 12 and 17. *Journal of Sustainable Tourism*, 30(10), 2448–2475. DOI: 10.1080/09669582.2020.1849232

Delgado, A., Rodriguez, R., & Staszewska, A. (2023). Tackling Food Waste in the Tourism Sector: Towards a Responsible Consumption Trend. *Sustainability (Basel)*, 15(17), 13226. DOI: 10.3390/su151713226

Delina, L. L., Perez, O. A., Afable, S. B., & Steuer, B. (2023). Do global public health crises change people's behaviors towards sustainability? Evidence of the COVID-19 and sustainability nexus from Hong Kong. *Cleaner and responsible consumption, 10*, 100132.

Despoudi, S., Bucatariu, C., Otles, S., & Kartal, C. (2021). Food waste management, valorization, and sustainability in the food industry. In *Food waste recovery* (pp. 3–19). Academic Press. DOI: 10.1016/B978-0-12-820563-1.00008-1

Dhir, A., Talwar, S., Kaur, P., & Malibari, A. (2020). Food waste in hospitality and food services: A systematic literature review and framework development approach. *Journal of Cleaner Production*, 270, 122861. DOI: 10.1016/j.jclepro.2020.122861

El Zein, B., Elrashidi, A., Dahlan, M., Al Jarwan, A., & Jabbour, G. (2024). Nano and Society 5.0: Advancing the Human-Centric Revolution.

ElShafei, R. (2022). Managers' risk perception and the adoption of sustainable consumption strategies in the hospitality sector: The moderating role of stakeholder salience attributes. *Smart and Sustainable Built Environment*, 11(1), 1–18. DOI: 10.1108/SASBE-03-2020-0024

Filimonau, V., Coşkun, A., Derqui, B., & Matute, J. (2022). Restaurant management and food waste reduction: Factors affecting attitudes and intentions in restaurants of Spain. *International Journal of Contemporary Hospitality Management*, 34(3), 1177–1203. DOI: 10.1108/IJCHM-07-2021-0899

Filimonau, V., Krivcova, M., & Pettit, F. (2019). An exploratory study of managerial approaches to food waste mitigation in coffee shops. *International Journal of Hospitality Management*, 76, 48–57. DOI: 10.1016/j.ijhm.2018.04.010

Filimonau, V., Matute, J., Kubal-Czerwińska, M., Krzesiwo, K., & Mika, M. (2020). The determinants of consumer engagement in restaurant food waste mitigation in Poland: An exploratory study. *Journal of Cleaner Production*, 247, 119105. DOI: 10.1016/j.jclepro.2019.119105

Filimonau, V., & Uddin, R. (2021). Food waste management in chain-affiliated and independent consumers' places: A preliminary and exploratory study. *Journal of Cleaner Production*, 319, 128721. DOI: 10.1016/j.jclepro.2021.128721

Fitchuk, C. (2021). Creating a Sustainable Business Strategy for a SME.

Font, X., Camilleri, M. A., Galati, A., & Vrontis, D. (2023). Responsible consumption and production of food: opportunities and challenges for hospitality practitioners.

Gao, J. Q., Li, D., Qiao, G. H., Jia, Q. R., Li, S. R., & Gao, H. L. (2024). Circular economy strategies in supply chains, enhancing resource efficiency and sustainable development goals. *Environmental Science and Pollution Research International*, 31(6), 8751–8767. DOI: 10.1007/s11356-023-31551-z PMID: 38180660

Gatto, A., & Chepeliev, M. (2024). Global food loss and waste estimates show increasing nutritional and environmental pressures. *Nature Food*, 5(2), 1–12. DOI: 10.1038/s43016-023-00915-6 PMID: 38287151

Glebova, E., Gerke, A., & Book, R. (2023). The transformational role of technology in sports events. In *Sports Management in an Uncertain Environment* (pp. 169–187). Springer Nature Singapore. DOI: 10.1007/978-981-19-7010-8_8

Goh, E., Okumus, B., Jie, F., Djajadikerta, H. G., & Lemy, D. M. (2022). Managing food wastage in hotels: Discrepancies between injunctive and descriptive norms amongst hotel food and beverage managers. *British Food Journal*, 124(12), 4666–4685. DOI: 10.1108/BFJ-05-2021-0513

Gupta, M., Choudhary, B., & Sharma, D. (2023). Emersion and Immersion of Technology in the Development of Smart Cities: A Bibliometric Analysis. In *Enabling Technologies for Effective Planning and Management in Sustainable Smart Cities* (pp. 303–318). Springer International Publishing. DOI: 10.1007/978-3-031-22922-0_12

Habib Ur Rehman, H. (2020). *Food Waste Reduction in Oslo'Restaurants: an approach toward Sustainability* (Master's thesis).

Han, H. (2021). Consumer behavior and environmental sustainability in tourism and hospitality: A review of theories, concepts, and latest research. *Journal of Sustainable Tourism*, 29(7), 1021–1042. DOI: 10.1080/09669582.2021.1903019

Han, H., Chua, B. L., Ariza-Montes, A., & Untaru, E. N. (2020). Effect of environmental corporate social responsibility on green attitude and norm activation process for sustainable consumption: Airline versus restaurant. *Corporate Social Responsibility and Environmental Management*, 27(4), 1851–1864. DOI: 10.1002/csr.1931

Hassoun, A., Aït-Kaddour, A., Abu-Mahfouz, A. M., Rathod, N. B., Bader, F., Barba, F. J., Biancolillo, A., Cropotova, J., Galanakis, C. M., Jambrak, A. R., Lorenzo, J. M., Måge, I., Ozogul, F., & Regenstein, J. (2023). The fourth industrial revolution in the food industry—Part I: Industry 4.0 technologies. *Critical Reviews in Food Science and Nutrition*, 63(23), 6547–6563. DOI: 10.1080/10408398.2022.2034735 PMID: 35114860

Hong, J., Jaegler, A., & Gergaud, O. (2024). Mobile applications to reduce food waste in supply chains: A systematic literature review. *British Food Journal*, 126(2), 509–530. DOI: 10.1108/BFJ-09-2022-0742

Hong, J., Jaegler, A., & Gergaud, O. (2024). Mobile applications to reduce food waste in supply chains: A systematic literature review. *British Food Journal*, 126(2), 509–530. DOI: 10.1108/BFJ-09-2022-0742

Huang, Y., Ma, E., & Wang, D. (2021). Message framing strategies, food waste prevention, and diners' repatronage intentions: The mediating role of corporate social responsibility. *Journal of Sustainable Tourism*, 29(10), 1694–1715. DOI: 10.1080/09669582.2020.1867151

Hunter, D., Monville-Oro, E., Burgos, B., Roel, C. N., Calub, B. M., Gonsalves, J., & Lauridsen, N. (Eds.). (2020). *Agrobiodiversity, school gardens and healthy diets: Promoting biodiversity, food and sustainable nutrition*. Routledge. DOI: 10.4324/9780429053788

Hwang, J., & Kim, J. J. (2021). Edible insects: How to increase the sustainable consumption behavior among restaurant consumers. *International Journal of Environmental Research and Public Health*, 18(12), 6520. DOI: 10.3390/ijerph18126520 PMID: 34204332

Islam, J. U., Nazir, O., & Rahman, Z. (2023). Sustainably engaging employees in food wastage reduction: A conscious capitalism perspective. *Journal of Cleaner Production*, 389, 136091. DOI: 10.1016/j.jclepro.2023.136091

Jauhar, S., Pratap, S., Lakshay, , Paul, S., & Gunasekaran, A. (2023). Internet of things based innovative solutions and emerging research clusters in circular economy. *Operations Management Research : Advancing Practice Through Research*, 16(4), 1968–1988. DOI: 10.1007/s12063-023-00421-9

Kalenjuk Pivarski, B., Šmugović, S., Tekić, D., Ivanović, V., Novaković, A., Tešanović, D., Banjac, M., Đerčan, B., Peulić, T., Mutavdžić, B., Lazarević, J., & Vukelić, N. (2022). Characteristics of Traditional Food Products as a Segment of Sustainable Consumption in Vojvodina's Hospitality Industry. *Sustainability (Basel)*, 14(20), 13553. DOI: 10.3390/su142013553

Kannan, H., Rodriguez, R. V., Paprika, Z. Z., & Ade-Ibijola, A. (Eds.). (2023). *Exploring Ethical Dimensions of Environmental Sustainability and Use of AI*. IGI Global. DOI: 10.4018/979-8-3693-0892-9

Kasavan, S., Mohamed, A. F., & Halim, S. A. (2019). Drivers of food waste generation: Case study of island-based hotels in Langkawi, Malaysia. *Waste Management (New York, N.Y.)*, 91, 72–79. DOI: 10.1016/j.wasman.2019.04.055 PMID: 31203944

Kasavan, S., Siron, R., Yusoff, S., & Fakri, M. F. R. (2022). Drivers of food waste generation and best practice towards sustainable food waste management in the hotel sector: A systematic review. *Environmental Science and Pollution Research International*, 29(32), 48152–48167. DOI: 10.1007/s11356-022-19984-4 PMID: 35579831

Katuwal, S., & Afzal, M. (2022). A framework for managing food wastage: Strategies to promote recycling and sustainability in the hotel industry.

Kilibarda, N., Djokovic, F., & Suzic, R. (2019). Food waste management–reducing and managing food waste in hospitality. *Scientific journal. Metals Technology*, 60(2), 134–142.

Koshariya, A. K., Krishnan, M. S., Jaisankar, S., Loganathan, G. B., Sathish, T., Ağbulut, Ü., Saravanan, R., Tuan, L. T., & Pham, N. D. K. (2024). Waste to energy: An experimental study on hydrogen production from food waste gasification. *International Journal of Hydrogen Energy*, 54, 1–12. DOI: 10.1016/j.ijhydene.2023.05.221

Koskela-Huotari, K., Svärd, K., Williams, H., Trischler, J., & Wikström, F. (2024). Drivers and hinderers of (un) sustainable service: A systems view. *Journal of Service Research*, 27(1), 106–123. DOI: 10.1177/10946705231176071

Kumar, M., Kim, C., Son, Y., Singh, S. K., & Kim, S. (2024). Empowering Cyberattack Identification in IoHT Networks With Neighborhood Component-based Improvised Long Short-Term Memory. *IEEE Internet of Things Journal*, 11(9), 16638–16646. DOI: 10.1109/JIOT.2024.3354988

Kumar, S., Singh, S. K., Varshney, S., Singh, S., Kumar, P., Kim, B. G., & Ra, I. H. (2023). Fusion of deep sort and Yolov5 for effective vehicle detection and tracking scheme in real-time traffic management sustainable system. *Sustainability (Basel)*, 15(24), 16869. DOI: 10.3390/su152416869

Labadi, S., Giliberto, F., Rosetti, I., Shetabi, L., & Yildirim, E. (2021). Heritage and the sustainable development goals: Policy guidance for heritage and development actors. *International Journal of Heritage Studies*.

Lagioia, G., Amicarelli, V., Strippoli, R., Bux, C., & Gallucci, T. (2023). Sustainable and circular practices in the hotel industry in Southern Italy: Opportunities, barriers and trends in food waste management. *British Food Journal*.

Lawal, I. M., Ndagi, A., Mohammed, A., Saleh, Y. Y., Shuaibu, A., Hassan, I., Abubakar, S., Soja, U. B., & Jagaba, A. H. (2024). Proximate analysis of waste-to-energy potential of municipal solid waste for sustainable renewable energy generation. *Ain Shams Engineering Journal*, 15(1), 102357. DOI: 10.1016/j.asej.2023.102357

Lee, H. H., & Huang, P. Y. (2023). Food Waste and Environmental Sustainability of the Hotel Industry in Taiwan. *Sustainability (Basel)*, 15(21), 15459. DOI: 10.3390/su152115459

Lee, S., Bae, G., & Kim, H. (2021). A study on the sustainable use intention of restaurant companies using the information attributes of sns: The dual process theory. *Sustainability (Basel)*, 13(22), 12922. DOI: 10.3390/su132212922

Legrand, W., Chen, J. S., & Laeis, G. C. (2022). *Sustainability in the hospitality industry: Principles of sustainable operations*. Taylor & Francis.

Legrand, W., Chen, J. S., & Laeis, G. C. (2022). *Sustainability in the hospitality industry: Principles of sustainable operations*. Taylor & Francis.

Legrand, W., Kuokkanen, H., & Day, J. (Eds.). (2023). *Critical Questions in Sustainability and Hospitality*. Taylor & Francis. DOI: 10.4324/9781003218425

Lund-Durlacher, D., & Antonschmidt, H. (2019). Towards a Framework for Sustainable and Responsible Food Operations in the Holiday Context and Implementation Approaches in the Travel Industry. *Corporate Sustainability and Responsibility in Tourism: A Transformative Concept*, 327-341.

Luu, T. T. (2020). Reducing food waste behavior among hospitality employees through communication: Dual mediation paths. *International Journal of Contemporary Hospitality Management*, 32(5), 1881–1904. DOI: 10.1108/IJCHM-09-2019-0779

Machado, M., & Hatakeyama, K. (2023, July). Digital Transformation as an Enabler of Sustainability in Supply Chain and Logistics: Evidence from the Field. In *2023 Portland International Conference on Management of Engineering and Technology (PICMET)* (pp. 1-6). IEEE. DOI: 10.23919/PICMET59654.2023.10216881

Malheiro, A., Sousa, B. B., Liberato, D., & Liberato, P. (2020, April). Sustainability in tourism and hospitality: Trends and challenges. In *35th IBIMA conference education excellence and innovation management: A 2025 vision to sustain economic development during global challenges* (pp. 15545-15555). IBIMA.

Martin-Rios, C., Demen Meier, C., & Pasamar, S. (2022). Sustainable waste management solutions for the foodservice industry: A Delphi study. *Waste Management & Research*, 40(9), 1412–1423. DOI: 10.1177/0734242X221079306 PMID: 35176923

Meier, T., von Borstel, T., Welte, B., Hogan, B., Finn, S. M., Bonaventura, M., Friedrich, S., Weber, K., & Dräger de Teran, T. (2021). Food waste in healthcare, business and hospitality catering: Composition, environmental impacts and reduction potential on company and national levels. *Sustainability (Basel)*, 13(6), 3288. DOI: 10.3390/su13063288

Menzies, J., Sabert, B., Hassan, R., & Mensah, P. K. (2024). Artificial intelligence for international business: Its use, challenges, and suggestions for future research and practice. *Thunderbird International Business Review*, 66(2), 185–200. DOI: 10.1002/tie.22370

Miconi, M. (2022). The approach to sustainability policies in family firms: how this phenomenon has impacted management remuneration.

Mirela, S., & Borza, M. (2014). Measures concerning sustainable consumption in hotels. *Acta Universitatis Danubius. Œconomica, 10*(3).

Mishra, P., & Singh, G. (2023). Introduction: Importance of Sustainable Smart City. In *Sustainable Smart Cities: Enabling Technologies, Energy Trends and Potential Applications* (pp. 1–37). Springer International Publishing.

Mishra, P., & Singh, G. (2023). *Sustainable Smart Cities: Enabling Technologies, Energy Trends and Potential Applications*. Springer Nature. DOI: 10.1007/978-3-031-33354-5

Munir, K. (2022). Sustainable food waste management strategies by applying practice theory in hospitality and food services-a systematic literature review. *Journal of Cleaner Production*, 331, 129991. DOI: 10.1016/j.jclepro.2021.129991

NaThalang, S. (2019). The environmental management approach for small hotel in Koh Chang (Chang Island), Trat province.

Nwokolo, S. C., Eyime, E. E., Obiwulu, A. U., & Ogbulezie, J. C. (2023). Africa's Path to Sustainability: Harnessing Technology, Policy, and Collaboration. *Trends in Renewable Energy*, 10(1), 98–131. DOI: 10.17737/tre.2024.10.1.00166

Okumus, B., Taheri, B., Giritlioglu, I., & Gannon, M. J. (2020). Tackling food waste in all-inclusive resort hotels. *International Journal of Hospitality Management*, 88, 102543. DOI: 10.1016/j.ijhm.2020.102543

Omar, I., Hassan, H., Jayaraman, R., Salah, K., & Omar, M. (2023). Using Blockchain Technology to Achieve Sustainability in the Hospitality Industry by Reducing Food Waste. *Available atSSRN* 4530352. DOI: 10.2139/ssrn.4530352

Ordóñez de Pablos, P., Almunawar, M. N., & Anshari, M. (Eds.). (2023). *Perspectives on the Transition Toward Green and Climate Neutral Economies in Asia*. IGI Global. DOI: 10.4018/978-1-6684-8613-9

Oriekhoe, O. I., Ashiwaju, B. I., Ihemereze, K. C., Ikwue, U., & Udeh, C. A. (2024). Review Of Technological Advancements In Food Supply Chain Management: A Comparative Study Between The Us And Africa. *International Journal of Management & Entrepreneurship Research*, 6(1), 132–149. DOI: 10.51594/ijmer.v6i1.713

Oriekhoe, O. I., Ashiwaju, B. I., Ihemereze, K. C., Ikwue, U., & Udeh, C. A. (2024). Review Of Technological Advancements In Food Supply Chain Management: A Comparative Study Between The Us And Africa. *International Journal of Management & Entrepreneurship Research*, 6(1), 132–149. DOI: 10.51594/ijmer.v6i1.713

Ozdemir, G., & Güçer, E. (2018). Food waste management within sustainability perspective: A study on five star chain hotels. *Journal of Tourism & Gastronomy Studies*, 6(1), 280–299. DOI: 10.21325/jotags.2018.187

Ozturkcan, S., & Kitapci, O. (2023). A sustainable solution for the hospitality industry: The QR code menus. *Journal of Information Technology Teaching Cases*, 20438869231181599.

Park, J. H., Singh, S. K., Salim, M. M., Azzaoui, A. E., & Park, J. H. (2022). Ransomware-based Cyber Attacks: A Comprehensive Survey. *Wangji Wanglu Jishu Xuekan*, 23(7), 1557–1564. DOI: 10.53106/160792642022122307010

Park, J. H., Yotxay, S., Singh, S. K., & Park, J. H. (2024). PoAh-Enabled Federated Learning Architecture for DDoS Attack Detection in IoT Networks. Hum.-Centric Comput. Inf. Sci, 14(3).

Principato, L., Pratesi, C. A., & Secondi, L. (2018). Towards zero waste: An exploratory study on restaurant managers. *International Journal of Hospitality Management*, 74, 130–137. DOI: 10.1016/j.ijhm.2018.02.022

Rahimi-Midani, A. (2023). Deep Technology for Sustainable Fisheries and Aquaculture.

Ramadass, R., Sathyanarayana, N., & Fathima, Y. (2023). Improving citizen lifestyle in green smart city. In *Green Blockchain Technology for Sustainable Smart Cities* (pp. 29–63). Elsevier. DOI: 10.1016/B978-0-323-95407-5.00007-4

Rane, N., Choudhary, S., & Rane, J. (2023). Metaverse as a cutting-edge platform for attaining Sustainable Development Goals (SDGs). *Available atSSRN* 4644035. DOI: 10.2139/ssrn.4644035

Riva, F., Magrizos, S., Rubel, M. R. B., & Rizomyliotis, I. (2022). Green consumerism, green perceived value, and restaurant revisit intention: Millennials' sustainable consumption with moderating effect of green perceived quality. *Business Strategy and the Environment*, 31(7), 2807–2819. DOI: 10.1002/bse.3048

Ruiz Molina, M. E., Belda-Miquel, S., Hytti, A., & Gil-Saura, I. (2022). Addressing sustainable food management in hotels: Proposing a framework and examining hotel groups. *British Food Journal*, 124(2), 462–492. DOI: 10.1108/BFJ-12-2020-1171

Said, Z., Sharma, P., Nhuong, Q. T. B., Bora, B. J., Lichtfouse, E., Khalid, H. M., & Hoang, A. T. (2023). Intelligent approaches for sustainable management and valorisation of food waste. *Bioresource Technology*, 377, 128952. DOI: 10.1016/j.biortech.2023.128952 PMID: 36965587

Sartor, D. (2021). An Exploratory Study of Woolworths Consumer's Post-Covid Food Waste Behaviour.

Scholz, P., Linderová, I., & Konečná, K. (2020). Green Management Tools as a Way to Sustainable Behaviour in the Hotel Industry: Case Study from Czechia. *Sustainability (Basel)*, 12(23), 10027. DOI: 10.3390/su122310027

Shams, S. R., Vrontis, D., Weber, Y., Tsoukatos, E., & Iaia, L. (Eds.). (2023). *Technology, Business and Sustainable Development: Advances for People, Planet and Profit*. Taylor & Francis. DOI: 10.4324/9781003293187

Sharma, A., Saulais, L., & Huang, Y. (2023). Sustainable consumer choices–critical reflection on hospitality and tourism. *International Journal of Contemporary Hospitality Management*.

Sharma, A., & Singh, B. (2022). Measuring Impact of E-commerce on Small Scale Business: A Systematic Review. *Journal of Corporate Governance and International Business Law*, 5(1).

Sharma, T. (2020). What a waste: confronting consumer food waste behavior in hospitality settings. In *Advances in hospitality and leisure* (pp. 167–176). Emerald Publishing Limited.

Sharma, Y., Sijariya, R., & Gupta, P. (2023). How Deep Learning Can Help in Regulating the Subscription Economy to Ensure Sustainable Consumption and Production Patterns (12th Goal of SDGs). In *Deep Learning Technologies for the Sustainable Development Goals: Issues and Solutions in the Post-COVID Era* (pp. 1–20). Springer Nature Singapore. DOI: 10.1007/978-981-19-5723-9_1

Singh, B. (2019). Profiling Public Healthcare: A Comparative Analysis Based on the Multidimensional Healthcare Management and Legal Approach. *Indian Journal of Health and Medical Law*, 2(2), 1–5.

Singh, B. (2022). COVID-19 Pandemic and Public Healthcare: Endless Downward Spiral or Solution via Rapid Legal and Health Services Implementation with Patient Monitoring Program. *Justice and Law Bulletin*, 1(1), 1–7.

Singh, B. (2022). Relevance of Agriculture-Nutrition Linkage for Human Healthcare: A Conceptual Legal Framework of Implication and Pathways. *Justice and Law Bulletin*, 1(1), 44–49.

Singh, B. (2023). Blockchain Technology in Renovating Healthcare: Legal and Future Perspectives. In *Revolutionizing Healthcare Through Artificial Intelligence and Internet of Things Applications* (pp. 177-186). IGI Global.

Singh, B. (2023). Federated Learning for Envision Future Trajectory Smart Transport System for Climate Preservation and Smart Green Planet: Insights into Global Governance and SDG-9 (Industry, Innovation and Infrastructure). *National Journal of Environmental Law*, 6(2), 6–17.

Singh, S. K., Azzaoui, A. E., Choo, K. K. R., Yang, L. T., & Park, J. H. (2023). ArticlesA Comprehensive Survey on Blockchain for Secure IoT-enabled Smart City beyond 5G: Approaches, Processes, Challenges, and Opportunities. Hum.-Centric Comput. Inf. Sci, 13(51).

Singh, S. K., Azzaoui, A. E., Kim, T. W., Pan, Y., & Park, J. H. (2021). DeepBlockScheme: A deep learning-based blockchain driven scheme for secure smart city. *Human-centric Computing and Information Sciences*, 11(12), 1–13.

Singh, S. K., Kumar, M., Tanwar, S., & Park, J. H. (2024). GRU-based digital twin framework for data allocation and storage in IoT-enabled smart home networks. *Future Generation Computer Systems*, 153, 391–402. DOI: 10.1016/j.future.2023.12.009

Singh, S. K., Lee, C., & Park, J. H. (2022). CoVAC: A P2P smart contract-based intelligent smart city architecture for vaccine manufacturing. *Computers & Industrial Engineering*, 166, 107967. DOI: 10.1016/j.cie.2022.107967 PMID: 35136280

Singh, S. K., Pan, Y., & Park, J. H. (2021). OTS scheme based secure architecture for energy-efficient iot in edge infrastructure. *Computers, Materials & Continua*, 66(3), 2905–2922. DOI: 10.32604/cmc.2021.014151

Singh, S. K., Pan, Y., & Park, J. H. (2022). Blockchain-enabled secure framework for energy-efficient smart parking in sustainable city environment. *Sustainable Cities and Society*, 76, 103364. DOI: 10.1016/j.scs.2021.103364

Singh, S. K., Pan, Y., & Park, J. H. (2022). Blockchain-enabled secure framework for energy-efficient smart parking in sustainable city environment. *Sustainable Cities and Society*, 76, 103364. DOI: 10.1016/j.scs.2021.103364

Singh, S. K., & Park, J. H. (2022). TaLWaR: Blockchain-based trust management scheme for smart enterprises with augmented intelligence. *IEEE Transactions on Industrial Informatics*, 19(1), 626–634. DOI: 10.1109/TII.2022.3204692

Singh, S. K., Park, J. H., Sharma, P. K., & Pan, Y. (2021). BIIoVT: Blockchain-based secure storage architecture for intelligent internet of vehicular things. *IEEE Consumer Electronics Magazine*, 11(6), 75–82. DOI: 10.1109/MCE.2021.3089992

Singh, S. K., Park, L., & Park, J. H. (2022, October). Blockchain-based Federated Approach for Privacy-Preserved IoT-enabled Smart Vehicular Networks. In *2022 13th International Conference on Information and Communication Technology Convergence (ICTC)* (pp. 1995-1999). IEEE. DOI: 10.1109/ICTC55196.2022.9952835

Singh, S. K., Yang, L. T., & Park, J. H. (2023). FusionFedBlock: Fusion of blockchain and federated learning to preserve privacy in industry 5.0. *Information Fusion*, 90, 233–240. DOI: 10.1016/j.inffus.2022.09.027

Spiess-Knafl, W. (2022). *Artificial Intelligence and Blockchain for Social Impact: Social Business Models and Impact Finance*. Taylor & Francis. DOI: 10.4324/9781003218913

Stewart, J., & Gorman, C. (2024). Authentic Assessment and Embedding Sustainable Literacy and Technology Skills in Tourism Education. *Irish Journal of Academic Practice*, 11(2), 10.

Stewart, J., & Gorman, C. (2024). Authentic Assessment and Embedding Sustainable Literacy and Technology Skills in Tourism Education. *Irish Journal of Academic Practice*, 11(2), 10.

Synani, K., Abeliotis, K., & Lasaridi, K. (2021). Food waste in the catering and hospitality sector: generation factors and innovative practices.

Taneja, A., Sharma, R., Khetrapal, S., Sharma, A., Nagraik, R., Venkidasamy, B., Ghate, M. N., Azizov, S., Sharma, S., & Kumar, D. (2023). Value Addition Employing Waste Bio-Materials in Environmental Remedies and Food Sector. *Metabolites*, 13(5), 624. DOI: 10.3390/metabo13050624 PMID: 37233665

Tapo, A. A., Traore, A., Danioko, S., & Tembine, H. (2024). Machine Intelligence in Africa: a survey. *arXiv preprint arXiv:2402.02218.*

Tice, P. H. (2024). *The Race to Zero: How ESG Investing will Crater the Global Financial System*. Encounter Books.

Toivonen, A. (2020). *Sustainable space tourism: an introduction* (Vol. 3). Channel view publications.

Tomaszewska, M., Bilska, B., Tul-Krzyszczuk, A., & Kołożyn-Krajewska, D. (2021). Estimation of the scale of food waste in hotel food services—A case study. *Sustainability (Basel)*, 13(1), 421. DOI: 10.3390/su13010421

Vargas-Sanchez, A. (2023). Toward a circular tourism industry: The importance of a start-up ecosystem. *Worldwide Hospitality and Tourism Themes*, 15(6), 625–632. DOI: 10.1108/WHATT-09-2023-0111

Verma, A., Grobbelaar, W., & Shukla, V. K. (2023). Smart Innovation in the Hospitality and Tourism Industry. In *Computational Intelligence in Urban Infrastructure* (pp. 94–116). CRC Press. DOI: 10.1201/9781003218715-9

Vind, D. L. (2023). *Danish Design Heritage and Global Sustainability*. Taylor & Francis. DOI: 10.4324/9781003261018

Voorhoeve, J. (2021). Fighting Poverty and Violence: How Basic Human Needs Could Be Met.

Voukkali, I., Papamichael, I., Loizia, P., & Zorpas, A. A. (2023). The importance of KPIs to calibrate waste strategy in hospitality sector. *Energy Nexus*, 11, 100211. DOI: 10.1016/j.nexus.2023.100211

Workie, E., Kumar, V., Bhatnagar, A., He, Y., Dai, Y., Tong, Y. W., & Fu, C. (2023). Advancing the bioconversion process of food waste into methane: A systematic review. *Waste Management (New York, N.Y.)*, 156, 187–197. DOI: 10.1016/j.wasman.2022.11.030 PMID: 36493662

Yoon, B., Chung, Y., & Jun, K. (2020). Restaurant industry practices to promote healthy sustainable eating: A content analysis of restaurant websites using the value chain approach. *Sustainability (Basel)*, 12(17), 7127. DOI: 10.3390/su12177127

Zhang, X., Sheng, Y., & Liu, Z. (2024). Using expertise as an intermediary: Unleashing the power of blockchain technology to drive future sustainable management using hidden champions. *Heliyon*, 10(1), e23807. DOI: 10.1016/j.heliyon.2023.e23807 PMID: 38226273

Chapter 18
Fostering a Circular Economy in Tourism:
Insights From Egypt

Tamer M. Elsawy
https://orcid.org/0000-0003-4852-6610
Pharos University in Alexandria, Egypt

ABSTRACT

This chapter examines the state of the Circular Economy (CE) in Egypt's tourism sector, revealing both challenges and opportunities. The chapter uses a multi-faceted approach, combining secondary data analysis with a literature review. It systematically reviews and synthesises existing data from government reports, industry statistics, and prior research studies. It analyses barriers, demographic influences, and organisational dynamics and explores how technology can aid in overcoming these challenges. The chapter also draws on best practices from successful transitions elsewhere. Findings highlight the need for robust policies, targeted educational campaigns, and technological adoption. Effective CE transition in Egypt will require structured approaches, international collaboration, and tailored practices, such as adapting strategies to the Red Sea and promoting agritourism in the Nile Delta. This chapter is novel for its detailed analysis of CE adoption in Egypt's tourism sector, offering strategic pathways for advancing circularity.

1. INTRODUCTION

The tourism industry plays a pivotal role in Egypt's economy, attracting millions of visitors yearly to its ancient monuments, pristine beaches, and vibrant cultural sites. However, the rapid growth of tourism has also brought about environmental

DOI: 10.4018/979-8-3693-6110-8.ch018

challenges, necessitating a shift towards more sustainable practices. According to Economist Impact (2022), the tourism industry in the Middle East lags in the sustainability transition, marked by high resource consumption, significant waste generation, and the absence of industry-wide sustainability mandates. Ghanem (2016) emphasises the need for Egypt to enhance its efforts towards a transition to sustainable development, considering the prevailing economic, social, and environmental challenges. Circular Economy (CE) principles offer a promising framework to address these challenges by promoting resource efficiency, minimising waste, and fostering sustainable consumption and production patterns. By adopting circular practices, tourism businesses can reduce their resource footprint, lower greenhouse gas emissions, and minimise waste generation, thereby contributing to broader sustainable development goals. Thus, in recent years, the concept of a Circular Economy has gained traction within the tourism sector as a pathway to mitigate environmental impacts and promote sustainable development (Vatansever et al., 2021). Similarly, the global tourism industry increasingly recognises the imperative to transition towards more sustainable and circular practices in response to pressing environmental challenges associated with traditional linear consumption patterns (Manniche et al., 2020).

Concepts such as sustainable development and green growth have arisen over the past thirty years to tackle worldwide challenges associated with the current production and consumption model centred on growth. These challenges include issues like resource scarcity, climate change, and pollution. While the Circular Economy (CE) aligns with these concepts, it provides unique and actionable principles (Martins, 2021). The concepts of CE recognise the need for our world's finite resources to be used efficiently and sustainably, as opposed to the dead-end system of the linear economy. Kirchherr et al. (2017) consolidated more than 100 definitions of the Circular Economy, presenting it as an economic framework that moves away from the traditional "end-of-life" approach to emphasising the minimisation, reuse, recycling, and recovery of materials throughout the stages of production, distribution, and consumption. Despite the predominant focus on manufacturing industries (Sorin & Einarsson, 2020), the tourism sector has received limited consideration in CE initiatives and analyses (Manniche et al., 2020). Nevertheless, the tourism industry has substantial potential to enhance sustainability and profitability through CE practices (Manniche et al., 2020), underscoring the importance of acknowledging travel and tourism in the global shift towards CE (UNWTO, 2020).

The concept of Circular Economy holds relevance for the tourism sector, given its reliance on natural resources and the generation of significant waste and emissions (UNWTO, 2020). Traditional linear consumption patterns, characterised by take-make-dispose practices, are increasingly recognised as unsustainable. In contrast, Circular Economy principles advocate for a regenerative approach to keep resources in use for as long as possible, extracting maximum value from them and minimising waste and pollution (Hernandez et al., 2020). According to Aryal (2020), adopting Circular Economy (CE) practices can enhance a nation's workforce by 4%, minimise waste, and potentially lower carbon dioxide emissions by 70%. As a result, numerous companies are incorporating circular principles to reduce costs, boost revenues, and manage risks more effectively. However, the global economy remains only 8.6% circular (WBCSD, 2022). According to Vatansever et al. (2021), companies frequently encounter significant obstacles in the implementation process despite the incentives provided for transitioning to a Circular Economy. Sorensen and Bærenholdt (2020) used the Delphi Method with 32 Danish experts to identify barriers to CE tourist practices, finding that lack of political action was the most cited obstacle. Martinez-Cabrera and López-del-Pino (2021), through a systematic review and interviews with 33 European experts, highlighted macro-environmental challenges such as insufficient government support and low social awareness. According to CEnTOUR (2020), tourism SMEs lack the resources and skills to transition to a CE model. Khan et al. (2022) found that while tourism SMEs show positive intentions towards CE, financial pressures, a lack of skilled personnel, and insufficient information on CE practices hinder their adoption.

To implement CE strategies, collaboration among all key players, including DMOs, tourism businesses, locals, and visitors, is essential (Florido et al., 2019). Robaina (2022) highlights the crucial role of tourists' attitudes and behaviours in engaging with green, sustainable, and circular practices for a successful CE transition in tourism. Destination efforts may be ineffective without tourists adopting circular behaviours (Robaina, 2022). Raising awareness among tourists about the impact of their consumption choices is vital (Giurea et al., 2018). Travel and tourism stakeholders can play a pivotal role in promoting circularity, benefiting from the collaborative creation of circular value within relevant supply chains (UNWTO, 2020). The rapid ascent of disruptive technologies, exemplified by platforms like Airbnb, Homestay, CouchSurfing, and Uber, has profoundly reshaped the tourism sector, illustrating its ongoing evolution (Dawadi, 2022). Manniche et al. (2020) argue that CE principles are pertinent for all tourism SMEs, regardless of their perceived environmental commitments.

Within the context of Egyptian tourism, Circular Economy principles present an opportunity to reduce the industry's ecological footprint while simultaneously promoting economic growth and social development (Elsawy, 2024a). By transitioning towards circular practices, Egyptian tourism stakeholders can create value through resource efficiency, innovation, and improved environmental stewardship (Elsawy, 2024b). A bibliometric review of CE in all research fields emphasises its importance (Rodríguez et al., 2020), making Egypt's adoption of CE initiatives in tourism a crucial step towards creating a more sustainable future for the sector. When applied to tourism in Egypt, CE initiatives can offer economic advantages such as increased growth and employment opportunities while reducing waste production and resource consumption (Rezk et al., 2023; UNIDO, 2023). According to Elsawy (2024a), the transition to a Circular Economy in Egypt offers a twofold advantage: bolstering economic growth and employment opportunities while making the best use of available resources. Examples of circularity in Egyptian tourism could include collecting and processing waste produced by tourists for recycling and repurposing, using renewable energy sources, and promoting sustainable tourism practices (Rezk et al., 2023). Adopting such initiatives is crucial for creating a more sustainable and eco-friendly future for Egypt's tourism sector.

Recent empirical studies focused on Circular Economy adoption within Egyptian tourism have yielded valuable insights into the barriers and facilitators of circularity. Elsawy (2024b) highlighted that approximately 91.69% of Egyptian travel agencies do not employ Circular Economy practices, indicating a significant need for improvement in adopting sustainable practices. The study underscores the challenges faced by SMEs in the tourism sector, with limited infrastructure and resource constraints hindering the transition to circular business models. Moreover, demographic factors among domestic tourists, including age, gender, and education, have been found to influence environmental consciousness and attitudes towards circular tourism practices in Egypt (Elsawy, 2024a). However, according to Elsawy (2024a), there appears to be a noticeable gap between environmental consciousness and the attitudes of domestic tourists in Egypt regarding circular practices. This insight is echoed by Budovska et al. (2019) and Rafiq et al. (2022), who emphasised that environmental awareness grows slowly in developing nations. Despite increased awareness, travellers' actions often do not align with their concerns, creating an attitude-behaviour gap and suggesting the need for targeted interventions to bridge this gap effectively.

The synthesis of these research findings provides a nuanced understanding of the complexities surrounding the adoption of a Circular Economy in Egyptian tourism. By examining both internal and external barriers, demographic influences, the potential of advanced technologies to support the CE within the tourism sector, and examples of successful transitions to CE, the study can identify strategic pathways for advancing circularity within the sector, which will fill a crucial gap in this area.

Despite the previously discussed importance of adopting a Circular Economy in the Egyptian tourism sector, no study has comprehensively examined this topic in Egypt, one of the most significant destinations in developing countries and the Middle East, using an integrated approach. Building on the preceding, this chapter aims to accomplish the following objectives:

- investigate the current implementation barriers to a CE within Egypt's tourism industry.

- examine how demographic factors, such as age, gender, and education, influence environmental consciousness and attitudes towards circular tourism among domestic tourists in Egypt.
- explore the potential of advanced technologies to support and enhance the CE within the tourism sector.
- discuss implications of identified barriers for tourism stakeholders and propose actionable strategies and interventions to overcome these challenges.
- offer policy recommendations based on empirical findings, outlining pathways for policymakers to promote Circular Economy practices in Egyptian tourism.

Through this exploration, the author aims to contribute to the discourse on sustainable tourism development in Egypt and inspire stakeholders to embrace Circular Economy principles as a pathway towards a more sustainable and resilient tourism industry. The following sections of this chapter will delve into the methodological framework, present the findings, and discuss their implications for promoting circularity in Egyptian tourism.

2. METHODOLOGY

This chapter employs a multi-faceted methodological approach that combines secondary data analysis and literature review to comprehensively understand the challenges and opportunities in promoting Circular Economy practices within the Egyptian tourism industry (Figure 1). By integrating secondary data analysis with a thorough literature review, the study can offer a nuanced understanding of macro-level trends and micro-level implementation strategies (Cheong et al., 2023; Taherdoost, 2021). This dual approach ensures that the recommendations are grounded in empirical data and theoretical insights, making them more robust and actionable. Additionally, this framework highlights the interconnectedness of various factors, such as policy, industry practices, and consumer behaviour, which are crucial for successfully implementing Circular Economy practices in tourism (Elsawy, 2024b).

Secondary data analysis involves systematically synthesising existing data from various sources, such as government reports, industry statistics, and previous research studies. This approach identifies trends, patterns, and insights that are not immediately apparent from primary data collection alone (Cheong et al., 2023). The literature review component further enriches this analysis by critically evaluating previous studies on the Circular Economy and sustainable tourism practices. This involves reviewing academic articles, case studies, and industry reports to synthesise knowledge about effective strategies and common barriers. For example, research by Geissdoerfer et al. (2017) outlines the theoretical underpinnings of the Circular Economy, providing a foundation for understanding how these principles can be applied within the tourism sector.

Figure 1. The methodological framework

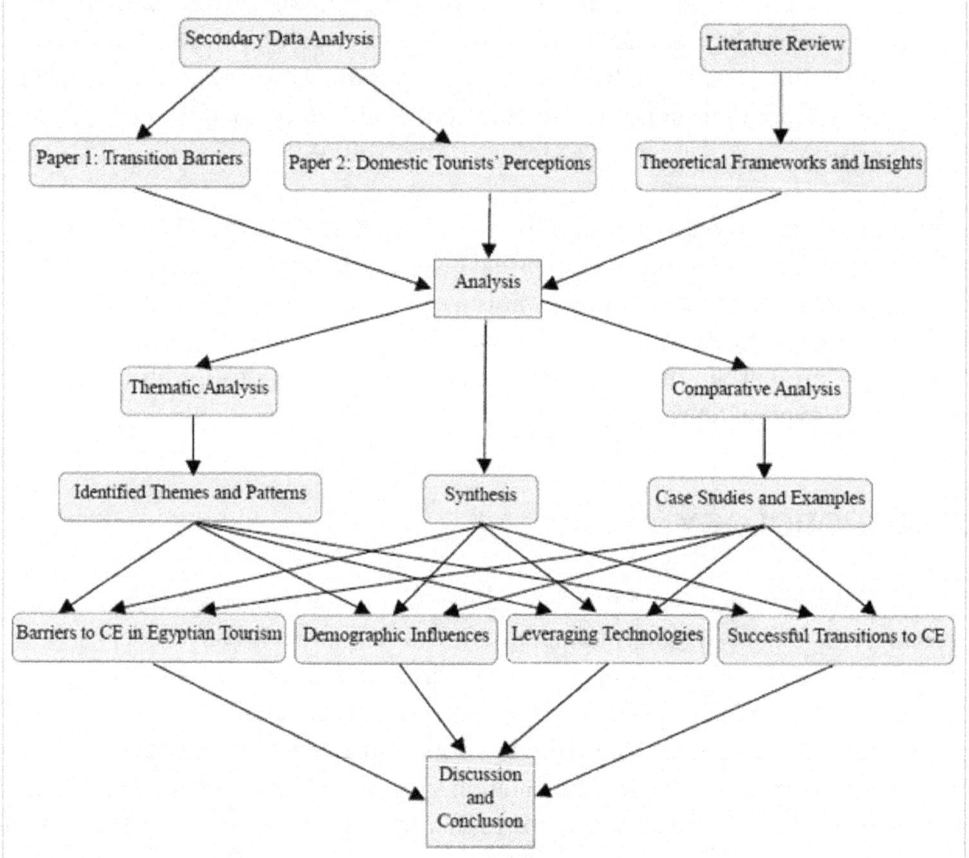

Source: Author's Compilation

2.1 Secondary Data Analysis

Two of the author's previously published papers are this chapter's primary data sources and foundational references for the current analysis.

- Paper 1: "The Transition Barriers Toward the Circular Economy for Travel Agencies in Egypt" (https://dx.doi.org/10.21608/ijtah.2024.258367.1114) (Elsawy, 2024b).

This study employed a quantitative approach to investigate the transition barriers to the Circular Economy for travel agencies in Egypt. Data was collected using structured questionnaires, targeting 448 category A travel agencies in the country.

- Paper 2: "Domestic Tourists' Perception and Attitude Towards Circularity in Egypt" (https://dx.doi.org/10.21608/pijth.2024.263493.1008) (Elsawy, 2024a)

This study explores the perceptions and attitudes of domestic tourists in Egypt towards Circular Economy practices. The aim is to gain insights into the preferences and behaviours of this group, filling a gap in the existing literature on this topic. Structured surveys were conducted among a random sample of 397 diverse domestic tourists. Using SmartPLS 4 for data analysis, a quantitative method was utilised to explore the relationships between environmental awareness, demographic characteristics, and attitudes toward tourism circularity.

The data extracted from these papers were analysed to identify recurring themes, challenges, and successful strategies for implementing Circular Economy practices in Egyptian tourism. This analysis provided a robust foundation for developing recommendations and insights for this chapter.

2.2 Literature Review

In addition to the secondary data analysis, a comprehensive literature review was conducted to contextualise the findings within the broader academic discourse on Circular Economy and sustainable tourism. The literature review mainly focused on recent studies published in the last five years, ensuring the inclusion of up-to-date perspectives and developments in the field. The key areas of focus in the literature review included:

- Theoretical frameworks of Circular Economy and their applicability to the tourism sector.
- Circular Economy challenges and facilitators in the tourism sector.
- Global best practices in Circular Economy initiatives within tourism.
- Policy and regulatory frameworks supporting the Circular Economy in tourism.

2.3 Data Collection and Analysis

Data collection involved sourcing articles, reports, and case studies from academic journals, industry publications, and reputable online databases. The selection criteria for the literature included relevance to the topic, recent publication date, and contributions to the understanding of Circular Economy practices in tourism. The analysis process included:

a) **Thematic Analysis**: Identify key themes and patterns related to Circular Economy practices in tourism. This involved coding the data to highlight significant themes such as barriers, demographic influences, technology use, and successful case studies. These themes were then categorised and analysed for common patterns and discrepancies (Braun & Clarke, 2006).
b) **Comparative Analysis**: Compare findings from different sources to highlight common challenges and successful strategies. This step involved a detailed examination of the data to identify similarities and differences across various studies and contexts. The comparative analysis identified best practices and potential pitfalls in implementing Circular Economy practices in tourism (Huberman & Miles, 2002).
c) **Synthesis**: Integrate secondary data and literature review insights to form a cohesive narrative and draw comprehensive conclusions. This synthesis aimed to combine the identified themes and comparative insights to develop a comprehensive understanding of the current state of Circular Economy practices in the Egyptian tourism sector and to propose actionable recommendations for fostering these practices (Thomas & Harden, 2008).

The analysis process resulted in four major themes, which were examined in detail:

a) Barriers to Circular Economy Adoption in Egyptian Tourism Businesses

This theme focused on the obstacles Egyptian tourism businesses face in adopting Circular Economy practices. The analysis included barriers such as lack of awareness, financial constraints, and inadequate regulatory support.

b) Demographic Influences on Environmental Consciousness and Attitudes Towards Circularity in Egypt

This theme explored how demographic factors such as age, gender, education, and income influence environmental consciousness and attitudes towards Circular Economy practices among Egyptian tourists and tourism professionals.

c) Leveraging Advanced Technologies for Circular Economy in Tourism

This theme investigated the role of advanced technologies, including blockchain, AI, IoT technologies, and big data analytics, in promoting Circular Economy practices in the tourism sector.

d) Successful Transitions to Circular Economy in Tourism: Lessons for Egypt

This theme analysed case studies and examples of successful transitions to Circular Economy practices in the global tourism sector, drawing lessons and best practices that can be applied to the Egyptian context.

2.4 Ethical Considerations

This research adhered to ethical guidelines for secondary data analysis and literature review. Proper citations and acknowledgements were provided for all sources, ensuring academic integrity and respecting intellectual property rights. This chapter provides a well-rounded exploration of the circular economy in the tourism sector in Egypt by combining the analysis of secondary data from previous research with a thorough literature review. The methodology ensures a robust and comprehensive understanding of the subject, laying the groundwork for practical recommendations and future research directions.

3. FINDINGS

3.1 Barriers to Circular Economy Adoption in Egyptian Tourism Businesses

Tourism circularity in Egypt refers to implementing Circular Economy (CE) principles in the country's tourism industry. This model keeps resources in use for as long as possible, and waste is reduced through recycling and reuse (OECD, 2023). Recent empirical studies highlight substantial barriers hindering the adoption of Circular Economy within Egyptian travel agencies, particularly among small- to medium-sized enterprises (SMEs). Despite the numerous initiatives and programmes related to CE and sustainable strategies, a quick look reveals that SME tourism businesses in Egypt are struggling. This slow progress is due to significant societal transformations affecting dominant business models, financing, communication, education systems, and energy provision. These changes extend beyond technology, encompassing shifts in regulations, laws, infrastructures, industrial networks, and consumption cultures (Manniche et al., 2020).

Karam (2020) argued that businesses in Egypt aiming to transition to CE might encounter potential weaknesses and threats. Weaknesses include bureaucracy and the disparity between major cities like Cairo and Alexandria and other regions regarding literacy rates, culture, infrastructure, and technology, impacting their capacity to implement CE. Threats include a lack of social awareness about CE, which could limit support for related initiatives or restrict them to clean energy production and waste management. According to Economist Impact (2022), Egypt's recycling industry, still in its early stages, faces capacity limitations.

Elsawy (2024b) examines Egyptian travel agencies, identifying barriers to adopting the Circular Economy (CE). The study revealed that 90% of tourism companies are small to medium-sized enterprises (SMEs) and that approximately 91.69% of agencies lack CE practices, placing them at the initial and least mature stage of CE implementation, mirroring the global economy's 8.6% circular status (WBCSD, 2022) and indicating challenges despite incentives (Vatansever et al., 2021), indicating a significant gap in sustainability implementation. The study underscores external and internal barriers that impede the transition towards circular business models. External barriers include political, economic, social, legal, external resources, value chains, and infrastructure, while internal impediments encompass organisational culture, internal resources, cost structure, and size (see Figure 2). Organisational characteristics, particularly size, significantly shape the adoption of the Circular Economy within Egyptian travel agencies. Elsawy (2024b) identified a positive relationship between organisational size and Circular Economy adoption, with larger agencies exhibiting a greater inclination towards circular practices, consistent with previous studies (e.g., Fernández-Robin et al., 2019). However, conflicting studies (e.g., Manniche et al., 2021) challenge the notion that small size hampers CE practices in tourism SMEs.

Furthermore, a negative moderating effect of organisational size between external factors and CE implementation was found, highlighting the nuanced impact of organisational characteristics on overcoming sustainability challenges (Martinez-Cabrera & López-del-Pino, 2021). Smaller travel agencies may face more significant difficulties in accessing financing, implementing technological solutions, and navigating regulatory frameworks, impeding their adoption of circular practices (Fernández-Robin et al., 2019). This explains the higher effect of external factors on CE transition in Egyptian SMEs (Elsawy, 2024b).

Figure 2. Barriers to circular economy adoption in egyptian tourism businesses

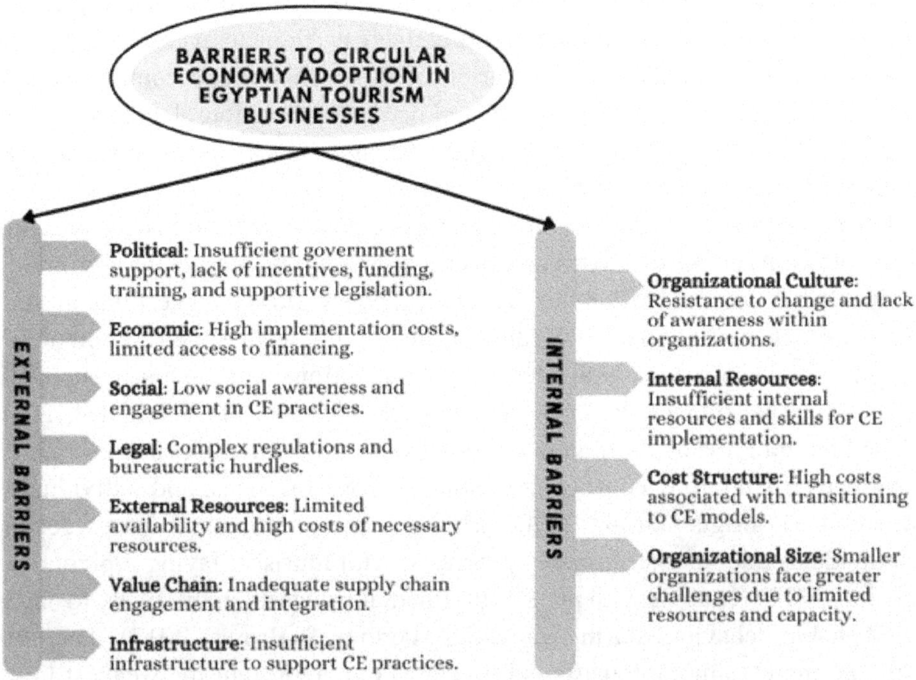

Source: Adapted from Elsawy (2024b), Martínez-Cabrera & López-del-Pino (2021)

These findings align with global observations highlighting financial constraints, lack of awareness, and resource limitations as primary barriers to Circular Economy adoption in the tourism sector (De Jesus & Mendonça, 2018; Vatansever et al., 2021). Limited access to financing and high implementation costs often deters tourism businesses, especially SMEs, from investing in circular practices and sustainability initiatives (CEnTOUR, 2020; Khan et al., 2022; Ranta et al., 2018). Moreover, a lack of supportive policies and government incentives further complicates adopting Circular Economy initiatives in tourism (Martinez-Cabrera & López-del-Pino, 2021).

3.2 Demographic Influences on Environmental Consciousness and Attitudes Towards Circularity in Egypt

While discussions on the Circular Economy (CE) often focus on production-side transitions, it is essential to recognise that this process occurs within a societal context, affecting both production and consumption practices (Manniche et al., 2020). Moreover, the growing number of tourists engaging in unsustainable consumption habits exacerbates solid waste issues, overburdening local waste management systems and contradicting CE principles (Dawadi, 2022). Educating tourists is a significant challenge in transitioning to a circular tourism model. While pressures from the supply side are essential, tourists actively participate in co-producing, co-performing, and co-creating their experiences. They choose and combine tourism elements, interact with local resources, and drive the behavioural changes needed for circular tourism. Therefore, consumer awareness is crucial for this transition (Sorensen & Bærenholdt, 2020; Aryal, 2020).

Understanding demographic influences on environmental consciousness and attitudes towards circular tourism practices among tourists is crucial for promoting sustainable behaviours. Hockenbury and Hockenbury (2011) suggest that attitudes directly influence behaviour. According to the Theory of Reasoned Action (TRA), these attitudes shape behavioural intent, impacting actions. Consequently, successful destination management relies on the dynamic attitudes of locals (Peters et al., 2018).

The literature review highlights the importance of promoting social awareness through advertising to encourage governments, tourism firms, and individuals to shift to CE (Rodríguez et al., 2020). Minimising environmental issues depends on individuals adopting sustainable practices, with tourists playing a pivotal role (Robaina, 2022). Tourists with pro-environmental values are more likely to engage in sustainable behaviour during vacations (Perkins & Brown, 2012). Therefore, raising awareness among tourists is crucial, as they represent the weakest link in the value chain (Robaina, 2022). However, Budovska et al. (2019) and Rafiq et al. (2022) emphasised the gap between environmental awareness and attitudes towards circular practices, particularly among tourists in developing nations like Egypt.

A study by Elsawy (2024a) found that while 56% of domestic tourists in Egypt demonstrate high environmental awareness, only 44.6% exhibit a positive attitude towards circularity. This highlights the complexity of translating awareness into sustainable behaviours. Although many tourists may be environmentally aware, this awareness does not always lead to support for circular practices. This highlights the need for targeted interventions and educational campaigns to close the gap between awareness and behaviour, promoting a more sustainable and circular tourism culture among domestic tourists in Egypt. Moreover, the study concluded that age, gender, and education emerged as key demographic factors influencing environmental consciousness, with older and more educated tourists demonstrating higher awareness and support for circular tourism practices. Female domestic tourists tend to strongly prefer environmentally friendly features and demonstrate greater engagement in sustainable practices than their male counterparts (Elsawy, 2024a). This highlights the role of gender in shaping attitudes and behaviours related to sustainability in tourism.

These results align with previous research, which highlights the importance of demographic differences, such as age, education, gender, nationality, and income, in understanding tourists' awareness and behaviours regarding sustainable and circular practices (Aryal, 2020; Holmes et al., 2019; Robaina, 2022; Sorensen & Bærenholdt, 2020). Higher education is linked to greater eco-friendly awareness and circular travel behaviour (Robaina, 2022). Additionally, females prefer green attributes and more circular behaviour than males (Millar & Baloglu, 2011; Robaina, 2022).

3.3 Leveraging Advanced Technologies for Circular Economy in Tourism

The tourism industry has increasingly turned to advanced technologies to foster a Circular Economy in recent years. This paradigm shift is essential for reducing environmental impacts and promoting sustainability within the sector. Recent studies highlight the pivotal role of advanced technologies such as blockchain (Treiblmaier, 2018; Kouhizadeh & Sarkis, 2018), artificial intelligence (AI) (Demirel et al., 2021; Gretzel et al., 2015), the Internet of Things (IoT) (Buhalis & Leung, 2018; Demirel et al., 2021), and big data (BD) analytics (Elsawy & Eltayeb, 2023; Li et al., 2018) in promoting Circular Economy practices within the tourism industry and effectively addressing the transition barriers towards it.

The integration of blockchain technology, for instance, offers a secure and efficient platform that enhances transparency and reduces the reliance on intermediaries, leading to more sustainable tourism practices (Halkiopoulos et al., 2023). In a post-pandemic context, smart contracts powered by blockchain have shown potential in streamlining transactions and eliminating intermediaries, facilitating more efficient resource management (Demirel et al., 2021). IoT devices further contribute by monitoring and analysing tourist activities, optimising resource use and minimising waste. AI technologies also play a significant role in predictive analytics, enabling better demand forecasting and resource allocation. Collectively, these technologies support a transition towards a Circular Economy in tourism by promoting sustainable practices and improving operational efficiencies.

By integrating blockchain, AI, IoT technologies, and big data analytics, Egyptian tourism businesses can address these barriers and enhance sustainability. These technologies support better resource management, transparency, and efficiency, facilitating the transition towards a Circular Economy, as follows:

a) **Blockchain**: Blockchain technology can address trust and transparency issues, providing secure and immutable records of transactions and resource flows. This reduces the need for intermediaries, lowers costs, and improves efficiency, which can mitigate resistance to CE adoption (Halkiopoulos et al., 2023). Blockchain enables tourism operators to ensure their supply chains are sustainable and ethically sourced. For instance, blockchain can track the origins of food and other products used in hotels and restaurants, verifying that they come from sustainable sources. This technology can also facilitate managing waste more effectively by providing a transparent record of waste generation and disposal and encouraging responsible waste management practices (Treiblmaier, 2018; Kouhizadeh & Sarkis, 2018).

b) **AI**: AI is transforming the tourism industry by enabling smarter decision-making and personalised experiences. AI-powered systems can analyse vast amounts of data to predict tourist behaviour, optimise resource allocation, and enhance the efficiency of tourism operations. For example, AI can forecast demand for services such as accommodation and transportation, allowing providers to adjust their offerings accordingly. This helps reduce overconsumption and ensures that resources are used efficiently. Moreover, AI can personalise tourist experiences by suggesting sustainable travel options and activities that align with their preferences, promoting eco-friendly tourism (Gretzel et al., 2015). According to Demirel et al. (2021), AI-driven predictive analytics aid strategic planning and decision-making, ensuring alignment with CE principles.

c) **IoT**: IoT devices offer real-time monitoring and management of resources, identifying inefficiencies and areas for improvement. This technology enhances operational efficiency and promotes sustainable practices by providing data on energy consumption, waste generation, and resource utilisation (Demirel et al., 2021). According to Buhalis and Leung (2018), IoT can optimise energy and water use in hotels through smart metres and sensors, which track consumption patterns and adjust settings to reduce wastage. For example, smart thermostats and lighting systems can adapt to occupancy patterns, ensuring energy is used efficiently. This not only reduces operational costs but also lowers the carbon footprint of tourism establishments.

d) **BD**: With technological advancements and the rapid increase in data generation by electronic devices, decision-makers and Destination Management Organisations (DMOs) are increasingly interested in applications that can capture and analyse big data. These tools help them leverage valuable information to enhance the effectiveness of their strategies, mitigate potential risks in their decisions, and differentiate their destination from competitors (Elsawy & Eltayeb, 2023). Big data analytics plays a pivotal role in understanding and optimising tourism operations for sustainability. By analysing data from various sources, tourism operators can gain insights into tourist behaviour, resource consumption, and waste generation. These insights can inform strategies to reduce environmental impacts and promote a Circular Economy. For instance, data analytics can identify energy and water use patterns, helping develop targeted conservation interventions. Additionally, big data can support the development of sustainable tourism policies by providing evidence-based insights into the effectiveness of different initiatives (Li et al., 2018).

By leveraging these technologies, the tourism sector can minimise its environmental impact, improve operational efficiency, and provide better tourist experiences. As the industry continues to evolve, the adoption of advanced technologies will be key to achieving a sustainable and Circular Economy in tourism (Buhalis & Leung, 2018; Elsawy & Eltayeb, 2023; Halkiopoulos et al., 2023; Kouhizadeh & Sarkis, 2018; Li et al., 2018; Treiblmaier, 2018).

3.4 Successful Transitions to Circular Economy in Tourism: Lessons for Egypt

Several countries have successfully transitioned towards a Circular Economy (CE) in tourism, focusing on sustainable practices and reducing environmental impact. A study in the Valencian Region of Spain revealed that tourism companies followed a structured journey from initial awareness to implementing actions based on the nine Rs framework, indicating a gradual and focused transition towards circularity. The research involved three focus groups in Alicante, Castellon, and Valencia, with eighteen executives from hotels, restaurants, and tourist industry suppliers. This study provides a detailed mapping of the transition process and highlights the significant strides made by these companies towards sustainable tourism practices (de-Miguel-Molina et al., 2024). This structured approach could be replicated for Egypt by organising similar focus groups and creating tailored transition plans for different tourism sectors.

Similarly, a survey of tourism SMEs across Cyprus, France, Italy, and Spain identified challenges such as lack of funds, information about potential partners, and lack of skilled personnel. Despite these obstacles, the study found that many tourism SMEs recognised the positive outcomes of adopting CE practices, such as improved sustainability and competitive advantage. This survey, which analysed a sample of 256 tourism SMEs, underscores the need for better support systems and resources to facilitate the transition to CE in the tourism sector (Khan et al., 2021). For Egypt, addressing these challenges could involve government and private sector collaboration to provide funding, information resources, and training programmes to boost the skills needed for CE adoption.

In the Nordic region, hotel chains have made strides in waste minimisation and reducing plastic use. However, the total adoption of CE principles requires more comprehensive efforts, including collaboration with industry partners and support from the public sector. The study analysed data from webpages, annual reports, and sustainability reports of three Nordic hotel chains, revealing that while individual actions are being taken, a more integrated approach is necessary for true circularity (Holmberg & Konttinen, 2023). For Egypt, hotels and other tourism businesses could benefit from forming alliances with local and international partners to share best practices and resources. At the same time, government policies could incentivise broader CE adoption.

On the small island of Guam, initiatives to transform linear tourism into circular tourism have focused on sustainable practices tailored to the island's unique context. This includes efforts to promote recycling, reduce waste, and encourage the use of local resources. The case of Guam highlights the importance of adapting CE principles to fit the specific needs and circumstances of small island tourism, providing a model that can be replicated in similar contexts (Schumann, 2020). With its unique environmental and cultural context, Egypt can adapt CE principles to fit its specific tourism needs, such as promoting local materials and sustainable practices in popular tourist destinations like the Red Sea and the Sinai Peninsula.

Lastly, agritourism has shown potential as a forerunner for the green transition in Sicily, driven by financial support from the European Union's Common Agricultural Policy (CAP) and promoting sustainable behaviours among tourists and farmers. The study involved surveys of Sicilian agritourism and their visitors, demonstrating that agritourism naturally supports the green transition by providing quality food and ecosystem services and promoting healthy behaviours and local consumption. This case illustrates how agritourism can serve as a vehicle for sustainable tourism and contribute to the broader goals of economic circularity (Ingrassia et al., 2023). For Egypt, developing agritourism in regions like the Nile Delta and Upper Egypt can promote sustainable tourism practices and support local economies by encouraging the consumption of local produce and fostering eco-friendly tourism activities.

4. DISCUSSION

This chapter seeks to thoroughly explore insights to foster a Circular Economy (CE) in Egyptian tourism. It includes analysing the obstacles to CE implementation in the tourism sector, assessing the impact of demographic factors on attitudes towards circular tourism, and investigating supportive technologies for CE. Additionally, the chapter proposes strategies to address these barriers and provides policy recommendations to advance CE practices in Egypt's tourism industry.

The low adoption rate of Circular Economy practices among Egyptian travel agencies, as revealed by Elsawy (2024b), highlights substantial barriers that hinder sustainability initiatives in the tourism sector. The synthesis of these empirical insights provides a comprehensive understanding of the barriers, demographic influences, and organisational dynamics shaping Circular Economy adoption within Egyptian tourism. External barriers, including political, economic, and social challenges, along with inadequate infrastructure, legal frameworks, and limited government support, present formidable obstacles to Circular Economy implementation (Elsawy, 2024b; Martinez-Cabrera & López-del-Pino, 2021). These external barriers underscore the need for policy interventions to create an enabling environment for sustainable practices.

Internally, resource constraints, organisational culture, and cost considerations influence the feasibility of adopting circular business models (Elsawy, 2024b; Rizos et al., 2016; Vatansever et al., 2021). Small to medium-sized enterprises (SMEs), which constitute a significant portion of the tourism sector in Egypt, often face challenges in accessing financing and implementing technological solutions required for circular practices (Rizos et al., 2016; Vatansever et al., 2021). Overcoming these barriers requires targeted intervention.

Organisational characteristics, particularly size, significantly influence the adoption of the Circular Economy within Egyptian travel agencies. Elsawy (2024b) identifies a positive relationship between organisational size and Circular Economy adoption, with larger agencies exhibiting a greater inclination towards sustainable practices. This finding aligns with global research emphasising resource constraints faced by SMEs in the tourism sector (Elsawy, 2023), limiting their capacity to invest in sustainability initiatives (CEnTOUR, 2020; Khan et al., 2022). The moderating effect of organisational size on the relationship between external barriers and Circular Economy implementation highlights the nuanced impact of organisational dynamics on sustainability efforts (Martinez-Cabrera & López-del-Pino, 2021). Smaller agencies may face greater challenges in accessing financing, implementing technological solutions, and navigating regulatory frameworks, thereby impeding their adoption of circular practices.

The relationship between demographic factors and environmental consciousness among domestic tourists in Egypt unveils nuanced insights into sustainable tourism behaviours. Age and education play pivotal roles in shaping environmental consciousness, with older and more educated tourists exhibiting higher awareness and support for circular tourism practices (Sorensen & Bærenholdt, 2020; Aryal, 2020). This demographic influence underscores the importance of targeted educational campaigns and interventions to bridge the gap between awareness and sustainable behaviours among domestic tourists. Moreover, gender moderates the relationship between environmental consciousness and attitudes towards circularity, with female tourists demonstrating higher preferences for green attributes and circular behaviours than males (Millar & Baloglu, 2011). Recognising these demographic influences is essential for developing tailored strategies that promote sustainable tourism practices among diverse traveller segments.

The tourism industry's growing adoption of advanced technologies is pivotal in fostering a CE essential for reducing environmental impacts and enhancing sustainability. Key technologies such as blockchain, AI, IoT, and BD analytics play crucial roles in promoting CE practices and addressing barriers to its implementation. The integration of blockchain technology addresses trust and transparency issues in the tourism industry by providing secure, immutable records of transactions and resource flows. This reduces intermediaries, lowers costs, and improves efficiency, mitigating resistance to CE adoption (Treiblmaier, 2018). Blockchain can also verify the sustainability of food and products in hotels and manage waste through transparent records (Kouhizadeh & Sarkis, 2018). AI revolutionises tourism by enabling smarter decision-making and personalised experiences. AI systems analyse vast data to predict tourist behaviour, optimise resources, and enhance efficiency, promoting eco-friendly tourism (Gretzel et al., 2015). IoT devices offer real-time monitoring and management of resources, enhancing efficiency and sustainability by providing data on energy consumption, waste generation, and resource use (Buhalis & Leung, 2018). Big data analytics help optimise tourism operations for sustainability by providing insights into tourist behaviour, resource consumption, and waste generation, informing strategies to reduce environmental impacts and promote a Circular Economy (Elsawy & Eltayeb, 2023).

Successful transitions to a Circular Economy (CE) in tourism provide useful insights for Egypt. The structured approach in Spain's Valencian Region—moving from awareness to action—can be mirrored in Egypt through focus groups tailored to different tourism sectors (de-Miguel-Molina et al., 2024). Challenges faced by SMEs in Cyprus, France, Italy, and Spain, such as funding and skill shortages, highlight the need for support systems in Egypt. Government and private sector collaboration could address these issues by providing resources and training (Khan et al., 2021). In the Nordic region, a more integrated approach involving industry collaboration and public support is crucial for full CE adoption (Holmberg & Konttinen, 2023). Egypt could form alliances and offer incentives to promote broader CE practices. Guam's adaptation of CE principles to local contexts, such as recycling and local resource use, suggests that Egypt should tailor practices to specific tourism areas like the Red Sea and the Sinai Peninsula (Schumann, 2020). Sicily's agritourism, supported by the EU, shows how niche sectors can drive green transitions. Egypt could develop agritourism in regions like the Nile Delta to promote sustainability and support local economies (Ingrassia et al., 2023).

These research findings have critical implications for tourism stakeholders in Egypt and beyond. Addressing the identified barriers requires targeted interventions, including financial incentives, capacity-building programs, and infrastructure development (De Jesus & Mendonça, 2018; Ranta et al., 2018). Stakeholders must collaborate to foster a culture of sustainability and innovation within the tourism sector, leveraging demographic insights to tailor strategies that promote Circular Economy practices among domestic tourists (Florido et al., 2019).

Policymakers, in turn, can play a pivotal role in facilitating the transition towards circular tourism by implementing supportive policies, providing incentives, and raising awareness about the benefits of sustainable practices (Economic Commission for Europe, 2022). By addressing these barriers and leveraging demographic trends, advanced technologies, and best practices, tourism stakeholders and policymakers can promote the adoption of Circular Economy principles in Egyptian tourism, contributing to the sector's long-term sustainability and resilience.

4.1 Implications for Stakeholders

The identified barriers to Circular Economy (CE) adoption within Egyptian tourism have profound implications for stakeholders across the industry, including tourism businesses, policymakers, and researchers. Addressing these challenges requires targeted strategies and interventions to foster sustainable practices and promote the transition towards circularity. The interventions must consider the demographic influences, the facilitating role of advanced technologies, and the lessons from the case studies towards a successful transition to a CE.

4.1.1 Tourism Agencies and Businesses

For tourism agencies and businesses in Egypt, CE practices' low adoption rate presents challenges and opportunities. The predominance of small to medium-sized enterprises (SMEs) within the sector, as highlighted by Elsawy (2024b), underscores the need for tailored support mechanisms. Resource constraints, including limited financial capacity and inadequate internal resources, pose significant barriers to CE adoption (Rizos et al., 2016; Vatansever et al., 2021). To overcome these challenges, tourism agencies can consider the following strategies:

- **Capacity Building:** Invest in training programs and capacity-building initiatives to enhance staff skills and knowledge related to Circular Economy principles (Elsawy, 2024b).
- **Strategic Partnerships:** Collaborate with industry associations, non-governmental organisations (NGOs), and academia to access resources, expertise, and funding opportunities that support CE implementation (Martinez-Cabrera & López-del-Pino, 2021).
- **Innovation and Technology**: Leverage innovative technologies, such as blockchain and artificial intelligence, to optimise resource use and minimise waste within tourism operations (Ranta et al., 2018).

- **Supply Chain Transparency:** Tourism operators should implement blockchain to ensure sustainable and ethical sourcing of their supply chains, thus gaining consumer trust.
- **Waste Management:** Establish blockchain-based systems to monitor and manage waste, promoting accountability and reducing environmental impact (Halkiopoulos et al., 2023).
- **Circular Supply Chains:** Promote circular supply chain practices by sourcing goods and services from suppliers that adhere to sustainable and circular principles (Elsawy, 2024b).
- **Customer Engagement:** Educate customers about the benefits of CE practices and encourage sustainable behaviours through targeted marketing and communication campaigns (Rizos et al., 2016).
- **Demand Forecasting:** Use AI to predict demand for services like accommodation and transportation, reducing overconsumption and ensuring efficient resource use (Demirel et al., 2021).
- **Personalized Experiences:** Develop AI-driven platforms to offer personalised, sustainable travel options, encouraging tourists to make eco-friendly choices.
- **Energy and Water Management:** Implement IoT solutions such as smart metres and hotel sensors to optimise energy and water use, reducing wastage and operational costs.
- **Adaptive Systems:** Use IoT-enabled smart thermostats and lighting systems to adjust settings based on occupancy patterns, improving energy efficiency and lowering carbon footprints.

4.1.2 Policymakers

Policymakers are crucial in creating an enabling environment for CE adoption within the tourism sector. The identified external barriers, including inadequate infrastructure and limited government support, require policy interventions that incentivise and support sustainable practices (Martinez-Cabrera & López-del-Pino, 2021). Critical strategies for policymakers in this matter include:

- **Financial Incentives:** Implement financial incentives, such as grants and subsidies, to encourage tourism businesses to invest in Circular Economy initiatives (De Jesus & Mendonça, 2018).
- **Regulatory Frameworks:** Develop and enforce regulations that promote sustainable tourism practices, including waste management and resource conservation (Martinez-Cabrera & López-del-Pino, 2021).

- **Capacity Development:** Invest in educational programs and training initiatives that raise awareness and build capacity among tourism stakeholders (Economic Commission for Europe, 2022).
- **Infrastructure Development:** Allocate resources for infrastructure development, including waste recycling facilities and renewable energy infrastructure, to support CE implementation (Martinez-Cabrera & López-del-Pino, 2021).
- **Stakeholder Engagement:** Facilitate multi-stakeholder dialogues and partnerships to develop collaborative solutions and drive collective action towards sustainability goals (De Jesus & Mendonça, 2018).
- **National Policies:** Many countries have integrated Circular Economy principles into their national policies. These policies promote resource efficiency and waste reduction in the tourism sector. National strategies often encourage tourism businesses to adopt sustainable practices through incentives and regulations to minimise environmental impacts (UNWTO, n.d.; UNECE, 2022).
- **Public-Private Collaboration:** An effective transition to a Circular Economy in tourism requires active collaboration between the public and private sectors. Governments, businesses, and other stakeholders work together to create policies that support sustainable practices. This includes developing standards and guidelines that promote circularity in tourism operations (Rodríguez et al., 2020; UNWTO, n.d.).
- **Global Initiatives:** International frameworks and initiatives are crucial in promoting Circular Economy principles in tourism. For instance, the Global Tourism Plastics Initiative aims to reduce plastic waste by encouraging tourism businesses to eliminate unnecessary plastics and adopt reusable alternatives. Similarly, the Global Roadmap for Food Waste Reduction in Tourism promotes strategies to minimise food waste, supporting sustainable food systems within the tourism industry (UNWTO, n.d.).
- **Sustainable Development Goals (SDGs):** The Circular Economy in tourism is closely linked to the United Nations' SDGs. Policies supporting circularity often align with SDG targets, such as responsible consumption and production (SDG 12) and climate action (SDG 13). This alignment helps integrate Circular Economy practices into broader sustainability agendas (UNWTO, n.d.; UNECE, 2022).
- **Local Supply Chains:** The importance of local supply chains has been emphasised, especially in light of the COVID-19 pandemic. Strengthening local production and consumption patterns supports Circular Economy goals by reducing dependency on global supply chains and enhancing resilience against disruptions (UNWTO, n.d.).

To fully realise the potential of advanced technologies in fostering a Circular Economy within Egypt's tourism industry, a multi-faceted approach led by policymakers and involving stakeholders at all levels is necessary. Key recommendations include:

- **Investment in Technology:** Encourage tourism businesses to invest in blockchain, AI, IoT, and big data analytics to enhance sustainability and operational efficiency.
- **Training and Capacity Building:** Training programmes for tourism professionals to use and manage these technologies effectively.
- **Collaboration and Partnerships:** Foster collaboration between tourism operators, technology providers, and government bodies to develop integrated solutions and share best practices.
- **Regulatory Support:** Develop and enforce regulations that promote the adoption of sustainable technologies and practices within the tourism industry.
- **Data-Driven Decision Making:** Utilise big data analytics to develop evidence-based strategies for energy and water conservation, waste reduction, and overall sustainability (Li et al., 2018).
- **Policy Development:** Support the creation of sustainable tourism policies using insights derived from big data to assess the effectiveness of different initiatives.

By adopting these frameworks, the tourism sector can significantly contribute to a more sustainable and resilient economy, minimising environmental impacts and enhancing resource efficiency.

4.1.3 Researchers

Researchers are critical in advancing knowledge and informing evidence-based policies and practices in sustainable tourism. The identified barriers and demographic influences underscore the importance of targeted research initiatives that address specific challenges and opportunities within Egyptian tourism. Critical strategies for researchers include:

- **Interdisciplinary Research:** Foster interdisciplinary collaborations integrating insights from environmental science, economics, sociology, and technology to address complex sustainability challenges (Rizos et al., 2016).
- **Longitudinal Studies:** Conduct longitudinal studies to track the progress of CE adoption and assess the effectiveness of interventions over time (Elsawy, 2024b).

- **Policy Evaluation:** Evaluate the impact of existing policies and interventions on CE adoption to inform evidence-based policy recommendations (Martinez-Cabrera & López-del-Pino, 2021).
- **Knowledge Sharing:** Disseminate research findings through academic publications, conferences, and workshops to facilitate knowledge exchange and promote best practices in sustainable tourism (De Jesus & Mendonça, 2018).
- **Community Engagement:** Engage local communities and stakeholders in research activities to ensure participatory and inclusive approaches to sustainability (Elsawy, 2024c; Economic Commission for Europe, 2022).

In addition, future research should focus on exploring interactions between different barriers, cultural influences, and community engagement to enhance the depth and breadth of knowledge on Circular Economy adoption in Egyptian tourism. Comparative analyses across different countries and cultural contexts can offer insights into common patterns and unique challenges tourism businesses face (Martinez-Cabrera & López-del-Pino, 2021). Additionally, exploring the role of technology in promoting circular practices and assessing the effectiveness of specific interventions can contribute to a more comprehensive understanding of sustainable tourism practices. Moreover, integrating advanced technologies such as blockchain, AI, IoT, and big data analytics transforms the tourism industry by promoting sustainable practices and fostering a Circular Economy (CE). Researchers play a crucial role in exploring these technologies' implications, uncovering their potential, and addressing associated challenges. The following section offers suggestions for future research on this critical topic.

- **Interdisciplinary Approaches:** Combine insights from technology, sustainability, and tourism studies to understand the impact of advanced technologies on CE comprehensively.
- **Case Studies and Pilot Projects:** Conduct comparative studies across regions to identify best practices and challenges in implementing these technologies in tourism.
- **Policy and Regulatory Frameworks:** Analyse existing frameworks to identify gaps and propose recommendations for encouraging the adoption of advanced technologies in tourism.
- **Stakeholder Collaboration:** Engage with industry stakeholders to co-create solutions and ensure the relevance of research findings.
- **Longitudinal Studies:** Monitor the long-term impacts of these technologies on tourism sustainability to gain deeper insights into their effectiveness.
- **Tourist Perceptions and Behaviour:** Study tourist responses to personalised and sustainable tourism options to enhance eco-friendly practices.

By addressing these, researchers can significantly advance the understanding and implementation of advanced technologies in fostering a Circular Economy within the tourism industry.

5. CONCLUSION

This chapter has explored the current state of Circular Economy (CE) adoption within Egypt's tourism sector, revealing a complex landscape of barriers and opportunities. Despite the growing global emphasis on sustainability, Egyptian tourism businesses, particularly small and medium-sized enterprises (SMEs), face significant challenges in transitioning to circular practices. The findings underscore the need for comprehensive policy and regulatory frameworks to support the adoption of Circular Economy (CE) practices in Egypt's tourism sector. The identified barriers, such as lack of awareness, bureaucratic inefficiencies, and inadequate infrastructure, highlight the necessity for targeted government interventions. Understanding the demographic influences on environmental consciousness and attitudes towards circularity is crucial for promoting sustainable tourism practices and implementing targeted educational campaigns for tourists to bridge the gap between environmental awareness and behaviour. These initiatives should focus on translating awareness into actionable, sustainable practices and tailoring strategies based on demographic factors such as age, gender, and education. For Egyptian tourism businesses, particularly SMEs, transitioning to a Circular Economy presents challenges and opportunities. Embracing technologies like blockchain, AI, IoT, and big data can help businesses overcome resource limitations and improve operational efficiency. Investing in these technologies can facilitate better resource management, waste reduction, and enhanced transparency in supply chains. Lessons from successful transitions in other regions, including Spain, Cyprus, and the Nordic countries, offer valuable insights for Egypt. These examples underscore the importance of structured approaches, international collaboration, and tailored strategies to overcome the challenges of CE adoption. Overall, this chapter's implications suggest a multifaceted approach to advancing Circular Economy practices in Egyptian tourism. Egypt can make significant strides towards a sustainable and circular tourism industry by addressing policy gaps, leveraging technology, educating consumers, and learning from international experiences.

REFERENCES

Aryal, C. (2020). Exploring circularity: A review to assess the opportunities and challenges to close loop in Nepali tourism industry. *Journal of Tourism & Adventure*, 3(1), 142–158. DOI: 10.3126/jota.v3i1.31362

Braun, V., & Clarke, V. (2006). Using thematic analysis in psychology. *Qualitative Research in Psychology*, 3(2), 77–101. DOI: 10.1191/1478088706qp063oa

Budovska, V., Torres Delgado, A., & Øgaard, T. (2019). Pro-environmental behaviour of hotel guests: Application of the theory of planned behaviour and social norms to towel reuse. *Tourism and Hospitality Research*, 20(1), 146735841983143. DOI: 10.1177/1467358419831431

Buhalis, D., & Leung, R. (2018). Smart hospitality—Interconnectivity and interoperability towards an ecosystem. *International Journal of Hospitality Management*, 71, 41–50. DOI: 10.1016/j.ijhm.2017.11.011

CEnTOUR. (2020). *Circular economy in tourism*. COSME Project Funded by the European Union. Available at: https://circulartourism.eu/ [Accessed 10 Jun. 2024].

Cheong, H., Lyons, A., Houghton, R., & Majumdar, A. (2023). Secondary qualitative research methodology using online data within the context of social sciences. *International Journal of Qualitative Methods*, 22(1), 1–19. DOI: 10.1177/16094069231180160

Dawadi, B. (2022). *Implication of circular economy in the tourism industry: Case study – Lahti, Finland* [Master's thesis]. Available at: https://www.theseus.fi/bitstream/handle/10024/743910/Dawadi_Bipin.pdf?sequence=2&isAllowed=y

De Jesus, A., & Mendonça, S. (2018). Lost in transition? Drivers and barriers in the eco-innovation road to the circular economy. *Ecological Economics*, 145, 75–89. DOI: 10.1016/j.ecolecon.2017.08.001

de-Miguel-Molina, B., de-Miguel-Molina, M., Miret-Pastor, L., & Silva-Cárdenas, M. B. (2024). The Transition Journey to the Circular Economy by Companies in the Valencian Region's Tourism Industry. Managing the Transition to a Circular Economy, 2.

Demirel, E., Karagöz Zeren, S., & Hakan, K. (2021). Smart contracts in the tourism industry: A model with blockchain integration for post-pandemic economy. *Current Issues in Tourism*, •••, 1–15. DOI: 10.1080/13683500.2021.1960280

Economic Commission for Europe. (2022). *Applying principles of circular economy to sustainable tourism*. Substantive thematic documents, Information Paper No. 3. The Ninth Environment for Europe Ministerial Conference. Geneva: Economic Commission for Europe, Committee on Environmental Policy.

Economist Impact. (2022). *In transit: Pathways to circular tourism in the Middle East*. Economist Impact. https://impact.economist.com/perspectives/sites/default/files/download/economist_impact_in_transit_english_nov22.pdf

Elsawy, T. M. (2023). Determinants of e-business usage by travel agencies in developing countries: A decision tree approach. *International Journal of Tourism Policy*, 13(1), 1. DOI: 10.1504/IJTP.2023.129173

Elsawy, T. M. (2024a). Domestic tourists' perception and attitude towards circularity in Egypt. *Pharos International Journal of Tourism and Hospitality*, 3(1), 0. Advance online publication. DOI: 10.21608/pijth.2024.263493.1008

Elsawy, T. M. (2024b). The transition barriers toward the circular economy for travel agencies in Egypt. *International Journal for Tourism. Archaeology and Hospitality*, 4(1), 60–88. DOI: 10.21608/ijtah.2024.258367.1114

Elsawy, T. M. (2024c). Harnessing appreciative inquiry in tourism planning: Towards a strategy for poverty alleviation in marginalized communities. *International Journal of Tourism Cities*. Advance online publication. https://doi.org/DOI: 10.1108/IJTC-10-2023-0208

Elsawy, T. M., & Eltayeb, N. (2023). Examining the adoption barriers of big data applications from the perspective of tourism authorities in Egypt. *Journal of the Faculty of Tourism and Hotels. Mansoura University*, 13(13), 81–124. DOI: 10.21608/mkaf.2023.299783

Fernández-Robin, C., Celemín-Pedroche, M. S., Santander-Astorga, P., & Alonso-Almeida, M. del M. (2019). Green practices in hospitality: A contingency approach. *Sustainability (Basel)*, 11(13), 3737. DOI: 10.3390/su11133737

Florido, C., Jacob, M., & Payeras, M. (2019). How to carry out the transition towards a more circular tourist activity in the hotel sector: The role of innovation. *Administrative Sciences*, 9(2), 47. DOI: 10.3390/admsci9020047

Geissdoerfer, M., Savaget, P., Bocken, N. M. P., & Hultink, E. J. (2017). The circular economy—A new sustainability paradigm? *Journal of Cleaner Production*, 143(1), 757–768. DOI: 10.1016/j.jclepro.2016.12.048

Ghanem, S. K. (2016). The relationship between population and the environment and its impact on sustainable development in Egypt using a multi-equation model. *Environment, Development and Sustainability*, 20(1), 305–342. DOI: 10.1007/s10668-016-9882-8

Giurea, R., Precazzini, I., Ragazzi, M., Achim, M., Cioca, L.-I., Conti, F., Torretta, V., & Rada, E. (2018). Good practices and actions for sustainable municipal solid waste management in the tourist sector. *Resources*, 7(3), 51. DOI: 10.3390/resources7030051

Gretzel, U., Sigala, M., Xiang, Z., & Koo, C. (2015). Smart tourism: Foundations and developments. *Electronic Markets*, 25(3), 179–188. DOI: 10.1007/s12525-015-0196-8

Halkiopoulos, C., Antonopoulou, H., & Kostopoulos, N. (2023). Integration of blockchain technology in the tourism industry: Opportunities and challenges. In V. Katsoni (Ed.), *Tourism, travel, and hospitality in a smart and sustainable world: IACuDiT 2022* (pp. 355–367). Springer Proceedings in Business and Economics. Springer. https://doi.org/DOI: 10.1007/978-3-031-26829-8_22

Hernandez, R. J., Miranda, C., & Goñi, J. (2020). Empowering sustainable consumption by giving back to consumers the "right to repair.". *Sustainability (Basel)*, 12(3), 850. DOI: 10.3390/su12030850

Hockenbury, D. H., & Hockenbury, S. E. (2011). *Discovering psychology*. Worth Publishers.

Hoffmann, T. C., Glasziou, P. P., Boutron, I., Milne, R., Perera, R., Moher, D., Altman, D. G., Barbour, V., Macdonald, H., Johnston, M., Lamb, S. E., Dixon-Woods, M., McCulloch, P., Wyatt, J. C., Chan, A.-W., & Michie, S. (2014). Better reporting of interventions: Template for Intervention Description and Replication (TIDieR) checklist and guide. *BMJ, 348*(mar07 3), g1687. https://doi.org/DOI: 10.1136/bmj.g1687

Holmberg, E., & Konttinen, A. (2023). Circular economy in tourism and hospitality – A Nordic perspective. In A. Köchling, S. Seeler, P. van der Merwe, & A. Postma (Eds.), *Towards sustainable and resilient tourism futures* (pp. 97–113). Schriftenreihe des Deutschen Instituts für Tourismusforschung, vol 1. Erich Schmidt Verlag GmbH & Co. KG. https://doi.org/DOI: 10.37307/b.978-3-503-21195-1.08

Holmes, M. R., Dodds, R., & Frochot, I. (2019). At home or abroad, does our behavior change? Examining how everyday behavior influences sustainable travel behavior and tourist clusters. *Journal of Travel Research*, 60(1), 004728751989407. DOI: 10.1177/0047287519894070

Huberman, A., & Miles, M. (2002). *The qualitative researcher's companion*. SAGE Publications, Inc., DOI: 10.4135/9781412986274

Ingrassia, M., Bacarella, S., Bellia, C., Columba, M. S., Altamore, L., & Chironi, S. (2023). Circular economy and agritourism: A sustainable behavioral model for tourists and farmers in the post-COVID era. *Frontiers in Sustainable Food Systems*, 7, 1174623. Advance online publication. DOI: 10.3389/fsufs.2023.1174623

Karam, A. (2020). *Egypt country profile*. Switchmed. https://switchmed.eu/wp-content/uploads/2021/02/Country-Profile-Egypt.pdf

Khan, O., Bellini, N., Daddi, T., & Iraldo, F. (2022). Effects of behavioral intention and dynamic capabilities on circular economy adoption and performance of tourism SMEs. *Journal of Sustainable Tourism*, •••, 1–20. DOI: 10.1080/09669582.2022.2066683

Khan, O., Marrucci, L., Daddi, T., & Bellini, N. (2021). Adoption of circular economy and environmental certifications: Perceptions of tourism SMEs. *Journal of Management and Sustainability*, 11(1), 218. DOI: 10.5539/jms.v11n1p218

Kirchherr, J., Reike, D., & Hekkert, M. (2017). Conceptualizing the circular economy: An analysis of 114 definitions. SSRN *Electronic Journal, 127*, 221–232. DOI: 10.2139/ssrn.3037579

Kouhizadeh, M., & Sarkis, J. (2018). Blockchain practices, potentials, and perspectives in greening supply chains. *Sustainability (Basel)*, 10(10), 3652. DOI: 10.3390/su10103652

Li, J., Xu, L., Tang, L., Wang, S., & Li, L. (2018). Big data in tourism research: A literature review. *Tourism Management*, 68, 301–323. DOI: 10.1016/j.tourman.2018.03.009

Manniche, J., Larsen, K., Broegaard, R., & Holland, E. (2020). *Destination: A circular tourism economy - A handbook for transitioning toward a circular economy within the tourism and hospitality sectors in the South Baltic Region* (Version 2.0). Centre for Regional & Tourism Research (CRT). https://circulareconomy.europa.eu/platform/sites/default/files/cirtoinno-handbook_eng-rev.-4.pdf

Martínez-Cabrera, J., & López-del-Pino, F. (2021). The 10 most crucial circular economy challenge patterns in tourism and the effects of COVID-19. *Sustainability (Basel)*, 13(9), 4940. DOI: 10.3390/su13094940

Martins, C. S. (2021). *Circular economy in the hospitality industry: Awareness and implementation in Europe* [Master's thesis]. https://repositorio.ucp.pt/bitstream/10400.14/36860/1/202894428.pdf

Millar, M., & Baloglu, S. (2011). Hotel guests' preferences for green guest room attributes. *Cornell Hospitality Quarterly*, 52(3), 302–311. DOI: 10.1177/1938965511409031

OECD. (2023). *Competition in the circular economy – Note by Egypt*. Organisation for Economic Co-operation and Development (OECD). https://one.oecd.org/document/DAF/COMP/WD(2023)43/en/pdf

Perkins, H. E., & Brown, P. R. (2012). Environmental values and the so-called true ecotourist. *Journal of Travel Research*, 51(6), 793–803. DOI: 10.1177/0047287512451133

Peters, M., Chan, C.-S., & Legerer, A. (2018). Local perception of impact-attitudes-actions towards tourism development in the Urlaubsregion Murtal in Austria. *Sustainability (Basel)*, 10(7), 2360. DOI: 10.3390/su10072360

Rafiq, F., Adil, M., & Sadiq, M. (2022). Does contextual factor influence travelers' towel reuse behavior? Insights from circular economy. *Sustainability (Basel)*, 14(10), 6155. DOI: 10.3390/su14106155

Ranta, V., Aarikka-Stenroos, L., Ritala, P., & Mäkinen, S. J. (2018). Exploring institutional drivers and barriers of the circular economy: A cross-regional comparison of China, the US, and Europe. *Resources, Conservation and Recycling*, 135, 70–82. DOI: 10.1016/j.resconrec.2017.08.017

Rezk, M. R. A., Kapiel, T. Y. S., Piccinetti, L., Salem, N., Khasawneh, A., Santoro, D., Montagnino, F. M., El-Bary, A. A., & Sakr, M. M. (2023). Circular economy in Egypt: An overview of the current landscape and potential for growth. *Insights into Regional Development*, 5(3), 45–57. DOI: 10.9770/IRD.2023.5.3(3)

Rizos, V., Behrens, A., van der Gaast, W., Hofman, E., Ioannou, A., Kafyeke, T., Flamos, A., Rinaldi, R., Papadelis, S., Hirschnitz-Garbers, M., & Topi, C. (2016). Implementation of circular economy business models by small and medium-sized enterprises (SMEs): Barriers and enablers. *Sustainability (Basel)*, 8(11), 1212. DOI: 10.3390/su8111212

Robaina, C. R. (2022). *Circular economy and tourism: Socioeconomic profile of tourists with a greater circular behaviour and factors affecting the implementation of circular practices in hotels in a mature island destination* [PhD thesis].

Rodríguez, C., Florido, C., & Jacob, M. (2020). Circular economy contributions to the tourism sector: A critical literature review. *Sustainability (Basel)*, 12(11), 4338. DOI: 10.3390/su12114338

Schumann, F. R. (2020). Circular economy principles and small island tourism. *Journal of Global Tourism Research*, 5(1), 13–20. DOI: 10.37020/jgtr.5.1_13

Sørensen, F., & Bærenholdt, J. O. (2020). Tourist practices in the circular economy. *Annals of Tourism Research*, 85, 103027. DOI: 10.1016/j.annals.2020.103027

Sorin, F., & Einarsson, S. (2020). *Circular economy in travel and tourism: A conceptual framework for a sustainable, resilient, and future-proof industry transition* [Online]. CE360 Alliance. https://www.oneplanetnetwork.org/sites/default/files/from-crm/circular_economy_paper_ce360_fabrice.pdf

Taherdoost, H. (2021). Data collection methods and tools for research: A step-by-step guide to choose data collection technique for academic and business research projects. *International Journal of Academic Research in Management (IJARM), 10*(1), 10–38. https://hal.science/hal-03741847/document

Thomas, J., & Harden, A. (2008). Methods for the thematic synthesis of qualitative research in systematic reviews. *BMC Medical Research Methodology*, 8(1), 1–10. DOI: 10.1186/1471-2288-8-45 PMID: 18616818

Treiblmaier, H. (2018). The impact of the blockchain on the supply chain: A theory-based research framework and a call for action. *Supply Chain Management*, 23(6), 545–559. DOI: 10.1108/SCM-01-2018-0029

UNECE. (2022). *Sustainable tourism: Policy and regulatory frameworks*. United Nations Economic Commission for Europe. Retrieved from https://unece.org/sites/default/files/2022-05/CEP-SS_Sustainable_Tourism.IP_.03.e.pdf

UNIDO. (2023, July 24). Embracing a strong circular economy in Egypt's ICT & electronics value chain. *Switch to Circular Economy Value Chains*. https://www.switchtocircular.eu/news-and-events/embracing-strong-circular-economy-egypts-ict-electronics-value-chain

UNWTO. (2020, December 16). *Circular economy in travel and tourism – A conceptual framework for a sustainable, resilient, and future-proof industry transition*. World Tourism Organization. https://www.unwto.org/covid-19-oneplanet-responsible-recovery-initiatives/circular-economy-in-travel-and-tourism-a-conceptual-framework-for-a-sustainable-resilient-and-future-proof-industry-transition

UNWTO. (n.d.). *Circular economy: Integrating circular economy principles in tourism*. World Tourism Organization. Retrieved from https://www.unwto.org/sustainable-development/circular-economy

Vatansever, K., Akarsu, H., & Kazançoğlu, Y. (2021). Evaluation of transition barriers to circular economy: A case from the tourism industry. *International Journal of Mathematical. Engineering and Management Sciences*, 6(3), 824–846. DOI: 10.33889/IJMEMS.2021.6.3.049

WBCSD. (2022). *Circular transition indicators v3.0: Metrics for business, by business*. World Business Council for Sustainable Development. https://www.wbcsd.org/contentwbc/download/14172/204337/1

Chapter 19
Turning the Waste Into Wealth:
Progressing Toward Global Sustainability Through the Circular Economy in Waste Management

Rejaul Karim
https://orcid.org/0000-0001-6175-8825
Varendra University, Bangladesh

Abdul Waaje
https://orcid.org/0009-0004-7207-0750
Michigan Technological University, USA

Md. Mustaqim Roshid
https://orcid.org/0000-0001-8517-6336
University of Rajshahi, Bangladesh

Md. Bony Yeamin
https://orcid.org/0009-0009-1791-2691
University of Rajshahi, Bangladesh

ABSTRACT

Increasing waste production presents substantial environmental and public health obstacles, necessitating inventive and sustainable waste management approaches. This chapter explores the fundamental concepts of circular economy and how it may be used to change waste management systems from linear disposal methods to sustainable and efficient models that make the most of resources. It explores the

DOI: 10.4018/979-8-3693-6110-8.ch019

influence of efficient waste management on diminishing environmental deterioration, fostering resource preservation, and bolstering economic development through waste materials' reuse, recycling, and repurposing. The chapter highlights the importance of adopting comprehensive strategies that combine technology advancements, policy restructuring, and community involvement to enable a shift toward circular waste management practices that will support attaining global sustainability.

1. INTRODUCTION

The world is stranded in an unusual situation in which the global population is growing, and the emergence of a never-ending amount of human and industrial waste is generating an imbalanced ecology, reducing environmental symmetry. This is the case in a world that is equipped with superior technology. From the beginning of craft manufacturing till the present day, in the fourth iteration of the Industrial Revolution, we have witnessed a variety of wastes that are the result of production. The release of vast quantities of waste into the environment, which occurs primarily in three different forms such as solid, liquid, and gaseous wastes, has occurred as a result of the mass production of a variety of chemicals and energy as well as other developmental activities such as industrialisation, urbanisation, digitalisation, and the utilisation of fertiliser with various machinery in agriculture and medical care at various locations around the world over the past four decades (Cheah et al., 2022; Musarat et al., 2022; Perey et al., 2018).

Waste disposal is a severe problem; by the year 2050, the amount of waste is anticipated to have doubled. Global greenhouse gas emissions are approximately 5% attributable to the decomposition of the organic components of solid waste, which accounts for around 11.2 billion tons of collected solid trash each year, according to UNEP (UNEP, 2024a). Global solid waste generation will triple by 2100 (The World Bank, 2013) due to the rising population and a linear industrialised system that is still being transformed into a circular system (Derhab & Elkhwesky, 2022; Minelgaitė & Liobikienė, 2019)Today, waste minimisation is crucial. The linear economic model of "take-make-dispose" is not sustainable because it cannot successfully reduce hazardous waste and repurpose garbage into beneficial items for society. This highlights the lack of a sustainable waste management system.

Circularity in the economy rises above the traditional economic model due to inadequate waste disposal and rising trash. Similarly, businesspeople are alarmed that efficient production lines, skilled personnel, and seamless but integrated supply chain operation supervision cannot alone generate profitability and create a win-win situation for all stakeholders (Karmakar et al., 2022; Puntillo, 2022; Sassanelli & Terzi, 2023), including the green initiatives by the financial institutions (Karim,

2020). One of the main obstacles to attaining economic circularity and creating a sustainable waste management system is the depletion of resources. Scarcity is a dilemma that everyone encounters, particularly when it comes to resources (Souza Piao et al., 2023a). Natural resources like fossil fuels are becoming increasingly depleted due to people's greed, lack of planning, and unproductive use. Thus, it is urgent to adopt circularity in the economy and create a sustainable waste management system that uses garbage as an end-user good or as an initial source of business earnings.

In 2015, the UN launched the 17 sustainable development goals (SDGs) to promote sustainable peace, advancement, and prosperity for ultimately. These goals provide a reinvigorated global outlook to address environmental issues and emphasise the need for multi-societal collaboration. Goal 11.6 of the SDGs is to lessen the negative impact of cities on the environment per person (González-Sánchez et al., 2023). By adopting a circular economy, enhancing the cleanliness of the air, and effectively dealing with commercial waste from communities, countries may do this. In the process of building a sustainable waste management system, an integrated approach to circularity in the economy can be useful not only in terms of lowering costs but also in terms of ensuring recyclability, reusability, reproducibility, and retrievability of trash (Ajwani-Ramchandani & Bhattacharya, 2022). In alignment with the change from the traditional waste management system to a sustainable waste management system with the vertical integration of circularity in the economy, it can also form a traditional economic model depicting the circular waste management system. By incorporating the ideas of circular economies in a cross-functional manner, this chapter investigates the adoption of sustainability techniques to develop a waste management system that is sustainable (Kadhila et al., 2023)Compared to conventional waste management, it examines the fundamentals of circular trash disposal and highlights its benefits and drawbacks. Examples from real-world situations are provided to show the application of circularity in constructing circular logistical chains. Additionally, eco-design principles push systemic transition toward a more robust and ecologically sound future. To achieve Target 6 under SDG 17, the primary focus is on the interconnection of financial, ecological, and social aspects.

The principal objective of this chapter is to delve into the revolutionary possibilities of the circular economy in trash management to advance toward a more sustainable world. It aims to shed light on how the circular economy's ideas and frameworks might be used for waste management to generate profit from trash, which in turn can promote economic growth and environmental sustainability. This chapter examines waste-to-wealth initiatives' best practices and case studies, lists the monetary, social, and ecological advantages of sustainable waste management, explains the difficulties of putting circular economy strategies into practice, and concludes with practical suggestions for how communities, businesses, and governments can improve their sustainability by adopting circular economy principles.

2. WASTE

Waste is just a result of insufficient cognitive processes. Scholars have endeavoured to precisely define the term "waste" by drawing upon their comprehension of diverse scenarios, scholarly expertise, imaginative foresight, and practical encounters. Economists commonly define waste as resources or materials resulting from economic operations deemed superfluous, underutilised, or abandoned without fully realising their potential value or utility. Both humans and machines can generate waste. Except for human waste, "waste" can be defined in various ways. Waste consists of materials that are abandoned after their first use and are judged to have no value, be defective, and be undesired (Silva et al., 2017; White et al., 1995). Despite the costs involved, this leads to the need for disposal. The waste is typically a result of inefficient production methods that continuously deplete important resources (Cheremisinoff, 2003).

The composition of garbage differs among different sectors. In the service industry, particularly in industries like hospitality, waste can take various forms, including extra packaging, superfluous documentation generated during guest check-in/check-out procedures, and printed materials such as brochures and menus. These can be rendered useless if employees do not exercise mindfulness during working hours. For example, if a guest asks explicitly for a hotel with two queen beds but the staff accidentally books a room with only one queen bed, this mistake might be classified as wasteful since it could cause the guest to rethink their choice. Determining the specific behaviours that qualify as waste and establishing a precise definition can be intricate and contingent on the particular sector.

However, the lack of a globally acknowledged definition of trash is uncommon because of many businesses' varied nature and distinct features. As a result, various international organisations have created precise definitions of garbage to establish a common understanding worldwide. This helps recognise different types of waste and promotes initiatives to create communities and enterprises that produce no waste. As per the European Union (EU), waste refers to any item or object mentioned in Annex I that is either thrown away or required to be thrown away by the holder (European Union Law, 1991). Similarly, the United Nations Environment Programme (UNEP) has defined waste as substances or items that are abandoned, planned for disposal, or required to be disposed of according to national regulations (European Union, 1993). Therefore, providing an unambiguous definition of waste is challenging. Table 1 presents a visual representation of Annex I, which outlines waste categories based on the EU Council Directive of 18 March 1991, which modified Directive 75/442/EEC on garbage.

Table 1. Categories of waste

Serial No.	Categories of Waste
1	Production of consumption residues not otherwise specified below
2	Off-specification products
3	Products whose date for appropriate use has expired
4	Materials spilled, lost, or having undergone other mishap, including any materials, equipment, etc., contaminated as a result of the mishap
5	Materials contaminated or soiled due to planned actions (e.g. residues from cleaning operations, packing materials, containers, etc.)
6	Unusable parts (e.g. reject batteries, exhausted catalysts, etc.)
7	Substances that no longer perform satisfactorily (e.g. contaminated acids, contaminated solvents, exhausted tempering salts, etc.)
8	Residues of industrial processes (e.g. slags, still bottoms, etc.)
9	Residues from pollution abatement processes (e.g. scrubber sludges, baghouse dusts, spent filters, etc.)
10	Machining/finishing residues (e.g. lathe turnings, mill scales, etc.)
11	Residues from raw materials extraction and processing (e.g. mining residues, oil field slops, etc.)
12	Adulterated materials (e.g. oils contaminated with PCBs, etc .)
13	Any materials, substances, or products that have been banned by law.
14	Products for which the holder has no further use (e.g. agricultural, household, office, commercial, shop discards, etc.)
15	Contaminated materials, substances, or products resulting from remedial action concerning land.
16	Any materials, substances, or products not contained in the above categories.

Source: European Union Law (1991)

The EU Council has established 16 categories of waste, which serve as a systematic framework for identifying and categorising different types of trash, regardless of whether they come from industrial or service-related activities. Implementing this categorisation system allows businesses and regulatory agencies to categorise garbage more efficiently, hence facilitating waste management methods such as minimising waste, minimising expenses, and improving resilience. By categorising themselves according to these criteria, organisations can simplify the process of identifying waste, implement specific strategies to reduce waste production, enhance production efficiency, and encourage the recovery and reuse of resources. Ultimately, this will contribute to adopting environmentally friendly habits and safeguarding the environment. Manufacturing waste includes surplus raw material inventory after demand fulfilment, surplus production, defective goods manufacturing, and other issues that interrupt production. These include lengthier cycle times, changeover intervals, work-in-process interruptions, and manufacturing expenses. Such actions, discarded products, and acts are waste that existed.

3. CONVENTIONAL WASTE MANAGEMENT

The phrase "conventional waste management" suggests a conventional or frequently utilised approach to trash management. Even though the form of this customary practice differs from region to region, it is characterised by significant similarities that indicate its traditional nature. Effective waste management is the primary idea behind the concept of waste management. The phrase may appear to be basic; nevertheless, it comprises a variety of processes that are aimed at waste management (Farooq et al., 2022). The selection of the term "management" rather than "disposal" suggests that disposal is but one component of the more comprehensive waste management process. Additionally, to successfully achieve the goal of proper disposal, it is vital to have a planned strategy for waste management.

Waste management techniques differ from nation to nation, depending on their development level (Awino & Apitz, 2024a). These practices might range from less developed to developing to developed nations. In addition, these behaviours are influenced by the particular characteristics of a nation's geography, such as the distinctions between urban and rural areas and residential and industrial sectors. It is the ultimate goal of waste management to eliminate various types of waste, including solid, liquid, and gaseous waste, which may cause environmental effects that are either detrimental or non-detrimental. Waste management encompasses the systematic assembling, conveyance, processing, overseeing, and governance of waste, all carried out following the law. These waste materials are produced by various sources, including mining, industrial, residential, and agricultural operations (Sawalkar et al., 2023). Conventional waste management is sometimes connected with conventional approaches emphasising waste disposal, typically accomplished through landfills or incinerators. Collecting garbage, temporarily storing it in specified places, preprocessing or separating it as required, and lastly, transporting it to disposal sites are the steps involved in these procedures. The disposal of garbage is typically given higher priority than environmental concerns in this method, and it sometimes ignores the potentially harmful impact of waste that is not segregated, which can have long-lasting consequences on soil and the ecosystem. Over several decades, it has become abundantly clear that a sustainable waste management system is necessary to achieve ecological viability and national prosperity through a variety of indices of well-being.

Conventional waste management procedures are widespread, but there is a rising awareness of the need for technical advances to preserve the environment and a healthier atmosphere for future generations. The UN's sustainable development goals have brought sustainability to the forefront. Sustainable waste management is indispensable for industrial and residential waste management and conservation (Ali & Shirazi, 2023). Challenges remain, especially in underdeveloped nations where willingness and economic limits prevent adopting sustainable practices, leaving conventional methods in use. Besides the aforementioned, conventional waste management methods have numerous constraints.

Figure 1. Conventional waste management system

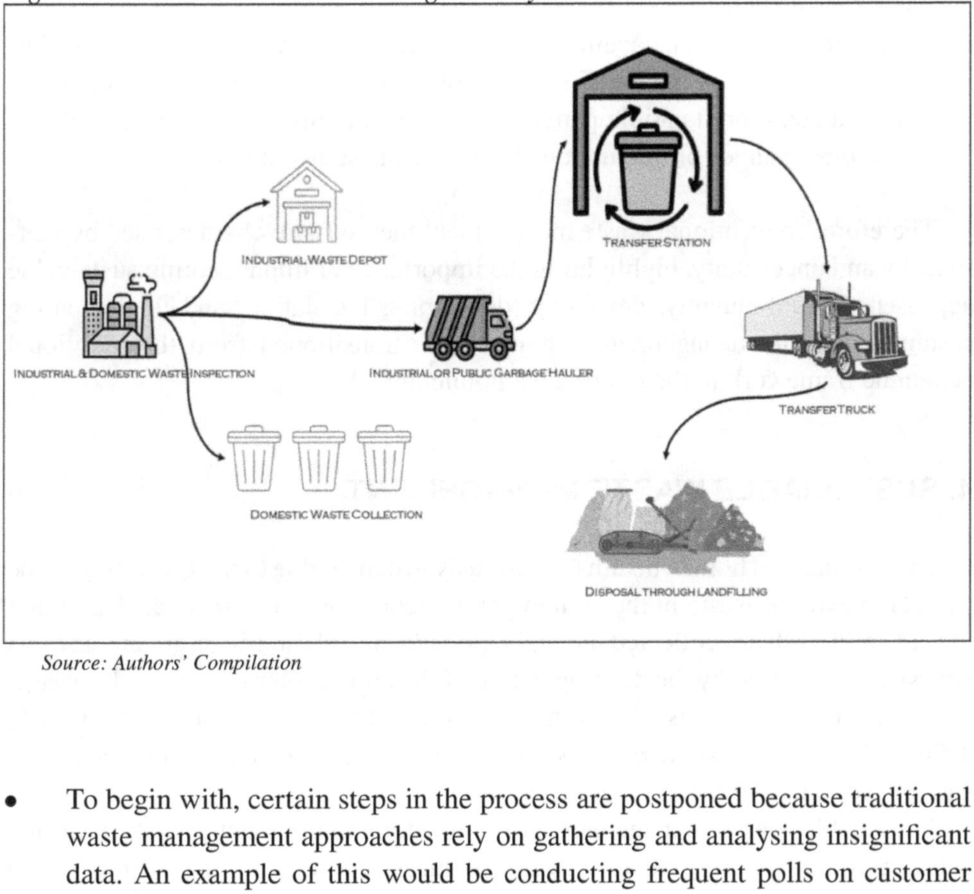

Source: Authors' Compilation

- To begin with, certain steps in the process are postponed because traditional waste management approaches rely on gathering and analysing insignificant data. An example of this would be conducting frequent polls on customer satisfaction in industries where the quality of products and functionality have remained the same throughout time.
- Conventional waste management measures may have irreversible repercussions. The focus is on performing irreversible actions rather than planning for unanticipated consequences. For example, Auckland City in New Zealand

did not anticipate the ensuing increase in garbage volumes. It did not develop suitable strategies for handling the extra waste when it increased the size of its waste collection containers from 40 to 240 litres (City Council, 2008).

- Lacking sustainability, this approach is mainly concerned with meeting immediate needs rather than planning. As a result, moving towards a more sustainable method is being heavily advocated.
- Due to its short-term concentration, conventional waste management is unorganised and less valuable. Procedure changes can prolong this interval of time.
- The potential repercussions of typical waste management measures are generally not considered. Intervention side effects are often overlooked or underestimated.
- Traditional waste management practices ignore more significant, more far-reaching issues in favour of making forecasts or forecasts concerning more immediate, more tangible patterns. This presumption ignores the possibility of future changes or fluctuations favouring the status quo.

Therefore, conventional waste management methods are characterised by inefficiency and uncertainty, highlighting the importance of implementing sustainable approaches. Consequently, designing, developing legislation, and implementing sustainable waste management techniques has transitioned from the traditional economic framework to the circular economic model.

4. SUSTAINABLE WASTE MANAGEMENT

Ecosystems and human health are seriously in danger due to the growing volume and complexity of waste in the contemporary economy. An estimated 11.2 billion tons of solid trash are collected annually globally, and about 5% of greenhouse gas emissions are caused by the decomposition of the organic element of solid garbage. Globally, 11.2 billion tons of solid trash are reportedly collected annually (UNEP, 2024b). Therefore, sustainable waste management ensures the natural ecosystem and reduces environmental deterioration.

Sustainable waste management is crucial for environmental protection and public health. Properly managing waste reduces the negative impact on natural ecosystems, decreases pollution, and conserves natural resources. Effective waste management practices, such as recycling, composting, and energy recovery, help minimise the amount of waste in landfills and incinerators, which are significant sources of greenhouse gases and other pollutants. Moreover, sustainable waste management is vital for economic and social sustainability. Implementing efficient waste management systems creates green jobs and stimulates economic growth by encouraging the development of recycling and waste-to-energy industries. It also fosters a culture of responsibility and awareness among communities, promoting sustainable consumption and production patterns. By managing waste sustainably, communities can reduce the costs associated with waste disposal and environmental remediation, ultimately leading to more resilient and sustainable societies.

Sustainable waste management is a critical aspect of environmental stewardship that seeks to minimise the negative impact of waste on the environment and human health. This approach prioritises waste reduction at the source, advocating for reducing, reusing, and recycling to lower the amount of waste generated. By implementing effective strategies such as composting organic waste, recycling materials like plastics, metals, and paper, and promoting reusable products, sustainable waste management aims to conserve natural resources, reduce greenhouse gas emissions, and decrease the burden on landfills. The majority of waste management models take into account economic and environmental factors, while only a small percentage take into account social factors (Morrissey & Browne, 2004). Advanced technologies, such as waste-to-energy plants and biodegradable materials, also play a crucial role in transforming waste into valuable resources, thereby contributing to a circular economy in which waste is seen not as a problem but a resource.

Moreover, sustainable waste management involves comprehensive planning and collaboration among governments, businesses, and communities. Policy frameworks and regulations need to be in place to support sustainable practices, such as extended producer responsibility (EPR), which holds manufacturers accountable for the lifecycle of their products. Public education and community engagement are essential to fostering a culture of sustainability, where individuals are informed and motivated to make environmentally responsible choices. By integrating economic, social, and environmental considerations, sustainable waste management addresses the immediate challenges of waste disposal. It contributes to long-term sustainability goals, ensuring a healthier planet for future generations.

5. PRINCIPLES OF SUSTAINABLE WASTE MANAGEMENT

Reduced adverse effects on the environment, preservation of natural resources, and improvement of public health are the overarching goals of sustainable waste management. Central to it all is the waste hierarchy, a structure for strategic planning that ranks waste prevention at the top, then minimisation, recovery, and appropriate disposal. Improving product design, changing consumption behaviour, and implementing lean production techniques are all examples of upstream interventions that are the focus of prevention efforts. These aim to decrease waste at its source. Reusing, repairing, and recycling are at the forefront of waste reduction initiatives (Katan, 2023; Schindler & Demaria, 2020), which aims to increase the useful life of objects and materials. Using techniques like composting, anaerobic digestion, and waste-to-energy technology, materials can be recovered from waste streams, preventing their disposal in landfills and ultimately lowering environmental impacts. The term "responsible disposal" refers to handling refuse in a way that does not harm the environment or people's health using facilities that follow all applicable regulations and industry standards. Therefore, sustainable waste management is an approach aimed at reducing waste's environmental impact by managing waste materials in an environmentally responsible, economically viable, and socially acceptable manner. Here are the fundamental principles governing this.

5.1 Waste Hierarchy

The waste hierarchy states that the first step in reducing garbage generation is prevention, which includes making items with less packing, advertising reusable solutions, and promoting habits that reduce waste, such as buying in bulk and utilising refillable containers. Reduced waste is the goal of minimisation, which is why many companies are embracing circular economy tactics to make their products last longer and be more recyclable (Wiprächtiger et al., 2022). Composting organic waste into soil amendments, producing biogas and fertiliser through anaerobic digestion, and converting non-recyclable trash into heat or power through gasification and incineration are all examples of waste-to-energy technologies that can be utilised for recovery, reducing landfill waste, and increasing renewable energy. Last but not least, disposal is the responsible treatment of trash that cannot be recycled or reused in any way (Bogusz et al., 2021). This includes techniques like landfilling and incineration with energy recovery, as well as the use of pollution control technology and advanced technical controls to lessen the adverse effects on the environment.

The waste hierarchy prioritises waste management practices from most to least preferred:

- Prevention: Reducing waste generation at the source.
- Minimization: Reducing the amount and toxicity of waste produced.
- Reuse: Using items as many times as possible before discarding them.
- Recycling: Converting waste into new materials and objects.
- Energy Recovery: Converting waste into energy, typically through incineration.
- Disposal: Safe and environmentally sound disposal methods for waste that cannot be reused or recycled.

5.2 Pollution Prevention

Reducing the use of hazardous materials, improving production processes and product design, optimising packaging to lessen waste, and implementing measures to prevent pollution, such as spill management, are all part of source reduction's focus on minimising waste and pollutants at their origin. By decreasing emissions, leachate, and other contaminants, pollution control methods in waste management facilities seek to safeguard public health and environmental quality (Lee & Bi, 2019; Zhou et al., 2022). Systems for treating landfill leachate purify the water, while filters and scrubbers remove particles and harmful gases from incinerator emissions to control air pollution. Plasma gasification and thermal depolymerisation are two examples of modern waste processing technologies that provide alternatives to conventional disposal methods that are less harmful to the environment.

5.3 Community Engagement and Education

Reducing, recycling, and composting are just a few examples of sustainable behaviours that may be promoted through public outreach, educational programs, and media campaigns to raise awareness and encourage responsible trash management. Communities are empowered to participate in local waste reduction activities when public awareness of the necessity of waste management is increased. By holding public hearings, consulting with stakeholders, and other forms of stakeholder engagement, participatory decision-making encourages communities to be actively involved in the process of creating, executing, and evaluating policies, programs, and infrastructure projects related to waste management (Hutner et al., 2018; Kountouris, 2022). This method guarantees that waste management plans are customised to meet the specific requirements of each community, taking into account the wide range of opinions and interests held by those people.

5.4 Lifecycle Assessment and Environmental Impact

Products and systems can affect the environment from the extraction of raw materials until their ultimate disposal. This effect can be measured through lifetime assessment (LCA). Throughout a product's existence, lifecycle assessments (LCAs) assess possible waste management solutions by considering energy use, greenhouse gas emissions, water consumption, and resource depletion. This allows for informed decisions that minimise environmental consequences and maximise resource efficiency. Efforts to lessen the environmental effect of waste management centres on implementing more efficient systems for garbage collection and transportation, which in turn reduce emissions from landfills and other waste treatment facilities

(Antunes et al., 2021; Koinig et al., 2022; Psomopoulos et al., 2022). Additionally, investments in renewable energy and energy efficiency projects are being considered. These efforts ensure that ecosystem services, biodiversity, and natural resources are maintained for the benefit of current and future generations.

6. THE CIRCULAR ECONOMY

In contrast to the conventional linear economy that operates on the 'take, make, dispose' principle, a circular economy signifies a revolutionary change. As an alternative, the goal of the circular economy is to reduce waste while extending the useful life of resources through a closed-loop system. The circular economy is an economic system aimed at eliminating waste and the continual use of resources. It contrasts with the traditional linear economy, which follows a "take, make, dispose" model of production (Antunes et al., 2021; Kang et al., 2023; Sassanelli & Terzi, 2023; Souza Piao et al., 2023a). In a circular economy, the focus is on creating a closed-loop system where resources are kept in use for as long as possible, extracting maximum value from them, then recovering and regenerating products and materials at the end of each service life.

This method reduces waste and encourages new ideas and business expansion. Resource efficiency, product stewardship, collaboration, and closed-loop systems are the fundamental elements of a circular economy based on some of the basic principles described below.

6.1 Resource Efficiency

The core principle of a circular economy is resource efficiency. It minimises adverse environmental effects by making the most efficient use of energy and materials at every stage of production and consumption. By adhering to this idea, businesses are motivated to create more valuable products and processes that reduce resource use and waste. For instance, by using lean manufacturing practices, both energy use and material waste can be significantly reduced. To further improve resource efficiency, cutting down on material input and waste can be achieved through the use of cutting-edge technologies and materials such as lightweight composites and additive manufacturing (3D printing) (Pearce & Turner, 1989; Huntington et al., 2023; Kirchherr et al., 2017). In reality, there are many ways to improve resource efficiency. One is through material substitution, which involves using more plentiful or sustainable materials instead of scarce or harmful ones. Another is process optimisation, which aims to increase manufacturing efficiency and decrease waste. Lastly, product design innovations can improve durability, reparability, and recyclability. By emphasising resource efficiency, businesses may lessen their impact on the environment, save money, and stay competitive.

6.2 Keep Products and Materials in Use

Extending the life of products through reuse, repair, and remanufacturing is a crucial strategy for reducing waste and conserving resources. Reuse minimises the need for new materials and reduces environmental impacts associated with production. Recycling plays a complementary role by recovering and processing materials to create new products, thereby maintaining the value of materials within the economy and reducing the demand for virgin resources. Additionally, remanufacturing involves rebuilding products to their original specifications using a combination of reused, repaired, and new parts, extending the product life cycle and enhancing sustainability by minimising the consumption of new materials and energy. These approaches - reuse, recycling, and remanufacturing - form a holistic strategy that supports a circular economy and reduces environmental footprints.

6.3 Closed-loop Systems

The circular economy relies on closed-loop systems, which seek to continuously reuse, refurbish, remanufacture, and recycle resources and products. By ensuring items can be readily dismantled, fixed, or improved, closed-loop systems ensure that products are designed with their end-of-life in mind. Reducing the demand for fresh resources and minimising waste, this strategy extends the lives of products and materials. One way to recycle and reuse resources is to make electrical devices that are easily disassembled. This way, we can recover valuable components and put them to use in other goods. Another example of a closed-loop system is industrial symbiosis, which involves reusing materials from one operation to feed another. Industries work together in this approach to build energy and material exchange networks, reusing and recycling materials and improving resource efficiency.

For closed-loop systems to work, reverse logistics—the process of collecting, transporting, and processing spent products and materials—must be strong (Brody, 2022; Ghazanfari, 2023; Seier et al., 2023). Efficient reverse logistics systems keep commodities moving through the economy by facilitating the return of used goods to the manufacturing cycle. Developing infrastructure and laws that facilitate closed-loop systems, including extended producer responsibility (EPR) programs and incentives for recycling and remanufacturing, is an essential obligation of both governments and corporations.

6.4 The Polluter Pays Principle

The principle is that those who generate pollution or waste should bear the costs of managing it to prevent damage to human health or the environment. This can be implemented through regulations, taxes, and extended producer responsibility (EPR) schemes.

- Responsibility: Ensuring that those who produce pollution bear the costs of managing it to prevent damage to human health or the environment.
- Extended Producer Responsibility (EPR): Producers are responsible for the entire lifecycle of their products, especially for take-back, recycling, and final disposal.

6.5 Product Stewardship

A concept that makes producers responsible for their products throughout their entire lifecycle, from design and production to end-of-life disposal, is product stewardship, also called extended producer responsibility (EPR). This strategy encourages companies to design products with a lower environmental impact, greater sustainability, and ease of recycling. Producing goods that adhere to the principles of the circular economy requires manufacturers to innovate while bearing the costs of waste management. The product take-back program is an example of product stewardship in action; it is a service that manufacturers provide to collect customers' old products so that they can be recycled or repaired. As a result, precious items can be more easily recovered, and correct disposal can be ensured. In addition, businesses can lessen their adverse effects on the environment by incorporating design-for-environment (DfE) principles into their product development processes. This involves making eco-friendly material choices, cutting down on energy usage in manufacturing and operation, and making it easier to disassemble and recycle (Kang et al., 2023; Seier et al., 2023). Companies are also encouraged to use recyclable or biodegradable materials and design waste-minimizing packaging as part of product stewardship. Policies and regulations like the European Union's Waste Electrical and Electronic Equipment (WEEE) Directive require producers to handle the collection, recycling, and disposal of electronic waste to promote more sustainable product designs and lower environmental effects.

6.6 Stakeholders' Collaboration

Stakeholders' collaboration is crucial for the circular economy concepts to be effectively implemented. Governments, corporations, academic institutions, and civil society must collaborate to create and spread circular practices. Governments can foster favourable conditions by enacting laws, rules, and incentives that promote circular economy projects. Businesses might be incentivised to adopt circular processes through tax incentives for recycling technology investments or fines for noncompliance with waste management legislation. Round supply chains can be created, and best practices can be shared when businesses work with their suppliers, consumers, and rivals. For instance, companies can work together in consortiums to standardise procedures and materials, facilitating product recycling and reusing. New circular technologies can be more easily researched and developed through public-private partnerships, and recycling and waste management infrastructure can be more easily built (Senaratne et al., 2023; Souza Piao et al., 2023a).

The advancement of the circular economy relies heavily on the research, innovation, and education conducted by academic institutions. Academic institutions might create novel materials, technology, and economic models to promote circular practices. The next generation of workers can be better equipped to adopt and promote sustainable practices if educational programs focus on the circular economy and help students learn about it. Advocating for the concepts of the circular economy and holding governments and corporations responsible for their environmental repercussions are two roles that civil society, which includes NGOs and community groups, may play. Businesses can be encouraged to embrace the ideas of the circular economy through public awareness campaigns and grassroots activities that increase consumer demand for sustainable products and practices.

7. THE CIRCULAR ECONOMY AND SUSTAINABLE WASTE MANAGEMENT

To begin with, in this conceptual area of transforming circular economy fundamentals to sustainable waste management, a person must have at least a tiny knowledge of sustainability. Sustainability might express consistency in a certain state of an event. To put it another way, sustainability satisfies the requirements of the present without jeopardising the capacity of successor generations to satisfy their unique needs (Keeble, 1988). The key to achieving sustainability in both micro and macroeconomics lies in a circular economy and sustainable waste management. Readers could be confused about the difference between uniqueness because of the terminology used in the heading. Some scholars have drawn attention to the essential distinctions between these ideas. Nevertheless, a sustainable approach to eliminating waste can be developed over time by utilising the idea of a Circular Economy. This entails extending product lifecycles through the implementation of thorough waste management policies that prioritise waste minimisation, reuse, and recycling even though it resource replenishing materials (Karim et al., 2024). Consistent with this, multiple academic groups have defined sustainable waste management (SWM) as an essential component of sustainability and the circular economy (CE) since our current way of existence is leading us away from, rather than towards, a more sustainable tomorrow.

The circular economy is an environmentally friendly, sustainable approach to waste administration, resulting in a transition away from the linear economic model and toward the circular economic paradigm. Therefore, the circular economy concept can significantly enhance sustainable waste management by transforming how we view and handle waste, shifting from a linear "take, make, dispose" approach to a regenerative system where resources are continuously reused and recycled. The circular economy can contribute to sustainable waste management in several ways.

7.1 The Circular Waste Management System

Circular waste management is an integrated approach to waste management based on the Circular Economy principles. This approach emphasises resource efficiency while simultaneously fulfilling the straightforward challenge of succeeding in sustainability in domestic, organisational, national, and cross-national boundaries to achieve ecological equilibrium. Integrated waste management is another term that may be used to describe this waste management technique. This is because it involves control over all different kinds of solid waste products and sources, such as trash from homes, businesses, factories, institutionalised ongoing construction, and farming operations. Although hazardous waste is managed within the system, it is dealt with in a special stream. The Circular Waste Management System, implemented directly or indirectly, is an integrated waste management approach that follows circular economy concepts to ensure long-term sustainability by efficiently handling and transporting garbage. Hence, although all Circular Waste Management Systems fall under the category of Integrated Waste Management Systems, not all Integrated Waste Management Systems may be classified as Circular Waste Management Systems (González-Sánchez et al., 2023).

Figure 2. Circular waste management system

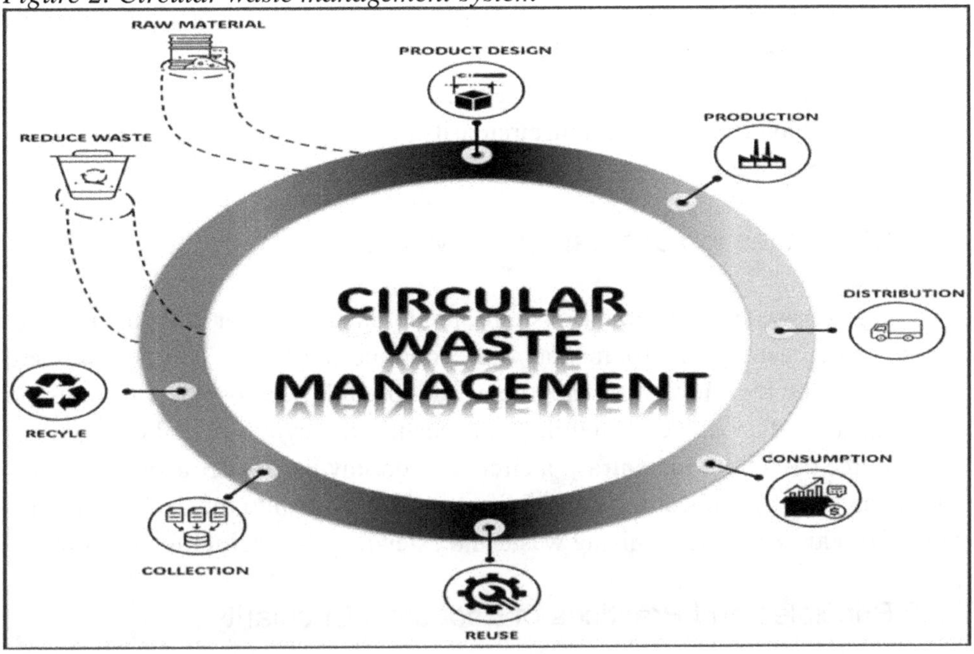

Source: Authors' Compilation

Cities and industries generate the majority of solid waste. In a circular economy framework, where an integrated waste management system may meet the fundamental principles of the economy while maximising sustainability, scholars and regulatory bodies are more inclined to focus on eliminating industrial waste (Awino & Apitz, 2024b; Ghazanfari, 2023). It should be acknowledged that the amount of undeniable and ecologically injurious manufacturing waste, electronics solid waste, and packaging from fast-moving consumer goods (FMCG) increases in direct correlation with the nation's overall level of development and poverty. Industries produce more dangerous waste than other sectors because they generate solid, liquid, and gaseous waste. A more sustainable climate, livable weather, and cleaner air can be ours when industries incorporate circularity into their current economic models.

7.2 The Transition of Waste Management into Economic Circularity

It is becoming increasingly clear that the transition towards a circular economy is a trend and a required evolution in our approach to waste management and resource sustainability. This realisation comes when we delve further into the complexities of altering economic paradigms. The change discussed here is premised on reconfiguring old 'take-make-dispose' models into intrinsically regenerative systems. These systems emphasise the conservation and reuse of resources everywhere in the economy (Meyer et al., 2020; Moradi & Sormunen, 2023; Peng et al., 2023; Souza Piao et al., 2023b; van Zanten et al., 2023).

The steps below are taken to attain circularity in the economy to attain sustainable waste management goals.

7.2.1 Challenges to Long-esteemed Models

Traditional waste management systems, predominantly linear in their structure, have led to escalating environmental and economic burdens. These systems are characterised by high levels of resource extraction, followed by short-term usage and subsequent disposal, contributing significantly to environmental degradation and resource depletion. In contrast, a circular economy introduces a systemic shift to sustainability by embedding end-of-life considerations into the design and life-cycle of products, thus minimising waste and extending the useful life of resources.

7.2.2 Principles and Practices of Economic Circularity

The concepts of eliminating waste, reusing resources, and recycling materials are the foundation upon which the circular economy is built! However, it goes beyond these principles and includes redesigning products and business processes to ensure they are durable and can be repaired. These activities ensure that products have a longer lifespan and make it easier to recycle resources back into the market once they have reached the end of their useful life. This strategy fosters a loop of continual resource efficiency, which reduces the amount of trash produced and the reliance on fresh materials.

7.2.3 Using Regenerative Methods in Practice

Promoting the durability of goods, straightforward repair, and material recyclability requires innovative company structures and regulatory frameworks. For example, right-to-repair policies and monetary rewards for product return and recycling are critical in promoting the reuse and refurbishment of electronics, which reduces e-waste and encourages sustainable consumer behaviours. This collaborative effort is necessary to successfully adopt circular strategies involving producers, consumers, waste administrators, and policymakers.

7.2.4 Resolving Restrictions and Emphasis on Potential Opportunities

There are many obstacles to achieving a circular economy, including changes in consumer behaviour and infrastructure, but there are also many opportunities. These include fewer environmental consequences, better resource efficiency, and economic gains from new markets and cost savings. The circular economy helps achieve larger sustainability objectives in both the economy and the environment, in addition to waste management.

7.2.5 Continuing Onward

Finally, one of the most important ways to achieve sustainable development is to move towards economic circularity in trash management. Our perspective on waste management must change dramatically to turn trash from an undesirable byproduct into a useful resource. This change necessitates a collective endeavour by all parties involved to revamp our economic systems, enhancing their sustainability, resilience, and inclusivity. To fully appreciate the advantages of a circular economy, it is necessary to take a holistic perspective, which is why this part delves into the mechanics of this change.

8 TURNING THE WASTE INTO WEALTH: THE ECONOMIC AND ENVIRONMENTAL BENEFITS

Transforming trash into treasure is a game-changing strategy that reimagines trash management. It intends to maximise financial gain while reducing adverse environmental effects. The ideas behind a circular economy—which promotes the reusing, recycling, and repurposing materials—are consistent with this philosophy. This method significantly impacts several industries and helps achieve larger sustainability goals, both economically and environmentally.

8.1 Economic Advantages

8.1.1 Boosting Employment and the Economy

Creating new jobs in recycling, manufacturing, and energy production are just a few benefits of repurposing trash. In the recycling industry, the gathering, sorting, processing, and producing of recycled goods necessitate a diversified workforce. Research shows that the recycling and reusing industry in the US alone is worth about

$37 billion per year and employs over 1.25 million people. Secondary markets, such as the refurbishing and remanufacturing industries and recycling facilities, generate additional economic activity and job opportunities (Karim et al., 2024; Zhang et al., 2018). The development of novel waste management methods and techniques has led to a high demand for engineers, researchers, and environmental scientists. This leads to the creation of more green jobs, which is crucial for the long-term viability of the economy. Investing in educational and training programs that educate the workforce for new waste management and resource recovery industries is an excellent way for the public and commercial sectors to take advantage of these opportunities.

8.1.2 Minimizing Waste and Cutting Costs

Businesses and economies can save money by reusing and recycling materials instead of buying new ones. For example, using recycled materials in manufacturing can reduce production costs by reducing raw material extraction, transportation, and processing expenses. An industry that greatly benefits from this is the aluminium sector. Recycling aluminium reduces energy consumption and costs significantly compared to producing new aluminium from raw ore. In addition to improving cost efficiency, firms can avoid landfill disposal fees by redirecting waste away from landfills and into recycling and reuse streams (Park et al., 2022). Industry can also benefit from substituting more sustainable or abundant materials for scarce or damaging ones. For instance, switching from petroleum-based plastics to bioplastics can reduce the use of fossil fuels and greenhouse gas emissions. Also, making your production processes more efficient and reducing waste can help you save and make more money. These practices can help businesses improve their environmental credentials and competitiveness.

8.1.3 Exploring Novel Approaches and Emerging Markets

There has been a proliferation of innovative technology and business models spurred by the need to find ways to recycle and repurpose garbage. More and more, businesses and entrepreneurs are looking for innovative ways to recycle and reuse resources, such as making biofuels from food scraps, building supplies from plastic trash, and manufacturing biodegradable packaging from food waste. In addition to attracting investments and collaborations, these breakthroughs open up new market opportunities, which drive economic growth. New business models, such as product-as-a-service, have emerged along with the circular economy. Corporations maintain ownership of products in this model but provide them as a service, ensuring their return for reuse, refurbishment, or recycling. Research and development are the engines behind these advancements. Technological, scientific, and engineering advancements in recent years have made it possible to recover valuable materials from various types of garbage (Neto et al., 2017). For example, scientists are working to turn plastic trash into valuable chemicals and fuels, turning a substantial environmental hazard into something useful. Bioprocessing advances are also opening the door to sustainable alternatives to conventional materials by allowing the manufacturing of bio-based products from agricultural byproducts.

8.1.4 Increased Business Credibility and Confidence

A corporation can improve its image and gain customers' trust by using eco-friendly waste management procedures and turning trash into valuable resources. Customers care more and more about how their purchases affect the environment and are more inclined to support businesses that demonstrate they care about sustainability. Companies may stand out, appeal to eco-conscious customers, and strengthen customer loyalty by clearly articulating their waste-to-wealth programs. A larger market share and an edge over the competition are two outcomes that may result from a better reputation (Huntington et al., 2022). To gain consumers' trust, it is essential to be open and responsible. Customers are more likely to remain loyal to businesses that are transparent about their sustainability efforts and show evidence of their success (Karmakar et al., 2023; Waaje et al., 2023). Better brand recognition and conscientious purchasing habits are two additional benefits of consumer engagement through sustainability efforts and marketing. When businesses take the lead in sustainability, they frequently discover that their initiatives connect with an expanding market sector, leading to higher sales and stronger consumer loyalty.

8.2 Environmental Advantages

8.2.1 A Decrease in Greenhouse Gas Emissions

Reusing and recycling materials instead of sending them to landfills or incinerators drastically reduces greenhouse gas (GHG) emissions. Organic waste in landfills produces methane, a potent greenhouse gas. One way to significantly reduce methane emissions is to send organic waste to composting or anaerobic digestion facilities. Additionally, recycling materials such as metals, plastics, and paper reduce the energy required for their production, reducing carbon dioxide emissions. One metric ton of recycled paper may prevent the felling of about 17 trees, and the reduction of greenhouse gas emissions is equal to taking one vehicle off the road for six months (Aracil et al., 2023). Limiting greenhouse gas emissions has environmental benefits beyond simply slowing climate change. Reduced emissions improve air quality, which is good for people's health (Tanchangya, T., et al., 2024).

Recycling and alternative energy sources can lessen our environmental impact by cutting down on our consumption of fossil fuels. Achieving global climate goals and establishing a low-carbon economy depends on these efforts.

8.2.2 Environmental Preservation

By decreasing the demand for virgin material extraction, recycling, and reusing waste products, we help preserve natural resources. Ecosystems, wildlife, and water supplies depend on this preservation effort. Recycling copper and gold, for instance, lessens the environmental impact of mining—a process that can lead to habitat loss, soil erosion, and water contamination—than mining would otherwise be necessary. Advocating for material reuse and recycling can reduce these negative impacts and advance the sustainable management of natural resources (Cheah et al., 2022). In addition to reducing resource use, eco-friendly garbage disposal methods shield wildlife and their natural environments from harm. We can protect ecosystems and biodiversity by making human activities such as resource extraction and trash disposal less environmentally harmful. Earth's natural systems provide vital services such as clean air and water, pollination, and climate management, which must remain healthy for these endeavours to continue.

8.2.3 Less Waste and Less Pollution at Landfills

Reusing or recycling materials instead of dumping them in landfills can prevent environmental pollution, such as reduced soil and water contamination from leachate and air pollution from landfill gases. Communities and ecosystems are vulnerable to the pollution that landfills produce. We can protect public health and reduce the environmental impact of garbage disposal by increasing recycling, composting, and energy recovery rates and decreasing the amount of trash delivered to landfills. Reducing landfill consumption also prolongs the life of current dump sites, which delays the necessity of developing new landfills and the land use implications that come with it (Ghazanfari, 2023). Implementing cutting-edge waste management technologies, such as biological processing and enhanced thermal treatment, can take additional steps to lessen the environmental toll of trash disposal. While reducing emissions and other pollutants, these technologies transform trash into valuable resources like electricity and compost. Waste management systems incorporating these technologies can lessen the adverse effects of trash disposal on the environment while improving resource recovery.

8.2.4 Advocate for a Regenerative Economic Model

Finding ways to transform waste into wealth is essential for promoting a circular economy, a regenerative system that continuously cycles products and materials back into the economy. This method lessens the linear depletion of resources and reduces trash production. The goal of a circular economy is to reduce waste and maximise efficiency in the use of resources in order to make production and consumption more sustainable. This systemic shift can have long-term environmental benefits, such as less resource depletion, less environmental damage, and more resilient ecosystems. The circular economy also promotes sustainable business practices and innovation. Companies adhering to the principles of the circular economy are more likely to develop new goods and services that lessen environmental effects and generate value from waste (Puntillo, 2022). This change has the potential to boost competitiveness and economic growth while also helping to achieve larger environmental targets. Stakeholders and policymakers can promote sustainable consumption and production by implementing public awareness campaigns, incentives, and legislation that facilitate the transition towards a circular economy.

8.2.5 Benefits to Soil and Agriculture

Fertilising soil and increasing crop yields. In addition to improving water retention and soil structure, compost also supplies plants with the nutrients they need to flourish. Chemical fertilisers have negative environmental impacts due to nutrient runoff and soil degradation; however, using compost in farming can lessen this demand. Composting organic waste has multiple benefits, including lowering landfill trash, bolstering sustainable agriculture, and enhancing soil health. Composting is beneficial not only in agriculture but also in other industries. One environmentally friendly way to deal with organic waste is to use compost in landscaping, erosion control, and habitat restoration initiatives (Ciccullo et al., 2021; Kör et al., 2022). Reusing organic waste instead of sending it to landfills and turning it into compost can improve soil health, sustainable agriculture, and environmental sustainability.

9 WASTE TO WEALTH INITIATIVES WORLDWIDE: FEW CASES

Many cities and countries are trying to manage their waste to ensure that it can be used as input to produce something treated as the nation's wealth. A few of the cases worldwide are described here.

9.1 Sweden's Waste-to-Energy Program

Sweden has established itself as a global leader in waste-to-energy initiatives, where over 99% of household waste is recycled or used for energy production. The country's state-of-the-art waste-to-energy plants convert waste into heat and electricity, significantly reducing landfill use and supplying power to numerous homes and businesses. Sweden's efficient system imports waste from other countries to fuel its plants. This initiative minimises environmental impact, provides a stable renewable energy source, generates economic revenue, and creates numerous job opportunities in the waste management and energy sectors. The Swedish waste-to-energy program has highlighted the issue of sustainable waste management. The country has established extensive recycling systems and public education campaigns to promote garbage sorting and consumption reduction. These initiatives have successfully implemented waste-to-energy technologies, resulting in high recycling rates.

9.2 The McCarthur Family Fund

The Ellen MacArthur Foundation is one of the most well-known groups advocating for a circular economy. The foundation's numerous projects and collaborations demonstrate the environmental and economic benefits of recycling and reusing materials. For example, their Circular Economy 100 program facilitates partnerships between academic institutions, government agencies, and private companies to address resource efficiency and waste. Redesigning products and waste management procedures have resulted in substantial cost reductions, new jobs, and environmental advantages, all thanks to projects funded by this initiative (Morcillo-Bellido et al., 2022). As a result of the foundation's work, new technology and business models have emerged to bolster the circular economy. For instance, they have worked with businesses to develop more durable and recyclable items, which helps cut down on trash and boost the economy. The Ellen MacArthur Foundation's efforts show how the circular economy can benefit from teamwork and new ideas.

9.3 Japan's Zero Waste Town - Kamikatsu

Kamikatsu, a small town in Japan, has committed to becoming zero waste by 2025 through meticulous waste separation and recycling. Residents separate their waste into 45 distinct categories, significantly enhancing recycling rates to over 80%. The town's success is driven by strong community involvement and education, alongside the establishment of reuse shops where residents can exchange used goods. Kamikatsu's initiative highlights the environmental benefits of reducing landfill waste, conserving resources, and empowering the community to take responsibility for their waste, fostering a culture of sustainability and environmental stewardship. The zero waste initiative in Kamikatsu began in 2003 when the town recognised the need to move away from incineration and landfill disposal methods, which were harmful to the environment and public health. Faced with the closure of their incinerators due to stricter emissions regulations, the town's leadership and its residents embarked on a transformative journey toward waste reduction and recycling. Kamikatsu's zero-waste efforts have attracted attention from around the world. They serve as a model for other communities and inspire global movements toward sustainable waste management. The town regularly hosts visitors from various countries who come to learn about its practices and replicate them in their regions.

9.4 Carpet Tiles from Interface

As part of their commitment to the circular economy concepts, the international carpet producer Interface has created a closed-loop recycling mechanism for their carpet tiles. As part of its ReEntry program, Interface gathers old carpet tiles, sorts them by substance, and then uses the recycled resources to make new goods. The project has successfully reduced the amount of garbage transported to landfills, the cost of raw materials and the company's carbon footprint. Interface's dedication to sustainability has improved the company's image and drawn in eco-conscious consumers. The Re-entry program at Interface emphasises how material choice and product design contribute to the success of the circular economy. The company has developed a sustainable business model that minimises waste and conserves resources by producing carpets that are easy to deconstruct and recycle. Businesses may learn from Interface's success and use creative waste management strategies to improve the environment and their bottom line.

9.5 The City of San Francisco's Zero Waste Initiative

Due to its extensive zero-waste project, San Francisco hopes to divert all its trash from landfills by 2030. By enforcing recycling and composting legislation and educating the public, the city aims to reduce trash and increase recycling and composting. Consequently, San Francisco has reduced landfill usage and greenhouse gas emissions, increased local economic activity, and created green jobs, ranking among the highest trash diversion rates in the US. The zero-trash effort in San Francisco is a prime example of how public involvement and integrated waste management may provide positive results. The city is launching several initiatives to combat trash, such as recycling and composting program partnerships, public awareness campaigns, and financial incentives for companies and individuals to reduce garbage. These initiatives have proven that communities can accomplish their zero-waste goals with substantial economic growth and positive environmental impacts.

9.6 The Company, TerraCycle

TerraCycle, a new startup, specialises in recycling challenging items such as cigarette butts, coffee capsules, and snack wrappers. Park benches and playground equipment are only two examples of the new items made from recycled materials that TerraCycle acquires and manufactures in collaboration with companies and individuals. Not only does this method keep trash out of landfills, but it also makes money off of things that would otherwise go to waste. The waste management sector has the potential for innovation and success, as shown by TerraCycle's business model. The success of TerraCycle shows how important it is to work together as a team to accomplish sustainability objectives. The company's programs collaborate with big brands, retailers, and customers to develop recycling solutions for complex waste streams. Businesses may learn from TerraCycle's example and find ways to find value in trash while also helping the environment. The company's creative mindset has resulted in new products and markets.

10. CHALLENGES OF TURNING THE WASTE INTO WEALTH

Despite the substantial economic and environmental benefits of converting trash into wealth, sustainable waste management strategies are not widely adopted. To address these challenges, we must concertedly address several social, technological, economic, and regulatory obstacles.

10.1 Technological Challenges

10.1.1 A Lack of Cutting-Edge Recycling Technologies

Although recycling technology has advanced significantly, numerous materials still pose substantial problems for recycling. Electronic devices and multi-layered packaging are examples of complex items with many elements that are notoriously difficult to deconstruct and recycle effectively. For example, recycling electronic trash (e-waste) goods becomes more difficult because many contain hazardous materials and a combination of metals and polymers. The absence of sophisticated recycling systems that manage these intricate waste streams limits the possibility of converting waste into riches. We must develop and scale up cutting-edge robotic sorting systems and chemical recycling to address these issues (Cheremisinoff,

2003; Ciccullo et al., 2021)Chemical recycling is a promising yet underdeveloped technique that holds the potential to transform the management of plastic waste into its constituent chemicals for reuse.

10.1.2 Complexity of Waste Streams

Waste materials are often heterogeneous, containing a mix of plastics, metals, organic matter, and other substances. Separating these efficiently can be technologically challenging and costly.

10.1.2 Recycling and Repurposing Technologies

Advanced technologies are required to recycle materials effectively and to repurpose them into valuable products. Developing these technologies can be expensive and time-consuming. Some materials, such as certain types of plastics, are challenging to recycle due to their chemical properties, requiring specialised processes.

10.1.3 Quality and Contamination

Contaminants in waste streams can reduce the quality of recycled materials, making them less desirable for manufacturers. Ensuring the purity of recycled materials often requires sophisticated sorting and cleaning technologies.

10.1.5 Lack of Technological Integration

Technological innovation could be one of the great reasons behind the expansion of growth of the business concerns as well as other technology-based initiatives (Ahmad et al., 2024). Integrating new waste management technology into current systems can be difficult and expensive. It takes a significant financial commitment and specialised knowledge to upgrade or convert current facilities to accept cutting-edge recycling or waste-to-energy technology. In addition, achieving smooth and effective operations depends on the compatibility of various waste management technologies and systems (Waaje et al., 2018). Working together, waste management firms, legislators, and technology suppliers can overcome these obstacles to technological integration. For example, switching to a smart waste management system that optimises waste collection routes using IoT sensors and data analytics requires significant infrastructure modifications and inter-system connections.

10.2 Economic Challenges

10.2.1 Exorbitant Outlay of Funds

High upfront investment costs are frequently associated with implementing sustainable waste management techniques and technologies. Building new recycling facilities, renovating existing infrastructure, and developing cutting-edge technology require significant financial resources. These expenses may be unaffordable, particularly for towns with tight budgets and small and medium-sized businesses (SMEs) (Gazi et al., 2024; Karim et al., 2023)For instance, establishing a new anaerobic digestion facility for organic waste can cost millions of dollars. Access to incentives and funding is required to reduce these obstacles and encourage the switch to sustainable waste management techniques. Governments and financial institutions can play a significant role by offering grants, low-interest loans, and subsidies to support waste management initiatives.

10.2.2 Recycled Products' Viability in the Market

The unpredictable and fluctuating nature of the market for recovered goods impacts the economic sustainability of recycling projects. Recycling businesses may not always be profitable due to changes in the cost of recycled goods and raw resources. For instance, the price of crude oil influences the cost of manufacturing virgin plastics, potentially influencing the price of recycled plastics. Furthermore, customer approval of recycled goods varies, affecting market stability and demand. Ensuring the economic sustainability of recycling projects requires creating stable markets for recycled goods and implementing regulations that promote the use of recycled materials. Green public procurement rules that prefer recycled content in goods and services are one way that governments can assist market development.

10.2.3 Virgin Materials as a Rival

Recycled materials frequently compete with virgin materials, which may be less expensive because of production cost reductions, economies of scale, or subsidies. Because of this rivalry, it could be challenging for recycled items to establish themselves in the market. For example, virgin plastic production is less expensive than recycled plastic because raw material prices are lower due to fossil fuel subsidies. Legislative measures such as mandated recycled content standards, taxes on virgin resources, or subsidies for recycled materials can level the playing field, address this economic barrier, and encourage using recycled products. By implementing extended producer responsibility (EPR) initiatives, we can encourage manufacturers to employ recycled materials and create products that are easier to recycle.

10.2.4 Limited Facilities

There are huge regional differences in the infrastructure for collecting, sorting, and processing waste; there are insufficient resources to handle and recycle waste in many places. Inadequate infrastructure can lead to environmental pollution, increased landfill usage, and ineffective waste management techniques (Minelgaitė & Liobikienė, 2019). For example, open dumping and burning waste lead to serious health and environmental issues in many underdeveloped nations due to a lack of basic waste management facilities. Investments in waste-to-energy plants, recycling facilities, and comprehensive waste management systems are essential to overcome these physical constraints and enable more effective waste management systems. Investing in sophisticated materials recovery facilities (MRFs) can significantly improve recycling processes' efficiency.

10.3 Social Challenges

10.3.1 Public Awareness and Involvement

Every Business enterprise has to think about the stakeholders of the society as their part of corporate social responsibility (Miah et al., 2019). Effective waste management requires public involvement to reduce consumption, recycle garbage, and sort waste. Public engagement and awareness, however, differ, with many people ignorant of the advantages of recycling and appropriate trash management techniques. One of the biggest challenges is getting the general public to embrace sustainable waste management practices. Public awareness campaigns, educational initiatives, and community projects can significantly increase the public's involvement and support for sustainable waste management procedures. Effective public education initiatives, for instance, have considerably raised recycling rates and decreased landfill waste in places like San Francisco.

10.3.2 Modification of Behavior

Breaking old habits and behaviours can be challenging when producing and disposing of waste. Adopting new habits like separating waste, consuming fewer throwaway items, or buying products made of recycled materials may encounter resistance from the public. It will require persistent training, rewards, and alterations to societal norms to overcome this resistance and promote lasting habits.

Behavioural science insights can influence the development of interventions that successfully advance sustainable waste management techniques. One way to affect behaviour change is through social marketing strategies that highlight the positive effects recycling has on society and the environment.

10.3.3 Socioeconomic Status Gaps

Socioeconomic inequality can impact participation in recycling programs and access to sustainable waste management services. Communities with low incomes might not have the infrastructure or resources to support efficient waste management, which would increase trash production and the degradation of the environment. Targeted interventions and financial expenditures for marginalised groups must address these gaps and provide fair access to waste management services and participation in recycling programs. Implementing policies that support community-based waste management projects and subsidise trash management services for low-income families can partially fill these gaps.

10.3.3 Community Involvement

Engaging communities in waste management practices is crucial but can be challenging. People may be resistant to change or lack the motivation to participate actively in recycling programs.

10.4 Regulatory Challenges

10.4.1 Incoherent Rules and Guidelines

Regional and national disparities in waste management laws and regulations may hinder the implementation of sustainable waste management methods. It can be challenging for businesses and consumers to implement sustainable practices with inconsistent policies because they can cause confusion, inefficiencies, and uneven enforcement. Harmonising laws and regulations at regional, national, and international levels is imperative to provide a cohesive framework that promotes sustainable waste management. For instance, the Circular Economy Action Plan of the European Union seeks to promote efficiency and uniformity by standardising waste management laws throughout its member states.

10.4.2 A Lack of Rewards

One major obstacle may be the lack of incentives for consumers and enterprises to use sustainable waste management techniques. Subsidies, tax breaks, grants, and other financial incentives can promote investment in infrastructure and recycling systems. Moreover, regulatory incentives such as mandated recycling objectives and extended producer responsibility (EPR) schemes may encourage businesses to prioritise sustainable waste management. Creating and implementing effective incentive programs is critical to encouraging sustainable practices. Germany's packaging rules, for example, incentivise waste reduction and the use of recyclable materials by requiring producers to pay to recycle their packaging.

10.4.2 Policy Support

Effective waste-to-wealth initiatives often require strong policy support, including subsidies, tax incentives, and other financial mechanisms to make recycling economically viable. A lack of comprehensive policies can hinder the development and implementation of waste management and recycling programs.

10.4.3 Compliance and Enforcement

Ensuring adherence to waste management policies and regulations poses a significant challenge. Insufficient enforcement methods and inadequate monitoring may compromise the efficacy of waste management strategies. To ensure that waste management procedures align with sustainability objectives, enforcement and compliance must be strengthened through frequent inspections, fines for non-compliance, and assistance for regulatory organisations. Furthermore, open reporting and accountability systems can improve stakeholder compliance and trust. For example, vital monitoring programs and regulations requiring public reporting can ensure that waste management firms follow environmental guidelines.

10.5 Environmental and Logistical Challenges

10.5.1 The Contamination of Recyclable Materials

Contamination of recyclable resources is one major issue that can lower the quality and value of recycled products. Allowing improper sorting, mixing non-recyclable materials with recyclables, and leftover food or chemical contamination can lead to more expensive and less effective recycling procedures. For example, food residues on cardboard and paper could render them unfit for recycling. Investing in cutting-edge sorting systems and educating the public about appropriate recycling techniques is crucial to lower contamination and raise the quality of recyclable materials. By encouraging customers to return clean, separated materials, deposit-return programs for beverage containers can also help prevent contamination.

10.5.2 Logistics and Transportation

Effective garbage management depends on efficient logistics and transportation, especially in rural or isolated places. Logistically challenging are the expenses and environmental effects of transporting waste to facilities for recycling or processing (Roshid et al., 2024). Creating decentralised waste management initiatives, such as neighbourhood composting projects and recycling facilities, can help address these issues while lowering transportation expenses and emissions. Community composting programs, for instance, can handle organic waste on-site and eliminate the need for long-distance shipping.

10.5.3 The Effect of Waste Management Procedures on the Environment

Although recycling and waste-to-energy methods suit the environment, they can sometimes have adverse effects, including pollution from burning waste or energy use in recycling centres. We must balance the environmental trade-offs of various waste management techniques to maintain overall sustainability. Strict environmental impact assessments, ongoing waste management technology, and practice improvement can reduce adverse environmental effects. One way to lessen their environmental impact is to invest in energy-efficient recycling technology and emissions control systems for waste-to-energy plants.

11. TURNING THE WASTE INTO WEALTH: A BLESSING FOR GLOBAL SUSTAINABILITY

Transforming waste into wealth represents a paradigm shift that can significantly contribute to global sustainability, serving as a profound blessing for the Earth. This approach mitigates the detrimental environmental impacts of waste, fosters economic growth, and promotes social well-being. We can drive a more sustainable future by recognising waste as a valuable resource rather than a disposal problem

11.1 Environmental Preservation

The most immediate benefit of converting waste into wealth is the substantial reduction in environmental degradation. Traditional waste disposal methods, such as landfilling and incineration, contribute to pollution, greenhouse gas emissions, and the depletion of natural resources. Recycling and repurposing waste materials can reduce the strain on landfills, lower carbon emissions, and conserve natural resources. This shift can lead to cleaner air and water, reduced soil contamination, and a decrease in the ecological footprint of human activities. In essence, waste transformation can help restore and maintain the planet's health.

11.2 Economic Growth and Job Creation

Economic opportunities abound in waste management and recycling. Establishing and expanding recycling industries can create numerous jobs in collection, sorting, processing, and manufacturing. These industries can thrive on a robust market for recycled materials, fostering new business ventures and stimulating local economies. Moreover, businesses that integrate sustainable practices can save money through reduced material expenses and waste disposal costs. The circular economy model, where products are continuously cycled back into use, enhances resource efficiency and drives innovation and economic resilience.

11.3 Social Benefits and Community Engagement

Transforming waste into wealth has significant social implications. Community-based recycling programs can strengthen social ties and foster a sense of collective responsibility. Educational campaigns and public awareness initiatives can lead to behavioural changes that prioritise sustainability and environmental stewardship. Moreover, communities can benefit from improved public health outcomes due to reduced pollution and cleaner environments. The participatory nature of effective waste management strategies can empower individuals and communities, making them active stakeholders in the journey toward sustainability.

12. FUTURE DIRECTIONS

Future directions should be pursued to fully realise the potential of waste transformation. These directions encompass technological advancements, policy frameworks, collaborative efforts, educational initiatives, and global cooperation. Each area is crucial in establishing a sustainable and effective waste-to-wealth ecosystem.

12.1 Innovation in Recycling Technologies

Significant investment in research and development is essential to create and refine advanced recycling technologies. These technologies should aim to efficiently process complex and mixed waste streams, which are often challenging to manage with conventional methods. Key areas of focus include improving separation and purification processes to enhance the quality of recovered materials and developing innovative techniques for recycling currently hard-to-recycle materials, such as multi-layer plastics and electronic waste. These advancements can help close the loop on various waste types, ensuring that a broader range of materials are effectively recycled and reintegrated into the production cycle. Implementing automation and artificial intelligence (AI) in recycling facilities can significantly enhance sorting accuracy, increase efficiency, and reduce labour costs. AI-powered systems can precisely identify and separate materials from traditional manual or mechanical sorting methods. This advanced sorting capability improves the quality of recycled outputs and streamlines the recycling process, making it more cost-effective and scalable. By integrating these technologies, recycling operations can achieve higher throughput and better quality control, driving overall efficiency and effectiveness.

12.2 Policy Support and Regulation

Governments should develop and enforce comprehensive waste management policies that promote recycling and waste reduction. Effective policies might include mandates for waste segregation at the source, which can significantly improve the quality and efficiency of recycling processes. Additionally, incentives for businesses that adopt sustainable practices can encourage wider participation in waste reduction efforts, while penalties for non-compliance can ensure adherence to waste management regulations. By creating a regulatory environment that supports sustainable waste practices, governments can drive significant improvements in recycling rates and overall waste management. Implementing Extended Producer Responsibility (EPR) policies can further enhance waste management by holding manufacturers accountable for the entire lifecycle of their products, including end-of-life disposal and recycling. EPR shifts the responsibility for waste management from consumers and governments to producers, encouraging them to design products that are easier to recycle and to invest in take-back programs. This approach reduces the environmental impact of products and incentivises innovation in sustainable product design and recycling infrastructure, ultimately contributing to a more circular economy.

12.3 Public-Private Partnerships

Public-private partnerships can play a crucial role in developing and implementing effective waste management solutions by leveraging the strengths of both sectors. Such collaborations enable the sharing of resources, expertise, and technologies, which can accelerate the transition to a circular economy. By combining the efficiency and innovation of the private sector with the regulatory and oversight capabilities of the public sector, these initiatives can more effectively address complex waste management challenges. Collaborative efforts can lead to the development of more integrated and scalable solutions, promoting sustainability across industries and communities. Establishing innovation hubs and centres of excellence can enhance collaboration between academia, industry, and government. These hubs can serve as incubators for new recycling technologies, business models, and sustainability practices, providing a platform for stakeholders to work together on cutting-edge solutions. By fostering an innovation and knowledge exchange environment, these centres can drive advancements in recycling efficiency, product design, and waste management strategies. This approach supports the development of more sustainable practices and encourages the continuous improvement and adaptation of waste management systems in response to emerging challenges and opportunities.

12.4 Education and Community Engagement

Continuous public education campaigns are crucial for raising awareness about recycling and waste reduction. These campaigns should emphasise the environmental, economic, and social benefits of transforming waste and offer practical guidance on how individuals can contribute to these efforts. By engaging the public through various media channels and community outreach programs, these initiatives can build a culture of sustainability and encourage widespread participation in recycling and waste reduction activities. Effective public awareness campaigns inform and motivate individuals to adopt more sustainable behaviours daily, thereby amplifying the impact of broader waste management strategies. Integrating sustainability and waste management topics into school curriculums can further support these efforts by educating future generations about the importance of environmental stewardship. By introducing these concepts early, students can develop a deeper understanding of the impact of waste on the planet and learn how they can make a difference. Hands-on activities, such as school recycling programs and sustainability projects, can reinforce these lessons and empower students to take action. This educational approach helps cultivate environmentally conscious citizens who are more likely to carry sustainable practices into adulthood, positively impacting waste management and environmental health.

12.5 Global Cooperation

Harmonising international recycling and waste management standards can significantly facilitate global trade in recycled materials by ensuring consistent quality and safety across borders. This cooperation helps to create a level playing field, making it easier for countries to implement and adhere to effective waste management practices. By aligning standards, nations can reduce barriers that hinder the flow of recycled materials, promoting a more integrated and efficient global recycling market. Consistent international standards can also foster innovation and improve environmental outcomes by encouraging the adoption of best practices worldwide. Promoting global knowledge sharing and best practices is another vital component of improving waste management globally. Through international conferences, workshops, and online platforms, countries can exchange ideas and learn from successful models, which can help them avoid common pitfalls and adapt effective strategies to their specific contexts. This knowledge exchange accelerates the adoption of innovative waste management solutions and builds a sense of global community and shared responsibility in tackling environmental challenges. Providing financial and technical support to developing countries is essential for building the infrastructure for effective waste management. Many developing nations face significant challenges, including limited resources and technical expertise, which can hinder their ability to manage waste sustainably. International assistance can help bridge these gaps by funding infrastructure projects, supplying necessary equipment, and offering capacity-building programs to train local stakeholders in sustainable practices. This support empowers developing countries to improve their waste management systems, contributing to global environmental health and economic development

13. CONCLUSION

The transition to a circular economy signifies a revolutionary method of handling trash, providing significant advantages for the environment, economics, and society. Communities and enterprises may reduce the adverse effects of waste, protect natural resources, and promote economic resilience by adopting the concepts of reuse, recycling, and recovery. Efficient adoption of circular waste management strategies requires governments, corporations, and consumers to work together to reconsider and reshape the lifespan of materials. Promoting education, innovation, and regulatory frameworks to expand circular economy practices will be essential as we progress. This will transform waste management from a necessity into a strategic resource for sustainable development. This chapter emphasises the significance of a comprehensive and proactive strategy for waste management, intending to transform trash into valuable resources and create a sustainable future.

REFERENCES

Ahmad, S., Saxena, C., Islam, S., & Karim, R. (2024). Impact of Insur-Tech on the Premium Performance of Insurance Business. *SN Computer Science*, 5(1), 1–7. DOI: 10.1007/s42979-023-02462-0

Ajwani-Ramchandani, R., & Bhattacharya, S. (2022). Moving towards a circular economy model through I4.0 to accomplish the SDGs. *Cleaner and Responsible Consumption*, 7, 100084–100084. DOI: 10.1016/j.clrc.2022.100084

Ali, S., & Shirazi, F. (2023). The Paradigm of Circular Economy and an Effective Electronic Waste Management. *Sustainability (Basel)*, 15(3), 1998–1998. DOI: 10.3390/su15031998

Antunes, A., Martins, R., Silvestre, J. D., Do Carmo, R., Costa, H., Júlio, E., & Pedroso, P. (2021). Environmental Impacts and Benefits of the End-of-Life of Building Materials: Database to Support Decision Making and Contribute to Circularity. *Sustainability (Basel)*, 13(22), 12659. DOI: 10.3390/su132212659

Aracil, C., Villanueva Perales, Á. L., Giuntoli, J., Cristóbal, J., & Haro, P. (2023). The Role of Renewable-Derived Plastics in the Analysis of Waste Management Schemes: A Time-Dependent Carbon Cycle Assessment. *Sustainability (Basel)*, 15(12), 9292. DOI: 10.3390/su15129292

Awino, F. B., & Apitz, S. E. (2024a). Solid waste management in the context of the waste hierarchy and circular economy frameworks: An international critical review. *Integrated Environmental Assessment and Management*, 20(1), 9–35. DOI: 10.1002/ieam.4774 PMID: 37039089

Awino, F. B., & Apitz, S. E. (2024b). Solid waste management in the context of the waste hierarchy and circular economy frameworks: An international critical review. *Integrated Environmental Assessment and Management*, 20(1), 9–35. DOI: 10.1002/ieam.4774 PMID: 37039089

Bogusz, M., Matysik-pejas, R., Krasnodębski, A., & Dziekański, P. (2021). The Concept of Zero Waste in the Context of Supporting Environmental Protection by Consumers. *Energies*, 14(18), 5964. DOI: 10.3390/en14185964

Brody, H. (2022). Building a circular economy. *Nature*, 611(7936), S1–S1. DOI: 10.1038/d41586-022-03643-2 PMID: 36385552

Cheah, C. G., Chia, W. Y., Lai, S. F., Chew, K. W., Chia, S. R., & Show, P. L. (2022). Innovation designs of industry 4.0 based solid waste management: Machinery and digital circular economy. *Environmental Research*, 213, 113619. DOI: 10.1016/j.envres.2022.113619 PMID: 35700763

Cheremisinoff, N. P. (2003). *Handbook of solid waste management and waste minimization technologies*. Butterworth-Heinemann., http://www.sciencedirect.com:5070/book/9780750675079/handbook-of-solid-waste-management-and-waste-minimization-technologies

Ciccullo, F., Cagliano, R., Bartezzaghi, G., & Perego, A. (2021). Implementing the circular economy paradigm in the agri-food supply chain: The role of food waste prevention technologies. *Resources, Conservation and Recycling*, 164, 105114. DOI: 10.1016/j.resconrec.2020.105114

City Council. A. (2008). *Auckland City Council Guideline for working towards zero waste events*. http://www.waitakere.govt.nz/abtcit/ec/clnprod/guidelines-events.asp

David, W. Pearce, & R. Kerry Turner. (1989). *Economics of Natural Resources and the Environment*. Johns Hopkins University Press. https://books.google.com.bd/books?hl=en&lr=&id=ex8vaG6m4RMC&oi=fnd&pg=PR11&dq=+Economics+of+natural+resources+and+the+environment.+Baltimore&ots=lB2SSxSCcK&sig=7rEpPVLKyja1xBKipbVOm0CauNc&redir_esc=y#v=onepage&q=Economics%20of%20natural%20resources%20and%20the%20environment.%20Baltimore&f=false

de Oliveira Neto, G. C., de Jesus Cardoso Correia, A., & Schroeder, A. M. (2017). Economic and environmental assessment of recycling and reuse of electronic waste: Multiple case studies in Brazil and Switzerland. *Resources, Conservation and Recycling*, 127, 42–55. DOI: 10.1016/j.resconrec.2017.08.011

Derhab, N., & Elkhwesky, Z. (2022). A systematic and critical review of waste management in micro, small and medium-sized enterprises: Future directions for theory and practice. *Environmental Science and Pollution Research International*, 30(6), 13920–13944. DOI: 10.1007/s11356-022-24742-7 PMID: 36543988

European Union. (1993). *Council Regulation (EEC) No. 259/93 on the supervision and control of shipments of waste within, into and out of the European Community*. FAOLEX. https://www.fao.org/faolex/results/details/en/c/LEX-FAOC038118/

European Union Law. (1991). *Council Directive 91/156/EEC of 18 March 1991 amending Directive 75/442/EEC on waste*. EUR-Lex. https://eur-lex.europa.eu/legal-content/EN/ALL/?uri=celex%3A31991L0156

Farooq, M., Cheng, J., Khan, N. U., Saufi, R. A., Kanwal, N., & Bazkiaei, H. A. (2022). Sustainable Waste Management Companies with Innovative Smart Solutions: A Systematic Review and Conceptual Model. *Sustainability 2022, Vol. 14, Page 13146, 14*(20), 13146. https://doi.org/DOI: 10.3390/su142013146

Gazi, M., Islam, H., Islam, M. A., Karim, R., Momo, S. M., & Senathirajah, A. R. B. S. (2024). Unlocking sustainable development in East Asia Pacific and South Asia: An econometric exploration of ESG initiatives. *Sustainable Environment*, 10(1), 2366558. Advance online publication. DOI: 10.1080/27658511.2024.2366558

Ghazanfari, A. (2023). An Analysis of Circular Economy Literature at the Macro Level, with a Particular Focus on Energy Markets. *Energies*, 16(4), 1779–1779. DOI: 10.3390/en16041779

González-Sánchez, R., Alonso-Muñoz, S., & Medina-Salgado, M. S. (2023). Circularity in waste management: A research proposal to achieve the 2030 Agenda. *Operations Management Research : Advancing Practice Through Research*, 16(3), 1–21. DOI: 10.1007/s12063-023-00373-0

Huntington, V. E., Coulon, F., & Wagland, S. T. (2022). Innovative Resource Recovery from Industrial Sites: A Critical Review. *Sustainability 2023, Vol. 15, Page 489, 15*(1), 489. https://doi.org/DOI: 10.3390/su15010489

Huntington, V. E., Coulon, F., & Wagland, S. T. (2023). Innovative Resource Recovery from Industrial Sites: A Critical Review. In *Sustainability (Switzerland)* (Vol. 15, Issue 1). MDPI. https://doi.org/DOI: 10.3390/su15010489

Hutner, P., Helbig, C., Stindt, D., Thorenz, A., & Tuma, A. (2018). Transdisciplinary Development of a Life Cycle–Based Approach to Measure and Communicate Waste Prevention Effects in Local Authorities. *Journal of Industrial Ecology*, 22(5), 1050–1065. DOI: 10.1111/jiec.12781

Kadhila, T., de Wit, M. P., & Schenck, R. (2023). A conceptual framework for sustainable waste management in small municipalities: The cases of Langebaan, South Africa and Swakopmund, Namibia. *Environmental Science and Pollution Research International*, 30(60), 1–16. DOI: 10.1007/s11356-023-26904-7 PMID: 37095215

Kang, S., Hong, Y. S., & Kwak, M. (2023). Evaluating the circularity and multi-lifecycle green profit of product family design. *Resources, Conservation and Recycling*, 197, 107106–107106. DOI: 10.1016/j.resconrec.2023.107106

Karim, R. (2020). Green Banking Practices in Bangladesh: Present Scenario and Future Challenges. *International Conference on Business and Sustainable Development (ICBSD)*, 34–44. https://www.researchgate.net/publication/355168961_Green_Banking_Practices_in_Bangladesh_Present_Scenario_and_Future_Challenges

Karim, R., Hasan, M. M., Waaje, A., Yesmin, M. S. H., & Roshid, M. M. (2023). Financial Performance of Bangladeshi Listed Commercial Banks: A Generation-based Analysis. International Journal of Finance & Banking Studies (2147-4486), 12(4), 46-56.

Karim, R., Roshid, M. M., Shamme, F. B., & Hasan, M. M. (2022). Non-performing loans and bank profitability: Evidence from Bangladesh. International Journal of Finance & Banking Studies (2147-4486), 11(4), 95-102.

Karim, R., Roshid, M., & Waaje, A. (2024). Circular Economy and Supply Chain Sustainability. In *Strategic Innovations for Dynamic Supply Chains* (pp. 1–30). https://doi.org/DOI: 10.4018/979-8-3693-3575-8.ch001

Karmakar, A., Hossain Khan, Z., Roshid, M., Hoshain Yesmin, S., & Bashar Shamme, F. (2023). Consumer Learning and Split-Brain Theory: Potential Usage in an Advertisement. *International Journal of Economics, Finance and Management Sciences*. https://doi.org/DOI: 10.11648/j.ijefm.20231103.21

Karmakar, A., Khan, Z. H., & Roshid, M. (2022, November). M., Shamme, F. B., & Rahman, Md. A. (2022). A Study of Farmers Direct Marketing: Assessing the Factors for the Home to Land Approach. *European Journal of Business and Management*. Advance online publication. DOI: 10.7176/EJBM/14-22-05

Katan, L. (2023). How the waste management system's materialised normativity influences engagement in sustainable waste practices. *Journal of Environmental Policy and Planning*, 25(4), 443–458. DOI: 10.1080/1523908X.2023.2171976

Keeble, B. R. (1988). The Brundtland report: 'Our common future.'. *Medicine and War*, 4(1), 17–25. DOI: 10.1080/07488008808408783

Kirchherr, J., Reike, D., & Hekkert, M. (2017). Conceptualizing the circular economy: An analysis of 114 definitions. *Resources, Conservation and Recycling*, 127, 221–232. DOI: 10.1016/j.resconrec.2017.09.005

Koinig, G., Grath, E., Barretta, C., Friedrich, K., Vollprecht, D., & Oreski, G. (2022). Lifecycle Assessment for Recycling Processes of Monolayer and Multilayer Films: A Comparison. *Polymers*, 14(17), 3620–3620. DOI: 10.3390/polym14173620 PMID: 36080696

Kör, B., Krawczyk, A., & Wakkee, I. (2022). Addressing food loss and waste prevention. *British Food Journal*, 124(8), 2434–2460. DOI: 10.1108/BFJ-05-2021-0571

Kountouris, Y. (2022). The influence of local waste management culture on individual recycling behavior. *Environmental Research Letters*, 17(7), 074017. Advance online publication. DOI: 10.1088/1748-9326/ac7604

Lee, S., & Bi, X. (2019). Can adoption of pollution prevention techniques reduce pollution substitution. *PLoS One*, 14(11), 1–18. DOI: 10.1371/journal.pone.0224868 PMID: 31697762

Meyer, D. E., Li, M., & Ingwersen, W. W. (2020). Analyzing economy-scale solid waste generation using the United States environmentally-extended input-output model. *Resources, Conservation and Recycling*, 157, 104795–104795. DOI: 10.1016/j.resconrec.2020.104795 PMID: 32831477

Miah, M., Saha, K. B., & Karim, R. (2019). The Role of Private Commercial Bank in Corporate Social Responsibility with Reference to Islami Bank Bangladesh Limited. *OAlib*, 6(05), 1–14. DOI: 10.4236/oalib.1105422

Minelgaitė, A., & Liobikienė, G. (2019). Waste problem in European Union and its influence on waste management behaviours. *The Science of the Total Environment*, 667, 86–93. DOI: 10.1016/j.scitotenv.2019.02.313 PMID: 30826684

Moradi, S., & Sormunen, P. (2023). Revisiting the concept of waste and its causes in construction from analytical and conceptual perspectives. *Construction Management and Economics*, 41(8), 621–633. DOI: 10.1080/01446193.2023.2189278

Morcillo-Bellido, J., Santos-Iscoa, E., & Isasi-Sánchez, L. (2022). Circular Economy Analysis Applying Ellen MacArthur Model: Spanish Glass Sector Case. In *Ensuring Sustainability* (pp. 13–21). Springer., DOI: 10.1007/978-3-030-95967-8_2

Morrissey, A. J., & Browne, J. (2004). Waste management models and their application to sustainable waste management. *Waste Management (New York, N.Y.)*, 24(3), 297–308. DOI: 10.1016/j.wasman.2003.09.005 PMID: 15016418

Musarat, M. A., Alaloul, W. S., Hameed, N., Dhinaharan, R., Qureshi, A. H., & Wahab, M. M. A. (2022). Efficient Construction Waste Management: A Solution through Industrial Revolution (IR) 4.0 Evaluated by AHP. *Sustainability (Basel)*, 15(1), 274–274. DOI: 10.3390/su15010274

Park, H., Nam, K., & Lee, J. (2022). Lessons from aluminum and magnesium scraps fires and explosions: Case studies of metal recycling industry. *Journal of Loss Prevention in the Process Industries*, 80, 104872. DOI: 10.1016/j.jlp.2022.104872

Peng, X., Jiang, Y., Chen, Z., Osman, A. I., Farghali, M., Rooney, D. W., & Yap, P. S. (2023). Recycling municipal, agricultural and industrial waste into energy, fertilizers, food and construction materials, and economic feasibility: A review. *Environmental Chemistry Letters*, 21(2), 1–37. DOI: 10.1007/s10311-022-01551-5

Perey, R., Benn, S., Agarwal, R., & Edwards, M. (2018). The place of waste: Changing business value for the circular economy. *Business Strategy and the Environment*, 27(5), 631–642. DOI: 10.1002/bse.2068

Psomopoulos, S., Leligou, H. C., Ponci, F., Guerrero, J. M., Peñalvo-López, E., & Thanos, A. C. (2022). A Comprehensive Assessment of Products Management and Energy Recovery from Waste Products in the United States. *Energies*, 15(18), 6581–6581. DOI: 10.3390/en15186581

Puntillo, P. (2022). Circular economy business models: Towards achieving sustainable development goals in the waste management sector—Empirical evidence and theoretical implications. *Corporate Social Responsibility and Environmental Management*, 30(2), 941–954. DOI: 10.1002/csr.2398

Roshid, M., Waaje, A., Meem, T. N., & Sarkar, A. (2024). Logistics 4.0: A Comprehensive Literature Review of Technological Integration, Challenges, and Future Prospects of Implementation of Industry 4.0 Technologies. *International Journal of Technology, Knowledge and Society*, 20(1), 65–85. DOI: 10.18848/1832-3669/CGP/v20i01/65-85

Sassanelli, C., & Terzi, S. (2023). Circular Economy and Sustainable Business Performance Management. *Sustainability (Basel)*, 15(11), 8619–8619. DOI: 10.3390/su15118619

Sawalkar, R. S., Undale, S., Muluk, S., Mude, G., Saxena, V. D., & Pasumarti, S. (2023). Strategic waste management practices for environmental sustainability – a case of Indian university. *Management of Environmental Quality: An International Journal, ahead-of-print*(ahead-of-print). https://doi.org/DOI: 10.1108/MEQ-07-2022-0201/FULL/XML

Schindler, S., & Demaria, F. (2020). Garbage is gold: Waste-based commodity frontiers, modes of valorization and ecological distribution conflicts. *Capitalism, Nature, Socialism*, 31(4), 52–59. DOI: 10.1080/10455752.2019.1694553

Seier, M., Roitner, J., Archodoulaki, V. M., & Jones, M. P. (2023). Design from recycling: Overcoming barriers in regranulate use in a circular economy. *Resources, Conservation and Recycling*, 196, 107052–107052. DOI: 10.1016/j.resconrec.2023.107052

Senaratne, S., Rodrigo, N., Almeida, L. M. M. C. E., Perera, S., & Jin, X. (2023). Systematic review on stakeholder collaboration for a circular built environment: Current research trends, gaps and future directions. *Resources, Conservation & Recycling Advances*, 19, 200169. DOI: 10.1016/j.rcradv.2023.200169

Silva, A., Rosano, M., Stocker, L., & Gorissen, L. (2017). From waste to sustainable materials management: Three case studies of the transition journey. *Waste Management (New York, N.Y.)*, 61, 547–557. DOI: 10.1016/j.wasman.2016.11.038 PMID: 27955907

Souza Piao, R., Vincenzi, T. B., Vazquez-Brust, D. A., Yakovleva, N., Bonsu, S., & Carvalho, M. M. (2023a). Barriers toward circular economy transition: Exploring different stakeholders' perspectives. *Corporate Social Responsibility and Environmental Management*, 31(1), 153–168. DOI: 10.1002/csr.2558

Souza Piao, R., Vincenzi, T. B., Vazquez-Brust, D. A., Yakovleva, N., Bonsu, S., & Carvalho, M. M. (2023b). Barriers toward circular economy transition: Exploring different stakeholders' perspectives. *Corporate Social Responsibility and Environmental Management*, 31(1), 153–168. DOI: 10.1002/csr.2558

Tanchangya, T., Mia, M. R., Roshid, M. M., Chakma, J., Islam, F., & Rahman, M. H. (2024). Role of green financing and energy production to mitigate climate change in emerging economy like Bangladesh. *International Journal of Sustainable Energy and Environmental Research*, 13(2), 114–139. DOI: 10.18488/13.v13i2.3935

The World Bank. (2013). *Global Waste on Pace to Triple by 2100*. The World Bank. https://www.worldbank.org/en/news/feature/2013/10/30/global-waste-on-pace-to-triple

UNEP. (2024a). *Solid waste management*. UN Environment Programme. https://www.unep.org/explore-topics/resource-efficiency/what-we-do/cities/solid-waste-management

UNEP. (2024b). *Solid waste management*. UNEP - UN Environment Programme. https://www.unep.org/explore-topics/resource-efficiency/what-we-do/cities/solid-waste-management

van Zanten, H. H. E., Simon, W., van Selm, B., Wacker, J., Maindl, T. I., Frehner, A., Hijbeek, R., van Ittersum, M. K., & Herrero, M. (2023). Circularity in Europe strengthens the sustainability of the global food system. *Nature Food*, 4(4), 320–330. DOI: 10.1038/s43016-023-00734-9 PMID: 37117548

Waaje, A., Islam, J. M. S., & Akter, T. (2018). Study on Gas Sensor and Its Application as a IOT Device. *EWU Institutional Repository*. http://dspace.ewubd.edu:8080/xmlui/handle/2525/3056

Waaje, A., Yong, C., & Hossain, S. (2023). Attributes Influence Customers towards Ice Cream Purchase with Brand Preference and Income Effect on Purchasing Ice Cream: An Empirical Study on Dhaka City. *Global Journal of Management and Business Research*, 23(5), 1. https://www.researchgate.net/publication/378968239

White, P., Franke, M., & Hindle, P. (1995). Integrated Solid Waste Management: A Lifecycle Inventory. *Integrated Solid Waste Management: A Lifecycle Inventory*. https://doi.org/DOI: 10.1007/978-1-4615-2369-7

Wiprächtiger, M., Rapp, M., Hellweg, S., Shinde, R., & Haupt, M. (2022). Turning trash into treasure: An approach to the environmental assessment of waste prevention and its application to clothing and furniture in Switzerland. *Journal of Industrial Ecology*, 26(4), 1389–1405. DOI: 10.1111/jiec.13275

Zhang, B., Lai, K., Wang, B., & Wang, Z. (2018). Financial benefits from corporate announced practice of industrial waste recycling: Empirical evidence from chemical industry in China. *Resources, Conservation and Recycling*, 139, 40–47. DOI: 10.1016/j.resconrec.2018.07.019

Zhou, Y., İnce, F., Teng, H., Kaabar, M. K. A., Xu, J., & Yue, X. G. (2022). Waste management within the scope of environmental public awareness based on cross-sectional survey and social interviews. *Frontiers in Environmental Science*, 10, 1030525. Advance online publication. DOI: 10.3389/fenvs.2022.1030525

Chapter 20
Waste-to-Energy Innovations:
Evidence From the Tourism Industry

Mohammad Badruddoza Talukder
https://orcid.org/0009-0008-1662-9221
International University of Business Agriculture and Technology, Bangladesh

Dil Afrin Swarna
https://orcid.org/0009-0000-9176-920X
Independent Researcher, Bangladesh

ABSTRACT

The convergence of waste-to-energy innovations and the tourism industry presents a dynamic frontier with significant potential for sustainable development. As tourism continues to grow globally, the environmental footprint of waste generated by this sector is becoming increasingly concerning. Waste-to-Energy (WtE) technologies, with the crucial role of technology providers, offer a promising solution by converting waste into usable energy, thereby reducing environmental impact and promoting a circular economy. This paper explores future trends and opportunities in integrating WtE systems within the tourism industry. By embracing WtE innovations, the tourism industry can significantly reduce its waste burden, lower energy costs, and contribute to a more sustainable future. This chapter underscores the necessity of multidisciplinary collaboration and supportive regulatory frameworks to fully realise the potential of WtE technologies in transforming the environmental landscape of tourism.

DOI: 10.4018/979-8-3693-6110-8.ch020

1. INTRODUCTION

An unparalleled boom in the tourist sector, a vital pillar of the world economy, contributes to a marked rise in waste production and heightened environmental concerns (Adeniyi et al., 2024). Conventional methods cannot manage the increasing volume of waste. Thus, creative and long-lasting solutions must be found. Technologies such as waste-to-energy (WtE) transform waste materials into valuable energy, offering a potential solution to these urgent problems (Zreik, 2024). The tourism industry may reduce its carbon footprint, modernise waste management procedures, and support a circular economy using WtE technologies (Bitok & Shirandula, 2024). Numerous causes, including stricter legal frameworks, the economic benefits of renewable energy, and growing environmental consciousness among visitors, drive this shift towards sustainability (Smolčić, 2024). Travelers are becoming more environmentally concerned, looking for locations and lodging that value environmental conservation. As a result, green practices are encouraged for tourism operators, and WtE innovations provide a workable solution to satisfy these demands (Ilieva et al., 2023). Using WtE solutions in the tourism industry involves various technologies, including biological techniques like anaerobic digestion and sophisticated thermal processes like gasification and pyrolysis. These solutions reduce operating expenses and dependency on fossil fuels by producing renewable energy that may be utilised to power tourism facilities and mitigate the environmental impact of garbage (Olshanska et al., 2021).

This chapter explores WtE innovations' potential and future developments in the travel and tourism sector. The most recent developments in waste-to-energy technologies are examined, together with the function of intelligent waste management systems and the possibility of decentralised energy production (Kariru, 2023). It also looks at the advantages of using WtE solutions, such as better economic results for tourism operators and more excellent sustainability credentials. The tourist sector can utilise WtE technology to promote a more resilient and sustainable future using a multidisciplinary strategy and a supporting regulatory framework (Kumar & Barua, 2024).

2. WASTE-TO-ENERGY (WtE) TECHNOLOGY

Waste-to-energy (WtE) technology integration is becoming increasingly popular in the tourism sector, with studies showing both the advantages and disadvantages of doing so (Chissom, 2024). This literature review examines the developments in WtE technologies, their environmental effects, and their financial ramifications for the travel and tourist industry (Haris & Ningsih, 2020). Improvements in biological and thermal processes have been the main focus of recent WtE technology developments. According to Dolynska et al. (2023), gasification and pyrolysis are becoming more popular as effective ways to turn trash into energy because they emit fewer emissions than conventional incineration. These methods work well with the many wastes tourism produces, including packaging materials and food trash (Kumar & Barua, 2024). Another promising technique is anaerobic digestion, which produces biogas to produce heat and power from organic waste (Talukder, 2024). These technologies provide essential byproducts like digestate and biochar, which are helpful in landscaping and agriculture and reduce the volume of garbage produced.

There are significant environmental advantages to integrating WtE technologies into the tourism industry (Lagioia et al., 2024). Through the diversion of garbage from landfills and the decrease in the usage of fossil fuels, these activities significantly reduce greenhouse gas emissions.As per Dighliya and Dahiya (2024). Adopting WtE treats trash as a resource instead of a burden, which aligns with the principles of the circular economy. (Prudeaux, 2020). This strategy is especially pertinent to the tourism sector, frequently chastised for its environmental impact. Tourism businesses can decrease their influence on nearby ecosystems and help achieve broader sustainability goals by turning garbage into electricity (Battour et al., 2024).

A critical research topic is the economic viability of WtE technology in the tourism industry. Long-term benefits frequently outweigh the costs of WtE infrastructure, even though initial investments can be substantial (Ke, 2024). Waste-to-energy (WtE) plants can lower tourism operators' energy and waste management expenses. By using the energy generated, tourism facilities can be powered, increasing economic efficiency. According to financial models, the tourism industry can be encouraged to embrace WtE technology by utilising public-private partnerships and creative funding structures such as carbon credits and green bonds (Killa & Tawa, 2024).

One crucial development in WtE technology applications is the rise of smart waste management systems (El Archi & Benbba, 2024). These systems maximise garbage collection, sorting, and processing through Internet of Things (IoT) devices and data analytics. Innovative solutions increase the effectiveness of Waste Engineering (WtE) processes and reduce operating costs by offering real-time data on waste generation and composition (Srinivasan et al., 2024). To successfully apply WtE technology in the tourism industry and create more economical and sustainable waste management practices, smart waste management is essential (Havrylova, 2019).

While WtE technologies have great potential, several issues need to be resolved. The main impediments to widespread adoption include technological constraints, regulatory restrictions, and public perception (Gorina et al., 2022). Nonetheless, noteworthy prospects exist via technological advancement, regulatory backing, and growing consciousness of environmental issues among stakeholders and tourists (Demir, 2024). Cooperation between governments, travel agencies, and technology companies is crucial to overcome these obstacles and realise the full potential of WtE solutions (Talukder, 2021).

The revolutionary potential of these technologies is highlighted in the literature on WtE advances in the tourism industry (Talukder & Kumar, 2024). Developments in biological and thermal processes and intelligent waste management systems offer a workable solution to lessen environmental effects and improve economic sustainability (Badruddoza et al., 2024). Although there are still obstacles to overcome, there are many chances to use WtE innovations to build a more sustainable tourism sector. To promote the use of technology, future research should concentrate on creating scalable WtE solutions, encouraging interdisciplinary collaboration, and creating supportive regulatory frameworks (Talukder et al., 2024).

3. THE HISTORICAL EVOLUTION OF WASTE MANAGEMENT IN TOURISM

How waste management has changed in the tourism sector reflects more significant cultural changes toward sustainability and environmental consciousness (Pasternak, 2024). Historically, landfilling and crude recycling have been the primary methods the tourism sector uses to handle garbage with little regard for the environmental effects. This method was primarily motivated by cost-effectiveness and convenience rather than sustainability (Vydaichuk et al., 2024). However, when these procedures' adverse environmental effects grew more prominent, the industry started looking for new environmentally friendly ways to manage garbage (Beltramo et al., 2024).

3.1 Early Practices and Challenges

Waste management was not a big concern in the early phases of tourism development. According to Muqsit et al. (2024), tourist destinations frequently depended on local municipal trash systems, which usually involved straightforward disposal techniques. The main option involved little to no recycling or garbage separation and disposal in landfills. Significant environmental deterioration resulted from this, especially in well-known tourist locations with heavy trash generation. According to Martínez-Navarro et al. (2024), inappropriate garbage disposal has resulted in severe pollution problems in coastal zones, affecting local ecosystems and marine life.

3.2 The Rise in Environmental Awareness

There was a dramatic change in the 1970s and 1980s as environmental consciousness increased everywhere. Waste management techniques have been reevaluated in light of the release of papers like "The Limits to Growth" by the Club of Rome and mounting scientific evidence of pollution and resource depletion (Herman et al., 2024). The first initiatives to encourage recycling and trash reduction in the tourism industry were launched at this time. According to Tsypko et al. (2024), hotels and resorts began introducing rudimentary recycling programs, frequently concentrating on easily recyclable items like wood, glass, and aluminium.

3.3 The 1990s: Towards Sustainable Tourism

In the 1990s, "sustainable tourism" gained popularity, highlighting the necessity of eco-friendly methods that reduce adverse effects on travel locations (Mavrin, 2024). Due to market demand for more environmentally friendly travel options and regulatory constraints, waste management became a primary priority. Sustainable waste management methods have been promoted by international organisations, including the International Hotel and Restaurant Association (IH&RA) and the United Nations World Tourism Organization (UNWTO) (Saarinen, 2021). During this time, programs like the Green Key and Blue Flag certifications were introduced, which pushed travel agencies to use more environmentally friendly waste management techniques. According to Brunet (1999), these certificates offered a framework for more efficient waste management, which included recommendations for recycling, trash separation, and cutting waste production at its source.

3.4 Early 2000s: Integration of Advanced Technologies

More sophisticated waste management methods were introduced to tourism in the early 2000s. Anaerobic digestion and composting are two examples of mechanical and biological treatment techniques gaining traction (Baranova, 2024). These technological advancements made it possible to treat organic garbage more effectively, decreasing the amount of waste dumped in landfills and producing valuable byproducts like compost and biogas (Cao, 2023). Furthermore, waste-to-energy (WtE) started to gain traction. Pilot projects and small-scale implementations proved the potential of waste-to-energy (WtE) technologies to generate renewable energy and reduce waste (Elsayed, 2024). These early initiatives established the foundation for a wider acceptance of WtE solutions in the tourism industry.

3.5 Recent Trends: Embracing Circular Economy Principles

With an emphasis on closing the resource use loop, the tourism sector has increasingly adopted the circular economy principles in recent years. According to Kiaušienė et al. (2024), this strategy strongly emphasises waste prevention, reuse, and recycling, with WtE technologies playing a critical role. The transition to circular economy models has been fueled by financial incentives, legal requirements, and rising consumer demand for eco-friendly travel options (Constance Obiuto et al., 2024). Top tourism operators are now incorporating modern anaerobic digestion systems, gasification, pyrolysis, and other cutting-edge WtE technologies into their waste management plans. According to Rahmawati and Novani (2024), these technologies support global sustainability goals by offering a renewable energy source and assisting with more sustainable waste management. According to Kumar et al. (2024), the development of waste management in the tourism industry shows a shift from basic disposal techniques to complex, technologically advanced solutions targeted at sustainability. An important turning point in this evolution is the growing use of waste-to-energy technologies, which provide a workable solution to the twin problems of waste management and renewable energy production (Džajić et al., 2024). Integrating WtE solutions will be crucial to encouraging sustainable practices and reducing environmental effects as tourism expands.

4. WASTE-TO-ENERGY (WtE) TECHNOLOGIES IN TOURISM

There is growing pressure on the tourism sector, recognised for making a significant economic contribution to the world, to address its environmental impact (Nanda et al., 2023a). There are considerable sustainability issues since as the number of tourists increases, so does the amount of waste produced. The industry's numerous and expanding waste streams make traditional waste management techniques—which frequently depend on landfilling and rudimentary recycling—increasingly insufficient (Wicaksono & Triwigati, 2023). As a result, there is an urgent need for creative solutions to revolutionise waste management and support a more environmentally friendly travel and tourism sector. In this context, waste-to-energy (WtE) technologies are a promising area. These technologies transform waste materials into valuable energy, such as fuel, heat, and electricity, through combustion, gasification, pyrolysis, and anaerobic digestion (Sharma, 2023). WtE technologies help achieve the aims of energy sustainability and trash reduction by turning garbage into a helpful resource that generates renewable energy and reduces the amount of waste sent to landfills (Jamilatun et al., 2023). The use of WtE technology in the travel and tourist sector is consistent with more significant global movements towards sustainability and the circular economy. Today's travellers are more concerned about the environment and expect their travel destinations to follow more sustainable policies (Talukder et al., 2023). WtE technologies provide a workable and significant solution to the pressing need for tourism operators to embrace greener practices due to the shift in customer preferences. Furthermore, WtE is a desirable alternative for tourism enterprises due to the financial advantages of lower energy prices and waste management expenses (Das et al., 2024).

An overview of WtE technologies and their applicability to the travel and tourist sector is given in this chapter. It examines the various WtE process types, their uses, and how tourism operators might profit from them (Themelis, 2023). The chapter also lays the groundwork for a thorough analysis of the regulatory, economic, and environmental issues of WtE adoption in later sections of this research (Mertzanakis et al., 2024). Stakeholders in the tourism sector might have a deeper appreciation for the transformative power of WtE technologies in promoting sustainable tourism practices by learning about its foundations and possible effects (Prebilič et al., 2024).

4.1 Technological Advancements In Waste-To-Energy Solutions

Significant technological developments have been made in the waste-to-energy (WtE) industry, providing creative ways to turn garbage into sound energy (Jaiswal et al., 2024). The tourism sector, which produces much garbage and is under increasing pressure to adopt sustainable practices, needs these developments. This chapter examines Several WtE technologies, emphasising their guiding principles, uses, and possible advantages (Sukla et al., 2024).

4.1.1 Thermal Processes

One of the earliest and most well-known WTE technologies is incineration. High temperatures are used to burn trash, producing heat, flue gas, and ash. According to Usman et al. (2024), we use the heat produced to create steam, which powers turbines to produce electricity. Modern incinerators are outfitted with sophisticated pollution control technologies to reduce the amount of hazardous pollutants they emit. These systems significantly lessen the discharge of dioxins, furans, and particulate matter, including scrubbers, filters, and catalytic converters (Kushwaha et al., 2024). According to Talukder et al. (2023), gasification is a sophisticated thermal process that converts carbonaceous materials from plants or fossils into syngas, a mixture of carbon dioxide, hydrogen, and monoxide, using a regulated amount of oxygen. Syngas can be used as a chemical feedstock to make synthetic fuels or power electric vehicles. Compared to conventional incineration, gasification has many benefits, such as increased energy efficiency, reduced emissions, and the capacity to manage various waste materials (Dawar & Negi, 2024). Pyrolysis is breaking down organic compounds by heat at high temperatures without oxygen. Char, syngas, and bio-oil are produced during this process. Char can be utilised as activated carbon or a soil amendment; syngas can generate energy; and bio-oil can be converted into renewable fuels. Pyrolysis is a flexible solution for the tourism industry since it works exceptionally well for processing trash with a high energy content, such as plastics (Manurung, 2024).

4.1.2 Biological Processes

Anaerobic digestion (AD) is a biological process that breaks down organic waste in the absence of oxygen, resulting in the production of digestive (a residue rich in nutrients) and biogas (a mixture of carbon dioxide and methane). While digestate can be fertiliser, biogas can generate heat and power (Khan, 2023). Food waste processing, prevalent in the tourism sector, is a perfect fit for AD. According to Raman et al. (2024), contemporary AD systems' architecture maximises biogas production and guarantees the effective use of both biogas and digestate. Microorganisms use a biological process called fermentation to turn organic resources like food scraps and agricultural waste into ethanol and other biofuels. Ethanol can be used as a chemical feedstock or as fuel. The development of more effective microbial strains and process optimisation approaches due to advancements in fermentation technology has increased the viability of this WtE strategy (Le et al., 2023).

4.1.3 Smart Waste Management Systems

Incorporating IoT technology into these systems has completely transformed waste management operations. IoT devices offer real-time data on garbage generation, composition, and bin fill levels. These devices include smart bins and sensors (Agre et al., 2024). Using this data, we can improve sorting effectiveness, optimise garbage collection routes, and ensure that the best WtE technology is used for waste processing. IoT integration improves the tourism sector's sustainability and efficiency of trash management (Abdullahi et al., 2024). AI and data analytics are becoming more and more crucial to WtE solutions. These technologies can analyse large amounts of data from waste management systems to spot trends, forecast waste creation, and improve waste-to-energy operations. Additionally, waste sorting and separation can be automated by AI-powered systems, maximising energy recovery and enhancing feedstock quality for WtE processes (Wirani et al., 2024).

4.1.4 Decentralized WTE Systems

One notable development in the industry is the rise of decentralised WtE systems. It is feasible to install these more compact, modular units near locations where trash is generated, such as resorts, hotels, and tourist destinations (Wirani et al., 2024). Decentralised waste-to-energy (WtE) systems facilitate local energy production, cut operating costs, and eliminate the need for waste transportation. According to Srivastava et al. (2024), they are especially well-suited for isolated or off-grid tourist locations where conventional waste management and electricity infrastructure may be missing.

4.1.5 WtE Integration and Intelligent Waste Management Systems

Tourism is one of the industries that innovative trash management systems (SWMS) are altering through cutting-edge technology to optimise garbage collection, sorting, and processing (Shah & Sarif, 2023). Waste-to-Energy (WtE) technology integration with SWMS increases operational efficacy, efficiency, and sustainability. The advantages, challenges associated with implementation, and prospective uses of the synergy between SWMS and WtE technologies are all covered in this chapter (Wicaksono & Triwigati, 2023). Because it connects garbage cans, sensors, and collection trucks to a single, central platform, IoT is crucial to SWMS. Real-time fill levels, temperature, and location data allow cities and businesses to improve overall service efficiency, optimise waste collection routes, and save costs. Trash production in the tourism industry varies geographically and seasonally. Prebilič et al. (2024) state that IoT-enabled SWMS provide vital information for timely waste management decisions. By optimising the collection and handling of waste, SWMS enhances the efficacy of Waste Management (WtE) technology. By monitoring fill levels and waste composition, SWMS can schedule maintenance for WtE facilities and prioritise collection routes based on actual demand. This proactive approach enhances operational sustainability and energy recovery by minimising downtime and maximising the utilisation of WtE assets (Nubi et al., 2024). To identify patterns, predict waste generation trends, and improve WtE operations, massive volumes of data collected by SWMS are analysed using data analytics and AI algorithms (Chandre et al., 2024). Predictive maintenance models help WtE facilities avoid equipment failures, reduce downtime, and save maintenance costs. AI-driven sorting systems enhance the quality of waste feedstock for WtE, maximising resource and energy recovery.

4.2 Future Trends In Waste-To-Energy (WtE) Technology and Innovations

Waste-to-Energy (WtE) innovations have great potential to change waste management procedures and solve sustainability issues in various industries, including tourism. According to Michael Becidan et al. (2015), this chapter examines WtE technology trends and opportunities, stressing developments, possible uses, and ramifications for the travel and tourism industry.

4.2.1 Advanced Gasification and Pyrolysis

Gasification and pyrolysis processes are continually developing, providing more effective and adaptable ways to turn various waste streams into valuable energy products. Catalyst science, reactor design, and process optimisation developments raise energy recovery rates while lowering emissions and leftover waste. Tourist locations can effectively handle polluted and mixed trash with this technology (Downmore et al., 2022).

4.2.2 Enhanced Anaerobic Digestion

Anaerobic digestion systems (AD) are evolving to become more complex and capable of processing organic waste products more effectively. Pre-treatment innovations like heat and enzymatic hydrolysis raise the yields of biogas production and enhance the quality of digestate. According to Agha et al. (2024), AD is still a feasible choice for tourism businesses looking for environmentally friendly ways to handle agricultural and food waste.

4.2.3 Integration of Advanced Sorting Technologies

Incorporating AI-driven robotic sorting systems improves the efficiency of waste management processes by increasing trash segregation and recycling rates. Automated sorting methods can recognise and collect valuable elements from mixed waste streams to maximise feedstock quality for WtE operations and encourage resource recovery (Tian et al., 2024).

5. THE EFFECTS OF WASTE-TO-ENERGY INNOVATIONS ON THE ENVIRONMENT

Waste-to-energy (WtE) technologies have enormous potential to drastically lessen the environmental effect of the tourism industry (Lesnicenonka & Blumberga, 2024). In addition to converting waste into valuable energy, these revolutionary technologies address several environmental problems (Jaghoub et al., 2024). With a focus on resource conservation, waste reduction, pollution control, and greenhouse gas emissions, this chapter explores the numerous environmental advantages of WtE breakthroughs. It inspires hope for a brighter future (Usmanetal., 2024). The capacity of WtE technologies to drastically cut down on trash going to landfills is one of its main environmental advantages. Conventional waste disposal techniques, including landfilling, deteriorate the soil, contaminate groundwater, and release methane, a potent greenhouse gas (Sukla et al., 2024). However, trash can be converted into energy and byproducts using WtE processes, including incineration, gasification, and anaerobic digestion, which reduces the amount of garbage that needs to be landfilled (Amissah et al.,2024). Now, let us explore the particular WtE procedures and how they help the environment. One way to reduce the amount of municipal solid waste that ends up in landfills is through incineration, which can reduce it by up to 90%.

On the other hand, gasification and pyrolysis minimise trash and generate syngas and bio-oil for energy production, lessening the need for landfills. Another method that produces biogas from organic waste and reduces the amount of garbage going to landfills is anaerobic digestion (AD) (Nwokediegwu & Ugwuanyi, 2024). It reduces greenhouse gas emissions. WtE technologies are essential for reducing greenhouse gas (GHG) emissions and provide a viable route to producing sustainable energy. These methods considerably minimise carbon dioxide emissions from conventional energy sources by stopping methane development in landfills. This greenhouse gas is 25 times more potent than carbon dioxide over 100 years. They also replace fossil fuels with energy created through WtE processes.

Reducing our dependency on fossil fuels is a significant step towards a more sustainable and environmentally friendly energy future, and WtE technologies are crucial in this regard. Anaerobic digestion, for example, creates biogas, which may be used to produce electricity and heat, eliminating the need for fossil fuels and reducing greenhouse gas emissions. Similarly, we can lessen our carbon footprint by feeding energy from gasification and incineration operations into the grid to replace power generated by burning coal, natural gas, and other fossil fuels (Do et al., 2024).

Contemporary WtE technologies are built with sophisticated pollution control systems to lower the emission of harmful pollutants. These technologies protect public health and air quality by guaranteeing that emissions from WtE plants remain subject to strict legal constraints (Phonthanukitithaworn et al., 2024). Gasification and incineration plants use sophisticated scrubbers and filters to absorb and neutralise harmful gases, including sulfur dioxide, nitrogen oxides, and volatile organic compounds. Catalytic converters: These devices ensure that WtE plants follow stringent environmental requirements by reducing the discharge of combustion byproducts such as dioxins and furans.

6. OPPORTUNITIES FOR WtE IN TOURISM

Sustainable waste management, mainly through Waste-to-Energy (WtE) technologies, offers notable environmental benefits, especially within the tourism sector. By converting waste into renewable energy, tourist destinations can significantly reduce the volume of waste sent to landfills, thereby lessening their overall environmental footprint. This approach helps preserve the natural attractions vital to the appeal of many tourist locations and also plays a critical role in lowering carbon emissions, contributing to broader environmental health. As a renewable energy source, WtE provides a dependable alternative for tourist facilities, reducing their reliance on fossil fuels. This is especially advantageous for remote or off-grid tourist destinations where conventional energy sources may be limited or expensive. These areas can improve their infrastructure by harnessing energy from waste, making them more appealing to visitors.

From an economic perspective, implementing WtE systems can result in significant cost savings for tourism businesses by lowering waste disposal and energy costs. Moreover, establishing and operating WtE plants generates local employment opportunities and supports the economy. Tourism businesses that integrate WtE technologies also stand to enhance their corporate social responsibility (CSR) profiles, bolstering their reputation among stakeholders and environmentally conscious tourists. Destinations that adopt WtE technologies can market themselves as eco-friendly, attracting visitors who prioritise sustainability. Additionally, WtE facilities can serve as educational attractions, offering tourists insights into sustainable practices and energy production, enriching their overall visitor experience and promoting green tourism.

The scalability and flexibility of decentralised WtE systems make them particularly suitable for isolated or island tourist locations that often struggle with inadequate waste management infrastructure. By customising modular WtE units to align with the specific waste volumes and energy needs of tourism facilities, these destinations can reduce their dependence on external energy and waste disposal sources (Romianingsih, 2023). WtE technologies also play a pivotal role in advancing circular economy principles within the tourism industry. By treating waste as a resource, tourism operators can minimise waste generation, optimise energy recovery, and close the loop on resource use, enhancing their sustainability credentials and meeting the growing demand for eco-friendly travel experiences (Wickasono & Triwigati, 2023).

Furthermore, integrating Smart Waste Management Systems (SWMS) enabled by the Internet of Things (IoT) can streamline real-time waste collection and processing. IoT-enabled sensors in WtE plants monitor waste composition, fill levels, and operational performance, facilitating data-driven maintenance and decision-making. This boosts operational efficiency, reduces costs, and improves environmental outcomes (Nubi et al., 2024). By embracing these innovative approaches, the tourism sector can significantly advance its commitment to sustainability while enhancing tourist destinations' attractiveness and economic viability.

7. APPROACHES TO WtE IN TOURISM

7.1 Policy and Regulatory Support

Incentives and supportive policy frameworks are essential for hastening the tourism sector's adoption of WtE technologies. Governments and regulatory agencies can encourage investment in sustainable waste management practices by offering grants, financial incentives, and expedited approval procedures for WtE projects. Clear environmental standards and emissions restrictions guarantee that WtE plants will function ethically and fulfil strict environmental targets (Verma et al., 2021).

7.2 Research and Development Initiatives

Enhancing the effectiveness, sustainability, and financial feasibility of WtE technologies is the primary goal of ongoing research and development projects (Kabir et al., 2022). Innovation is fueled by cooperative relationships between government organisations, industrial players, and academic institutions in reactor design, waste pre-treatment, energy conversion efficiency, and environmental performance measures. Emerging technologies such as microbial fuel cells and plasma arc gasification could expand the applicability of WtE in tourism and other fields (Nanda et al., 2023b).

7.3 Collaborative Efforts and Multidisciplinary Approaches in Waste-to-Energy Innovations

Multidisciplinary methods and cooperative efforts are necessary to effectively integrate Waste-to-Energy (WtE) developments in the tourism sector (Ramanathan et al., 2024). This chapter emphasises the value of collaboration among stakeholders from different industries and disciplines to develop sustainable WtE solutions. It highlights difficulties, best practices, and synergies (Nwokediegwu & Ugwuanyi, 2024).

7.4 Stakeholder Approach in WtE Adoption

Effective collaboration involves active participation and coordination among diverse stakeholders:

- Government and Regulatory Bodies: Creating laws, rules, and rewards that encourage using WtE, simplifying project approvals, and guaranteeing adherence to environmental norms (Abdulai et al., 2024).
- Industry and tourism operators must integrate WtE technologies into operational procedures. Amoo et al. (2020) suggest investing in infrastructure and technological changes for sustainable waste management and energy generation.
- Research and academic institutions are involved in advancing WtE technology through research and development, offering proficiency in energy management, environmental science, and engineering (Srivastava et al., 2024).
- Technological innovators and providers: creating and promoting WtE solutions. Working with end users to adapt technology for travel uses (Gray et al., 2024).

7.5 A Multidisciplinary Approach to WtE Solutions

WtE innovations benefit from a multidisciplinary approach that combines expertise from various fields:

- Engineering and Technology: Creating and refining WtE systems to ensure dependability and efficiency. Creating cutting-edge procedures like gasification, pyrolysis, and anaerobic digestion that are specific to the demands of the travel and tourism sector (Kromp et al., 2023).
- Environmental Science: Evaluating how WtE technologies affect the environment. To ensure sustainability targets are being met, waste residues, water quality, and air emissions are monitored (Mi, 2024).
- Economics and Business: Financial modelling and cost-benefit analysis. Assessing WtE projects' economic feasibility and locating sources of funding and investment opportunities (Amissah et al., 2024).
- Policy and Governance: Developing laws and other structures that support the implementation of WtE. Removing institutional and legal ob-

stacles to enable project operation and implementation (Hertati & Arif, 2022).

Collaborative efforts in implementing Waste-to-Energy (WtE) technologies face several challenges that require proactive solutions. One key challenge is interdisciplinary communication, which involves encouraging stakeholders from various fields of expertise to communicate and share knowledge effectively. This requires clearly defining roles, responsibilities, and communication pathways to ensure a smooth flow of information among all parties involved (Phonthanukitithaworn et al., 2024). Establishing these clear lines of communication can help bridge gaps between different disciplines, fostering a more cohesive and effective collaboration.

Another significant challenge is resource allocation, which involves securing the necessary funding and materials to support WtE initiatives. This can be achieved by tapping into various financial sources such as grants, international funding agencies, and investments from both public and private sectors. Effective resource management ensures that WtE projects have the support they need to succeed and can sustain operations over the long term. Technological integration is also a critical hurdle, as it involves overcoming technical obstacles to incorporate WtE technologies within existing infrastructures. Ensuring the interoperability, scalability, and operational efficiency of WtE systems across diverse tourism settings is essential for their successful implementation (Nadeau & Koebele, 2023). Addressing these technological challenges requires careful planning and the development of adaptable solutions that can fit the specific needs of different destinations, ultimately enabling WtE technologies to thrive within the tourism industry.

8. FUTURE DIRECTIONS

To promote the sustainable adoption of Waste-to-Energy (WtE) technologies in the tourism sector, stakeholders must navigate a complex landscape of challenges and opportunities. WtE presents a valuable solution for managing waste while generating renewable energy, making it a crucial component of sustainable tourism development. However, successful implementation requires careful consideration of various factors. The following suggestions provide actionable steps for stakeholders to overcome these challenges and foster the effective adoption of WtE in tourism destinations.

8.1 Capacity Building

Investing in skill development and workforce training for Waste-to-Energy (WtE) maintenance and operations is crucial for successfully implementing and sustaining these technologies in the tourism sector. According to Kurhayadi (2021), building the capacity of local communities and workers ensures a skilled workforce capable of operating and maintaining WtE facilities efficiently. This enhances the reliability and effectiveness of WtE systems and empowers local populations, providing them with valuable skills and employment opportunities. By prioritising capacity building, tourism destinations can ensure the long-term success and sustainability of their WtE initiatives.

8.2 Innovation Hubs

Creating innovation hubs and living labs to test and display WtE technology in authentic tourism environments is an effective way to demonstrate the potential and benefits of these systems. Theilen et al. (2022) suggest that these hubs serve as experimental platforms where new technologies can be developed, tested, and showcased. In the context of tourism, such hubs can provide real-world insights into how WtE systems perform and integrate with existing infrastructure. They also offer an educational opportunity for tourists and local stakeholders to learn about WtE technologies and their environmental benefits, fostering greater acceptance and support for sustainable practices.

8.3 Policy Advocacy

Promoting favourable policies and regulatory frameworks is essential for the widespread adoption of WtE technologies. Fredericks (2023) emphasises the need for local, national, and international advocacy to ensure that WtE is recognised and prioritised in sustainable development plans. By urging governments to incorporate WtE into their policy agendas, stakeholders can help create an enabling environment that supports the growth and scalability of WtE initiatives. Effective policy advocacy can lead to the establishment of incentives, subsidies, and regulations that facilitate the adoption of WtE technologies, making them more accessible and economically viable for tourism businesses and communities.

8.4 Platforms for Integrated Waste Management

Provide unified digital platforms with user-friendly interfaces integrating Smart Waste Management Systems (SWMS). These platforms may provide real-time information on waste generation, collection schedules, and energy output from WtE plants. Tourism businesses and local governments can use this data to streamline processes, boost productivity, and encourage visitors to adopt sustainable behaviours.

8.5 Eco-Tourism Bundles Including WtE Activities

Make educational workshops on sustainable waste management or guided visits to WtE facilities part of eco-tourism packages. Visitors can learn firsthand the waste-to-energy conversion process, promoting environmental stewardship awareness and respect. These events encourage tourists to adopt sustainable practices while educating them.

8.6 WtE Initiatives Based in the Community

Implement participatory planning and decision-making techniques to encourage community participation in WtE projects. Create neighbourhood WtE cooperatives where local companies and citizens can provide organic waste for composting or anaerobic digestion. Involving communities in WtE management and benefits can promote sustainability and a sense of ownership in waste reduction initiatives.

8.7 Transportable WtE Units for Organizing Events

Install mobile WtE devices in tourist areas to manage garbage produced during significant events and festivals. These operations can employ Advanced technologies like compact anaerobic digestion systems or mobile incinerators. They lower transportation emissions, offer on-site trash management options, and encourage environmentally friendly event-hosting techniques.

8.8 Green Accommodation Certification Initiatives

Implement or improve green certification schemes that incentivise lodging establishments to use sustainable waste management techniques and WtE technologies. A certification can showcase energy conservation, waste minimisation, and accomplishments in environmental performance. This encourages travel agencies to invest in WtE solutions and promote their eco-friendly reputations to environmentally aware customers.

8.9 Joint Research and Development Initiatives

To develop WtE technologies suited explicitly to tourism environments and promote cooperative research projects, including academic institutions, industry partners, and local governments. Concentrate on creating scalable and affordable solutions for various waste streams while considering regional environmental circumstances, seasonality, and visitor behaviour. These collaborations quicken the pace of technical development and real-world applications in the travel industry.

8.10 Reward Programs for WtE Infrastructure Purchases

Provide travel agencies and local governments who invest in WtE infrastructure subsidies, tax exemptions, or other financial incentives. By generating jobs and reducing energy use, these incentives can accelerate the adoption of WtE technology, minimise initial capital costs, and boost economic growth. Public-private partnerships can use these incentives to create long-term, sustainable trash management initiatives.

9. CONCEPTUAL FRAMEWORK

Figure 1. Conceptual framework

Source: Authors' Compilation

Figure 1 depicts a conceptual framework for incorporating Waste-to-Energy (WtE) technologies within the tourism industry, emphasising various essential elements contributing to sustainable practices. This model's heart is WtE, a crucial solution for transforming waste into renewable energy, addressing the escalating demand for eco-friendly waste management in tourism. This integration enhances waste management and aligns with the broader objectives of diminishing environmental impact and fostering sustainability in tourism destinations. One of the primary relationships in the model is between WtE and sustainable tourism. WtE technologies are pivotal in bolstering tourism operations' sustainability by minimising waste and generating clean energy. This supports environmentally responsible tourism practices, which are increasingly crucial to environmentally conscious travellers seeking destinations that prioritise sustainability. By adopting WtE, tourism businesses can decrease their reliance on conventional waste disposal methods, such as landfilling, and diminish their carbon footprint, thereby making a positive environmental impact.

The framework places WtE within a circular economy, where waste is considered a valuable resource rather than a disposal issue. This integration encourages tourism businesses to embrace sustainable waste management practices and close the loop on resource use. By treating waste as a valuable asset, WtE contributes to a more sustainable approach to resource management, which benefits both the environment and the economy.

Smart waste management is another critical aspect of the framework, emphasising cutting-edge technologies, such as IoT-enabled systems, to optimise waste collection, sorting, and processing. These advanced technologies improve the efficiency and effectiveness of WtE operations, allowing for real-time monitoring and data-driven decision-making. This enhances operational performance and reduces costs and environmental impact, making WtE a more viable and attractive option for various tourism settings.

Lastly, the framework highlights WtE's potential to generate renewable energy, showcasing its dual role as a waste management and energy production solution. By harnessing energy from waste, WtE reduces reliance on fossil fuels, supports energy resilience, especially in remote or off-grid destinations, and aligns with global renewable energy objectives. This comprehensive approach to waste management and energy production makes WtE a solution that addresses multiple sustainability challenges in the tourism sector.

Overall, this conceptual framework illustrates how WtE can be effectively integrated into the tourism industry, linking it with sustainable tourism practices, environmental impact reduction, circular economy principles, Smart waste management, and renewable energy generation. It provides a holistic approach to sustainable tourism development, showcasing the multifaceted benefits of WtE technologies.

10. CONCLUSION

In summary, waste-to-energy (WtE) technologies in the tourism sector offer a way forward for environmentally conscious and sustainable development. By adopting cooperative endeavours, incorporating interdisciplinary methodologies, and utilising technological progress, stakeholders can efficaciously tackle waste management predicaments while mobilising renewable energy resources. The suggestions, which range from integrated waste management systems to community-based programs and eco-tour packages, highlight the potential of WtE to improve visitor engagement and operational efficiency in tourist areas while reducing environmental impact.

Governments, travel agencies, tech companies, and local communities need to collaborate and make investments in creative solutions in the future. Environmental responsibility, cost-effectiveness, and scalability should be prioritised in these endeavours. Furthermore, encouraging WtE infrastructure investments and creating a welcoming regulatory framework would be essential to hastening the global acceptance and realisation of sustainable tourism practices. By doing this, the tourism sector may set an example for others in advancing the circular economy, reducing the effects of climate change, and guaranteeing a more environmentally friendly future for coming generations.

REFERENCES

Abdulai, I. A., Fuseini, M. N., & File, D. J. M.-B. (2024). Making cities clean with collaborative governance of solid waste infrastructure in Ghana. *Cleaner Waste Systems*, 8, 100150. DOI: 10.1016/j.clwas.2024.100150

Abdullahi, A., Mohammed, A., Bonet, M. U., El-Suleiman, A., Ahmad, R. B., & Chollom, T. D. (2024). Development of a Smart Waste Management System with Automatic Bin Lid Control for Smart City Environment. *EAI Endorsed Transactions on Smart Cities*, 7(3). Advance online publication. DOI: 10.4108/eetsc.4385

Adeleke, A. K., Danny, J. P. M., Olu-lawal, K. A., & Olajiga, O. K. (2024). Process development in mechanical engineering: Innovations, challenges, and opportunities. *Engineering Science & Technology Journal*, 5(3), 901–912. DOI: 10.51594/estj.v5i3.945

Agatha, N. P., Harvan, A. A., & Prasetya, F. R. (2024). Examining waste-to-energy technology potential through the pilot project of Bantargebang Waste-to-Energy Power Plant. *E3S Web of Conferences, 485*, 05007. DOI: 10.1051/e3sconf/202448505007

Agre, V., Barve, A., Gadekar, S., Gangurde, A., & Pawar, D. A. B. (2024). Smart Waste Management and Awareness System. *International Journal for Research in Applied Science and Engineering Technology*, 12(5), 5474–5479. DOI: 10.22214/ijraset.2024.61457

Alotaibi, S., Martinez-Vazquez, P., & Baniotopoulos, C. (2024). Waste Generation Factors and Waste Minimisation in Construction. In V. Ungureanu, L. Bragança, C. Baniotopoulos, & K. M. Abdalla (Eds.), *4th International Conference 'Coordinating Engineering for Sustainability and Resilience' & Midterm Conference of CircularB "Implementation of Circular Economy in the Built Environment"* (Vol. 489, pp. 552–559). Springer Nature Switzerland. DOI: 10.1007/978-3-031-57800-7_51

Amissah, E. K., Nduro, K., & Quansah, P. A. (2024). Towards Greener Printing: Examining Waste Disposal Innovations in the Ghanaian Printing Industry. *International Journal of Research and Innovation in Social Science*, VIII(III), 1244–1264. DOI: 10.47772/IJRISS.2024.803092

Amoo, M., Bringardner, J., Chen, J.-Y., Coyle, E., Finnegan, J., Kim, C., Koman, P., Lagoudas, M., Llewellyn, D., Logan, L., Sonnenberg-Klein, J., Trent, N., Strachan, S., & Ward, B. (2020). Breaking Down the Silos: Innovations for Multidisciplinary Programs. *2020 ASEE Virtual Annual Conference Content Access Proceedings*, 34224. DOI: 10.18260/1-2--34224

Badruddoza Talukder, M., Kumar, S., Misra, L. I., & Firoj Kabir, . (2024). Determining the role of eco-tourism service quality, tourist satisfaction, and destination loyalty: A case study of Kuakata beach. *Acta Scientiarum Polonorum. Administratio Locorum*, 23(1), 133–151. DOI: 10.31648/aspal.9275

Baranova, T. V. (2024). Digitalization Of Document Management System In Medical Organizations: Theoretical Aspect. *City Healthcare*, 5(1), 122–128. DOI: 10.47619/2713-2617.zm.2024.v.5i1;122-128

Battour, M., Mady, K., Salaheldeen, M., Ratnasari, R. T., Sallem, R., & Al Sinawi, S. (2024). Halal tourism and ChatGPT: An overview of current trends and future research directions. *Journal of Islamic Marketing*. Advance online publication. DOI: 10.1108/JIMA-11-2023-0379

Becidan, M., Wang, L., Fossum, M., Midtbust, H. O., Stuen, J., Bakken, J. I., & Evensen, E. (2015). Norwegian waste-to-energy (wte) in 2030: Challenges and opportunities. *Chemical Engineering Transactions*, 43, 2401–2406. DOI: 10.3303/CET1543401

Beltramo, R., Duglio, S., & Mascadri, G. (2024). Mountain Tourism: Environmental Management for Fostering Tourism Destinations: From Context Analysis to Tourism Products. In Lagioia, G., Paiano, A., Amicarelli, V., Gallucci, T., & Ingrao, C. (Eds.), *Innovation, Quality and Sustainability for a Resilient Circular Economy* (pp. 509–515). Springer International Publishing., DOI: 10.1007/978-3-031-28292-8_62

Bitok, K., & Shirandula, D. (2024). Continuous Improvement and Innovation During Covid-19 in East Africa: Implications for Future Tourism Policy and Planning. In Chihwai, P. (Ed.), *COVID-19 Impact on Tourism Performance in Africa* (pp. 161–182). Springer Nature Singapore., DOI: 10.1007/978-981-97-1931-0_8

Brunet, S. (1999). Is Sustainable Tourism Really Sustainable?: Protecting the Icon in the Commodity at Sites of Invasion. *M/C Journal, 2*(2). DOI: 10.5204/mcj.1745

Cao, L. (2023). *Advanced Fanout Embedded Bridge Packaging Technology for Chiplets Integration. IMAPSource Proceedings, 2022*. DPC., DOI: 10.4071/001c.91168

Chandre, V., Gharat, O., Ghonge, R., Kulkarni, S., & Jadhav, V. (2024). Intelligent Waste Management System using IOT. *International Journal of Innovative Science and Research Technology (IJISRT)*, 2467–2472. DOI: 10.38124/ijisrt/IJISRT24APR2236

Chissom, D. (2024). Post-Pandemic Trends in the Hospitality Industry. *Journal of Modern Hospitality*, 3(3), 1–13. DOI: 10.47941/jmh.1961

Constance Obiuto, N., Ninduwezuor-Ehiobu, N., Chigozie Ani, E., Andrew Olu-lawal, K., & David Ugwuanyi, E. (2024). Implementing Circular Economy Principles to Enhance Safety and Environmental Sustainability in Manufacturing. *International Journal of Advanced Multidisciplinary Research and Studies*, 4(2), 22–29. DOI: 10.62225/2583049X.2024.4.2.2432

Das, I. R., Talukder, M. B., & Kumar, S. (2024). Implication of Artificial Intelligence in Hospitality Marketing. In Kumar, S., Talukder, M. B., & Pego, A. (Eds.), *Advances in Hospitality, Tourism, and the Services Industry* (pp. 291–310). IGI Global., DOI: 10.4018/979-8-3693-1978-9.ch014

Dawar, I., & Negi, S. (2024). In In, R. K. K. V., Kannan, H., Spodarets, D., Khan, P. A., & Pradhan, B. K. (Eds.), *Sustainable E-Waste Management: Present Situation, Emerging Solutions, and Future Trends* (pp. 254–274). Practice, Progress, and Proficiency in Sustainability. IGI Global., DOI: 10.4018/979-8-3693-1018-2.ch017

Demir, Ş. Ş. (2024). Innovations With ChatGPT in the Tourism Industry: Current Challenges and Opportunities. In Demir, M., & Dalgic, A. (Eds.), *Advances in Hospitality, Tourism, and the Services Industry* (pp. 29–51). IGI Global., DOI: 10.4018/979-8-3693-1239-1.ch002

Dighliya, B., & Dahiya, A. (2024). Tourism and Hospitality Education in India: Innovation, Trends, and Status. In Sharma, A. (Ed.), *International Handbook of Skill, Education, Learning, and Research Development in Tourism and Hospitality* (pp. 1–24). Springer Nature Singapore., DOI: 10.1007/978-981-99-3895-7_36-1

Do, Q., Mishra, N., Correia, F., & Eldridge, S. (2024). The role of a boundary object in legitimacy-making strategies for food waste innovation: The perspective of emergent circular supply chains. *Supply Chain Management*, 29(3), 523–539. DOI: 10.1108/SCM-01-2023-0056

Dolynska, O., 24. Shorobura, I., & Binytska, O. (2023). Innovations in tourism. *the scientific issues of ternopil volodymyr hnatiuk national pedagogical university. series: geography, 55*(2), 58–64. DOI: 10.25128/2519-4577.23.2.7

Downmore, M., Chihobo Chido, H., & Zvikomborero, G. (2022). Emerging Trends for Biomass and Waste to Energy Conversion. In N. Kumar & Prabhansu (Eds.), *Renewable Energy Technologies* (1st ed., pp. 245–293). Wiley. DOI: 10.1002/9781119827634.ch8

Džajić Uršič, E., Fric, U., & Rončević, B. (2024). The Circular Economy: Recent debates and research trends. *Journal of Infrastructure. Policy and Development*, 8(3). Advance online publication. DOI: 10.24294/jipd.v8i3.2855

El Archi, Y., & Benbba, B. (2024). New Frontiers in Tourism and Hospitality Research: An Exploration of Current Trends and Future Opportunities. In Negru a, A. L., & Coroş, M. M. (Eds.), *Sustainable Approaches and Business Challenges in Times of Crisis* (pp. 149–166). Springer Nature Switzerland., DOI: 10.1007/978-3-031-48288-5_9

Elsayed, E. E. (2024). *Advancing Photonics Communications through the Integration of Advanced Functional Materials*. DOI: 10.13140/RG.2.2.35966.19520/1

Fredericks, S. (2023). *A multidisciplinary collaborative web site for cardiovascular surgery*. DOI: 10.32920/14640216.v2

Gorina, G. O., Romanukha, O. M., Lakomova, O. Y., & Snihur, K. V. (2022). Innovation development trends in international tourism: content and bibliometric analysis. *Visnyk of Donetsk National University of Economics and Trade named after Mykhailo Tugan-Baranovsky, 2 (77)*, 94–102. DOI: 10.33274/2079-4819-2022-77-2-94-102

Gray, P., Gubb, S., Butler, J., & Walton, B. (2024). A recycling quality improvement project to engage the multidisciplinary team with sustainability efforts. *Nursing in Critical Care*, nicc.13054. DOI: 10.1111/nicc.13054

Haris, A., & Ningsih, N. H. I. (2020). Impact Of Tourism On Community Development And Income In Kuta Mandalika Beach Kuta Village, Pujut District, Central Lombok. [Jurnal Ilmu Sosial Dan Pendidikan]. *JISIP*, 4(3). Advance online publication. DOI: 10.58258/jisip.v4i3.1225

Havrylova, H. (2019). Tourism product of mykolaiv region: opportunity and innovations. *geography and tourism, 49*, 56–64. DOI: 10.17721/2308-135X.2019.49.56-64

Herman, G. V., Blaga, L., Filimon, C., Caciora, T., Filimon, L., Herman, L. M., & Wendt, J. A. (2024). Spatial Distribution of Relationship between Historical Monuments and Tourism: The Case Study of Bihor County in Romania. *Land (Basel)*, 13(5), 668. DOI: 10.3390/land13050668

Hertati, D., & Arif, L. (2022). *Collaborative Governance in the Management of a Waste Bank*. KnE Social Sciences., DOI: 10.18502/kss.v7i9.10923

Ilieva, L., Petrova, M., & Todorova, L. (2023). Application of technological innovations in the tourism industry. *E3S Web of Conferences, 408*, 01003. DOI: 10.1051/e3sconf/202340801003

Jaiswal, A., Roy, P., Bugalia, N., Varghese, K., & Ha, Q. P. (2024). Framework to Identify Directions for Future Construction and Demolition Waste Management Technologies. *E3S Web of Conferences, 496*, 04005. DOI: 10.1051/e3sconf/202449604005

Jamilatun, S., Pitoyo, J., & Setyawan, M. (2023). Technical, Economic, and Environmental Review of Waste to Energy Technologies from Municipal Solid Waste. *Jurnal Ilmu Lingkungan*, 21(3), 581–593. DOI: 10.14710/jil.21.3.581-593

Kabir, Z., Kabir, M., & Sultana, N. (2022). Waste-to-energy (WtE): A potential renewable source for future electricity generation in the developing world. In *Renewable Energy and Sustainability* (pp. 203–226). Elsevier. DOI: 10.1016/B978-0-323-88668-0.00015-2

Kariru, D. (2023). Contemporary Trends and Issues in The Hospitality and Tourism Industry. *International Journal of Research and Innovation in Social Science*, VII(IV), 970–986. DOI: 10.47772/IJRISS.2023.7481

Ke, W. (2024). Tourism Demand Forecast and Future Market Trend Research. *Advances in Politics and Economics*, 7(2), 202. DOI: 10.22158/ape.v7n2p202

Khan, M. O. (2023). *Manufacturing Waste for Sustainable Energy Generation: A Comprehensive Review of Current Methods and Future Trends*. DOI: 10.20944/preprints202311.0935.v1

Kharmawphlang, I. M., Paul, J., Doloi, S., & Hussain, N. (2024). Waste to Energy Technologies: A Sustainable Key Approach Towards Circular Economy and Energy Recycling. In Singh, P., & Borthakur, A. (Eds.), *Trash or Treasure* (pp. 143–181). Springer Nature Switzerland., DOI: 10.1007/978-3-031-55131-4_6

Khomenko, L., Rosokhata, A., & Nesterenko, V. (2023). Waste Management in European Countries: Indicators, Clustering, and Socioeconomic Implications. *E3S Web of Conferences, 456*, 05004. DOI: 10.1051/e3sconf/202345605004

Kiaušienė, I., Hladkova, V., & Makūnaitė, G. (2024). Application of circular economy principles in the tourism sector. *Management Theory and Studies for Rural Business and Infrastructure Development*, 46(1), 31–44. DOI: 10.15544/mts.2024.04

Killa, M. F., & Tawa, A. T. (2024). Model for Small Enterprises Innovation on Climate Change and Tourism Industry: Learning from the Traditional Weaving Industry in Sumba. *Journal of Law and Sustainable Development*, 12(1), e2939. DOI: 10.55908/sdgs.v12i1.2939

Kromp, M. G., Zambetakis, F. P., & Wood, W. (2023). Innovative approaches to nuclear energy density optimization for enhanced power generation and waste minimization. *Vertex (Buenos Aires, Argentina)*, 12(2), 89–98. DOI: 10.35335/5jd0hm75

Kumar, A., & Barua, Dr. C. S. (2024). Leveraging Digital Innovations in Tourism Marketing: A Study of Destination Promotion Strategies. *International Journal of Business and Management Research, 12*(1), 08–12. DOI: 10.37391/IJBMR.120102

Kumar, S., Talukder, M. B., & Pego, A. (Eds.). (2024). *Utilizing Smart Technology and AI in Hybrid Tourism and Hospitality*. IGI Global., DOI: 10.4018/979-8-3693-1978-9

Kurhayadi. (2021). Collaborative Governance in Waste Management in Bandung City. *International Journal of Science and Society, 3*(4), 236–244. DOI: 10.54783/ijsoc.v3i4.482

Kushwaha, A. K. Raha, Dr. S., Kushwaha, M., & Kurup, Dr. P. S. (2024). *Environmental engineering* (1st ed.). Amkcorp Research Technologies Private Limited. DOI: 10.61909/AMKEDTB032415

Lagioia, G., Amicarelli, V., Strippoli, R., Bux, C., & Gallucci, T. (2024). Sustainable and circular practices in the hotel industry in Southern Italy: Opportunities, barriers and trends in food waste management. *British Food Journal*, 126(1), 428–452. DOI: 10.1108/BFJ-12-2022-1144

Le, M.-H., Lu, W.-M., & Chang, J.-C. (2023). Recycling E-Waste and the Sustainable Economy: A Bibliometric Exploration. *Sustainability (Basel)*, 15(22), 16108. DOI: 10.3390/su152216108

Ma, W., Boré, A., Cui, J., Huang, Z., Chen, G., & Themelis, N. J. (2024). *Growth of the WTE industry in China and environmental performance*. DOI: 10.21203/rs.3.rs-4474907/v1

Manurung, J. S. (2024). Green Tech and Human Dynamics: Transforming Indonesia's Waste Industry with VR, AR, and Renewable Energy Innovations. *International Journal of Energy Economics and Policy*, 14(2), 603–617. DOI: 10.32479/ijeep.15650

Martínez-Navarro, J., Campayo-Sánchez, F., & Ostrovskaya Fedorova, L. (2024). Mapping the evolution of tourism research: The Spanish case. *Journal of Historical Research in Marketing*, 16(1), 96–121. DOI: 10.1108/JHRM-07-2023-0030

Mavrin, I. (2024). European Capital of Culture and Sustainable Tourism: Challenges, Trends and Perspectives. *Tourism (Zagreb)*, 72(1), 20–34. DOI: 10.37741/t.72.1.2

Mertzanakis, C., Vlachokostas, C., Toufexis, C., & Michailidou, A. V. (2024). *Closing the Loop between Waste-to-Energy Technologies: A Holistic Assessment Based on Multiple Criteria*. https://doi.org/DOI: 10.20944/preprints202405.1023.v1

Mi, J. (2024). Making Open Resources Discoverable: Collaborative Approaches for Enhanced Access. *International Journal of Librarianship*, 8(4), 17–29. DOI: 10.23974/ijol.2024.vol8.4.350

Mohammad, B. T., & Iva, R. D. (2024). The technology impacts and AI solutions in hospitality. *I-Manager's Journal on Artificial Intelligence &. Machine Learning*, 2(1), 56. DOI: 10.26634/jaim.2.1.20291

Muqsit, F. A., Setiadi, R., & Lo, A. (2024). Waste Management in Heritage Tourism Area: Perspectives from Visitors and Waste Management Operators. *The Journal of Indonesia Sustainable Development Planning*, 5(1), 15–26. DOI: 10.46456/jisdep.v5i1.524

Nadeau, N., & Koebele, E. A. (2023). Collaborating to reduce food waste: Building collaborative advantage in local food systems. *Renewable Agriculture and Food Systems*, 38, e32. DOI: 10.1017/S1742170523000285

Nanda, M. A., Sugandi, W., Wijayanto, A. K., Imantho, H., Sutawijaya, A., Nelwan, L. O., Budiastra, I. W., & Seminar, K. B. (2023a). The Waste-to-Energy (WtE) Technology to Support Alternative Fuels for Agriculture in the Context of Effective Solid Waste Management in the Jabodetabek Area, Indonesia. *Energies*, 16(24), 7980. DOI: 10.3390/en16247980

Nanda, M. A., Sugandi, W., Wijayanto, A. K., Imantho, H., Sutawijaya, A., Nelwan, L. O., Budiastra, I. W., & Seminar, K. B. (2023b). The Waste-to-Energy (WtE) Technology to Support Alternative Fuels for Agriculture in the Context of Effective Solid Waste Management in the Jabodetabek Area, Indonesia. *Energies*, 16(24), 7980. DOI: 10.3390/en16247980

Nubi, O., Murphy, R., & Morse, S. (2024). Life Cycle Sustainability Assessment of Waste to Energy Systems in the Developing World: A Review. *Environments (Basel, Switzerland)*, 11(6), 123. DOI: 10.3390/environments11060123

Olshanska, O. V., Khmelevskyi, M. O., & Hryhorchuk, D. V. (2021). Innovation-driven opportunities and solutions to promote europe-oriented tourism development in ukraine. *Journal of Strategic Economic Research*, 2(2), 31–40. DOI: 10.30857/2786-5398.2021.2.3

Pasternak, K. (2024). Manor tourism as an indicator of the promotion of the cultural and historical heritage of small towns and rural settlements. *moscow economic journal*, 366–390. DOI: 10.55186/2413046X_2024_9_3_154

Phonthanukitithaworn, C., Maitree, N., & Naruetharadhol, P. (2024). Eco-innovation policies for food waste management: A European Union-ASEAN comparison. *Journal of Open Innovation*, 10(2), 100295. DOI: 10.1016/j.joitmc.2024.100295

Prebilič, V., Može, M., & Golobič, I. (2024). Waste-to-Energy Processes as a Municipality-Level Waste Management Strategy: A Case Study of Kočevje, Slovenia. *Processes (Basel, Switzerland)*, 12(5), 1010. DOI: 10.3390/pr12051010

Prideaux, B. (2020). Drive and car tourism: A perspective article. *Tourism Review*, 75(1), 109–112. DOI: 10.1108/TR-05-2019-0162

Rahmawati, D. M., & Novani, S. (2024). *Embracing Circular Economy Principles: How Indonesian MSMEs Food Services Drive Sustainability Through Local Sourcing*. DOI: 10.31219/osf.io/d4fyx

Ram, C., Kumar, A., & Rani, P. (2021). Municipal solid waste management: A review of waste to energy (WtE) approaches. *BioResources*, 16(2), 4275–4320. DOI: 10.15376/biores.16.2.Ram

Raman, A. P. S., Aslam, M., Singh, P., Pandey, G., Jain, P., & Kumari, K. (2024). Solid and Liquid Waste Management in Urban and Rural Areas: Current Practices and Future Perspectives. In Gupta, A., Kumar, R., & Kumar, V. (Eds.), *Integrated Waste Management* (pp. 141–157). Springer Nature Singapore., DOI: 10.1007/978-981-97-0823-9_7

Ramanathan, R., Duan, Y., Condell, J., Ramanathan, U., & Ajmal, T. (Eds.). (2024). *New Multidisciplinary Approaches for Reducing Food Waste in Agribusiness Supply Chains*. MDPI., DOI: 10.3390/books978-3-7258-0029-2

Romianingsih, N. P. W. (2023). Waste to energy in Indonesia: Opportunities and challenges. *Journal of Sustainability, Society, and Eco-Welfare*, 1(1). Advance online publication. DOI: 10.61511/jssew.v1i1.2023.180

Saarinen, J. (2021). Is Being Responsible Sustainable in Tourism? Connections and Critical Differences. *Sustainability (Basel)*, 13(12), 6599. DOI: 10.3390/su13126599

Shah, S. Q., & Sarif, S. M. (2023). Exploring the Economic Dimensions of Human Resource Management: A Systematic Literature Review. *International Journal of Business and Economic Affairs*, 8(3). Advance online publication. DOI: 10.24088/IJBEA-2023-830012

Sharma, R. (2023). Leveraging AI and IoT for Sustainable Waste Management. In Whig, P., Silva, N., Elngar, A. A., Aneja, N., & Sharma, P. (Eds.), *Sustainable Development through Machine Learning, AI and IoT* (Vol. 1939, pp. 136–150). Springer Nature Switzerland., DOI: 10.1007/978-3-031-47055-4_12

Singh, P., Gupta, A., & Raj, S. (2020). IoT Based Smart Waste Management System Using Arduino. International Journal of Research in Engineering. *Science and Management*, 3(5), 569–571.

Smolčić Jurdana, D. (2024). Evolution of trends affecting the tourism industry: Business challenges in the future. *Journal of Innovations in Business and Industry*, 3(3), 193–204. DOI: 10.61552/JIBI.2024.03.008

Srinivasan, S., SHerkar, A., Jayamani, J., Indora, A., & Mukherjee, R. (2024). Tourism Innovation And The Role Of Technology In Enhancing Visitor Experiences. *Educational Administration: Theory and Practice*, 30(4), 1506–1513.

Sukla, B., Jakhar, M. R., Begum, S., Jasmine, M. M., & Sundaram, S. (2024). Innovations In Sustainable Engineering: A Review Of Green Technologies. *Educational Administration: Theory and Practice*, 30(5), 3667–3675.

Talukder, M. B. (2021). An assessment of the roles of the social network in the development of the Tourism Industry in Bangladesh. *International Journal of Business, Law, and Education*, 2(3), 85–93. DOI: 10.56442/ijble.v2i3.21

Talukder, M. B. (2024). Implementing Artificial Intelligence and Virtual Experiences in Hospitality. In Manohar, S., Mittal, A., Raju, S., & Nair, A. J. (Eds.), *Advances in Hospitality, Tourism, and the Services Industry* (pp. 145–160). IGI Global., DOI: 10.4018/979-8-3693-2019-8.ch009

Talukder, M. B., Kabir, F., Muhsina, K., & Das, I. R. (2023). Emerging Concepts of Artificial Intelligence in the Hotel Industry: A Conceptual Chapter. *International Journal of Research Publication and Reviews*, 4(9), 1765–1769. DOI: 10.55248/gengpi.4.923.92451

Talukder, M. B., & Kumar, S. (2024). The Development of ChatGPT and Its Implications for the Future of Customer Service in the Hospitality Industry. In Derbali, A. M. S. (Ed.), (pp. 100–126). Advances in Information Security, Privacy, and Ethics. IGI Global., DOI: 10.4018/979-8-3693-1511-8.ch005

Talukder, M. B., Kumar, S., Kaiser, F., & Mia, Md. N. (2024). Pilgrimage Creative Tourism: A Gateway to Sustainable Development Goals in Bangladesh. In M. Hamdan, M. Anshari, N. Ahmad, & E. Ali (Eds.), *Advances in Public Policy and Administration* (pp. 285–300). IGI Global. DOI: 10.4018/979-8-3693-1742-6.ch016

Theilen, T.-M., Braun, Y., Bochennek, K., Rolle, U., Fiegel, H. C., & Friedmacher, F. (2022). Multidisciplinary Treatment Strategies for Wilms Tumor: Recent Advances, Technical Innovations and Future Directions. *Frontiers in Pediatrics*, 10, 852185. DOI: 10.3389/fped.2022.852185 PMID: 35911825

Themelis, N. J. (2023). Energy and materials recovery from post-recycling wastes: WTE. *Waste Disposal & Sustainable Energy*, 5(3), 249–257. DOI: 10.1007/s42768-023-00138-2

Tian, X., Xie, L., & Ai, H. (2024). Exploring the Path of Sustainable Cultural Tourism under the Context of Rural Revitalization: Taking Yantianzi, Hong Kong as an Example. *Journal of Humanities and Social Sciences Studies*, 6(2), 25–29. DOI: 10.32996/jhsss.2024.6.2.4

Tsypko, V., Andrusenko, S., Podpisnov, V., & Podpisnov, D. (2024). The potential of the global automotive industry as a tourist attraction. *E3S Web of Conferences, 508*, 08026. DOI: 10.1051/e3sconf/202450808026

Usman, M., Li, A., Huang, Y., Zhang, T., Zheng, Y., Li, S., & Yao, H. (2024). Evaluation and Perspective of the Thermal Treatment Technologies of Medical Waste for Energy and Value-Added Products. In Wan, Z. (Ed.), *Water Resources Management and Water Pollution Control* (pp. 179–189). Springer Nature Switzerland., DOI: 10.1007/978-3-031-53456-0_15

Verma, P., Jain, O., & Gupta, A. (2021). Waste to Energy (WTE) by Incineration: Current and Future Practices in India. In Muzammil, M., Chandra, A., Kankar, P. K., & Kumar, H. (Eds.), *Recent Advances in Mechanical Engineering* (pp. 369–375). Springer Singapore., DOI: 10.1007/978-981-15-8704-7_46

Vydaichuk, T., Rusachenko, N., Ladonia, K., Korotun, O., & Lakhno, N. (2024). Semantic evolution of a lexeme in historical context: Analysis of changes in its meaning and usage. *Multidisciplinary Science Journal, 6*, 2024ss0213. DOI: 10.31893/multiscience.2024ss0213

Wicaksono, P. B., & Triwigati, P. T. (2023). Potential Integration of Waste to Energy (WtE) and Carbon Mineralization Technology in Indonesia. *IOP Conference Series. Earth and Environmental Science*, 1199(1), 012008. DOI: 10.1088/1755-1315/1199/1/012008

Wirani, Y., Eitiveni, I., & Sucahyo, Y. G. (2024). Framework of Smart and Integrated Household Waste Management System: A Systematic Literature Review Using PRISMA. *Sustainability (Basel)*, 16(12), 4898. DOI: 10.3390/su16124898

Zamathula, Q. S. N., & Ugwuanyi, E. D. (2024). Implementing AI-driven waste management systems in underserved communities in the USA. *Engineering Science & Technology Journal*, 5(3), 794–802. DOI: 10.51594/estj.v5i3.903

Zreik, M. (2024). Resurgence and transformation: The impact of COVID-19 on Myanmar's tourism industry and future trends. *Southeast Asia: A Multidisciplinary Journal*. DOI: 10.1108/SEAMJ-12-2023-0085

About the Contributors

Reshma Sucheran is a distinguished academic and researcher who has significantly shaped the field of Hospitality and Tourism at the Durban University of Technology (DUT) in South Africa. Holding the prestigious position of Research Chair, Prof. Sucheran has been an esteemed faculty member since 1997. She earned her Ph.D. in Geography and Environmental Science from the University of KwaZulu-Natal in 2012, with her doctoral research focusing on Environmental Management in the Hotel Sector. Renowned for her proficiency in lecturing and guiding postgraduate research, Prof. Sucheran has been pivotal in developing tourism and hospitality curricula at DUT. Her expertise encompasses environmental management, climate change in tourism, crisis and disaster management, and education and training in tourism. With a keen focus on the global tourism and hospitality industry, she has led numerous independent research projects, including critical studies on the impacts of the COVID-19 pandemic on the tourism sector. Prof. Sucheran actively engages in international collaborative research, partnering with prestigious institutions worldwide. She collaborates with the University of Coimbra in Portugal on Creative Tourism, Western Sydney University in Australia on innovative business strategies in response to COVID-19's impact on tourism, the Federal University of Rio De Janeiro in Brazil on cultural integration through air transport and tourism, and the Guangzhou International Sister-City Universities Alliance on research exchange. Additionally, she has shared her expertise as a guest lecturer at the Munich University of Applied Sciences, Germany, and Chandigarh University, India. She is a member of the BRICS Research Institute at DUT. Her scholarly work includes publications in leading tourism and hospitality journals and presentations at international conferences. Prof. Sucheran is actively involved in academic societies, serving as a member and reviewer for organisations such as the Global Research and Development Services (GRDS), the Global Association for Humanities and Social Science Research (GAHSSR), the Social Science and Humanities Research Association (SSHRA), and the International Federation for

IT and Travel and Tourism (IFITT). She also contributes as a reviewer and editorial board member for esteemed journals like the Tourism Review International Journal and the African Journal of Hospitality and Tourism. At DUT, she is a key member of the Faculty Research Committee and the Faculty Research Ethics Committee. Prof. Sucheran's outstanding contributions have been recognized with multiple awards, including the Women in Research and Innovation Program Grant from DUT for three consecutive years (2020-2022). She has also been awarded a scholarship from the Human Science Research Council (SA) and the Mellon Foundation (USA) for the Summer School Program on Research Methodology at the University of Michigan (1997-1999). She was honoured as a Fellow at the Munich University of Applied Sciences, Department of Tourism, in 2017 and 2022 and received the Golden Key International Honour Society Award in 2013. Prof. Sucheran remains actively involved in the Golden Key International Honour Society, where she has served as President of the Alumni Committee for KwaZulu-Natal. Throughout her distinguished career, Prof. Sucheran has demonstrated an unwavering commitment to advancing the field of tourism and hospitality through education, research, and international collaboration, earning her a respected place in the global academic community.

***

Abdallah Mohamed Kahly Abd-Elsattar is an Assistant Lecturer in the Hotel Management Department at the Faculty of Tourism and Hotel Management, Pharos University in Alexandria, Egypt. He holds a Master's degree in Hotel Management and has extensive experience in the hospitality industry. Mr. Kahly's research interests include customer loyalty, service innovation, and hospitality marketing.

Anu Chandran is the Professor and Head of the Department of Tourism Studies at Pondicherry University. With 18 years of academic experience and over a year of administrative and industry experience, he specializes in destination branding, travel journalism, and cultural tourism. He can be contacted via email at anu.chandran48@gmail.com, and by phone at +919843244547.

Suman Chaudhury is a distinguished faculty member at the Department of Business Administration of NAAC "A" Graded Berhampur University in the Eastern State of Odisha, India. He is a versatile teacher with areas of interest in Finance & Accounts with core management subjects. He is an ardent researcher who have guided and produced 10 Ph.D. theses and 01 M.Phil. thesis which have been adjudicated as the best theses in the area concerned. He is also a renowned author who have written 04 books and edited 46 volumes published by reputed houses. Added to this, he has published nearly 100 articles in various national and international journals of repute in academia. To his credit, he is a personification of 19 years of teaching experience and 10 years of research experience with an industry experience of over 3 years from the beginning of his professional career.

Girish Das is a Faculty in the department of Humanities and social sciences, Parala Maharaja Engineering college, Odisha. He holds number of publications to his credit in various management related topics. He also a decent researcher continuously engaged in research in the field of rural development. He also a very good teacher teaching various subject to the post-graduate students. He holds five years of teaching and 12 years of research experience.

Tamer M. Elsawy is a lecturer at Pharos University's Faculty of Tourism and Hotel Management, and holds a PhD in tourism planning and development. With a decade of academic experience and eight years as Assistant Manager at F.S. Hotels, he brings expertise in hospitality and intercultural skills. Currently leading committees and serving as editorial manager of the Pharos International Journal of Tourism and Hospitality, he's also involved in various roles, including head of academic advising, marketing committee, and international relations. As an IR coordinator, he establishes partnerships globally and secures funding for projects, collaborating with academics worldwide to drive research and enterprise activities. Tamer M. Elsawy can be contacted at: tamer.elsawy@pua.edu.eg

B. Udaya Bhaskara Ganesh is an esteemed academic professional currently serving as an Assistant Professor at Mittal School of Business (MHRD NIRF India Rank 34; ACBSP USA, Accredited), Lovely Professional University, Phagwara, Punjab, India. He has done his M.B.A and Ph.D. from Andhra University, Visakhapatnam. With an impressive 14 years of experience in academia, research, and teaching, His teaching interests encompass a wide array of subjects, including Investment Banking, Financial Modelling, Financial Derivatives, Corporate Finance etc. he has showcased his research prowess at esteemed National and International conferences across different regions of the country. Dr.B. Udaya Bhaskara Ganesh research contributions are noteworthy, with sixteen research papers to his credit in esteemed databases such as Scopus, Web of Science (Wos), and UGC. and four books and two patents got published. Dr. B. Udaya Bhaskara Ganesh holds two professional memberships with reputable national and international organizations. He appointed as a BSE resource person and for the last 8 years he conducted various capital Market awareness programmes in reputed institutions both in offline & online.

Sarah Hussain is a Professor (of Hospitality) at the Department of Tourism and Hospitality Management, Jamia Millia Islamia (A Central University) New Delhi.

Previously she has served as the Head, Department of Tourism and Hospitality Management, Jamia Millia Islamia. She has also served as the Principal of Banrsidas Chandiwala Institute of Hotel Management, New Delhi. She has more than 25 years of stellar experience in Industry and Academia including Taj Group of Hotels, Times of India, Times Warner Group, Berggruen Education, BCIHMCT and Jamia Millia Islamia. An alumnus of prestigious Mayo College Girls School - Ajmer and Institute of Hotel Management – Aurangabad (1998 batch), she holds a Master's Degree in Hotel Management and a Ph.D. degree in Hospitality. Prof. Sarah has been imparting Hospitality Education since 2008. Prof. Sarah is also member of Board of Studies of various Universities in India (State and Central). Beyond her contribution as an educator, she is also an accomplished researcher, published in academic journals, books and hospitality industry publications. She has served as the editor of Indian Journal of Applied Hospitality and Tourism Research. She is an Editorial Board Member of Journal of Services Research and Transdisciplinary International Journal of Academic Research. She teaches Research Methodology to PhD scholars and has taught various courses in Hotel Management programmes including Revenue Management, Front Office Operations and Management, Events Management, Hospitality Sales and Marketing, Housekeeping Operations, and Business Communication. Her research interests include Guest Satisfaction, Service Quality, Expectation and Perception of Guests, Environment Management, and Women Guests/ Travelers. She is a Life Member of India Hospitality Congress and also Advisory Member of Women's Indian Chamber of Commerce & Industry (WICCI) Hospitality & Tourism Council NCR.

Rejaul Karim is an Assistant Professor of Finance and Banking in the Department of Business Administration at Varendra University, Rajshahi, Bangladesh. He received his PhD in finance from the Institute of Bangladesh Studies, University of Rajshahi, Bangladesh. He completed the Bachelor of Business Administration and Master of Business Administration in Finance and Banking from the Pabna University of Science and Technology, Pabna, Bangladesh with outstanding performance. Dr. Karim has experience in teaching and research for about a decade in undergrad and Master's level tertiary business education. He has a great passion for teaching and research in finance and economics. His research focused on financial performance, financial structure, circular economy, and sustainable finance with other contemporary issues in finance and economics. Dr. Karim is continuously contributing through his research in his field and publishing several papers in reputed national and international peer-reviewed journals, including conference proceedings. Dr. Karim founded and headed BusinessXcel Lab, a center for quantitative business research, a voluntary organization that aims to facilitate collaboration among domestic and international undergraduate, graduate, and PhD students.

Christian Kaunert is Professor of International Security at Dublin City University, Ireland. He is also Professor of Policing and Security, as well as Director of the International Centre for Policing and Security at the University of South Wales. In addition, he is Jean Monnet Chair, Director of the Jean Monnet Centre of Excellence and Director of the Jean Monnet Network on EU Counter-Terrorism (www.eucter.net).

Vipan Kumar is an independent researcher in the field of tourism and hospitality.

Narayana Maharana, B.Sc., MBA, M.Com. UGC NET (JRF)-Management, UGC NET-Commerce, Seasoned Researcher and a full time Faculty in the Dept. of Management Studies, GVP College of Engineering, Visakhapatnam, A.P., India, with Seven years of experience in teaching Post-Graduate courses in Management and allied subjects. Actively participated in different Faculty Development Programmes as well as presented research paper in several national and international conferences. Besides, having more than 13 articles published in different UGC CARE, Scopus and Web of Science listed journals as well as seven chapters in edited books in last five years.

Kevin Mearns is a Full Professor in the Department of Environmental Sciences at the University of South Africa (UNISA). Professor Mearns specialises in the research areas of sustainability, sustainable tourism, community-based tourism and ecotourism. Professor Mearns has supervised more than 43 postgraduate Master's and Doctoral candidates to completion and published more than 120 peer-reviewed scholarly journal publications, book chapters and conference proceedings

Jitendra Mohan Mishra is an accomplished academic with nineteen years of teaching experience. He currently serves as Dean of the School of Management Studies and is the Head of the Department of Tourism & Hospitality Management at Central Tribal University of Andhra Pradesh. His contributions include 10 completed and 5 ongoing Ph.D. studies, 60 paper presentations, and leadership in major research projects. Prof. Mishra's passion for outdoor activities, including trekking and yoga, complements his academic pursuits. He has authored 40 research articles and five books, including the one, Tourism Principles and Practices, published by Oxford University Press.

Shreeansh Mishra is an Assistant Tourism Professor at Central University of Tamil Nadu, India. His research highlights the transformative role of self-help groups, microfinance, and community readiness in advancing community-based tourism and fostering sustainable development. His work examines factors that synergize marginalized communities and foster sustainable development. With a commitment to global collaboration and continuous efforts, Shreeansh Mishra continues his mission to create a more inclusive and equitable tourism landscape where gender disparities are addressed, and women's empowerment takes center stage, catalyzing a brighter future for all.

Lingaraj Prasad Patnaik is a research scholar in the Dept. of Business Administration, Utkal University, Bhubaneswar, Odisha, India. He is also serving the Govt of Odisha in the Urban planning and Civil engineering department. He is pursuing his PhD in Utkal University. His research interest includes, urban development, rural infrastructure development, sustainable urban planning, urban waste management.

Chandra Sekhar Pattnaik (MBA, PGDPPM, UGC-NET, AP-SET, Ph.D, FIP-NITTTR-B) is a Professor at Omega P G College-MBA, Hyderabad. He has a rare combination of Teaching, Industry, Administration, Research, Project, and research Guidance experience of about three decades. He has completed many Courses from SWAYAM-NPTEL & MMTTP with all of India's Top scores and is well versed in Ranking, NAAC Assessment & Accreditation Processes including NEP-2020. He has published over 60 articles bearing ISBN/ISSN Numbers to his credit. He has been an Editorial & Review Board Member for various National & International Journals besides presenting papers at various National and International Conferences. Further, he has co-authored four textbooks on Marketing Management, Financial Services in India, Start-Up Opportunity & Feasibility, and Research Methodology. His areas of interest are Marketing, Banking & Financial Services, and Research Methodology.

Md. Mustaqim Roshid is a Bangladeshi citizen who has worked as a research associate at BusinessXcel Lab. Mustaqim has a Bachelor of Business Administration in Marketing from Varendra University and is currently pursuing an MBA in Management Studies at Rajshahi University. His research interests include various aspects of business operations, supply chain management, financial performance, agribusiness, neuromarketing, industry revolution, environmental economics, circular economy, and sustainable development goals.

Rajan Sharma is a Research Scholar with Center for Promotion of Ecological, Adventure, Health and Culture Tourism, Central University of Himachal Pradesh INDIA

Amrik Singh is working as Professor in the School of Hotel Management and Tourism at Lovely Professional University, Punjab, India. He obtained his Ph.D. degree in Hotel Management from Kurukshetra University, Kurukshetra. He started his academic career at Lovely Professional University, Punjab, India in the year 2007. He has published more than 40 research papers in UGC and peer-reviewed and Scopus/Web of Science) journals. He has published 12 patents and 01 patent has been granted in the inter-disciplinary domain. Dr. Amrik Singh participated and acted as a resource person in various national and international conferences, seminars, research workshops, and industry talks. His area of research interest is accommodation management, ergonomics, green practices, human resource management in hospitality, waste management, AR VR in hospitality, etc. He is currently guiding 8Ph.D. scholars and 2 Ph.D. scholars have been awarded Ph.D.

Amrik Singh is an Assistant Professor at the Centre for Promotion of Ecology, Adventure, Health & Cultural Tourism at Central University Himachal Pradesh in Dharamshala. With a Ph.D. in Tourism Management, a Master's degree in Tourism Administration, and a Master's in Economics, Dr. Singh's academic journey is complemented by a strong foundation in economics and geography from Himachal Pradesh University, Summer hill Shimla. His dynamic career has included roles with Travel Corporation (I) Pvt. Ltd, Cox & King (I) Pvt. Ltd, and the Dalmia Group's in-house travel vertical, Sun Shine Tourism Services (I) Pvt. Ltd. Dr. Singh's dedication to the tourism industry extends to his free lance consultancy work, where he promotes off-site experiences in Himachal, both in the international and Indian markets. He has also shared his Vocational expertise with students as part of the faculty at Centre of Excellence Government Degree College, Sanjauli, Shimla. Beyond his professional roles, Dr. Singh actively contributes to the field through his association with the Global Agree Network (GAN), participating in various committees such as Research, Definition & Standards, Education, and Communications. His commitment to outreach and extension activities, as well as his involvement in the CUHP Outreach, Extentions & UGC Notification Committee at Central University Himachal Pradesh, reflects his dedication to advancing the realms of ecology, adventure, health, and cultural tourism.

Bhupinder Singh is working as Professor at Sharda University, India. Also, Honorary Professor in University of South Wales UK and Santo Tomas University Tunja, Colombia. His areas of publications as Smart Healthcare, Medicines, fuzzy logics, artificial intelligence, robotics, machine learning, deep learning, federated learning, IoT, PV Glasses, metaverse and many more. He has 3 books, 139 paper publications, 163 paper presentations in international/national conferences and seminars, participated in more than 40 workshops/FDP's/QIP's, 25 courses from international universities of repute, organized more than 59 events with international and national academicians and industry people's, editor-in-chief and co-editor in journals, developed new courses. He has given talks at international universities, resource person in international conferences such as in Nanyang Technological University Singapore, Tashkent State University of Law Uzbekistan; KIMEP University Kazakhstan, All'ah meh Tabatabi University Iran, the Iranian Association of International Criminal law, Iran and Hague Center for International Law and Investment, The Netherlands, Northumbria University Newcastle UK,

Gagandeep Soni has over 22 years of expertise in the hospitality industry and 18 years dedicated to academic and teaching roles. Soni currently serves as the Incharge of Rooms Division Management and Assistant Professor (Senior Grade) at Banarsidas Chandiwala Institute of Hotel Management and Catering Technology (A NAAC A+ Institute affiliated by Guru Gobind Singh Indraprastha University and AICTE) She actively undertakes key institutional duties, including Assessment and Ranking Coordinator (Internal Quality Assurance Cell), serving as the Joint Editor for the Indian Journal of Applied Hospitality and Tourism Research (IJAHTR) a UGC indexed journal, and overseeing research and consultancy activities as the Deputy Coordinator for Research and Consultancy.Additionally, she chairs committees for Grievance Redressal, and Sexual Harassment, while also leading initiatives in environmental and corporate social responsibility.

Dil Swarna is an independent researcher Dil Afrin Swarna is a renowned expert in education and research. She holds a research-based MBA in Human Resource Management (HRM) from the National University of Bangladesh. Swarna's passion for HRM drives her to explore various themes, showcasing her understanding of business dynamics and interconnections. Her research benefits both academia and industry. Swarna has delved into organizational behavior, employee development, and strategic management, demonstrating the critical role of HRM in corporate performance and offering practical solutions to current managerial challenges. She actively participates in conferences, publishes articles, and collaborates with other specialists in education and research. Her innovative approaches to HRM challenges are highly regarded and push the boundaries of existing knowledge. Through her dedication to research and teaching, Dil Afrin Swarna continues to make significant contributions to academia and her industry. Her work exemplifies how committed and insightful research can enhance organizational practices and deepen our understanding of human resource management.

Mohammad Badruddoza Talukder is an Associate Professor, College of Tourism and Hospitality Management, IUBAT - International University of Business

Agriculture and Technology, Dhaka-1230, Bangladesh. He holds PhD in Hotel Management from Lovely Professional University, India. He has been teaching various courses in the Department of Tourism and Hospitality at various universities in Bangladesh since 2008. His research areas include tourism management, hotel management, hospitality management, food & beverage management, and accommodation management, where he has published research papers in well-known journals in Bangladesh and abroad. Mr. Talukder is one of the executive members of the Tourism Educators Association of Bangladesh. He has led training and consulting for a wide range of hospitality organizations in Bangladesh. He just became an honorary facilitator at the Bangladesh Tourism Board's Bangabandhu international tourism and hospitality training institution.

Abdul Waaje, a dedicated Master's student in Engineering Management at Michigan Technological University, seamlessly combines his expertise in both business and technical fields. Abdul completed his undergraduate studies in Electronics and Telecommunication Engineering at East West University, Bangladesh. He is currently pursuing a Master's degree in Engineering Management with a focus on Quality Engineering, which he is set to complete in December 2024. Abdul has a wealth of experience as a sales planner and market trend analyst. He has held the position of Business Development Officer, where he has honed his skills in these areas. During my time at Marico Bangladesh Ltd. With a background as a Research Associate at BusinessXcel Lab and a Teaching Assistant at Michigan Tech, his areas of focus include Business and Psychology, Circular Economy and SDG, Six Sigma and Industry Revolution, and Agri-Economy. Abdul actively participates in various associations and community programmes outside of his academic pursuits.

Michelle Whitmore is currently in the middle of a part-time PhD thesis with the Institute for Culture and Society at Western Sydney University. Her research topic, Imagineering a world with less plastic waste: A Disney case study, combines her interests in environmental management and tourism with her passion for all things Disney. Michelle brings to her research over 20-years' experience in government and research sectors, leading complex projects and applying deep understanding of government systems, policies, and processes. Her work has spanned sustainable, environmental, waste, event, and tourism management. She has a Bachelor of Applied Science in Environmental Management and Tourism from Western Sydney University and a Bachelor of Antarctic Studies (Honours) from the University of Tasmania.

Index

A

B

C

D

E

F

G

H

S

Z

www.ingramcontent.com/pod-product-compliance
Lightning Source LLC
LaVergne TN
LVHW080355011026
849161LV00009B/2310